# Texts and Monographs in Physics

**Series Editors:**

R. Balian, Gif-sur-Yvette, France
W. Beiglböck, Heidelberg, Germany
H. Grosse, Wien, Austria
E. H. Lieb, Princeton, NJ, USA
N. Reshetikhin, Berkeley, CA, USA
H. Spohn, München, Germany
W. Thirring, Wien, Austria

Springer
*Berlin*
*Heidelberg*
*New York*
*Barcelona*
*Hong Kong*
*London*
*Milan*
*Paris*
*Tokyo*

Physics and Astronomy ONLINE LIBRARY

http://www.springer.de/phys/

Vincenzo Barone   Enrico Predazzi

# High-Energy Particle Diffraction

With 188 Figures

Springer

Dr. Vincenzo Barone
Dipartimento di Scienze
Università del Piemonte Orientale
and INFN, Gruppo di Alessandria
Corso Borsalino 54
15100 Alessandria
Italy

Professor Dr. Enrico Predazzi
Dipartimento di Fisica Teorica
Università di Torino
and INFN, Sezione di Torino
Via P. Giuria 1
10125 Torino
Italy

Library of Congress Cataloging-in-Publication Data

Barone, Vincenzo.
High-energy particle diffraction / Vincenzo Barone, Enrico Predazzi.
p. cm. -- (Texts and monographs in physics, ISSN 0172-5998)
Includes bibliographical references and index.
ISBN 3540421076 (alk. paper)
1. Deep inelastic collisions. 2. Hadron interactions. 3. Particles (Nuclear physics)--Diffraction. I. Predazzi, Enrico. II. Title. III. Series.

QC794.6.C6 B37 2002
539.7'57--dc21

2001049856

ISSN 0172-5998
ISBN 3-540-42107-6 Springer-Verlag Berlin Heidelberg New York



Springer-Verlag Berlin Heidelberg New York
a member of BertelsmannSpringer Science+Business Media GmbH

http://www.springer.de


Printed in Germany



Typesetting by the authors using a Springer TEX makro package
Final processing by Steingraeber Satztechnik GmbH Heidelberg
Cover design: *design & production* GmbH, Heidelberg

Printed on acid-free paper SPIN: 10720945 55/3141/mf - 5 4 3 2 1 0

To our families

# Preface

High-energy diffraction has become a hot and fashionable subject in recent years due to the great interest triggered by the HERA and Tevatron data. These data have helped to show the field from a different perspective paving the road to a hopefully more complete understanding than hitherto achieved. The forthcoming data in the next few years from even higher energies (LHC) promise to sustain this interest for a long time.

We believe that it is therefore necessary to summarize the main developments which have marked the growth of high-energy diffractive physics in recent decades, and to assess the present state of the art. This is the purpose of the present book, which is especially aimed at the young researchers who are entering the field and want to get acquainted with the relevant results and the main theoretical techniques. The "new" diffraction has started to bridge the gap between the hard and soft regimes of strong interactions. A modern account of the subject, in our opinion, should reflect this situation, covering both the traditional approaches to soft processes which are still alive and useful, and the modern treatment of hard dynamics in the framework of perturbative QCD.

The book is divided into three parts. The first part (Chaps. 1–3) contains some introductory material: the systematics of diffractive processes, some historical remarks, the optical analogy, the eikonal approximation of quantum mechanics, and high-energy kinematics.

In the second part (Chaps. 4–7), the "old" diffraction is reviewed, with a focus on the theoretical and experimental results which are still of some relevance nowadays. Topics covered in this part include: $S$-matrix properties, Regge theory, $s$-channel models, and the phenomenology of soft diffractive reactions.

Finally, the third part (Chaps. 8–11) is devoted to the "new" diffraction. Our present understanding of the pomeron in perturbative QCD (the celebrated BFKL theory) is presented in some detail. Deep inelastic scattering at low $x$, diffractive deep inelastic scattering and diffractive hadron–hadron processes are analysed phenomenologically, and their QCD description is outlined. One of the main theoretical approaches, the color dipole picture, is discussed at various stages.

This book grew out of lectures that the authors delivered at the Centro Brasileiro de Pesquisas Fisicas (Rio de Janeiro, Brazil) in February 1998 (V.B.) and at the Florianopolis Workshop on Hadron Physics (Florianopolis, Santa Catarina, Brazil) in March 1998 (E.P.). We thank our Brazilian friends, Ignacio Bediaga, Francisco Caruso, Alberto Santoro, Erasmo Ferreira, for their hospitality and for giving us the opportunity to talk to an interested and stimulating audience.

We are deeply indebted to the countless number of colleagues and friends who have been discussing with us various aspects of DIS and diffraction in the past decade, and from whom we learned much of what we know on these subjects. Apologizing for any possible omission, we mention, in alphabetical order: Halina Abramowicz, Michele Arneodo, Bj. Bjorken, John Dainton, Umberto D'Alesio, Vittorio Del Duca, Victor Fadin, Roberto Fiore, Stefano Forte, Marco Genovese, Maurice Giffon, Dino Goulianos, Laszlo Jenkovszky, Boris Kopeliovich, Eduard Kuraev, Elliot Leader, Genya Levin, Aharon Levy, Lev Lipatov, Uri Maor, Genya Martynov, Giorgio Matthiae, Kolya Nikolaev, Francesco Paccanoni, Alessandro Papa, Cristiana Peroni, Irina Potashnikova, Bogdan Povh, Misha Ryskin, Alberto Santoro, Ada Solano, Amedeo Staiano, Slava Zakharov and Fabian Zomer. It is a pleasure to acknowledge a longlasting and fruitful collaboration with Boris Kopeliovich, Marco Genovese, Kolya Nikolaev and Slava Zakharov.

We are grateful to Umberto D'Alesio, Alexei Prokudin and Ada Solano for a careful reading of some chapters of the book, to Vittorio Del Duca, Dino Goulianos, Elliot Leader and Kolya Nikolaev for useful suggestions and comments, to Giorgio Matthiae and Erasmo Ferreira for providing some figures, and to Igor Pesando for his computer expertise which proved very helpful in an emergency situation.

We dedicate this book to our families.

Torino,
December 2001

*Vincenzo Barone,*
*Enrico Predazzi*

# Contents

# 1. Introduction

Hadronic processes are traditionally classified in two distinct classes: *soft processes* and *hard processes* (for a general discussion see, e.g., Abramowicz et al. 1996).

- The *soft processes* are characterized by an energy scale of the order of the hadron size $R$ ($\sim 1$ fm). This is the only typical scale of such processes. The momentum transfer squared is generally small: $|t| \sim 1/R^2 \sim$ (few hundred MeV)$^2$. The $t$-dependence of the cross sections is exponential, $\mathrm{d}\sigma/\mathrm{d}t \sim \mathrm{e}^{-R^2\,|t|}$, and large-$|t|$ events are highly suppressed.
  Classical examples of soft processes are elastic hadron–hadron scattering and diffractive dissociation.
  From a theoretical point of view, perturbative quantum chromodynamics (QCD) is inadequate to describe these processes. The presence of a large length scale ($R$) makes them intrinsically non-perturbative. The approach which has been adopted since the 60's to describe soft processes is Regge theory (Regge 1959; Chew and Frautschi 1961; Gribov 1961). According to this theory, soft hadronic phenomena at high energies are universally dominated by the exchange of an enigmatic object, the *pomeron*.

- The *hard processes* are characterized by two (or more) energy scales: one is still the hadron size, the other is a "hard" energy scale. The momentum transfer is of the order of this scale and therefore large ($\gtrsim 1$ GeV$^2$). The dependence of the cross sections on the momentum transfer is typically powerlike, modulo logarithms.
  Two examples of hard processes are deep inelastic scattering (the momentum transfer squared is $q^2$, the virtuality of the exchanged photon or vector boson) and large-$p_T$ jet production (the momentum transfer squared is $-\boldsymbol{p}_T^2$).
  The high value of the momentum transfer allows one to use perturbative QCD. Part of the process, however, is still of non-perturbative origin. This component is embodied in the quark and gluon distribution (or fragmentation) functions of hadrons. The so-called "factorization theorems" (Collins, Soper and Sterman 1989) ensure that the perturbative part can be separated from the non-perturbative one. The latter is universal: it can be extracted from one process and used to predict another one.

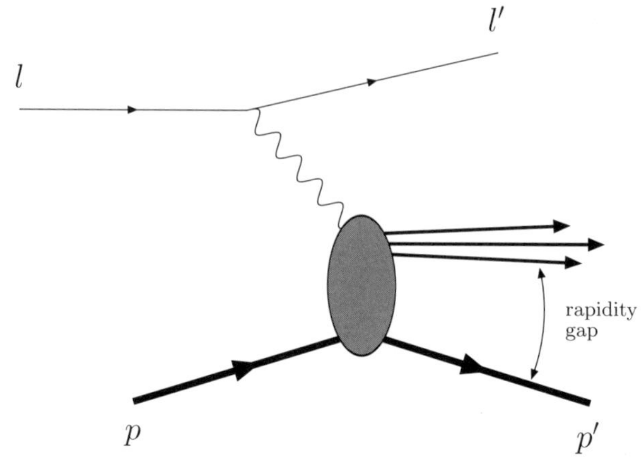


**Fig. 1.1.** Diffractive deep inelastic scattering.

Hadronic diffraction has always found its place in the first class, that of soft phenomena. The main novelty of the last years is the discovery and investigation of diffractive processes which have soft and hard properties at the same time.

A typical process of this type is *diffractive deep inelastic scattering* (DDIS), depicted in Fig. 1.1.

DDIS is simply a deep inelastic scattering reaction with a particular final state configuration, characterized by a large rapidity gap between the proton remnant and the products of the hadronization of the photon. As we shall see in the next section, this rapidity gap is a typical signature of diffraction. It implies that there is no exchange of quantum numbers (except those of the vacuum) between the virtual photon and the proton. DDIS has been first observed at HERA in 1993 by the ZEUS and the H1 Collaborations (Derrick et al. 1993; Ahmed et al. 1994) and amounts to about 10% of the total deep inelastic scattering events. DDIS is a diffractive process in which two different energy scales coexist: a soft one, $R$, and a hard one, the photon virtuality $Q^2$. That is why DDIS is often referred to as a *hard diffractive process.*

Another example of hard diffraction, initiated by hadrons only, is jet production in $pp$ collisions with a leading proton in the final state, i.e. a proton carrying a large fraction of the initial proton's momentum (Fig. 1.2). The first evidence for this process was found by the UA8 experiment at CERN (Bonino et al. 1988; Brandt et al. 1992). Events with large rapidity gaps between two jets were discovered by the CDF and D0 Collaborations at the Tevatron (Abe et al. 1995; Abachi et al. 1996) shortly after the observation of hard diffraction at HERA.

A further interesting manifestation of the interplay of soft and hard interactions is the phenomenon of *nuclear shadowing* in low-$x$ DIS. As shown by Gribov (1969), nuclear shadowing at large $Q^2$ is closely related to DDIS.

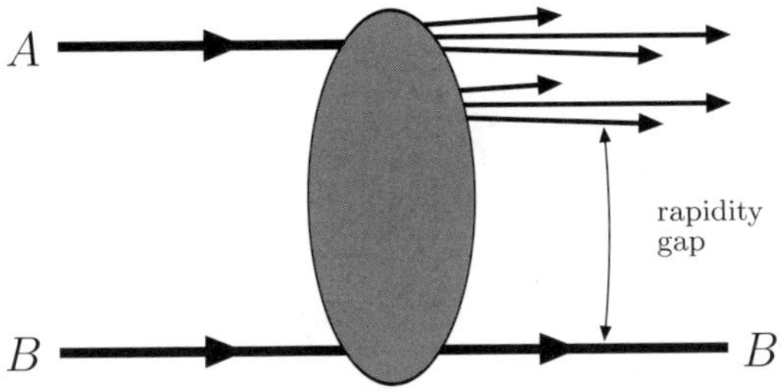


**Fig. 1.2.** Dijet production in hadron–hadron scattering with a leading proton and a rapidity gap in the final state.

Thus the observation of this phenomenon at CERN in the late 80's (for a review see Arneodo 1994) can be regarded as the first (indirect) evidence of diffraction in DIS.

What is exciting about hard diffractive processes is that they open up the possibility of studying diffraction, at least to some extent, in a perturbative QCD framework. Said otherwise, they allow investigating the QCD nature of the pomeron. The ultimate goal is to translate Regge theory (the somehow misterious but quite successful phenomenology of soft phenomena) into the language of QCD, the theory of strong interactions. This dream might come true in the next future.

## 1.1 Diffractive Phenomena: a Definition

The term *diffraction* was introduced in nuclear high-energy physics in the 50's. The very first to use it were Landau and his school (Landau and Pomeranchuk 1953; Feinberg and Pomeranchuk 1956; Akhiezer and Pomeranchuk 1958; Sitenko 1959). The term is used in strict analogy with the familiar optical phenomenon that occurs when a beam of light meets an obstacle or crosses a hole whose dimensions are comparable to its wavelength (so long as the wavelengths are much smaller than these dimensions, we have geometrical shadowing). To the extent that the propagation and the interaction of extended objects like the hadrons are nothing but the absorption of their wave function caused by the many inelastic channels open at high energy, the use of the optical terminology seems indeed appropriate.

As we shall see, in optics the intensity of the diffracted light at small angles and large wave numbers $k$ is given by

$$I(\vartheta) \simeq I(0)\,(1 - Bk^2\vartheta^2)\,, \tag{1.1}$$

where $B \propto R^2$ (the squared radius of the obstacle, or of the hole in the screen) and $q \simeq k\vartheta$ is the momentum transfer. Thus the intensity has a forward peak and a rapid decrease (eventually followed by other secondary maxima).

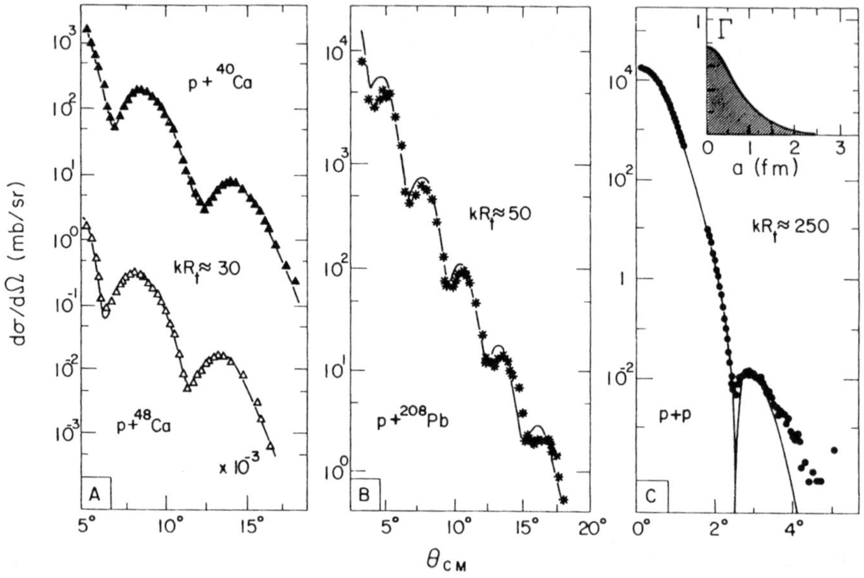


**Fig. 1.3.** Diffraction in nuclear and hadronic physics (from Amaldi, Jacob and Matthiae 1976).

Diffractive hadronic processes have a similar behavior. Their cross sections behave as

$$\frac{\mathrm{d}\sigma}{\mathrm{d}t} = \left.\frac{\mathrm{d}\sigma}{\mathrm{d}t}\right|_{t=0} \mathrm{e}^{-B\,|t|} \simeq \left.\frac{\mathrm{d}\sigma}{\mathrm{d}t}\right|_{t=0} (1 - B\,|t|)\ , \tag{1.2}$$

at small $|t|$ ($|t| \propto \vartheta^2$ at high energies). In this case the slope parameter $B$ is proportional to the squared radius of the target hadron. At larger $|t|$ other secondary maxima appear. The series of diffractive maxima and minima have become familiar in nuclear physics first and in hadronic physics later (see Fig. 1.3).

Optics and hadronic physics may, at first sight, appear very distant fields. In the latter, the collision of very high energy particles produces an almost arbitrary number of large varieties of different particles and this makes the situation considerably more confused than in optics where Huygens principle dictates the solution to every problem. In practice, however, the two fields are very much alike; in both cases we have high wave numbers in the game and in both cases it is the ondulatory character of the phenomenon which is responsible of what we observe[1].

Thus, diffraction covers a large span of phenomena, from optics to quantum mechanics, and from nuclear to hadron physics. The next task is to

[1] At the phenomenological level, however, the optical analogy should not be pushed too far: for instance, the observed shrinkage of the forward peak with increasing energy in hadronic diffraction has no optical analogue.

define diffraction in purely particle physics terms. The first authors to give a description of *hadronic diffraction* in perfectly modern terms were Good and Walker (1960) who, more than 40 years ago, wrote:

> A phenomenon is predicted in which a high energy particle beam undergoing diffraction scattering from a nucleus will acquire components corresponding to various products of the virtual dissociations of the incident particle [...] These diffraction-produced systems would have a characteristic extremely narrow distribution in transverse momentum and would have the same quantum numbers of the initial particle.

A general definition of hadronic diffractive processes can be formulated as follows:

1) *A reaction in which no quantum numbers are exchanged between the colliding particles is, at high energies, a diffractive reaction.*

In other words, diffraction is the phenomenon which takes place *asymptotically* (i.e., as the energy increases) whenever the particles (or ensembles of particles) diffused have the same quantum numbers as the incident particles.

There is an element of ambiguity in the prescription above. The request alone of no exchange of quantum numbers (other than those of the vacuum) is a *necessary* condition for the process to be diffractive, but not a *sufficient* one. It would not allow to recognize and eliminate a possible contamination of non-diffractive origin. The latter, however, is expected to become asymptotically smaller and smaller compared to the former. That is why, in the definition above, we explicitly demanded the process to be a *high-energy* process.

On the other hand, the big advantage of Definition 1 is that it is simple and general enough to cover all cases:

I *elastic scattering*, when exactly the same incident particles come out after the collision (Fig 1.4a)

$$1 + 2 \rightarrow 1' + 2' \;; \tag{1.3}$$

II *single diffraction*, when one of the incident particles comes out unscathed after the collision while the other gives rise to a bunch of final particles (or to a resonance) with the same quantum numbers (Fig. 1.4b)

$$1 + 2 \rightarrow 1' + X_2 \;, \tag{1.4}$$

and, finally,

III *double diffraction*, when each incident particle gives rise to a bunch of final particles (or to a resonance) with exactly the same quantum numbers of the two initial particles (Fig. 1.4c)

$$1 + 2 \rightarrow X_1 + X_2 \;. \tag{1.5}$$

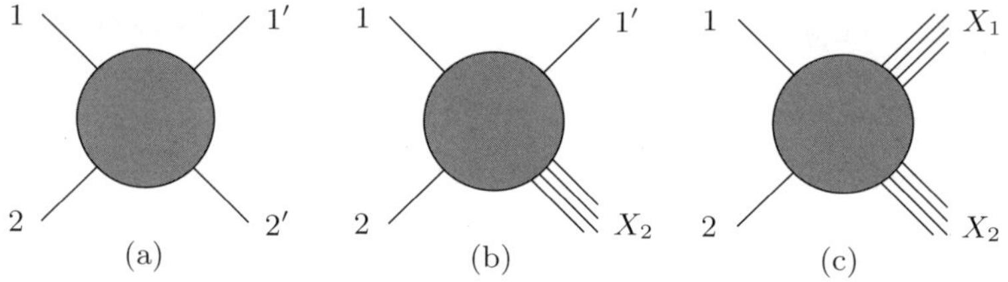


**Fig. 1.4.** (**a**) Elastic scattering. (**b**) Single diffraction. (**c**) Double diffraction.

In practice, while it is possible to recognize immediately that an elastic reaction such as (1.3) belongs to the class of diffractive processes defined above, it is difficult, when the final state is not fully reconstructed, to know if the outgoing systems have the same quantum numbers as the incoming particles. Hence it is convenient to provide an operational definition of diffraction (Bjorken 1993) which is equivalent to the one given above:

2) *A diffractive reaction is characterized by a large, non exponentially suppressed, rapidity gap*[2] *in the final state.*

This means, for instance, that a reaction such as (1.5) is diffractive if a large rapidity gap (i.e., a large angle separation) is observed between $X_1$ and $X_2$. However, there may be (few) events of type (1.5) which display a large rapidity gap, although they are of non-diffractive nature. The number of these events is expected to be exponentially suppressed. If we denote by $\Delta\eta$ the final state rapidity gap, the distribution of diffractive events is

$$\frac{\mathrm{d}N}{\mathrm{d}\Delta\eta} \sim \text{constant} , \tag{1.6}$$

to be compared with that of non-diffractive events which is

$$\frac{\mathrm{d}N}{\mathrm{d}\Delta\eta} \sim \mathrm{e}^{-\Delta\eta} . \tag{1.7}$$

Notice that the request alone of large final state rapidity gaps is again not sufficient to characterize diffraction. In order to avoid a contamination from non-diffractive events, one has to demand also that the rapidity gaps be non exponentially suppressed. True diffraction can be distinguished only asymptotically from non-diffraction contributions, as it is known that the latter decrease with energy.

It should be clear from the discussion so far that the two definitions of diffraction we have given are not fully empirical, but rather theory-laden. In fact, we have to resort to some theoretical input to discriminate between diffractive and non diffractive events and to show that the latter are asymptotically negligible.

[2] See Sect. 3.3.2 for the definition of rapidity.

The traditional theoretical framework for diffraction is Regge theory. This describes hadronic reactions at high energies in terms of the exchange of 'objects' (not particles !) called *reggeons*. The reggeon with vacuum quantum numbers which dominates asymptotically is the so-called *pomeron*[3]. The exchange of other scalars with vacuum quantum numbers, contributing to non-diffractive events, is suppressed at high energy. Thus in Regge theory the diffractive reactions are those dominated by pomeron exchange.

To see how Regge theory provides a bridge between Definitions 1 and 2, let us consider for definiteness the single-diffraction process (1.4). Anticipating some kinematics, when the center-of-mass energy $\sqrt{s}$ is much higher than the invariant mass $M$ of the $X_2$ system, particle 1 comes out with (almost) unmodified longitudinal momentum and its rapidity gap with $X_2$ is large, being $\Delta\eta \sim \ln(s/M^2)$. Regge theory describes this process as due to the exchange of a pomeron between particle 1 and particle 2, and hence no quantum numbers are exchanged in the reaction. As $s$ decreases, other reggeons contribute and the non-diffractive contamination gets larger.

In the Regge language "diffraction" is equivalent to "pomeron physics". Many consider the concept of pomeron misleading, illdefined or even meaningless. Here, by default, we shall refer to *pomeron exchange* simply as synonimous of *exchange of vacuum quantum numbers*, and elaborate, in different contexts, on the physical meaning to attach to this term.

## 1.2 Brief Historical Survey

**Sixties.** We can take the early 60's as the beginning of diffraction in hadronic physics with the advent of the first high-energy accelerators. On a purely empirical ground, the distinctive features of hadronic diffraction were soon recognized to be the following:

- Steep angular (or momentum transfer) distributions, as predicted by Good and Walker (1960).
- A slow increase with energy of the total and integrated cross sections[4].
- Shrinkage of the forward peak, i.e. an increase with energy of the slopes of the angular distributions.

---

[3] In honor of the Russian physicist I.Ya. Pomeranchuk, one of the founding fathers of hadronic diffractive physics.

[4] As a matter of fact it was only after the data from the Serpukhov accelerator at $p_{\rm lab} \approx 50 - 60$ GeV became available (in 1969) that total cross sections were recognized to increase with energy; until then, the prevalent belief was that they would, eventually, approach a constant at asymptotically high energies (this was wrongly considered to be a prediction of one of the Pomeranchuk theorems). The increase of the cross sections with energy was predicted theoretically by Cheng and Wu (1970a) and by Frolov, Gribov and Lipatov (1970a).

The interest focused on *elastic* reactions or, more generally, on *exclusive* reactions, i.e. on reactions in which the kinematics of all particles in the final state is fully reconstructed.

From a theoretical point of view, diffraction was a blend of various fairly heterogeneous items.

Let us mention, first of all, the rigorous theorems, like the optical theorem, the Pomeranchuk theorems (Pomeranchuk 1956, 1958), the Froissart-Martin theorem (Froissart 1961, Martin 1963), plus a variety of highly sophisticated theorems derived by several authors (A. Martin, N.N. Khuri, H. Cornille, J.D. Bessis, T. Kinoshita, and many others). While we shall discuss some of these theorems later on (Sect. 4.9), it is impossible to give the proper credit to all those who have contributed to this field. The interested reader is referred to some basic reviews and textbooks (for instance, Eden 1967; Roy 1972; Collins 1977) where details and original references can be found.

General properties of the $S$-matrix, such as crossing and analyticity were postulated and used. Various kinds of approximations and of intuitive or empirical properties were proposed, especially tailored to describe specific (or general) aspects of the problem (examples are the geometrical scaling, duality, etc.) Finally, the powerful techniques of dispersion relations were employed often together with the most naive (in retrospective) intuitions to provide a framework in which to describe the data.

Probably, the most longlasting development of this decade was the introduction of complex angular momenta pioneered in quantum mechanics by T. Regge (1959, 1960) and extended to relativistic particle physics by G.F. Chew and S.C. Frautschi (1961), V.N. Gribov (1961) and others, with the consequent use of the so-called Watson-Sommerfeld transform to represent the scattering amplitude. This formulation proved most useful to describe high energy data: basically all data on angular distributions were easily reproduced with a fairly limited number of Regge trajectories (and of parameters).

It is in this context that, among other things, the pomeron was first introduced as the leading trajectory, i.e. as the dominant singularity in the complex angular momentum plane at high energy.

However, not all the hadronic phenomenology fitted the simple Regge pole scheme. Thus, more complicated singularities (Regge cuts) were introduced and their properties explored.

To summarize, diffractive physics in the 60's was characterized by clean and clever mathematics, but the physical issues remained rather muddy and illdefined.

**Seventies.** During these years the interest shifted gradually from exclusive to *inclusive* reactions, where only one (or, perhaps, two) of the produced particles is detected, while the others are 'summed over'. A very remarkable extension of the optical theorem was formulated by A.H. Mueller (1970) through a clever use of analyticity and crossing. This theorem enabled to extend Regge theory to inclusive processes and even nowadays remains an important tool of the theory of hadronic diffraction.

But, in general, diffractive physics in the 70's lost ground. The main reasons for this were the explosion of interest for deep inelastic scattering (DIS) and the birth of QCD, the gauge field theory of strong interactions. The focus of research in hadronic physics thus moved from *soft* to *hard* processes, which provide tests for the QCD predictions.

It is both enlightening and surprising that diffraction was, eventually, rejuvenated due, to a large extent, to the developments of DIS.

**Eighties.** This decade witnessed a renewal of interest in diffractive physics, due to some important theoretical results.

A. Donnachie and P.V. Landshoff (1987, 1988, 1988/89) proposed an empirically simple Regge-based phenomenology of a variety of high-energy processes, which contributed to bringing the attention back to diffraction.

On the other hand, a perturbative QCD picture of the pomeron, pioneered by the two-gluon model of F.E. Low (1975) and S. Nussinov (1975, 1976), was developed by Y.Y. Balitsky, V.S. Fadin, E.A. Kuraev and L.N. Lipatov (for the original references see Chap. 8; other noteworthy contributions are due to J. Bartels 1979, 1980 and L.V. Gribov, E.M. Levin and M.G. Ryskin 1983).

At the same time, the first discussion of *hard diffraction* and the anticipation of diffractive phenomena in jet production and DIS were offered in a paper by G. Ingelman and P. Schlein (1985).

**Nineties.** In the last decade diffraction made a grandiose comeback. It was observed in DIS at HERA and in jet physics at the Tevatron.

J.D. Bjorken (1993) (see also Dokshitzer, Khoze and Troyan 1987) pointed out an interesting signature of diffractive events, the presence of large rapidity gaps in the final state. These rapidity gaps are an important signature of diffraction, both at HERA and at the Tevatron. At HERA, diffraction is intertwined with fully inclusive deep inelastic scattering at very low-$x$, which provides another window on pomeron physics.

Most important for the resurrection of diffraction in the 90's is that, for the first time, one goes beyond a purely qualitative understanding of the origin and properties of the pomeron both perturbatively and non-perturbatively. Generally speaking, our present understanding of hadronic physics marks a great improvement over the past mostly because of the much larger flexibility of experimental and theoretical tools by which we can approach the problem. A remarkable example, on the theoretical side, is the color dipole approach developed by Nikolaev and Zakharov (1991b, 1992, 1994b) and Mueller (1994), which establishes a link between the quantum mechanical picture of diffraction and QCD.

Some of the basic questions of the present days' debate are: *i)* What is the precise relation between hard and soft diffraction? *ii)* Are there several pomerons (as some believe) or is there only one (as others advocate)? *iii)*

Even more fundamental, is the notion of pomeron meaningful at all? *iv)* What does QCD tell us about diffraction and the pomeron?

It is quite clear that, ultimately, a complete understanding of these matters will require a much deeper mastering of highly complex perturbative *and* non-perturbative techniques. A full understanding of the origin and nature of diffraction, and of the pomeron, is the task for the present century.

# 2. Preliminaries

Since the 50's, the optical analogy has played an important rôle in shaping the research program of high-energy particle diffraction. Kirchhoff's diffraction theory is indeed formally similar to the quantum theory of potential scattering at high energies, the so-called "eikonal[1] approximation", and the latter is the basis of the geometrical and $s$-channel approaches to hadronic diffraction.

The analogy between optical and quantum mechanical diffraction is complete in the case of elastic scattering, where the internal structure of the interacting particles does not come into play. Inelastic diffraction excitations, on the other hand, are related to the compositeness of the particles and to the peculiar fluctuations of quantum fields. In this sense, hadronic diffraction represents a highly non-trivial extension of a macroscopic phenomenon into the microscopic world (for the similarities and differences between optical and hadronic diffraction see Amaldi, Jacob and Matthiae 1976; Alberi and Goggi 1981).

## 2.1 Optics

The first to introduce the notion and to coin the word "diffraction" was Francesco Maria Grimaldi (1618-1663), a Jesuit Professor of Mathematics at Bologna. Grimaldi collected his optical observations and speculations in a book which appeared shortly after his death under the title *Physico-Mathesis de lumine, coloribus, et iride.* In this work, he made the explicit statement

> *Lumen propagatur seu diffunditur non solum directe, refracte ac reflexe etiam quodam quarto modo diffracte*[2].

Diffraction owes its systematization to Joseph Fraunhofer (1787-1826) and Augustin Fresnel (1788-1827) who gave their names to the two main diffractive regimes. Finally, Gustav Kirchhoff (1824-1887) solved the problem of reconstructing the diffraction figure on the basis of Maxwell electromagnetic theory.

---

[1] "Eikonal", from a greek word meaning "image", is a term borrowed from optics.

[2] Light propagates and diffuses not only directly, refractively and reflectively, but also, somehow, in a fourth manner, that is diffractively.

For an exhaustive presentation of diffraction in optics, see the classical treatise by Born and Wolf (1959).

### 2.1.1 Kirchhoff Theory

Consider a plane wave which hits (perpendicularly, for simplicity) a screen $\Sigma$ with a hole $\Sigma_0$ of dimension $R$ (the case of a plane wave hitting an opaque obstacle will be treated later on). Let us suppose that the wave number $k = 2\pi/\lambda$ is sufficiently large that the short wavelength condition

$$kR \gg 1 \tag{2.1}$$

is satisfied.

The wave is represented by a function

$$\varphi(x, y, z, t) = U(x, y, z)\, \mathrm{e}^{-\mathrm{i}\omega t} , \tag{2.2}$$

where $\omega = ck$ and $U(x, y, z)$ satisfies Helmholtz's equation

$$(\nabla^2 + k^2)\, U = 0 . \tag{2.3}$$

The modulus squared of $U$ is the *intensity* of the light, that is the energy carried by the wave per second and unit perpendicular area.

Before the screen, $U$ is simply given by

$$U(x, y, z) = U_0\, \mathrm{e}^{\mathrm{i}kz} , \tag{2.4}$$

with $U_0$ constant.

According to the Huygens-Fresnel principle, each point of the hole becomes the center of a spherical wave. The envelope of these waves will give the resulting deflected wave. Let us put at a distance $D$ from $\Sigma$ another plane where we collect the image (the detector plane). Because of the varying distances and varying angles with the original direction of the beam, the amplitudes and phases of the wavelets collected at each point of the detector plane will be different. As a consequence, cancellations and reinforcements occur, which give rise to the phenomenon of diffraction.

The amplitude of the wave, which is $U_0$ immediately before the screen, is mapped to a value $U(x, y, z)$ at the point $P \equiv (x, y, z)$ of the detector plane. Mathematically, $U(x, y, z)$ is given by the Fresnel-Kirchhoff formula (see, for instance, Born and Wolf 1959)

$$U(x, y, z) = -\frac{\mathrm{i}k}{4\pi}\, U_0 \int_{\Sigma_0} \mathrm{d}^2\boldsymbol{b}\, (1 + \cos\chi)\, \frac{\mathrm{e}^{\mathrm{i}ks}}{s} , \tag{2.5}$$

where $\boldsymbol{b} \equiv (b_1, b_2, 0)$ is a vector lying on the plane $\Sigma$, $\boldsymbol{s}$ is the vector connecting $P$ to the point of $\Sigma_0$ individuated by $\boldsymbol{b}$, $\cos\chi$ is the inclination of $\boldsymbol{s}$ with respect to the normal to $\Sigma$ (see Fig. 2.1).

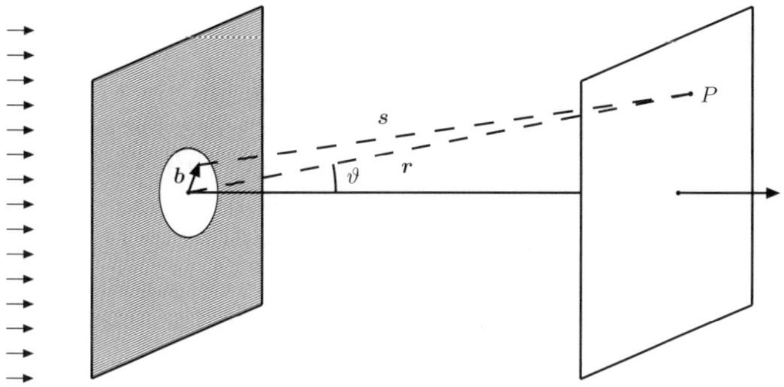


**Fig. 2.1.** Diffraction of a plane wave by a hole in a screen.

We define also the vector $\boldsymbol{r}$ as

$$\boldsymbol{r} = \boldsymbol{b} + \boldsymbol{s} \, . \tag{2.6}$$

The short wavelength condition (2.1) implies that the wave is diffracted at small angles, hence $\cos\chi \simeq 1$ and the Fresnel-Kirchhoff integral becomes

$$U(x, y, z) = -\frac{\mathrm{i}k}{2\pi} U_0 \int_{\Sigma_0} \mathrm{d}^2\boldsymbol{b} \, \frac{\mathrm{e}^{\mathrm{i}ks}}{s} \, , \tag{2.7}$$

What we have to do now is to calculate the integral (2.7). The problem is greatly simplified when the *large distance condition* holds

$$R/D \ll 1 \, . \tag{2.8}$$

In this case we can expand the distance $s$ in (2.7) as a power series of $b/r$. There are two diffraction regimes:

- *Fraunhofer diffraction*, when

$$kR^2/D \ll 1 \, ; \tag{2.9}$$

- *Fresnel diffraction*, when

$$kR^2/D \approx 1 \, . \tag{2.10}$$

The *geometrical optics* limit is obtained when

$$kR^2/D \gg 1 \, . \tag{2.11}$$

Note that there are three parameters with dimension of length: the wavelength, the size of the disk and the distance of the detector. Keeping the first two fixed in such a way that (2.1) holds, one observes, near the obstacle, the geometrical shadow, then, moving away from it, the Fresnel diffraction figure, and finally, at large distance, the Fraunhofer diffraction pattern.

### 2.1.2 Fraunhofer Diffraction

Fraunhofer diffraction is what we are interested in, in view of the application of optical concepts to hadronic phenomena. This can be seen by a simple back-of-the-envelope evaluation. If $R \sim 1$ fm, as in the case of a hadronic target, $D \gtrsim 1$ cm and $k \sim \sqrt{s} \sim 200$ GeV (a typical accelerator energy), then one has $kR \sim 10^3$ and $D/R \gtrsim 10^{13}$, and therefore $kR^2 \ll D$.

In the Fraunhofer regime (which corresponds to the genuine limit $D \to \infty$) we can neglect the terms quadratic in $b/r$ in the expansion of the exponent of $\mathrm{e}^{iks}$ in (2.7), that is

$$\begin{aligned} ks &= k\,|\boldsymbol{r} - \boldsymbol{b}| \\ &\simeq kr - k\,\frac{\boldsymbol{r}\cdot\boldsymbol{b}}{r}\,, \end{aligned} \tag{2.12}$$

and set $s \simeq r$ in the denominator of the integrand of (2.7).

In the second term on the r.h.s. of Eq. (2.12), $\boldsymbol{b}$ selects the transverse part of $k\,\boldsymbol{r}/r$, i.e.

$$\boldsymbol{q} \equiv \frac{k}{r}\,\boldsymbol{r}_\perp = \left(\frac{kx}{r}, \frac{ky}{r}, 0\right) . \tag{2.13}$$

Eq. (2.12) then reads

$$ks \simeq kr - \boldsymbol{q}\cdot\boldsymbol{b}\,, \tag{2.14}$$

The modulus of $\boldsymbol{q}$ is

$$|\boldsymbol{q}| = k\sin\vartheta \simeq k\vartheta\,, \tag{2.15}$$

where the last approximate equality is valid for small angles. In this limit $\boldsymbol{q}$ is the momentum transfer

$$\boldsymbol{q} \simeq \boldsymbol{k}' - \boldsymbol{k}\,, \tag{2.16}$$

where $\boldsymbol{k}'$ is the outgoing wave vector and $|\boldsymbol{k}| = |\boldsymbol{k}'| = k$. It should be kept in mind that the momentum transfer $\boldsymbol{q}$ is a transverse two-vector.

We can now write the wave function at the point $P(x, y, z)$ in the Fraunhofer limit as

$$U(x,y,z) = -\frac{ik}{2\pi}\,U_0\,\frac{\mathrm{e}^{ikr}}{r}\int_{\Sigma_0} \mathrm{d}^2\boldsymbol{b}\;\mathrm{e}^{-\mathrm{i}\boldsymbol{q}\cdot\boldsymbol{b}}\,. \tag{2.17}$$

In Eq. (2.17) the integration covers the surface $\Sigma_0$ of the hole. We can introduce a function $\Gamma(\boldsymbol{b})$, called *profile function*, such that

$$\Gamma(\boldsymbol{b}) = \begin{cases} 1 \text{ on } \Sigma_0 \\ 0 \text{ outside } \Sigma_0 \end{cases} \tag{2.18}$$

Thus the integral (2.17) can be rewritten as

$$U(x,y,z) = -\frac{\mathrm{i}k}{2\pi}\,U_0\,\frac{\mathrm{e}^{ikr}}{r}\int \mathrm{d}^2\boldsymbol{b}\;\Gamma(\boldsymbol{b})\,\mathrm{e}^{-\mathrm{i}\boldsymbol{q}\cdot\boldsymbol{b}}\,, \quad (\text{hole } \Sigma_0)\,, \tag{2.19}$$

where now the integration is extendend over the whole plane containing the aperture.

So far we have considered the diffraction by a hole in a screen. However, the optical analogue of high-energy particle scattering is the diffraction by an opaque obstacle. Babinet's principle (see Born and Wolf 1959) states that a hole and an obstacle of identical form and dimension produce the same diffraction figure. In fact, as a consequence of the Huygens-Fresnel principle, the waves diffracted by the hole and by the obstacle must combine in such a way to reconstruct the incident wave front. Thus the amplitude of the wave diffracted by an opaque disk $\Sigma_0$, identical to the hole in Fig. 2.1, is obtained subtracting the plane wave (2.19) from (2.4). This is equivalent to replacing $\Gamma(\boldsymbol{b})$ in (2.19) by

$$S(\boldsymbol{b}) = 1 - \Gamma(\boldsymbol{b}) \,, \tag{2.20}$$

where $S(\boldsymbol{b})$ is the analogue of the $S$-matrix of particle physics and describes the modification of the wave to the diffracting obstacle.

The wave amplitude beyond the obstacle is then

$$U(x, y, z) = -\frac{\mathrm{i}k}{2\pi}\, U_0\, \frac{\mathrm{e}^{\mathrm{i}kr}}{r} \int \mathrm{d}^2\boldsymbol{b}\; S(\boldsymbol{b})\, \mathrm{e}^{-\mathrm{i}\boldsymbol{q}\cdot\boldsymbol{b}} \,, \qquad (\text{disk } \Sigma_0) \,. \tag{2.21}$$

The unity in (2.20) represents the unperturbed wave. The second term, $\Gamma(\boldsymbol{b})$, yields the diffracted wave.

The profile function $\Gamma(\boldsymbol{b})$ is not necessarily a simple step function such as (2.18). It can be a more complicated function of $\boldsymbol{b}$, as in the case, for instance, of partially transmitting disks, filters, etc. (see Born and Wolf 1959).

The two extreme cases are:

- $\Gamma(\boldsymbol{b}) = 1$, i.e. $S(\boldsymbol{b}) = 0$, everywhere, which corresponds to a completely opaque screen (no trasmission).
- $\Gamma(\boldsymbol{b}) = 0$, i.e. $S(\boldsymbol{b}) = 1$, everywhere, which corresponds to a completely transparent screen (no obstacle, no diffraction).

The $S$-matrix satisfies a normalization condition due to energy conservation. This implies

$$A\,|U_0|^2 = \int \mathrm{d}x\,\mathrm{d}y\,|U(x, y, z)|^2 \,, \tag{2.22}$$

where $A$ is the area of the opening and the integral on the r.h.s. is extended over the detector's plane. On the other hand, applying Parseval's theorem to Eq. (2.21) and remembering the definition (2.13), we get

$$\int \mathrm{d}x\,\mathrm{d}y\,|U(x, y, z)|^2 = |U_0|^2 \int \mathrm{d}^2\boldsymbol{b}\;|S(\boldsymbol{b})|^2 \,, \tag{2.23}$$

and thus we must have

$$\int \mathrm{d}^2\boldsymbol{b}\;|S(\boldsymbol{b})|^2 = A \,. \tag{2.24}$$

This is the constraint on $S(\boldsymbol{b})$ arising from the conservation of energy.

Equation (2.21) can be written as (we neglect the constant phase shift of the plane wave)

$$\begin{aligned} U(x,y,z) &= U_{\text{inc}} + U_{\text{scatt}} \\ &= U_0 \left( \mathrm{e}^{\mathrm{i}kz} + f(\boldsymbol{q}) \frac{\mathrm{e}^{\mathrm{i}kr}}{r} \right) , \end{aligned} \tag{2.25}$$

where $U_{\text{inc}}$ is the incident plane wave and $U_{\text{scatt}}$ is the scattered wave.

The factor $f(\boldsymbol{q})$ multiplying the spherical wave $\mathrm{e}^{\mathrm{i}kr}/r$ is the physically relevant quantity, the *scattering amplitude*, given by

$$f(\boldsymbol{q}) = \frac{\mathrm{i}k}{2\pi} \int \mathrm{d}^2\boldsymbol{b}\, \Gamma(\boldsymbol{b})\, \mathrm{e}^{-\mathrm{i}\boldsymbol{q}\cdot\boldsymbol{b}} , \tag{2.26}$$

The scattering amplitude is the Fourier transform of the profile function. Inverting the Fourier integral (2.26), one gets[3]

$$\Gamma(\boldsymbol{b}) = \frac{1}{2\pi \mathrm{i}k} \int \mathrm{d}^2\boldsymbol{q}\, f(\boldsymbol{q})\, \mathrm{e}^{\mathrm{i}\boldsymbol{q}\cdot\boldsymbol{b}} . \tag{2.27}$$

If the profile function is azimuthally symmetric, i.e. if it depends on $b \equiv |\boldsymbol{b}|$ only, we can perform the integration over the azimuthal angle $\phi$ in (2.26) by means of

$$\frac{1}{2\pi} \int_0^{2\pi} \mathrm{e}^{\mathrm{i}x\,\cos\phi}\, \mathrm{d}\phi = j_0(x) , \tag{2.28}$$

and write the scattering amplitude as the Bessel transform of $\Gamma(b)$

$$f(q) = \mathrm{i}k \int_0^{\infty} \mathrm{d}b\, b\, J_0(qb)\, \Gamma(b) . \tag{2.29}$$

### 2.1.3 Cross-sections

The intensity of the incident and of the scattered light are

$$I_{\text{inc}} = |U_{\text{inc}}|^2 = |U_0|^2 , \tag{2.30}$$

and

$$I_{\text{scatt}} = |U_{\text{scatt}}| = |U_0|^2 \, \frac{|f(\boldsymbol{q})|^2}{r^2} , \tag{2.31}$$

respectively. We define the *differential cross section* $\mathrm{d}\sigma$ as the ratio of the outgoing energy in the solid angle $\mathrm{d}\Omega$ to the incident energy flux, namely

$$\mathrm{d}\sigma = \frac{I_{\text{scatt}}\, r^2 \mathrm{d}\Omega}{I_{\text{inc}}} . \tag{2.32}$$

---

[3] Here we ignore subtleties related to the (finite) range of variation of $\boldsymbol{q}$ on the presumption that the integrand vanishes fast enough.

Using Eqs. (2.31,2.32) we find that the differential cross section is the modulus squared of the scattering amplitude

$$\frac{d\sigma}{d\Omega} = |f(\boldsymbol{q})|^2 . \tag{2.33}$$

The integrated *scattering cross section* is, by definition

$$\sigma_{\text{scatt}} \equiv \int \frac{d\sigma}{d\Omega} \, d\Omega = \frac{1}{k^2} \int |f(\boldsymbol{q})|^2 \, d^2\boldsymbol{q} , \tag{2.34}$$

where in the second equality we have used the relation

$$d^2\boldsymbol{q} = \pi \, dq^2 \simeq \pi(2k^2\vartheta \, d\vartheta) \simeq k^2 \, d\Omega , \tag{2.35}$$

which follows from (2.15). Parseval's theorem then leads to

$$\sigma_{\text{scatt}} = \int d^2\boldsymbol{b} \, |\Gamma(\boldsymbol{b})|^2 = \int d^2\boldsymbol{b} \, |1 - S(\boldsymbol{b})|^2 . \tag{2.36}$$

The *absorption cross section* $\sigma_{\text{abs}}$, that is the ratio of the absorbed energy to the incident energy flux, is given by

$$\begin{aligned} \sigma_{\text{abs}} &= \int d^2\boldsymbol{b} \, (1 - |S(\boldsymbol{b})|^2) \\ &= \int d^2\boldsymbol{b} \, [2 \, \text{Re} \, \Gamma(\boldsymbol{b}) - |\Gamma(\boldsymbol{b})|^2] . \end{aligned} \tag{2.37}$$

The *total cross section*, finally, is obtained by summing the scattering cross section (2.36) and the absorption cross section (2.37)

$$\begin{aligned} \sigma_{\text{tot}} &= \sigma_{\text{scatt}} + \sigma_{\text{abs}} \\ &= 2 \int d^2\boldsymbol{b} \, (1 - \text{Re} \, S(\boldsymbol{b})) = 2 \int d^2\boldsymbol{b} \, \text{Re} \, \Gamma(\boldsymbol{b}) . \end{aligned} \tag{2.38}$$

Combining (2.38) with (2.26) we derive the so-called *optical theorem*

$$\sigma_{\text{tot}} = \frac{4\pi}{k} \, \text{Im} \, f(\vartheta = 0) , \tag{2.39}$$

which states that the total cross section equals (apart from a factor) the scattering amplitude in the forward direction (remember that $\vartheta = 0$ corresponds to $\boldsymbol{q} = 0$). The optical theorem is a consequence of the conservation of energy. We shall see that an analogous relation holds in particle physics, as a consequence of the conservation of probability.

### 2.1.4 Examples

We illustrate the results of the previous section with two examples.

**Black disk.** A profile function of the type

$$\Gamma(b) = \begin{cases} 1 & b \leq R\,, \\ 0 & b > R\,. \end{cases} \tag{2.40}$$

corresponds to diffraction by a black disk of radius $R$. Using Eq. (2.29) we get for the scattering amplitude

$$f(q) = \mathrm{i}kR^2\,\frac{J_1(qR)}{qR}\,. \tag{2.41}$$

The differential cross section (2.33) is, thus

$$\begin{aligned} \frac{\mathrm{d}\sigma}{\mathrm{d}\cos\vartheta} &= \frac{2\pi R^2}{\vartheta^2}\,[J_1(kR\vartheta)]^2 \\ &\underset{\vartheta\to 0}{\simeq} \frac{2\pi R^2}{\vartheta^2}\left[\frac{1}{4}\,(kR\vartheta)^2 - \frac{1}{16}\,(kR\vartheta)^4\right] \\ &= \frac{\pi}{2}\,k^2R^4\left[1 - \frac{1}{4}\,(kR\vartheta)^2\right]\,, \end{aligned} \tag{2.42}$$

where use has been made of the expansion of the Bessel functions for small arguments

$$J_1(x) \underset{x\to 0}{\sim} \frac{1}{2}\,x - \frac{1}{16}\,x^3 + \ldots\,. \tag{2.43}$$

The total cross section can be obtained by the optical theorem and is given by

$$\sigma_{\mathrm{tot}} = \frac{4\pi}{k}\,\mathrm{Im}\,f(0) = 2\pi R^2\,. \tag{2.44}$$

This is the sum of the elastic and the absorption cross sections, each of which is equal to the geometrical area of the disk

$$\sigma_{\mathrm{scatt}} = \sigma_{\mathrm{abs}} = \pi R^2\,. \tag{2.45}$$

**Gaussian profile.** For a Gaussian profile

$$\Gamma(b) = \exp\left(-\frac{b^2}{R^2}\right)\,, \tag{2.46}$$

one easily finds

$$f(q) = \mathrm{i}\,\frac{kR^2}{2}\,\exp\left(-\frac{q^2R^2}{4}\right)\,, \tag{2.47}$$

and

$$\begin{aligned} \frac{\mathrm{d}\sigma}{\mathrm{d}\cos\vartheta} &= \frac{\pi}{2}\,k^2R^4\,\exp\left(-\frac{q^2R^2}{2}\right) \\ &\underset{\vartheta\to 0}{\simeq} \frac{\pi}{2}\,k^2R^4\left[1 - \frac{1}{2}\,k^2R^2\vartheta^2\right]\,. \end{aligned} \tag{2.48}$$

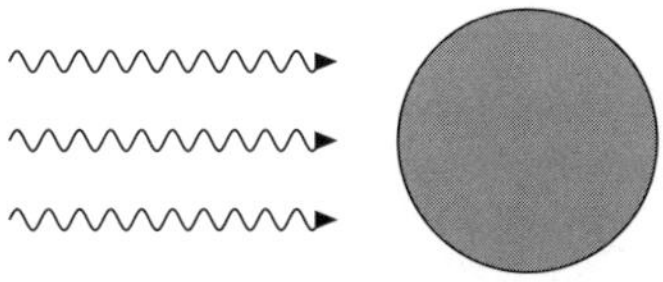

**Fig. 2.2.** Scattering of light by a sphere.

As in the case of the black disk the total cross section is

$$\sigma_{\text{tot}} = 2\pi R^2 \,, \tag{2.49}$$

but the scattering cross section and the absorption cross section are now different

$$\sigma_{\text{scatt}} = \frac{\pi R^2}{2} \,, \tag{2.50}$$

$$\sigma_{\text{abs}} = \frac{3\pi R^2}{2} \,, \tag{2.51}$$

In both examples the differential cross section behaves at small angles as

$$\frac{\mathrm{d}\sigma}{\mathrm{d}\cos\vartheta} = \left.\frac{\mathrm{d}\sigma}{\mathrm{d}\cos\vartheta}\right|_{\vartheta=0} \simeq 1 - Bk^2\vartheta^2 \,, \tag{2.52}$$

where $B \propto R^2$, namely it exhibits a rapid fall as we move away from the forward direction.

### 2.1.5 Scattering of Light by a Sphere

In view of the study of particle scattering at high energies, we now draw another analogy with electromagnetism. Let us consider the scattering of light by a perfectly conducting sphere of radius $R$ (Van de Hulst 1957, King and Wu 1959, Wu 1995). Imposing $E_\vartheta = E_\varphi = 0$ on the surface of the sphere, and using a partial wave expansion, the cross section is found to be ($k$ is the wave number)

$$\sigma_{\text{tot}} = \frac{2\pi}{k^2}\,\mathrm{Re}\sum_{n=0}^{\infty}(2n+1)\left\{\frac{j_n(\kappa)}{h_n^{(1)}(\kappa)} + \frac{(\partial/\partial\kappa)\,[\kappa\, j_n(\kappa)]}{(\partial/\partial\kappa)\,[\kappa\, h_n^{(1)}(\kappa)]}\right\} \,, \tag{2.53}$$

where $\kappa \equiv kR$ and $j_n, h_n^{(1)}$ are spherical Bessel functions. In the limit of short wavelengths ($\kappa \gg 1$), which is similar to the Kirchhoff domain of diffraction, the asymptotic expansion of (2.53) gives

$$\sigma_{\text{tot}} \underset{\kappa\gg 1}{\sim} \frac{2\pi}{k^2}\,\mathrm{Re}\sum_{n=0}^{[\kappa]}(2n+1)\left(\frac{1}{2}+\frac{1}{2}\right) \sim \frac{2\pi}{k^2}\,[\kappa]^2 \sim 2\pi R^2 \,, \tag{2.54}$$

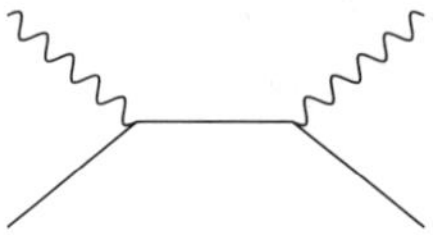
(a)

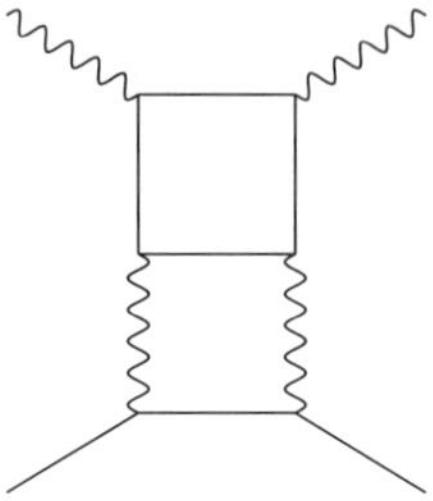
(b)

**Fig. 2.3.** Feynman diagrams contributing to Compton scattering. (**a**) order $\alpha_{\rm em}$; (**b**) order $\alpha_{\rm em}^3$.

where $[\kappa]$ is the largest integer $\leq \kappa$. We find again that, in the limit $k \to \infty$ with fixed $R$, the total cross section does not depend on $k$ (i.e., on the energy of light) and is twice the geometrical projected area of the sphere.

Now, what happens if we consider the scattering of very energetic photons by elementary particles? Obviously, classical electromagnetism must be replaced by quantum electrodynamics. It turns out that the excitation of the vacuum, which is a quantum field theory phenomenon with no classical analog, has a striking effect on the energy dependence of the cross section. Consider for instance Compton scattering ($\gamma e \to \gamma e$). While the lowest-order diagram (Fig. 2.3a) gives a contribution that goes as $s^0$, where $s$ is the center-of-mass squared energy, the diagrams with particle production in the intermediate state (Fig. 2.3b), which have no classical counterpart, contribute $\mathcal{O}(s)$ terms (modulo logarithms) and dominate at high energies. The total cross section then grows with the energy, as shown long ago by Cheng and Wu (1970a) and by Frolov, Gribov and Lipatov (1970a). Essentially, the same mechanism is at work in quantum chromodynamics and leads to increasing cross sections in photon–proton and hadron–hadron scattering as well. This has led T.T. Wu (1995) to the extraordinary conclusion that Maxwell equations (and their extension to non-abelian fields) "describe electromagnetic waves correctly over at least 18 orders of magnitude from the Edison-Hertz to the HERA wavelengths".

## 2.2 Potential Scattering

The second basic tool that we introduce very concisely on the road to particle diffraction is the quantum theory of non-relativistic potential scattering.[4]

There are two approaches: one starts from the Schrödinger equation with a given potential, the other is based on quite general properties of the scattering

[4] Although throughout the book we adopt natural units, in this section we write explicitly the $\hbar$ factors.

amplitudes. The former approach is more transparent, the latter is more widely applicable.

### 2.2.1 The Schrödinger Equation Approach

The relative motion of two spinless particles interacting via the potential $V(\boldsymbol{r})$ is governed by the Schrödinger equation[5]

$$-\frac{\hbar^2}{2\mu}\nabla^2\psi(\boldsymbol{r}) + V(\boldsymbol{r})\,\psi(\boldsymbol{r}) = E\,\psi(\boldsymbol{r})\,, \tag{2.55}$$

where $\mu$ is the reduced mass of the system. The problem is equivalent to that of the scattering of a particle of mass $\mu$ off a fixed potential $V(\boldsymbol{r})$ which will be supposed to have a finite range (Fig. 2.4).

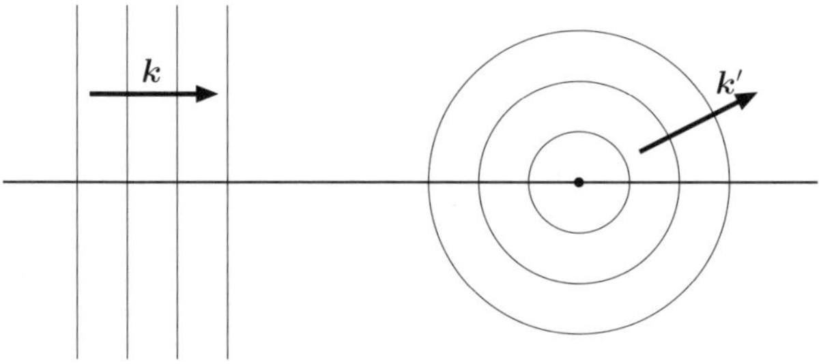


**Fig. 2.4.** Potential scattering.

Let us simplify the writing of Eq. (2.55) by introducing

$$k^2 = \frac{2\mu}{\hbar^2}E\,, \quad U(\boldsymbol{r}) = \frac{2\mu}{\hbar^2}V(\boldsymbol{r})\,, \tag{2.56}$$

so that Eq. (2.55) becomes

$$[\nabla^2 - U(\boldsymbol{r}) + k^2]\,\psi(\boldsymbol{r}) = 0\,. \tag{2.57}$$

We assume that the initial state is a plane wave. Then, asymptotically (i.e. as $r \to \infty$), the wave function of the system is a superposition of this incoming plane wave and the outgoing spherical wave emerging from the scattering center, that is

$$\psi(\boldsymbol{r}) \underset{r\to\infty}{\sim} \mathrm{e}^{\mathrm{i}\boldsymbol{k}\cdot\boldsymbol{r}} + f(\boldsymbol{k},\boldsymbol{k}')\,\frac{\mathrm{e}^{\mathrm{i}kr}}{r}\,, \tag{2.58}$$

[5] For simplicity, we consider here only time-independent spinless scattering. For a more general treatment the reader may consult Goldberger and Watson (1964), or Newton (1966).

where $\boldsymbol{k}$ and $\boldsymbol{k}'$ are the wave vectors of the incoming and outgoing wave respectively, and energy conservation requires

$$|\boldsymbol{k}| = |\boldsymbol{k}'| \equiv k \,. \tag{2.59}$$

In Eq. (2.58) $f(\boldsymbol{k},\boldsymbol{k}')$ is the *scattering amplitude*, which contains all the dynamical information on the collision process. In general, it is a function of the vectors $\boldsymbol{k},\boldsymbol{k}'$, or, equivalently, of $k$ and of the polar and azimuthal angles $(\vartheta,\phi)$ of $\boldsymbol{k}'$ relative to $\boldsymbol{k}$.

The solution of the Schrödinger equation (2.57) satisfying the boundary condition (2.58) is given by the Lippman-Schwinger integral equation (in the configuration space)

$$\psi(\boldsymbol{r}) = \mathrm{e}^{\mathrm{i}\boldsymbol{k}\cdot\boldsymbol{r}} - \frac{1}{4\pi}\int \frac{\mathrm{e}^{\mathrm{i}k\,|\boldsymbol{r}-\boldsymbol{r}'|}}{|\boldsymbol{r}-\boldsymbol{r}'|}\, U(\boldsymbol{r}')\,\psi(\boldsymbol{r}')\,\mathrm{d}^3\boldsymbol{r}' \,. \tag{2.60}$$

If the potential goes to zero at infinity more rapidly than $1/r$, that is $r\,U(r) \underset{r\to\infty}{\longrightarrow} 0$, we can expand $|\boldsymbol{r}-\boldsymbol{r}'|$ as

$$|\boldsymbol{r}-\boldsymbol{r}'| \simeq r - \boldsymbol{r}'\cdot\frac{\boldsymbol{r}}{r} \,, \tag{2.61}$$

and comparing (2.60) with (2.58) we find an expression for the scattering amplitude in terms of the particle wave function (remember that $\boldsymbol{r}$ and $\boldsymbol{k}'$ are parallel)

$$f(\boldsymbol{k},\boldsymbol{k}') = -\frac{1}{4\pi}\int \mathrm{e}^{\mathrm{i}\boldsymbol{k}'\cdot\boldsymbol{r}}\, U(\boldsymbol{r}')\,\psi(\boldsymbol{r}')\,\mathrm{d}^3\boldsymbol{r}' \,. \tag{2.62}$$

A vector that we shall often use in the following is the three-momentum transfer

$$\boldsymbol{q} = \boldsymbol{k}' - \boldsymbol{k} \,, \tag{2.63}$$

whose modulus is

$$q = 2\,k\sin\frac{\vartheta}{2} \,. \tag{2.64}$$

The *differential cross section* of the scattering process at hand is simply the squared modulus of the amplitude, i.e. $(\mathrm{d}\Omega = \mathrm{d}\phi\,\mathrm{d}\cos\vartheta)$

$$\frac{\mathrm{d}\sigma_{\mathrm{el}}}{\mathrm{d}\Omega} = |f(k,\vartheta,\phi)|^2 \,. \tag{2.65}$$

Notice that (2.65) is an *elastic cross section* (this is the reason of the subscript 'el'). So far, in fact, we have considered only elastic scattering (assuming, implicitly, $V(\boldsymbol{r})$ to be real). This means that, denoting by $|\boldsymbol{k}\,\lambda\rangle$ and $|\boldsymbol{k}'\,\lambda'\rangle$ the initial and final state respectively ($\lambda$ is any quantum number, besides the momentum, characterizing these states), we have (tacitly) supposed $\lambda = \lambda'$.

Absorption effects can be accounted for in the Schrödinger approach by means of the so-called *optical model*. This approach assumes that the potential is complex and, consequently, that the phase shifts develop an imaginary part. We do not pursue this model here as we shall discuss inelastic scattering in full generality within the $S$-matrix approach.

### 2.2.2 Partial Wave Expansion

For a *spherically symmetric potential* (this is the case we shall stick to hereafter) $f$ does not depend on $\phi$: $f(k,\vartheta)$, and, integrating over the azimuthal angle, Eq. (2.65) reduces to

$$\frac{\mathrm{d}\sigma_{\mathrm{el}}}{\mathrm{d}\cos\vartheta} = 2\pi\,|f(k,\vartheta)|^2 \,. \tag{2.66}$$

The scattering amplitude for central potentials can be expanded in a partial wave series

$$f(k,\vartheta) = \sum_{\ell=0}^{\infty} (2\ell+1)\, a_\ell(k)\, P_\ell(\cos\vartheta) \,, \tag{2.67}$$

where the sum is made over all possible values of the angular momentum $\ell$ and $P_\ell$ is a Legendre polynomial containing the angular dependence. For the partial wave amplitudes $a_\ell(k)$ one finds the structure (see, for instance, Goldberger and Watson 1964; Newton 1966)

$$a_\ell(k) = \frac{\mathrm{e}^{2\mathrm{i}\delta_\ell(k)} - 1}{2\mathrm{i}k} \equiv \frac{S_\ell(k) - 1}{2\mathrm{i}k} \,. \tag{2.68}$$

In Eq. (2.68) $\delta_\ell(k)$ is known as the *phase shift* of the $\ell$-th wave and we have called $S_\ell(k)$ the quantity $\mathrm{e}^{2\mathrm{i}\delta_\ell(k)}$ to remind that it is the element of the scattering matrix $S$ in a state of definite orbital angular momentum $\ell$ (see below). The meaning of $\delta_\ell$ can be understood comparing the asymptotic behaviors of the free radial wave function $u_\ell$, which is a Bessel function

$$u_\ell(r) \underset{r\to\infty}{\sim} \frac{1}{kr} \sin\left(kr - \frac{\pi\ell}{2}\right) \,, \tag{2.69}$$

and of the radial wave function $\varphi_\ell$ of the outgoing state

$$\varphi_\ell(r) \underset{r\to\infty}{\sim} \frac{\mathrm{e}^{\mathrm{i}\delta_\ell}}{kr} \sin\left(kr - \frac{\pi\ell}{2} + \delta_\ell\right) \,. \tag{2.70}$$

The only difference between the wave functions before and after the scattering (or, which is the same, without or with the potential $V(r)$) is the phase $\delta_\ell$, which contains all the dynamical information on the process.

Substituting Eq. (2.67) with (2.68) in (2.66) and using the orthogonality relation for Legendre polynomials

$$\int_{-1}^{1} \mathrm{d}\cos\vartheta\, P_\ell(\cos\vartheta) P_{\ell'}(\cos\vartheta) = \frac{2\,\delta_{\ell\ell'}}{2\ell+1} , \tag{2.71}$$

to integrate over the polar angle, we get the *integrated elastic cross section*[6]

$$\begin{aligned} \sigma_{\mathrm{el}}(k) &= \int \frac{\mathrm{d}\sigma_{\mathrm{el}}}{\mathrm{d}\cos\vartheta}\,\mathrm{d}\cos\vartheta \\ &= \frac{\pi}{k^2}\sum_{\ell=0}^{\infty}(2\ell+1)\,|S_\ell - 1|^2 \\ &= \frac{4\pi}{k^2}\sum_{\ell=0}^{\infty}(2\ell+1)\,\sin^2\delta_\ell\,. \end{aligned} \tag{2.72}$$

### 2.2.3 The $S$-Matrix Approach

We describe now the second, and more general, approach to quantum scattering, the $S$-matrix formalism. This formalism can be used to extend the treatment to inelastic scattering, spinning particles, non-central fields, time-dependent processes, multi-channel reactions, relativistic collisions etc. (a detailed presentation of the relativistic $S$-matrix theory will be offered in Chap. 4).

First of all, we rederive in the $S$-matrix language the results already obtained within the Schrödinger approach, then we shall extend our analysis to inelastic reactions.

The $S$-matrix is a unitary operator which transforms the initial state $|i\rangle$ of a scattering process into the final state $|f\rangle$

$$|f\rangle = S\,|i\rangle\,. \tag{2.73}$$

$S$ is related to the unitary time-evolution operator $U(t_2, t_1)$ connecting a state at time $t_1$ to the state at time $t_2$ by

$$S = U(-\infty, +\infty)\,, \tag{2.74}$$

and is given in terms of the interaction Hamiltonian $H'_{\mathrm{int}}$ of the system (in the interaction representation) by Dyson's formula

$$S = \mathcal{T}\,\mathrm{e}^{-\frac{\mathrm{i}}{\hbar}\int_{-\infty}^{+\infty} H'_{\mathrm{int}}(t)\,\mathrm{d}t}\,, \tag{2.75}$$

where $\mathcal{T}$ denotes time ordering. The unitarity of $S$

$$SS^\dagger = S^\dagger S = 1\!\!1\,, \tag{2.76}$$

implies that (2.73) can be inverted to give

[6] We reserve the name 'total cross section' to the integrated elastic + inelastic cross section (see below).

$$|i\rangle = S^\dagger |f\rangle \,. \tag{2.77}$$

Standard manipulations (that can be found in most quantum mechanics books and for the relativistic case will be presented in more detail in Chap. 4) lead to the following expression of the differential cross section for the process $|\boldsymbol{k}\,\lambda\rangle \to |\boldsymbol{k}'\,\lambda'\rangle$ in terms of the $S$-matrix elements

$$\mathrm{d}\sigma = \frac{4\pi^2}{k^2}\,|\langle \hat{\boldsymbol{k}}'\,\lambda'|S - \mathbb{1}|\hat{\boldsymbol{k}}\,\lambda\rangle|^2\,\mathrm{d}\Omega \,. \tag{2.78}$$

Here energy conservation is enforced by setting $k = k'$ (the $S$-matrix is taken on the energy shell). Since $k$ is fixed, we label the 'in' and 'out' states only by the directions $\hat{\boldsymbol{k}}, \hat{\boldsymbol{k}}'$ of their momenta, and we denote collectively all other quantum numbers by $\lambda, \lambda'$. These states are normalized as

$$\langle \hat{\boldsymbol{k}}'\,\lambda'|\hat{\boldsymbol{k}}\,\lambda\rangle = \delta_{\lambda\lambda'}\,\delta(\hat{\boldsymbol{k}} - \hat{\boldsymbol{k}}') = \delta_{\lambda\lambda'}\,\delta(\Omega_k - \Omega_{k'}) \,. \tag{2.79}$$

In the coordinate space the states $|\hat{\boldsymbol{k}}\rangle$ are plane waves

$$\psi_{\hat{\boldsymbol{k}}}(\boldsymbol{r}) \equiv \langle \boldsymbol{r}|\hat{\boldsymbol{k}}\rangle = \frac{\sqrt{\mu k}}{(2\pi)^{3/2}\hbar}\,\mathrm{e}^{\mathrm{i}\boldsymbol{k}\cdot\boldsymbol{r}} \,, \tag{2.80}$$

whose normalization follows from (2.79).

Equation (2.78) is a general formula, valid for any scattering process. We can introduce the scattering amplitude $f(k, \vartheta, \phi; \lambda, \lambda')$, such that

$$\frac{\mathrm{d}\sigma}{\mathrm{d}\Omega} = |f(k, \vartheta, \phi; \lambda, \lambda')|^2 \,. \tag{2.81}$$

In terms of the $S$-matrix elements the scattering amplitude is

$$f(k, \vartheta, \phi; \lambda, \lambda') = -\frac{2\pi\mathrm{i}}{k}\,\langle \hat{\boldsymbol{k}}'\,\lambda'|S - \mathbb{1}|\hat{\boldsymbol{k}}\,\lambda\rangle \,, \tag{2.82}$$

where the phase of $f(k, \vartheta, \phi; \lambda, \lambda')$ has been chosen in order to regain the formulas of Sect. 2.2.1 in the case of elastic central scattering (see below).

When the scattering is *elastic* one has $\lambda = \lambda'$. The quantum numbers lumped into $\lambda$ are untouched by the collision process and we can omit them in the expression for the scattering amplitude, which becomes then

$$f(k, \vartheta, \phi) = -\frac{2\pi\mathrm{i}}{k}\,\langle \hat{\boldsymbol{k}}'|S - \mathbb{1}|\hat{\boldsymbol{k}}\rangle \,. \tag{2.83}$$

### 2.2.4 Born Approximation

Retaining only the first order term in the expansion of the Dyson formula (2.75) for the $S$-matrix, we find what is known as the *Born approximation*

$$S \simeq \mathbb{1} - \frac{\mathrm{i}}{\hbar} \int_{-\infty}^{+\infty} H'_{\mathrm{int}}(t)\, \mathrm{d}t \; . \tag{2.84}$$

The matrix elements of (2.84) on the energy shell are

$$\langle \hat{\boldsymbol{k}}' \, \lambda' | S - \mathbb{1} | \hat{\boldsymbol{k}} \, \lambda \rangle \simeq -2\pi \mathrm{i} \, \langle \hat{\boldsymbol{k}}' \, \lambda' | H' | \hat{\boldsymbol{k}} \, \lambda \rangle \; , \tag{2.85}$$

where $H'$ is the interaction Hamiltonian in Schrödinger representation. Hence, in Born approximation, the scattering amplitude (2.82) is given by

$$f(k, \vartheta, \phi; \lambda, \lambda') = -\frac{4\pi}{k} \langle \hat{\boldsymbol{k}}' \, \lambda' | H' | \hat{\boldsymbol{k}} \, \lambda \rangle \; . \tag{2.86}$$

If the interaction Hamiltonian $H'$ is purely spatial, that is a potential $V(\boldsymbol{r})$ (see Eq. (2.55)), the r.h.s. of (2.86) reduces to the Fourier transform of $V(\boldsymbol{r})$

$$f(k, \vartheta, \phi) = -\frac{\mu}{2\pi\hbar^2} \int V(\boldsymbol{r}) \, \mathrm{e}^{-\mathrm{i}(\boldsymbol{k}' - \boldsymbol{k}) \cdot \boldsymbol{r}} \, \mathrm{d}^3 \boldsymbol{r} \; , \tag{2.87}$$

where we have used (2.80). In terms of the potential $U(\boldsymbol{r})$ defined in (2.56) Eq. (2.87) reads

$$f(\boldsymbol{q}) = -\frac{1}{4\pi} \int U(\boldsymbol{r}) \, \mathrm{e}^{-\mathrm{i}\boldsymbol{q} \cdot \boldsymbol{r}} \, \mathrm{d}^3 \boldsymbol{r} \; . \tag{2.88}$$

Note that in the Born approximation the scattering amplitude does not depend on $\boldsymbol{k}$ and $\boldsymbol{k}'$ independently but only on the momentum transfer $\boldsymbol{q} = \boldsymbol{k}' - \boldsymbol{k}$.

The Born approximation (2.88) can also be obtained from the general expression (2.62) of the scattering amplitude by approximating the exact scattered wave function $\psi(\boldsymbol{r})$ by the incoming plane wave $\mathrm{e}^{\mathrm{i}\boldsymbol{k} \cdot \boldsymbol{r}}$.

### 2.2.5 Central Fields

In the important case of central fields one can project the initial and final states onto angular momentum eigenvectors $|\ell \, m\rangle$, thus obtaining the spherical harmonics $Y_\ell^m$

$$\langle \hat{\boldsymbol{k}} \, \lambda | \ell \, m \, \lambda \rangle = Y_\ell^m(\Omega) \; . \tag{2.89}$$

Introducing the matrix elements of $S$ in the $|\ell \, m \, \lambda\rangle$ states

$$\langle \ell' \, m' \, \lambda' | S | \ell \, m \, \lambda \rangle = \delta_{\ell\ell'} \, \delta_{mm'} \, S_\ell^{\lambda\lambda'} \; , \tag{2.90}$$

and making use of (2.89) and of the addition theorem of spherical harmonics

$$\sum_{m=-\ell}^{\ell} Y_\ell^m(\Omega_k) \, Y_\ell^{m*}(\Omega_{k'}) = \frac{2\ell + 1}{4\pi} P_\ell(\cos\vartheta) \; , \tag{2.91}$$

where $\vartheta$ is the angle between $\hat{\boldsymbol{k}}$ and $\hat{\boldsymbol{k}}'$, that is the scattering angle, we get

$$\langle \hat{\boldsymbol{k}}'\,\lambda' | S - \mathbb{1} | \hat{\boldsymbol{k}}\,\lambda \rangle = \frac{1}{4\pi} \sum_{\ell=0}^{\infty} (2\ell+1)\,(S_\ell^{\lambda\lambda'} - \delta_{\lambda\lambda'})\,P_\ell(\cos\vartheta)\,. \tag{2.92}$$

Inserting (2.92) in Eq. (2.78) we obtain the differential cross section for central field scattering

$$\frac{\mathrm{d}\sigma}{\mathrm{d}\Omega} = \frac{1}{4k^2} \left| \sum_{\ell=0}^{\infty} (2\ell+1)\,(S_\ell^{\lambda\lambda'} - \delta_{\lambda\lambda'})\,P_\ell(\cos\vartheta) \right|^2 . \tag{2.93}$$

The integrated cross section can be computed using the orthogonality relation for Legendre polynomials, and reads

$$\sigma(k) = \int \frac{\mathrm{d}\sigma}{\mathrm{d}\Omega}\,\mathrm{d}\Omega = \frac{\pi}{k^2} \sum_{\ell=0}^{\infty} (2\ell+1)\,|S_\ell^{\lambda\lambda'} - \delta_{\lambda\lambda'}|^2\,. \tag{2.94}$$

In the case of *elastic scattering*, $\lambda = \lambda'$, and setting $S_\ell \equiv S_\ell^{\lambda\lambda}$ we get from Eq. (2.94) the integrated elastic cross section

$$\sigma_{\mathrm{el}}(k) = \frac{\pi}{k^2} \sum_{\ell=0}^{\infty} (2\ell+1)\,|S_\ell - 1|^2\,. \tag{2.95}$$

The integrated *inelastic cross section* is given by

$$\sigma_{\mathrm{in}}(k) = \frac{\pi}{k^2} \sum_{\ell=0}^{\infty} (2\ell+1) \sum_{\lambda'\neq\lambda} |S_\ell^{\lambda\lambda'}|^2\,, \tag{2.96}$$

where we summed over all possible final states characterized by the quantum numbers $\lambda'$, with $\lambda' \neq \lambda$.

At a first look it may seem that the calculation of this quantity would require the knowledge of a huge number of $S$-matrix elements. It is not so. The unitarity of the $S$-matrix (2.76) helps to simplify Eq. (2.96). Since $|\lambda\,\ell\,m\rangle$ is an orthonormal basis, the condition $S^\dagger S = \mathbb{1}$ implies

$$\sum_{\lambda'} |S_\ell^{\lambda\lambda'}|^2 = |S_\ell|^2 + \sum_{\lambda'\neq\lambda} |S_\ell^{\lambda\lambda'}|^2 = 1\,, \tag{2.97}$$

so that the inelastic cross section becomes

$$\sigma_{\mathrm{in}}(k) = \frac{\pi}{k^2} \sum_{\ell=0}^{\infty} (2\ell+1)\,(1 - |S_\ell|^2)\,. \tag{2.98}$$

If we define the *total cross section* as the sum of the elastic and the total inelastic cross section, i.e.

$$\sigma_{\mathrm{tot}} = \sigma_{\mathrm{el}} + \sigma_{\mathrm{in}}\,, \tag{2.99}$$

Eqs. (2.95) and (2.97) lead to

$$\sigma_{\text{tot}}(k) = \frac{2\pi}{k^2} \sum_{\ell=0}^{\infty} (2\ell + 1)\,(1 - \operatorname{Re} S_\ell)\ . \tag{2.100}$$

A very important sum rule can be derived by comparing Eqs. (2.67), taken at $\vartheta = 0$, and (2.100). One easily finds that the total cross section is proportional to the elastic amplitude in the forward direction ($\vartheta = 0$)

$$\operatorname{Im} f(0) = \frac{k}{4\pi}\,\sigma_{\text{tot}}\ , \tag{2.101}$$

a result known as the *optical theorem.* The derivation of Eq. (2.100) for the total cross section shows that the relation (2.101) is a direct consequence of unitarity (i.e., of the conservation of probability). This will be even more evident when we shall derive the optical theorem in the relativistic case. In optics we saw that a similar relation holds, which reflects the conservation of energy.

Let us now see the effects of unitarity on the partial wave amplitudes (2.68).

For elastic collisions Eq. (2.97) prescribes

$$|S_\ell| = 1\ . \tag{2.102}$$

Hence we can set

$$S_\ell(k) = \mathrm{e}^{2\mathrm{i}\delta_\ell(k)}\ , \tag{2.103}$$

where $\delta_\ell(k)$ are the phase shifts introduced before (see Eq. (2.68)). Note that for elastic scattering $\delta_\ell(k)$ are *real* quantities. From (2.102) and the definition (2.68) we derive the *elastic unitarity* conditions

$$\frac{1}{k} \operatorname{Im} a_\ell(k) = |a_\ell(k)|^2\ , \tag{2.104}$$

These equations admit a general solution of the form

$$a_\ell(k) = \frac{1}{g_\ell(k) - \mathrm{i}\,k}\ , \tag{2.105}$$

where $g_\ell(k)$ is an arbitrary real function.

In the more general case when the inelastic channels are open, Eq. (2.102) is no longer true and the unitarity condition (2.97) implies only

$$|S_\ell| \leq 1\ . \tag{2.106}$$

We can still write a relation such as (2.103), but now the phase shifts develop an imaginary part and we can rewrite (2.103) as

$$S_\ell(k) = \eta_\ell(k)\,\mathrm{e}^{2\mathrm{i}\zeta_\ell(k)}\ , \tag{2.107}$$

where $\eta_\ell \equiv \mathrm{e}^{-2\,\mathrm{Im}\,\delta_\ell}$ and $\zeta_\ell \equiv \mathrm{Re}\,\delta_\ell$ are real quantities. In view of the unitarity constraint (2.106), we must have

$$\mathrm{Im}\,\delta_\ell \geq 0\,, \tag{2.108}$$

and the *absorption* (or *inelasticity*) *coefficients* $\eta_\ell$ obey

$$0 \leq \eta_\ell(k) \leq 1\,. \tag{2.109}$$

The unitarity relation satisfied by the partial wave amplitudes reads now

$$\frac{1}{k}\,\mathrm{Im}\,a_\ell(k) - |a_\ell(k)|^2 = \frac{1}{4k^2}\left(1 - \eta_\ell^2(k)\right)\,, \tag{2.110}$$

and its r.h.s. reduces to zero in the elastic limit $\eta_\ell = 1$. In terms of $a_\ell(k)$ the condition (2.106) – or, equivalently, (2.109) – translates into

$$0 \leq k^2\,|a_\ell(k)|^2 \leq k\,\mathrm{Im}\,a_\ell(k) \leq 1\,. \tag{2.111}$$

## 2.3 The Eikonal Approximation

We shall now discuss the *high-energy limit* of non-relativistic potential scattering. This is described by the so-called *eikonal approximation*, valid when the particle energy largely dominates over the interaction potential

$$E \gg |V(\boldsymbol{r})|\,, \tag{2.112}$$

and the particle wavelength is much smaller than the interaction range $a$, i.e.

$$ka \gg 1\,. \tag{2.113}$$

In this case we can try a solution of the Schrödinger equation (2.57) of the form (here, we follow the method of Glauber 1958)

$$\psi(\boldsymbol{r}) = \varphi(\boldsymbol{r})\,\mathrm{e}^{\mathrm{i}\boldsymbol{k}\cdot\boldsymbol{r}}\,, \tag{2.114}$$

namely a plane wave in the forward direction modulated by some unknown function $\varphi(\boldsymbol{r})$, to be determined later. The trial wave function of the type (2.114) is suggested by the consideration that, when (2.112) and (2.113) are satisfied, we expect that the particle should be scattered predominantly at small angles.

Substituting (2.114) in Eq. (2.57) leads to an equation for $\varphi(\boldsymbol{r})$

$$\left(\nabla^2 + 2\mathrm{i}\,\boldsymbol{k}\cdot\nabla - U\right)\varphi(\boldsymbol{r}) = 0\,, \tag{2.115}$$

with the boundary condition

$$\varphi(x, y, -\infty) = 1 \,. \tag{2.116}$$

At high energies we can suppose that $\varphi$ and $U$ are smooth functions of $\boldsymbol{r}$ compared to $\mathrm{e}^{\mathrm{i}\boldsymbol{k}\cdot\boldsymbol{r}}$, i.e. that they vary only over a distance much larger than $1/k$. Therefore we can neglect in Eq. (2.115) the term $\nabla^2\varphi$ and get

$$\left[2\mathrm{i}k \frac{\partial}{\partial z} - U(x, y, z)\right] \varphi(x, y, z) = 0 \,, \tag{2.117}$$

where we have used the fact that $\boldsymbol{k}$ is directed along $z$.

The solution of Eq. (2.117) with the condition (2.116) is

$$\varphi(x, y, z) = \exp\left[-\frac{\mathrm{i}}{2k}\int_{-\infty}^{z} U(x, y, z')\,\mathrm{d}z'\right] \tag{2.118}$$

and the wave function of the scattered particle is then

$$\psi(x, y, z) = \exp\left[\mathrm{i}kz - \frac{\mathrm{i}}{2k}\int_{-\infty}^{z} U(x, y, z')\,\mathrm{d}z'\right] \,. \tag{2.119}$$

Notice that this wave function does not have a spherical wave behavior. This is no problem since we obtained it in the high-energy limit and hence it is valid in the small-angle region.

Decomposing $\boldsymbol{r}$ as

$$\boldsymbol{r} = \boldsymbol{b} + z\hat{\boldsymbol{u}}_z \,, \tag{2.120}$$

where $\boldsymbol{b}$ is a vector perpendicular to the propagation axis (its length is known as the *impact parameter*, see Fig. 2.5), and using $\boldsymbol{k} = k\hat{\boldsymbol{u}}_z$, we can rewrite (2.119) as

$$\psi(\boldsymbol{r}) = \exp\left[\mathrm{i}\boldsymbol{k}\cdot\boldsymbol{r} - \frac{\mathrm{i}}{2k}\int_{-\infty}^{z} U(\boldsymbol{b}, z')\,\mathrm{d}z'\right] \,. \tag{2.121}$$

Let us calculate now the scattering amplitude. Inserting the wave function (2.121) in the general expression (2.62) yields ($\mathrm{d}^3\boldsymbol{r} = \mathrm{d}^2\boldsymbol{b}\,\mathrm{d}z$)

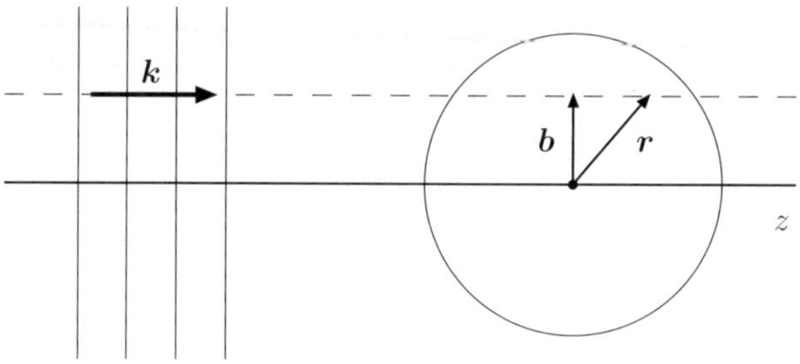


**Fig. 2.5.** Plane wave scattering off a potential.

$$f(\boldsymbol{k},\boldsymbol{k}') = -\frac{1}{4\pi}\int_{-\infty}^{+\infty} \mathrm{d}z \int \mathrm{d}^2\boldsymbol{b}\, \mathrm{e}^{-\mathrm{i}(\boldsymbol{k}'-\boldsymbol{k})\cdot(\boldsymbol{b}+z\,\hat{\boldsymbol{u}}_z)}\, U(\boldsymbol{b},z') \times \exp\left[-\frac{\mathrm{i}}{2k}\int_{-\infty}^{z} U(\boldsymbol{b},z)\,\mathrm{d}z'\right] . \tag{2.122}$$

Since $k = k'$, for small scattering angles the momentum transfer vector $\boldsymbol{q} \equiv \boldsymbol{k}' - \boldsymbol{k}$ is almost orthogonal to $\boldsymbol{k}$ and $\boldsymbol{q}\cdot\boldsymbol{r} \simeq \boldsymbol{q}\cdot\boldsymbol{b}$, so that we can approximate (2.122) by (recall that $q = 2\,k\,\sin\vartheta/2$)

$$f(k,\vartheta,\varphi) = -\frac{1}{4\pi}\int_{-\infty}^{+\infty} \mathrm{d}z \int \mathrm{d}^2\boldsymbol{b}\, \mathrm{e}^{-\mathrm{i}\boldsymbol{q}\cdot\boldsymbol{b}}\, U(\boldsymbol{b},z) \times \exp\left[-\frac{\mathrm{i}}{2k}\int_{-\infty}^{z} U(\boldsymbol{b},z')\,\mathrm{d}z'\right] . \tag{2.123}$$

We now perform the $z$ integration using the identity

$$2\mathrm{i}\,k\,\frac{\partial}{\partial z}\exp\left[-\frac{\mathrm{i}}{2k}\int_{-\infty}^{z} U\,\mathrm{d}z'\right] = U\,\exp\left[-\frac{\mathrm{i}}{2k}\int_{-\infty}^{z} U\,\mathrm{d}z'\right] , \tag{2.124}$$

and finally obtain

$$f(k,\vartheta,\varphi) = \frac{\mathrm{i}k}{2\pi}\int \mathrm{d}^2\boldsymbol{b}\,\mathrm{e}^{-\mathrm{i}\,\boldsymbol{q}\cdot\boldsymbol{b}}\,(1-\mathrm{e}^{\mathrm{i}\,\chi(\boldsymbol{b})}) , \tag{2.125}$$

where the *eikonal function* $\chi$ is defined as

$$\chi(\boldsymbol{b}) = -\frac{1}{2k}\int_{-\infty}^{+\infty} U(\boldsymbol{b},z)\,\mathrm{d}z . \tag{2.126}$$

The quantity

$$\Gamma(\boldsymbol{b}) \equiv 1 - \mathrm{e}^{\mathrm{i}\,\chi(\boldsymbol{b})} \tag{2.127}$$

is often called *profile function* in strict analogy to optics (see Sect. 2.1).

Equation (2.125) exhibits the scattering amplitude as the two-dimensional Fourier transform of the profile function. The latter can be obtained from $f(k,\vartheta,\varphi) = f(\boldsymbol{q})$ inverting Eq. (2.125) (see the footnote before Eq. (2.27))

$$\Gamma(\boldsymbol{b}) = -\frac{2\pi\mathrm{i}}{k}\int \mathrm{d}^2\boldsymbol{q}\; f(\boldsymbol{q})\,\mathrm{e}^{\mathrm{i}\,\boldsymbol{q}\cdot\boldsymbol{b}} . \tag{2.128}$$

The meaning of the eikonal function can be inferred from Eq. (2.121). Asymptotically, i.e. as $z \to \infty$, the outgoing wave function is

$$\psi(\boldsymbol{r}) \underset{z\to\infty}{\sim} \mathrm{e}^{\mathrm{i}\,\chi(\boldsymbol{b})}\,\mathrm{e}^{\mathrm{i}\,\boldsymbol{k}\cdot\boldsymbol{r}} . \tag{2.129}$$

Thus $\chi(\boldsymbol{b})$ represents the phase shift due to the scattering.

If $\chi(\boldsymbol{b}) \ll 1$, we can approximate the exponential $\mathrm{e}^{\mathrm{i}\,\chi(\boldsymbol{b})}$ as

$$\mathrm{e}^{\mathrm{i}\,\chi(\boldsymbol{b})} \simeq 1 + \mathrm{i}\,\chi(\boldsymbol{b}) \;, \tag{2.130}$$

and from Eq. (2.125) we reobtain the Born approximation for the scattering amplitude

$$f(\boldsymbol{q}) = -\frac{1}{4\pi} \int \mathrm{d}^3\boldsymbol{r}\, \mathrm{e}^{-\mathrm{i}\boldsymbol{q}\cdot\boldsymbol{r}}\, U(\boldsymbol{r}) \;. \tag{2.131}$$

Note that, if the potential has an azimuthal symmetry, we can perform the angular integration in (2.125) using

$$\boldsymbol{q}\cdot\boldsymbol{b} = q\, b\, \cos\varphi \simeq k\, b\, \vartheta\, \cos\varphi \;, \tag{2.132}$$

and

$$\frac{1}{2\pi} \int_0^{2\pi} \mathrm{e}^{\mathrm{i}\, x \cos\varphi}\, \mathrm{d}\varphi = J_0(x) \;. \tag{2.133}$$

Equation (2.125) thus reduces to

$$f(k,\vartheta) = \mathrm{i}k \int_0^\infty b\,\mathrm{d}b\, J_0(qb)\,(1 - \mathrm{e}^{\mathrm{i}\chi(b)}) \;, \tag{2.134}$$

which, in view of (2.127), is the quantum mechanical analogue of the optical expression (2.29).

It is possible to derive the eikonal representation directly from the partial wave expansion (Molière 1947; see also Predazzi 1966). At high energies ($k \to \infty$) and small scattering angles ($\vartheta \to 0$) large values of $\ell$ are important (hereafter we consider central potentials). Thus we can use in the partial wave series (2.67) the approximation

$$P_\ell(\cos\vartheta) \simeq J_0(\ell\,\vartheta) \;, \tag{2.135}$$

valid for $\ell\,\vartheta \sim 1$, and convert the sum $\sum_\ell$ into an integral over $\ell$ regarded as a continuous variable

$$f(k,\vartheta) \simeq \int_0^\infty \mathrm{d}\ell\,(2\ell+1)\, J_0(\ell\,\vartheta)\, a(\ell,k) \;. \tag{2.136}$$

Since the impact parameter $b$ is related to $\ell$ by $k\,b = \ell + 1/2$, we can rewrite (2.136) as an integral representation in the $b$-space (using $\ell\,\vartheta \simeq q\,b$)

$$f(k,\vartheta) = 2\,k^2 \int \mathrm{d}b\, b\, J_0(kb\vartheta)\, a(k,b) \;, \tag{2.137}$$

and finally

$$f(k,\vartheta) = \mathrm{i}\,k \int \mathrm{d}b\, b\, J_0(kb\vartheta)\, \Gamma(k,b) \;, \tag{2.138}$$

where the profile function $\Gamma(k,b)$ is given by

$$\Gamma(k,b) = -2\,\mathrm{i}\,k\,a(k,b) = 1 - \mathrm{e}^{2\,\mathrm{i}\,\delta(k,b)} \; . \tag{2.139}$$

If we identify the eikonal function $\chi(k, b)$ with $2\,\delta(k, b)$, (2.138) is the result already obtained in a different way, see (2.134). Aa a matter of fact, the derivation we have just outlined is more general, as it is based only on the partial wave expansion, and not on the Schrödinger equation. Therefore, it applies also to relativistic scattering. We shall make full use of it in Chap. 6.

To conclude this section we give the expressions for the elastic, inelastic and total cross sections in the eikonal limit. From what we have been saying so far it should be clear that they are the same as in optics, that is

$$\sigma_{\mathrm{el}} = \int \mathrm{d}^2\boldsymbol{b}\, |\Gamma(b)|^2 \; , \tag{2.140}$$

$$\sigma_{\mathrm{in}} = \int \mathrm{d}^2\boldsymbol{b}\, [2\,\mathrm{Re}\,\Gamma(b) - |\Gamma(b)|^2] \; , \tag{2.141}$$

$$\sigma_{\mathrm{tot}} = 2 \int \mathrm{d}^2\boldsymbol{b}\, \mathrm{Re}\,\Gamma(b) \; . \tag{2.142}$$

We shall make full use of the eikonal representation to describe high-energy particle diffraction in Chap. 6.

# 3. Kinematics

As anticipated in the Introduction, the experimental signatures of diffraction consist in particular kinematic configurations of the final states. In the remainder of the book we shall also see that some of the theoretical approaches to diffraction are based on the identification of the relevant kinematic regimes. Therefore, kinematics is by no means a secondary ingredient in diffractive physics, and so it requires a careful consideration.

In this chapter we review the kinematics of some scattering processes. The presentation will not be confined to the case of diffractive reactions. We start by considering very general processes and then apply our discussion to the specific case of diffraction.

## 3.1 Scattering Processes: Generalities

Consider a generic production process of the type

$$1 + 2 \to 3 + 4 + \ldots + N \,. \tag{3.1}$$

The number of independent Lorentz-invariant variables for this reaction is $3N - 10$. In fact, due to:

- the conservation of four-momentum

$$p_1 + p_2 = p_3 + p_4 + \ldots + p_N \,, \quad (4 \text{ constraints}) \,; \tag{3.2}$$

- the mass shell conditions

$$p_i^2 = m_i^2 \,, \quad i = 1, 2, \ldots, N \,, \quad (N \text{ constraints}) \,; \tag{3.3}$$

- the arbitrariness in fixing a 4-dimensional reference frame (6 constraints);

of the *a priori* available $4N$ variables (the components of the four-vectors $p_i$), only $3N - 10$ remain as truly independent variables. In practice, one of these variables (e.g., the beam energy, the center-of-mass energy of the colliding particles, etc.) is fixed by preparing the initial state.

Throughout this book we shall mainly be concerned with two kinds of processes:

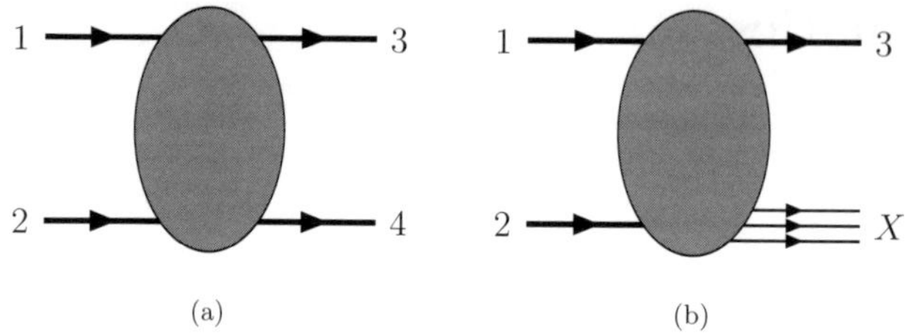


**Fig. 3.1.** (**a**) Two-body exclusive scattering $1 + 2 \to 3 + 4$; (**b**) Single-particle inclusive scattering $1 + 2 \to 3 + X$.

- *two-body exclusive scattering* (Fig. 3.1a)

$$1 + 2 \to 3 + 4 \,, \tag{3.4}$$

- and *semi-inclusive* (or, more precisely, *single-particle inclusive*) *scattering* (Fig. 3.1b)

$$1 + 2 \to 3 + X \,, \tag{3.5}$$

  where $X$ denotes an unresolved system of particles (or a resonance).

According to the general discussion above, reaction (3.4) is described by two independent variables, whereas reaction (3.5) requires three independent variables, since the system $X$ is not on mass shell.

A special case of (3.4) is the *elastic scattering*

$$1 + 2 \to 1' + 2' \,, \tag{3.6}$$

where the two particles remain unaltered (but in a different kinematic configuration).

*Single-diffractive dissociation*, instead, falls into class (3.5), being characterized only by a particular final state configuration: one of the two colliding particles (say 1) emerges unaltered while the other produces a system carrying the same quantum numbers

$$1 + 2 \to 1' + X_2 \,. \tag{3.7}$$

## 3.2 Two-Body Processes

### 3.2.1 Mandelstam Variables

We have seen that the kinematics of two-body reactions

$$1 + 2 \to 3 + 4 \,, \quad (s-\text{channel}) \,, \tag{3.8}$$

is described by two variables. These variables are usually chosen among the three *Mandelstam invariants*, defined as

$$s = (p_1 + p_2)^2 = (p_3 + p_4)^2 , \tag{3.9}$$

$$t = (p_1 - p_3)^2 = (p_2 - p_4)^2 , \tag{3.10}$$

$$u = (p_1 - p_4)^2 = (p_2 - p_3)^2 . \tag{3.11}$$

The Mandelstam variables obey the identity

$$s + t + u = \sum_{i=1}^{4} m_i^2 , \tag{3.12}$$

which can be easily derived from the definitions (3.9–3.11) and the energy-momentum conservation $p_1 + p_2 = p_3 + p_4$, and therefore only two of them are independent. In general, we take $s$ and $t$ as the independent variables[1].

In the reaction (3.8), see Fig. 3.2a, $s$ is the square of the total center-of-mass (CM) energy and $t$ is the squared momentum transfer. We refer to (3.8), or its time reversed, as to the *s-channel process*.

Analogously, *t-channel* (*u-channel*) means that $t$ ($u$, respectively), defined by Eqs. (3.10, 3.11), is the square of the total CM energy. The $t$-channel and $u$-channel reactions are then (see Fig. 3.2b,c)

$$1 + \bar{3} \to \bar{2} + 4 , \quad (t - \text{channel}) , \tag{3.13}$$

$$1 + \bar{4} \to \bar{2} + 3 , \quad (u - \text{channel}) , \tag{3.14}$$

together with their time reversed processes. Here $\bar{3}$, for instance, means that the momentum of particle 3 has been reversed and all additive quantum numbers have changed sign, that is $\bar{3}$ is the antiparticle of 3 with opposite momentum.

Just to make a specific example, suppose we start from the $s$-channel reaction (in brackets the particle momenta)

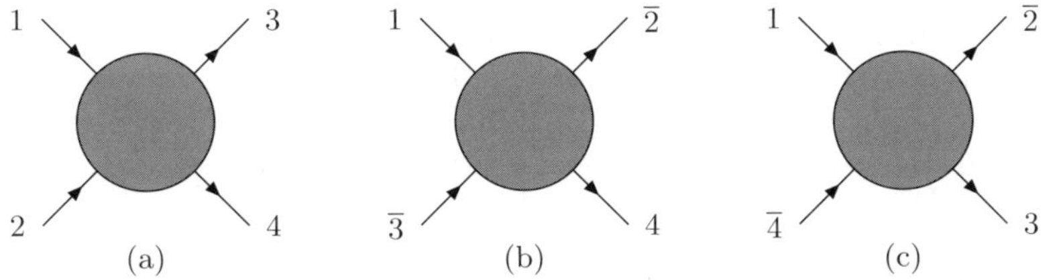


**Fig. 3.2.** (**a**) $s$-channel. (**b**) $t$-channel. (**c**) $u$-channel.

[1] In some cases it may be convenient to use a different pair of variables. For instance, if we look at backward scattering, $s$ and $u$ are a more practical choice, since the backward direction corresponds, as we shall see, to $u = 0$ for equal-mass particles (whereas the forward direction corresponds to $t = 0$).

$$p\,(p_1) + \pi^+\,(p_2) \to p\,(p_3) + \pi^+\,(p_4)\;; \tag{3.15}$$

its $t$-channel is

$$p\,(p_1) + \bar{p}\,(-p_3) \to \pi^-\,(-p_2) + \pi^+\,(p_4)\;; \tag{3.16}$$

its $u$-channel is

$$p\,(p_1) + \pi^-\,(-p_4) \to \pi^-\,(-p_2) + p\,(p_3)\;. \tag{3.17}$$

We shall see that the $s, t, u$-channel processes (3.8, 3.13, 3.14) have different physical domains in the Mandelstam plane, but the *crossing symmetry* of the $S$-matrix tells us that they are described by the same scattering amplitude (or by appropriate combinations of the same amplitudes).

We work out now the kinematics of the two-body reactions in two commonly used reference systems, the center-of-mass system (CM) and the laboratory system (Lab). These systems are characterized by the vanishing of some combinations of three-momenta.

### 3.2.2 The Center-of-Mass System

Consider the $s$-channel reaction (3.8). In the center-of-mass (CM) system (Fig. 3.3) we have, by definition

$$\boldsymbol{p}_1 + \boldsymbol{p}_2 = 0\;. \tag{3.18}$$

Hence, assuming that 1 and 2 travel along the $z$ axis, the four-momenta of the particles can be written as

$$p_1 = (E_1, \boldsymbol{p}) = (E_1, 0, 0, p_z)\;, \tag{3.19}$$
$$p_2 = (E_2, -\boldsymbol{p}) = (E_2, 0, 0, -p_z)\;, \tag{3.20}$$
$$p_3 = (E_3, \boldsymbol{p}') = (E_3, \boldsymbol{p}_\perp, p'_z)\;, \tag{3.21}$$
$$p_4 = (E_4, -\boldsymbol{p}') = (E_4, -\boldsymbol{p}_\perp, -p'_z)\;, \tag{3.22}$$

where we have used (3.18) and the conservation of three-momentum. In Eqs. (3.21, 3.22) $\boldsymbol{p}_\perp$ is a transverse two-vector.

Only two of the variables appearing in Eqs. (3.19–3.22) are independent. We choose them to be the CM momentum $|\boldsymbol{p}| = p_z$ (by convention, particle 1 moves in the $+z$-direction), and the scattering angle (in the $s$-channel) $\vartheta$, defined by

$$p'_z = |\boldsymbol{p}'|\cos\vartheta\;, \tag{3.23}$$
$$|\boldsymbol{p}_\perp| = |\boldsymbol{p}'|\sin\vartheta\;, \tag{3.24}$$

Let us now relate the CM variables to the Mandelstam invariants.

The energies $E_1, E_2, E_3, E_4$ can be expressed in terms of the CM energy squared $s = (p_1 + p_2)^2$ as

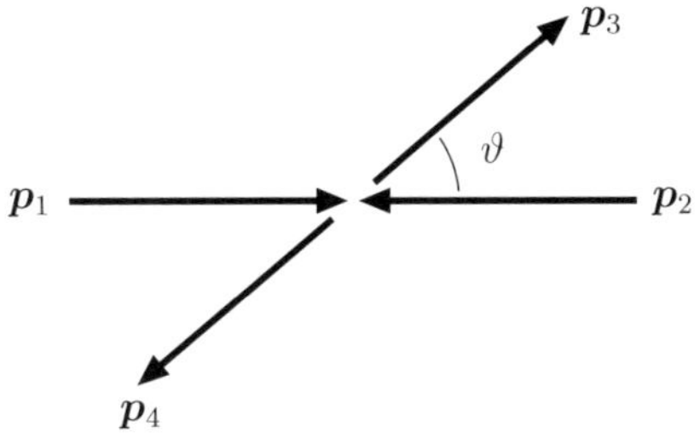


**Fig. 3.3.** The center-of-mass system.

$$E_1 = \frac{1}{2\sqrt{s}}\,(s + m_1^2 - m_2^2)\ , \tag{3.25}$$

$$E_2 = \frac{1}{2\sqrt{s}}\,(s + m_2^2 - m_1^2)\ , \tag{3.26}$$

$$E_3 = \frac{1}{2\sqrt{s}}\,(s + m_3^2 - m_4^2)\ , \tag{3.27}$$

$$E_4 = \frac{1}{2\sqrt{s}}\,(s + m_4^2 - m_3^2)\ . \tag{3.28}$$

From the mass-shell conditions, we get the following relations between $|\boldsymbol{p}|$, $|\boldsymbol{p}'|$ and $s$

$$\begin{aligned}\boldsymbol{p}^2 &= p_z^2 = E_1^2 - m_1^2\\ &= \frac{1}{4s}\,[s-(m_1+m_2)^2][s-(m_1-m_2)^2]\\ &= \frac{1}{4s}\,\lambda(s, m_1^2, m_2^2)\ ,\end{aligned} \tag{3.29}$$

$$\begin{aligned}\boldsymbol{p}'^2 &= \boldsymbol{p}_\perp^2 + p_z^2 = E_3^2 - m_3^2\\ &= \frac{1}{4s}\,[s-(m_3+m_4)^2][s-(m_3-m_4)^2]\\ &= \frac{1}{4s}\,\lambda(s, m_3^2, m_4^2)\ ,\end{aligned} \tag{3.30}$$

where the triangle function $\lambda(x, y, z)$ is defined as

$$\lambda(x, y, z) = x^2 + y^2 + z^2 - 2xy - 2yz - 2xz\ , \tag{3.31}$$

and is invariant under permutations of its arguments. A noteworthy special case is

$$\lambda(x, y, y) = x^2 - 4xy\ . \tag{3.32}$$

In the high-energy limit ($s \to \infty$) the masses can be neglected and one has

$$E_1, E_2, E_3, E_4 \underset{s\to\infty}{\simeq} \frac{\sqrt{s}}{2} \,. \tag{3.33}$$

and

$$|\boldsymbol{p}|,\, |\boldsymbol{p}'| \underset{s\to\infty}{\simeq} \frac{\sqrt{s}}{2} \,. \tag{3.34}$$

Let us now express the Mandelstam invariant $t$ in terms of the CM variables

$$\begin{aligned} t &= (p_1 - p_3)^2 \\ &= m_1^2 + m_3^2 - 2E_1 E_3 + 2\,|\boldsymbol{p}|\,|\boldsymbol{p}'|\,\cos\vartheta \,. \end{aligned} \tag{3.35}$$

Combining Eqs. (3.25), (3.27), (3.29), (3.30) and (3.35), we can extract $\cos\vartheta$

$$\cos\vartheta = \frac{s^2 + s\,(2t - \sum_i m_i^2) + (m_1^2 - m_2^2)(m_3^2 - m_4^2)}{\lambda^{\frac{1}{2}}(s, m_1^2, m_2^2)\,\lambda^{\frac{1}{2}}(s, m_3^2, m_4^2)} \,. \tag{3.36}$$

When all particles have *equal masses*[2] ($m_1 = m_2 = m_3 = m_4 \equiv m$), the relations between the CM variables $[|\boldsymbol{p}|, \vartheta]$ and the Mandelstam invariants $[s, t]$ are much simpler

$$|\boldsymbol{p}| = \frac{1}{2}\sqrt{s - 4m^2} \,, \tag{3.37}$$

$$\cos\vartheta = 1 + \frac{2t}{s - 4m^2} \,. \tag{3.38}$$

The inverse relations are

$$s = 4\,(\boldsymbol{p}^2 + m^2) \,, \tag{3.39}$$

$$t = -2\boldsymbol{p}^2\,(1 - \cos\vartheta) \,, \tag{3.40}$$

and to these we can add, using $s + t + u = 4m^2$,

$$u = -2\boldsymbol{p}^2\,(1 + \cos\vartheta) \,. \tag{3.41}$$

For massless particles, or when masses can be neglected (i.e. for $s \to \infty$) Eq. (3.37) reduces to (3.34) and Eq. (3.38) becomes

$$\cos\vartheta = 1 + \frac{2t}{s} \,. \tag{3.42}$$

In this limit one easily finds, from (3.24) and (3.42)

$$t \underset{s\to\infty}{\simeq} -\boldsymbol{p}_\perp^2 \,. \tag{3.43}$$

[2] Technically, this would be the case of elastic scattering of identical particles. At high energies, however, mass differences become irrelevant so long as all particles are on their mass shell. Thus, the assumption of equal masses is not so restrictive in the high-energy limit we are interested in.

Proceeding along similar lines, we can obtain the relations between the CM variables and the Mandelstam invariants for the $t$-channel and the $u$-channel. In the $t$-channel one finds for the scattering angle $\vartheta_t$

$$\cos\vartheta_t = \frac{t^2 + t\,(2s - \sum_i m_i^2) + (m_1^2 - m_3^2)(m_2^2 - m_4^2)}{\lambda^{\frac{1}{2}}(t, m_1^2, m_3^2)\,\lambda^{\frac{1}{2}}(t, m_2^2, m_4^2)} \,. \tag{3.44}$$

In the equal-mass case this simplifies to

$$\cos\vartheta_t = 1 + \frac{2s}{t - 4m^2} \,. \tag{3.45}$$

Similarly, in the $u$-channel we have

$$\cos\vartheta_u = \frac{u^2 + u\,(2t - \sum_i m_i^2) + (m_1^2 - m_4^2)(m_2^2 - m_3^2)}{\lambda^{\frac{1}{2}}(u, m_1^2, m_4^2)\,\lambda^{\frac{1}{2}}(u, m_2^2, m_3^2)} \,. \tag{3.46}$$

and

$$\cos\vartheta_u = 1 + \frac{2t}{u - 4m^2} \,. \tag{3.47}$$

for equal masses.

### 3.2.3 The Laboratory System

We now briefly illustrate the kinematics of a two-body process in the laboratory system (Lab), focusing on the $s$-channel configuration. In the Lab frame, by definition, one of the initial particles – say particle 2 – is at rest (Fig. 3.4)

$$\boldsymbol{p}_2 = 0 \,. \tag{3.48}$$

The four-momenta of the colliding particles, assuming, as usual, that particle 1 moves along the $z$ direction, can be written as

$$p_1 = (E_L, 0, 0, p_L) \,, \tag{3.49}$$

$$p_2 = (m_2, 0, 0, 0) \,. \tag{3.50}$$

where $p_L$ is the laboratory total momentum. The four-momenta of the outgoing particles are generically

$$p_3 = (E_3, \boldsymbol{p}_3) \,, \tag{3.51}$$

$$p_4 = (E_4, \boldsymbol{p}_4) \,. \tag{3.52}$$

In terms of the laboratory variables, the Mandelstam invariants $s$, $t$ and $u$ are given by

$$s = (p_1 + p_2)^2 = m_1^2 + m_2^2 + 2m_2 E_L \,, \tag{3.53}$$

$$t = (p_2 - p_4)^2 = m_2^2 + m_4^2 - 2m_2 E_4 \,, \tag{3.54}$$

$$u = (p_2 - p_3)^2 = m_2^2 + m_3^2 - 2m_2 E_3 \,. \tag{3.55}$$

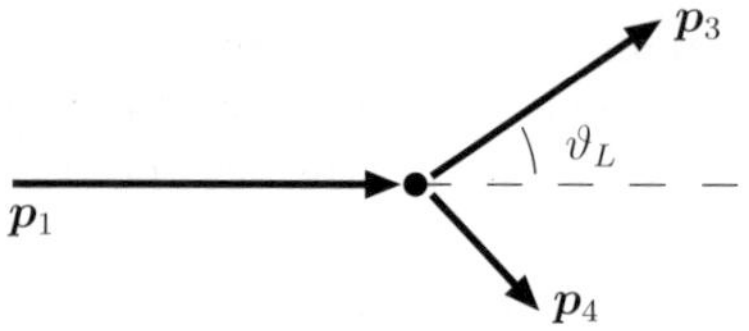


**Fig. 3.4.** The laboratory system.

Therefore the total Lab momentum $p_L$ and the energy of one of the produced particles represent a complete set of kinematic variables describing a two-body reaction in the Lab system.

From (3.53–3.55) one gets

$$E_L = \frac{1}{2m_2}(s - m_1^2 - m_2^2) , \tag{3.56}$$

$$E_4 = \frac{1}{2m_2}(m_2^2 + m_4^2 - t) , \tag{3.57}$$

$$E_3 = \frac{1}{2m_2}(m_2^2 + m_3^2 - u) , \tag{3.58}$$

and

$$p_L^2 = E_L^2 - m_1^2 = \frac{1}{4m_2^2}\lambda(s, m_1^2, m_2^2) , \tag{3.59}$$

$$\boldsymbol{p}_4^2 = E_4^2 - m_4^2 = \frac{1}{4m_2^2}\lambda(t, m_2^2, m_4^2) , \tag{3.60}$$

$$\boldsymbol{p}_3^2 = E_3^2 - m_3^2 = \frac{1}{4m_2^2}\lambda(u, m_2^2, m_3^2) . \tag{3.61}$$

$$\tag{3.62}$$

where $\lambda$ is the triangle function defined in (3.31).

At high energies, when masses can be neglected, Eqs. (3.53–3.55) become

$$s \simeq 2m_2 E_L \simeq 2m_2 p_L , \tag{3.63}$$

$$t \simeq -2m_2 E_4 , \tag{3.64}$$

$$u \simeq -2m_2 E_3 . \tag{3.65}$$

Instead of $E_3$ or $E_4$ one can use, as a final-state variable, the Lab scattering angle $\vartheta_L$, i.e. the angle formed by the direction of the outgoing particle 3 with the collision axis, which coincides with the direction of 1. Using

$$t = (p_1 - p_3)^2 = m_1^2 + m_3^2 - 2\,E_L E_3 + 2\,p_L|\boldsymbol{p}_3|\,\cos\vartheta_L , \tag{3.66}$$

the expressions (3.56, 3.58, 3.59, 3.61) for $E_L, E_3, p_L, |\boldsymbol{p}_3|$, and the constraint (3.12) it is possible to express $\cos\vartheta_L$ as a function of $s$ and $t$, or $s$ and $u$. The formula is quite complicated and we shall present it here only for the equal mass case (for a general treatment see Byckling and Kajantie 1973)

$$\cos\vartheta_L = \frac{s(s+t-4m^2)}{\lambda^{\frac{1}{2}}(s,m^2,m^2)\,\lambda^{\frac{1}{2}}(s+t,m^2,m^2)} \,. \tag{3.67}$$

### 3.2.4 Physical Domains of the $s$, $u$ and $t$ Channels

We derive now the physical domains of the $s$, $t$ and $u$ channels, limiting ourselves to the case of equal-mass scattering. Using Eqs. (3.39–3.41) one sees that the kinematical limits $p \geq 0$ and $-1 \leq \cos\vartheta \leq 1$ translate into the following physical ranges of the Mandelstam variables for the $s$-channel

$$s \geq 4m^2 \,, \quad t \leq 0 \,, \quad u \leq 0 \,. \tag{3.68}$$

This is the physical region of the $s$-channel in the Mandelstam plane. Note that $s$ has a threshold value corresponding to the production of two particles of mass $m$, and $t$ and $u$ are always negative.

By a similar reasoning, starting from Eqs. (3.45,3.47), one gets the physical domain of the $t$-channel

$$t \geq 4m^2 \,, \quad s \leq 0 \,, \quad u \leq 0 \,. \tag{3.69}$$

and of the $u$-channel

$$u \geq 4m^2 \,, \quad s \leq 0 \,, \quad t \leq 0 \,. \tag{3.70}$$

The important point to retain is that the physical domains of the $s$, $t$ and $u$ channels are *different* and *non overlapping*. We shall see in Sect. 4.5 what the crossing property of the $S$-matrix prescribes to this respect. The physical regions of the three channels are shown in the Mandelstam plot of Fig. 3.5.

For unequal masses the conditions specifying the physical regions of the three channels are much more complicated. They can be found in Kibble (1960).

## 3.3 Single-Inclusive Processes

Single-particle inclusive reactions

$$1 + 2 \to 3 + X \,, \tag{3.71}$$

are described by three independent variables. It is customary to use $s$, $t$ and the invariant mass $M^2 = (p_1 + p_2 - p_3)^2$ of the $X$ system, also called the "missing mass". The Mandelstam variables are defined as in (3.9–3.11), with $p_4$ replaced by $p_X$ (with the proviso that $X$ is not a real particle on the mass shell and hence $M^2$ is not a fixed number).

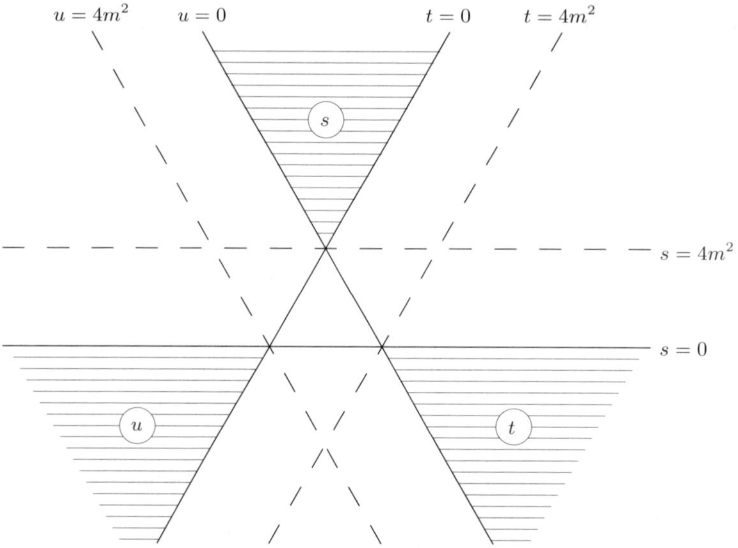


**Fig. 3.5.** The Mandelstam plot showing the physical regions of the $s$, $t$, and $u$ channels.

In the CM system the four-momenta of the three *bona fide* particles can be written as

$$p_1 = (E_1, \boldsymbol{p}) = (E_1, 0, 0, p_z) , \tag{3.72}$$

$$p_2 = (E_2, -\boldsymbol{p}) = (E_2, 0, 0, -p_z) , \tag{3.73}$$

$$p_3 = (E_3, \boldsymbol{p}') = (E_3, \boldsymbol{p}_\perp, p'_z) , \tag{3.74}$$

The relations between the CM variables appearing in Eqs. (3.72–3.74) and the Mandelstam invariants can be derived as we did in Sect. 3.2.2 for the exclusive two-body reactions, or simply obtained from those written in Sect. 3.2.2 by the replacement $m_4 \to M$. Limiting ourselves to the asymptotic case when $s$ and $M^2$ are much larger than the masses of the particles, we have

$$|\boldsymbol{p}| = p_z \simeq \frac{\sqrt{s}}{2} , \quad E_1, E_2 \simeq \frac{\sqrt{s}}{2} , \tag{3.75}$$

for $s \gg m_1^2, m_2^2$, and

$$|\boldsymbol{p}'| \simeq \frac{s - M^2}{2\sqrt{s}} , \quad E_3 \simeq \frac{s - M^2}{2\sqrt{s}} , \tag{3.76}$$

for $s, M^2 \gg m_3^2$.

From

$$t = (p_1 + p_3)^2 = m_1^2 + m_3^2 - 2\, E_1 E_3 + 2\, |\boldsymbol{p}|\, |\boldsymbol{p}'| \cos\vartheta , \tag{3.77}$$

using Eqs. (3.75, 3.76) we get, for $s, M^2 \gg m_1^2, m_3^2$,

$$\cos\vartheta \simeq 1 + \frac{2t}{s - M^2} , \tag{3.78}$$

In the same limit, the transverse momentum of the outgoing detected particle is

$$\boldsymbol{p}_\perp^2 = \boldsymbol{p}'^2 \sin^2\vartheta \simeq -t\left(1 - \frac{M^2}{s}\right) . \tag{3.79}$$

### 3.3.1 Feynman's $x_F$ Variable

A kinematic variable of common use in discussing inclusive processes is Feynman's $x_F$ variable, defined as

$$x_F \equiv \frac{|p_z'|}{p_z} . \tag{3.80}$$

In high-energy hadron scattering the transverse momentum of the produced particle is quite small on average: typically $|\boldsymbol{p}_\perp| \lesssim 0.5$ GeV. Hence $|p_z'| \simeq p'$ and, from (3.76), we get

$$|p_z'| \simeq |\boldsymbol{p}'| \simeq \frac{s - M^2}{2\sqrt{s}} . \tag{3.81}$$

Recalling (3.75), the Feynman variable then becomes (for $s, M^2 \gg m_i^2, \boldsymbol{p}_\perp^2$)

$$x_F \simeq 1 - \frac{M^2}{s} , \tag{3.82}$$

and the relation (3.79) gives

$$t \simeq -\frac{\boldsymbol{p}_\perp^2}{x_F} . \tag{3.83}$$

The triplet of variables $[s, x_F, \boldsymbol{p}_\perp^2]$ is often used to describe single inclusive processes in alternative to $[s, t, M^2]$.

While $|\boldsymbol{p}_\perp|$ is generally small, $p_z'$ can take values ranging from $p_z$ (when particle 3 is produced as a fragment of 1, see Fig. 3.6a), to $-p_z$ (when particle 3 is produced as a fragment of 2, Fig. 3.6b). Therefore, the range of $x_F$ is $[0, 1]$.

The kinematic domain where $|p_z'| \simeq 0$, i.e. $x_F \simeq 0$, is called *central region* (Fig. 3.6c). The domain where $|p_z'| \simeq p_z$, i.e. $x_F \simeq 1$, is the *fragmentation region*[3] (Fig. 3.6a,b). Diffraction dissociation processes fall into the latter region.

If we use light-cone variables (App. A.1) and write the four-momenta of particles 1 and 3 as

[3] In the Lab system one talks of *target* fragmentation region if $p_z' \simeq p_z$ and *beam* fragmentation region if $p_z' \simeq -p_z$.

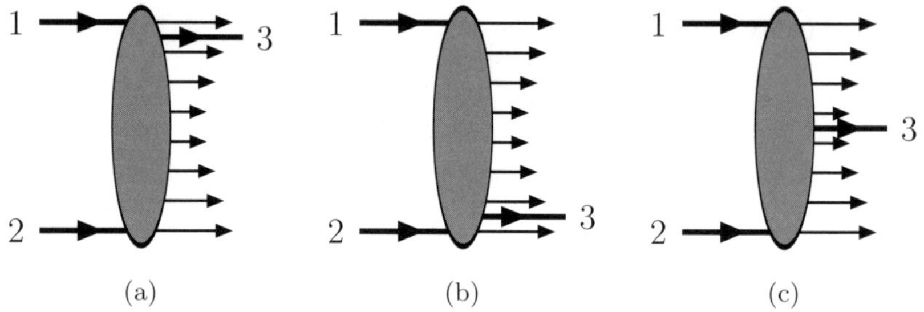


**Fig. 3.6.** (**a,b**) Fragmentation region, $x_F \simeq 1$: (**a**) particle 3 is produced as a fragment of 1; (**b**) particle 3 is produced as a fragment of 2. (**c**) Central region, $x_F \simeq 0$.

$$p_1 = \left(p_1^+, \frac{m_1^2}{2p_1^+}, \mathbf{0}\right) , \tag{3.84}$$

$$p_3 = \left(p_3^+, \frac{\boldsymbol{p}_\perp^2 + m_3^2}{2p_3^+}, \boldsymbol{p}_\perp\right) , \tag{3.85}$$

it is easy to see, using (3.75, 3.76) and assuming that particle 3 is produced as a fragment of 1 (that is $p_z' \geq 0$), that in the limit $s \gg m_i^2, \boldsymbol{p}_\perp^2$ one has

$$x_F \simeq \frac{p_3^+}{p_1^+} , \tag{3.86}$$

and thus the four-momentum of the outgoing particle can be rewritten as

$$p_3 = \left(x_F p^+, \frac{\boldsymbol{p}_\perp^2 + m_3^2}{2x_F p^+}, \boldsymbol{p}_\perp\right) . \tag{3.87}$$

In view of the study of diffraction dissociation of virtual photons, let us obtain a formula for $x_F$ valid when the restriction $s \gg m_2^2$ is not imposed. In $\gamma^* p$ collisions, in fact, the role of $m_2^2$ is played by the virtuality $q^2$ of the photon, which is generally non negligible (in absolute value) with respect to $s$. For $m_1^2 \ll s$ and any $m_2^2$, Eqs. (3.75) are replaced by (see Eqs. (3.25, 3.30))

$$E_1 \simeq \frac{s - m_2^2}{2\sqrt{s}} , \quad p_z \simeq \frac{s - m_2^2}{2\sqrt{s}} , \tag{3.88}$$

and using (3.76) we find

$$x_F \simeq \frac{s - M^2}{s - m_2^2} . \tag{3.89}$$

In the specific case of $\gamma^* p$ scattering, with $m_2^2 \to q^2 \equiv -Q^2$, we have

$$x_F \simeq \frac{s - M^2}{s + Q^2} . \tag{3.90}$$

### 3.3.2 Rapidity and Rapidity Gaps

Finally, we introduce another frequently used kinematic variable, the *rapidity*. For a particle of energy $E$ and momentum component $p_z$ along the $z$-axis, the rapidity $y$ is defined as

$$y = \frac{1}{2} \ln \frac{E + p_z}{E - p_z} . \tag{3.91}$$

In the non-relativistic limit, $v \ll 1$, $E \simeq m$, $p_z \simeq m v_z$, and $y$ reduces to the velocity of the particle.

From (3.91) we get

$$\frac{p_z}{E} = \tanh y . \tag{3.92}$$

Introducing the transverse mass $m_\perp$

$$m_\perp = \sqrt{m^2 + \boldsymbol{p}_\perp^2} , \tag{3.93}$$

so that

$$E = \sqrt{m_\perp^2 + p_z^2} , \tag{3.94}$$

we can disentangle $p_z$ and $E$ in (3.92)

$$p_z = m_\perp \sinh y , \tag{3.95}$$

$$E = m_\perp \cosh y . \tag{3.96}$$

What makes the rapidity a useful variable is the fact that it transforms additively under a Lorentz boost along $z$. Using

$$(E, \boldsymbol{p}_\perp, p_z) \underset{\text{boost}}{\rightarrow} (\gamma(E + \beta p_z), \boldsymbol{p}_\perp, \gamma(p_z + \beta E)) , \tag{3.97}$$

one finds indeed

$$y \underset{\text{boost}}{\rightarrow} y + \frac{1}{2} \ln \frac{1+\beta}{1-\beta} . \tag{3.98}$$

Hence any rapidity difference is invariant under longitudinal boosts and remains the same in all collinear frames.

For massless particles one has $E \simeq |\boldsymbol{p}|$ and the rapidity is directly related to the angle $\vartheta$ specifying the direction of motion with respect to the $z$-axis

$$y = \frac{1}{2} \ln \frac{1 + \cos\vartheta}{1 - \cos\vartheta} = - \ln \tan \frac{\vartheta}{2} , \tag{3.99}$$

One defines the *pseudorapidity* $\eta$ as

$$\eta \equiv y|_{m=0} = - \ln \tan \frac{\vartheta}{2} , \tag{3.100}$$

Thus for particles of zero mass, rapidity and pseudorapidity coincide. This is also the case when $|\boldsymbol{p}_\perp| \gg m$.

For very fast particles ($p_z \to \infty$) the rapidity takes the form

$$y \simeq \ln \frac{2p_z}{m_\perp} , \qquad (3.101)$$

immediately derivable from (3.91).

Coming back to the single-inclusive processes, from (3.101), (3.76) and (3.81) one finds that the rapidity of particle 3 at large $s$ is (we suppose that particle 3 is a fragment of 1, i.e. $p'_z \geq 0$, and, for simplicity, we take all particle masses to be equal to $m$)

$$y_3 = \frac{1}{2} \ln \frac{E_3 + p'_z}{E_3 - p'_z} \simeq \ln \frac{\sqrt{s}}{m_\perp} . \qquad (3.102)$$

The maximum value of $y_3$ is attained when $\boldsymbol{p}_\perp = 0$, that is

$$(y_3)_{\rm max} = \ln \frac{\sqrt{s}}{m} . \qquad (3.103)$$

As for the system $X$, it obviously exhibits a distribution of rapidity, due to its compositeness. The average value of $y_X$ corresponds approximately to a momentum $(p_X)_z \simeq -\sqrt{s}/2$ and to a transverse mass $(m_X)_\perp \simeq M$, hence

$$\langle y_X \rangle \simeq - \ln \frac{\sqrt{s}}{M} . \qquad (3.104)$$

The maximum rapidity (in absolute value) of the $X$ system is the rapidity of a particle with momentum $\sim \sqrt{s}/2$ and transverse mass $\sim m$,

$$|y_X|_{\rm max} \simeq \ln \frac{\sqrt{s}}{m} , \qquad (3.105)$$

whereas the minimum value of $|y_X|$ pertains to a particle with momentum $\sim (m/M)\sqrt{s}/2$ and transverse mass $\sim M$,

$$|y_X|_{\rm min} \simeq \ln \frac{m\sqrt{s}}{M^2} , \qquad (3.106)$$

The situation is depicted in Fig. 3.7.

The final-state rapidity gap between particle 3 and the edge of the rapidity distribution of the $X$ system is roughly given by

$$\Delta y \simeq \ln \frac{\sqrt{s}}{m} + \ln \frac{m\sqrt{s}}{M^2} \simeq \ln \frac{s}{M^2} . \qquad (3.107)$$

To the extent that we do not make a large error in identifying the rapidity $y$ with the pseudorapidity $\eta$, we can write

$$\Delta\eta \simeq \ln \frac{s}{M^2} , \qquad (3.108)$$

and, from (3.82), the relation between Feynman's variable and the pseudorapidity gap is found to be

$$1 - x_F \simeq {\rm e}^{-\Delta\eta} . \qquad (3.109)$$

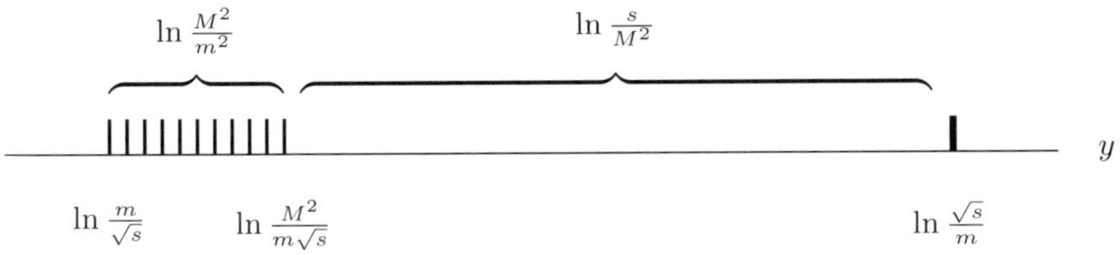


**Fig. 3.7.** Rapidity distribution in the final state.

### 3.3.3 Diffractive Dissociation

We now possess all the information necessary to characterize kinematically the subclass of single-particle inclusive reactions that we are most interested in, the diffractiive dissociation processes.

From (3.80), (3.82) and (3.75) we see that the longitudinal momentum transfer (when $s, M^2 \gg m_i^2$) is

$$|\Delta p_z| \simeq \frac{M^2}{2\sqrt{s}} . \tag{3.110}$$

The coherence condition between the outgoing and the incoming waves, which defines diffraction according to Good and Walker (1960) is

$$|\Delta p_z| \lesssim \frac{1}{R} , \tag{3.111}$$

where $R$ is the size of the target. In view of (3.110) this condition is equivalent to requiring $M^2/s$ to be very small, that is, remembering (3.82),

$$x_F \simeq 1 . \tag{3.112}$$

Thus diffractive events fall in the fragmentation region. The particle produced with $p_z' \simeq p_z$ is called *leading particle.* Its transverse momentum is related to $t$ by

$$t \simeq -\boldsymbol{p}_\perp^2 . \tag{3.113}$$

The same conclusion can be drawn when the mass of the particle 2 cannot be neglected. In this case the longitudinal momentum transfer is

$$|\Delta p_z| \simeq \frac{M^2 - m_2^2}{2\sqrt{s}} . \tag{3.114}$$

and the coherence condition still prescribes $x_F \simeq 1$, where now $x_F = (s - M^2)/(s - m_2^2)$. Notice that this means that in order for diffraction to occur the mass $M$ of the produced system $X$ should not be much different from the mass of the diffracted particle.

Another important signature of diffraction is the presence of large rapidity gaps in the final state. From Eq. (3.108) we see in fact that the condition $M^2/s \ll 1$, or equivalently $x_F \simeq 1$, corresponds to demanding a large $\Delta\eta$.

To understand qualitatively the relationship between diffraction and large rapidity gaps, consider the case of Fig. 3.1b (single diffraction). If the particle emitted at the upper vertex coincides with the incoming particle 1 and if the event is diffractive, by our Definition 1 (see Sect. 1.1) this means that no quantum numbers are exchanged in the $t$ channel. No other particles are emitted at the upper vertex: the emission of any particle, in fact, would immediately produce an exchange of non-vacuum quantum numbers. Thus, diffraction, according to Definition 1 implies (and is signaled by) the existence of large rapidity gaps (which is the content of Definition 2).

# 4. The Relativistic $S$-Matrix

The $S$-matrix program (see, e.g., Chew 1961; Frautschi 1963; Eden et al. 1966) was an ambitious attempt to achieve a complete description of strong interactions, alternative – to some extent – to quantum field theory. The program, in its original form, failed, but some of its methods are still useful and have found an important place in the modern approaches to hadronic phenomena. Unitarity relations, analytical properties of scattering amplitudes, dispersive techniques, exact theorems: all that still belongs to the background of hadronic physicists. In the following we present the main concepts and tools of relativistic $S$-matrix theory.

## 4.1 General Definitions

The scattering matrix (or, simply, $S$-matrix) is the linear operator which transforms the initial state $|i\rangle$ of a scattering process into the final state $|f\rangle$

$$S\,|i\rangle = |f\rangle\ . \tag{4.1}$$

The 'in' states $|i\rangle$ and the 'out' states $|f\rangle$ are defined at the time $-\infty$ and $+\infty$, respectively. They represent free particles and form complete sets of states.

The probability that, starting from $|i\rangle$, the system will be found, after the scattering, in the state $|f\rangle$ is, by definition,

$$P_{i\to f} = |\langle f|S|i\rangle|^2\ . \tag{4.2}$$

Clearly, the full knowledge of the $S$-matrix would allow in principle to reconstruct the entire dynamics of the interaction process.

The $S$-matrix coincides with the time evolution operator $U$ from $t = -\infty$ to $t = +\infty$

$$S \equiv U(-\infty, +\infty)\ , \tag{4.3}$$

and in quantum field theory is given by the Dyson series

$$S = 1\!\!1 + \sum_{n=1}^{\infty} \frac{\mathrm{i}^n}{n!} \int \mathrm{d}^4x_1 \ldots \mathrm{d}^4x_n\, \mathcal{T}\left(H'_{\mathrm{int}}(x_1) \ldots H'_{\mathrm{int}}(x_n)\right)\ , \tag{4.4}$$

where $H'_{\rm int}$ is the interaction Hamiltonian (in the interaction representation) and $\mathcal{T}$ denotes the time-ordered product.

The *linearity* of $S$ reflects the superposition principle of quantum mechanics. It is also natural to require $S$ to be *relativistically invariant.* That means that its matrix elements must depend on Lorentz-invariant combinations of the kinematic variables.

Three other general properties of the $S$-matrix are usually assumed. They are:

- *unitarity,*
- *analyticity,*
- *crossing.*

We shall analyze in detail these properties in the next sections. For the moment we just recall that the first of them (unitarity) is a direct consequence of the conservation of probability, while the other two (crossing and analyticity) are essentially *postulated,* although they are supported, to some extent, by perturbative quantum field theory.

We first proceed to show how the $S$-matrix elements are related to the most important observables of particle physics, the scattering cross sections.

## 4.2 Scattering Amplitudes and Cross-sections

Subtracting from $S$ the identity operator ($S = 1\!\!1$ means that nothing happens), we obtain the *transition* matrix $T$

$$S = 1\!\!1 + \mathrm{i}\, T \, . \tag{4.5}$$

The $S$-matrix elements are then

$$S_{if} \equiv \langle f|S|i\rangle = \delta_{if} + \mathrm{i}\,\langle f|T|i\rangle = \delta_{if} + \mathrm{i}\, T_{if} \, . \tag{4.6}$$

If we extract from $T$ the $\delta$ function which enforces the four-momentum conservation $p_i = p_f$, Eq. (4.6) can be rewritten as

$$S_{if} \equiv \langle f|S|i\rangle = \delta_{if} + \mathrm{i}\,(2\pi)^4\,\delta^4(p_f - p_i)\,A(i \to f) \, , \tag{4.7}$$

where $A(i \to f)$ is the relativistic *scattering amplitude.*

The normalization of the momentum eigenstates $|p\rangle$ is

$$\langle p|p'\rangle = (2\pi)^3\, 2E\, \delta^3(\boldsymbol{p} - \boldsymbol{p}') \, . \tag{4.8}$$

Hereafter we use natural units ($\hbar = c = 1$).

Let us write the cross section for a scattering process of the type $1+2 \to n$ particles. The initial state is a two-particle state

$$|i\rangle = |p_1\, p_2\rangle \, , \tag{4.9}$$

with total four-momentum $p_i = p_1 + p_2$. The final state is composed of $n$ particles

$$|f_n\rangle = |p_1' \dots p_n'\rangle \,. \tag{4.10}$$

We omit all other quantum numbers.

The differential cross section for this reaction is given by

$$\mathrm{d}\sigma = \frac{1}{\Phi} \, |A(i \to f_n)|^2 \, \mathrm{d}\Pi_n \,, \tag{4.11}$$

where $\Phi$ is the incident flux and $\mathrm{d}\Pi_n$ is the Lorentz-invariant phase space for $n$ particles in the final state. The total cross section $\sigma_{\mathrm{tot}}$ is obtained by integrating (4.11) and summing over all possible numbers of particles produced in the final state, i.e.

$$\sigma_{\mathrm{tot}} = \frac{1}{\Phi} \sum_n \int \mathrm{d}\Pi_n \, |A(i \to f_n)|^2 \,. \tag{4.12}$$

The explicit expression of the phase space $\mathrm{d}\Pi_n$ is

$$\begin{aligned} \mathrm{d}\Pi_n &= \prod_{j=1}^{n} \frac{\mathrm{d}^4 p_j'}{(2\pi)^3} \, \delta(p_j'^2 - m_j^2) \, (2\pi)^4 \, \delta^4(p_1 + p_2 - \sum_{j=1}^{n} p_j') \\ &= \prod_{j=1}^{n} \frac{\mathrm{d}^3 \boldsymbol{p}_j'}{(2\pi)^3 \, 2E_j'} \, (2\pi)^4 \, \delta^4 \left( p_1 + p_2 - \sum_{j=1}^{n} p_j' \right) \,, \end{aligned} \tag{4.13}$$

where in the second equality we have used

$$\int \mathrm{d}^4 p \, \delta(p^2 - m^2) = \int \frac{\mathrm{d}^3 \boldsymbol{p}}{2p_0} \, \theta(p_0) \,. \tag{4.14}$$

The incident flux is

$$\Phi = 2E_1 \, 2E_2 \, |\boldsymbol{v}_1 - \boldsymbol{v}_2| \,, \tag{4.15}$$

where $\boldsymbol{v}_1$ and $\boldsymbol{v}_2$ are the velocities of the colliding particles 1 and 2, respectively, i.e.

$$\boldsymbol{v}_1 = \frac{\boldsymbol{p}_1}{E_1} \,, \quad \boldsymbol{v}_2 = \frac{\boldsymbol{p}_2}{E_2} \,. \tag{4.16}$$

The flux (4.15) is invariant for boosts along the collision axis. To see this we write

$$\begin{aligned} \Phi &= 4E_1 E_2 \left| \frac{\boldsymbol{p}_1}{E_1} - \frac{\boldsymbol{p}_2}{E_2} \right| \\ &= 4 \, |E_2 \boldsymbol{p}_1 - E_1 \boldsymbol{p}_2| \\ &= 4 \, \left[ E_2^2 \boldsymbol{p}_1^2 + E_1^2 \boldsymbol{p}_2^2 - 2E_1 E_2 \, |\boldsymbol{p}_1| \, |\boldsymbol{p}_2| \right]^{\frac{1}{2}} \,. \end{aligned} \tag{4.17}$$

In any collinear frame, i.e. in any frame where $\boldsymbol{p}_1$ is parallel to $\boldsymbol{p}_2$, we also have

$$\begin{aligned}\left[(p_1 \cdot p_2)^2 - m_1^2 m_2^2\right]^{\frac{1}{2}} &= \left[(E_1 E_2 - |\boldsymbol{p}_1|\,|\boldsymbol{p}_2|)^2 - m_1^2 m_2^2\right]^{\frac{1}{2}} \\ &= \left[E_1^2\,\boldsymbol{p}_2^2 + E_2^2\,\boldsymbol{p}_1^2 - 2E_1E_2\,|\boldsymbol{p}_1|\,|\boldsymbol{p}_2|\right]^{\frac{1}{2}} ,\end{aligned} \tag{4.18}$$

and, comparing Eqs. (4.17) and (4.18), we obtain

$$\Phi = 4\,\left[(p_1 \cdot p_2)^2 - m_1^2 m_2^2\right]^{\frac{1}{2}} , \tag{4.19}$$

which shows explicitly the Lorentz invariance (limited to collinear frames) of the flux factor. Since $2p_1 \cdot p_2 = s - m_1^2 - m_2^2$ the flux (4.19) is given by

$$\Phi = 2\,\lambda^{\frac{1}{2}}(s, m_1^2, m_2^2) , \tag{4.20}$$

and in the large-$s$ limit reduces to

$$\Phi \simeq 2s . \tag{4.21}$$

Putting everything together, the differential cross section for the reaction $1 + 2 \to n$ particles is

$$\begin{aligned}\mathrm{d}\sigma = &\frac{1}{2\,\lambda^{\frac{1}{2}}(s, m_1^2, m_2^2)} \prod_{j=1}^{n} \frac{\mathrm{d}^3\boldsymbol{p}_j'}{(2\pi)^3\,2E_j'} \\ &\times (2\pi)^4\,\delta^4\left(p_1 + p_2 - \sum_{j=1}^{n} p_j'\right) |A(i \to f_n)|^2 .\end{aligned} \tag{4.22}$$

### 4.2.1 Two-Body Exclusive Reactions

Let us focus now on the important case of a two-body process $1 + 2 \to 3 + 4$. The differential cross section then reads

$$\begin{aligned}\mathrm{d}\sigma = &\frac{1}{2\,\lambda^{\frac{1}{2}}(s, m_1^2, m_2^2)} |A(12 \to 34)|^2 \\ &\times (2\pi)^4\,\delta^4(p_1 + p_2 - p_3 - p_4)\,\frac{\mathrm{d}^3\boldsymbol{p}_3\,\mathrm{d}^3\boldsymbol{p}_4}{(2\pi)^3 2E_3\,(2\pi)^3 2E_4} .\end{aligned} \tag{4.23}$$

Using $\delta^3(\boldsymbol{p}_1 + \boldsymbol{p}_2 - \boldsymbol{p}_3 - \boldsymbol{p}_4)$ to perform the integration over $\mathrm{d}^3\boldsymbol{p}_4$ and decomposing the remaining volume element as $\mathrm{d}^3 p_3 = \boldsymbol{p}_3^2\,\mathrm{d}|\boldsymbol{p}_3|\,\mathrm{d}\Omega$, where $\mathrm{d}\Omega = \mathrm{d}\cos\vartheta\,\mathrm{d}\phi$, we get

$$\begin{aligned}\frac{\mathrm{d}\sigma}{\mathrm{d}\Omega} = &\frac{1}{2\,\lambda^{\frac{1}{2}}(s, m_1^2, m_2^2)\,(2\pi)^2} \\ &\times \int \frac{\boldsymbol{p}_3^2\,\mathrm{d}|\boldsymbol{p}_3|}{2E_3\,2E_4}\,\delta(E_1 + E_2 - E_3 - E_4)\,|A(12 \to 34)|^2 .\end{aligned} \tag{4.24}$$

We can calculate (4.24) in any frame. Choosing for instance the CM system, where $\boldsymbol{p}_1 = -\boldsymbol{p}_2$ and $\boldsymbol{p}_3 = -\boldsymbol{p}_4$, and using

$$E_1 + E_2 = \sqrt{s}\ , \tag{4.25}$$

$$E_3 + E_4 = \sqrt{\boldsymbol{p}_3^2 + m_3^2} + \sqrt{\boldsymbol{p}_3^2 + m_4^2}\ , \tag{4.26}$$

the integration over $|\boldsymbol{p}_3|$ is easily performed and we obtain

$$\frac{\mathrm{d}\sigma}{\mathrm{d}\Omega} = \frac{\lambda^{\frac{1}{2}}(s, m_3^2, m_4^2)}{64\pi^2\, s\, \lambda^{\frac{1}{2}}(s, m_1^2, m_2^2)}\, |A(s,t)|^2\ . \tag{4.27}$$

At high energies, (4.27) reduces to

$$\frac{\mathrm{d}\sigma}{\mathrm{d}\Omega} \simeq \frac{1}{64\pi^2 s}\, |A(s,t)|^2\ . \tag{4.28}$$

This is also the exact expression for the equal-mass case.

Comparing Eq. (4.28) with Eq. (2.81) we see that the correspondence between the relativistic scattering amplitude $A(s,t)$ and the quantum-mechanical scattering amplitude $f(k, \vartheta)$ defined in Sect. 2.2.3 is

$$\frac{1}{8\pi\sqrt{s}}\, A(s,t) \to f(k,\vartheta)\ , \tag{4.29}$$

with

$$s = 4(k^2 + m^2)\ . \tag{4.30}$$

The differential cross section is often presented as a function of the momentum transfer $t$, which is related to $\cos\vartheta$ by Eq. (3.36). At fixed $s$ we have

$$\mathrm{d}\cos\vartheta = \frac{2s}{\lambda^{\frac{1}{2}}(s, m_1^2, m_2^2)\, \lambda^{\frac{1}{2}}(s, m_3^2, m_4^2)}\, \mathrm{d}t \underset{s\to\infty}{\simeq} \frac{2}{s}\, \mathrm{d}t\ . \tag{4.31}$$

If the amplitude does not depend on the azimuthal angle (this is the case for spinless particle scattering), the integration over $\phi$ is trivial and Eq. (4.27) becomes

$$\frac{\mathrm{d}\sigma}{\mathrm{d}t} = \frac{1}{16\pi\, \lambda(s, m_1^2, m_2^2)}\, |A(s,t)|^2\ , \tag{4.32}$$

and for large $s$

$$\frac{\mathrm{d}\sigma}{\mathrm{d}t} \simeq \frac{1}{16\pi s^2}\, |A(s,t)|^2\ . \tag{4.33}$$

**Partial-Wave Expansion.** Analogously to what we have done in quantum mechanics, we can expand the scattering amplitude $A(s,t)$ in partial waves. Using the relation (3.38) between $t$ and $\cos\vartheta$ in the $s$-channel, we can reexpress the amplitude in terms of $s$ and $\vartheta$ (calling it, with a slight abuse of notation, $A(s, \cos\vartheta)$) and write the expansion (in the $s$-channel)

$$A(s, \cos\vartheta) = \sum_{\ell=0}^{\infty} (2\ell + 1)\, A_\ell(s)\, P_\ell(\cos\vartheta) \; . \tag{4.34}$$

The partial-wave amplitudes $A_\ell(s)$ can be extracted from (4.34) by means of the orthogonality relation of Legendre polynomials (2.71). We obtain

$$A_\ell(s) = \frac{1}{2} \int_{-1}^{1} \mathrm{d}\cos\vartheta \, P_\ell(\cos\vartheta)\, A(s, \cos\vartheta) \; . \tag{4.35}$$

Similar partial-wave series can be written for the $t$ and $u$ channels. For instance, in the $t$ channel we have

$$A(s, \cos\vartheta_t) = \sum_{\ell=0}^{\infty} (2\ell + 1)\, A_\ell(t)\, P_\ell(\cos\vartheta_t) \; , \tag{4.36}$$

and

$$A_\ell(t) = \frac{1}{2} \int_{-1}^{1} \mathrm{d}\cos\vartheta_t \, P_\ell(\cos\vartheta_t)\, A(s, \cos\vartheta_t) \tag{4.37}$$

where $\vartheta_t$ is the $t$-channel scattering angle.

A final remark: comparing Eqs. (2.67) and (4.34), and using the correspondence (4.29) one sees that the relation between the $S$-matrix partial-wave amplitudes $A_\ell(s)$ and the quantum mechanical partial-wave amplitudes $a_\ell(k)$ is

$$\frac{1}{8\pi\sqrt{s}} A_\ell(s) \to a_\ell(k) \; . \tag{4.38}$$

### 4.2.2 Single-Particle Inclusive Reactions

In a single-inclusive reaction of the type $1 + 2 \to 3 + X$, an arbitrary number of particles (besides the identified particle 3) is produced in the final state. In general $n$ particles of type 3 (with $n \geq 1$) and $N$ particles of a different type (with $N \geq 1$) are generated in a scattering event. For brevity, in this section we shall drop the subscript 3 from the variables related to particle 3.

The cross section for producing at least one particle of type 3 is (see Eq. (4.22))

$$\begin{aligned} \mathrm{d}\sigma_{12\to 3X} &= \sum_{n=1}^{\infty} \mathrm{d}\sigma^{(n)} \\ &= \frac{1}{2\,\lambda^{\frac{1}{2}}(s, m_1^2, m_2^2)} \sum_{n=1}^{\infty} \sum_{N=1}^{\infty} |A(1 + 2 \to n + N)|^2 \, \mathrm{d}\Pi_{n+N} \; , \end{aligned} \tag{4.39}$$

where $\mathrm{d}\sigma^{(n)}$ is the differential cross section for producing $n$ particles of type 3. The Lorentz-invariant phase space factor in (4.39) is

$$\mathrm{d}\Pi_{n+N} = \prod_{i=1}^{n} \frac{\mathrm{d}^3\boldsymbol{p}_i'}{2E_i'\,(2\pi)^3} \prod_{j=1}^{N} \frac{\mathrm{d}^3\boldsymbol{k}_j}{2\varepsilon_j\,(2\pi)^3}$$

$$\times (2\pi)^4\,\delta^4\left(p_1 + p_2 - \sum_{i=1}^{n} p_i' - \sum_{j=1}^{N} k_i\right) , \qquad (4.40)$$

where we have called $p_i' = (E_i', \boldsymbol{p}_i')$ the four-momenta of the particles of type 3 and $k_j = (\varepsilon_j, \boldsymbol{k}_j)$ the four-momenta of the other particles.

The single-inclusive differential cross section, written in an invariant form, reads

$$f(\boldsymbol{p}, s) \equiv (2\pi)^3\, 2E\, \frac{\mathrm{d}^3\sigma}{\mathrm{d}^3\boldsymbol{p}}$$

$$= \frac{1}{2\,\lambda^{\frac{1}{2}}(s, m_1^2, m_2^2)} \sum_{n=1}^{\infty}\sum_{N=1}^{\infty} \int \mathrm{d}\Pi_{n+N}$$

$$\times \sum_{l=1}^{n} E_l'\,\delta^3(\boldsymbol{p} - \boldsymbol{p}_l')\,|A(1+2 \to n+N)|^2 . \qquad (4.41)$$

It is useful to recall a chain of equalities (that we write in the large-$s$ limit) which allow to reexpress the differential cross section in terms of the other sets of variables introduced in Sect. 3.3:

$$E\,\frac{\mathrm{d}^3\sigma}{\mathrm{d}^3\boldsymbol{p}} = \frac{s}{\pi}\,\frac{\mathrm{d}^2\sigma}{\mathrm{d}t\,\mathrm{d}M^2} = \frac{x_F}{\pi}\,\frac{\mathrm{d}^2\sigma}{\mathrm{d}x_F\,\mathrm{d}\boldsymbol{p}_\perp^2} = \frac{1}{\pi}\,\frac{\mathrm{d}^2\sigma}{\mathrm{d}y\,\mathrm{d}\boldsymbol{p}_\perp^2} . \qquad (4.42)$$

The total single-inclusive cross section is obtained from (4.41) by integrating over $\boldsymbol{p}$, that is

$$\sigma_{\mathrm{incl}} = \int \frac{\mathrm{d}^3\boldsymbol{p}}{(2\pi)^3\,2E}\, f(\boldsymbol{p}, s)$$

$$= \frac{1}{4\,(2\pi)^3\,\lambda^{\frac{1}{2}}(s, m_1^2, m_2^2)} \sum_{n=1}^{\infty}\sum_{N=1}^{\infty} \int \mathrm{d}\Pi_{n+N}$$

$$\times \sum_{l=1}^{n} \int \mathrm{d}^3\boldsymbol{p}\,\delta^3(\boldsymbol{p} - \boldsymbol{p}_l')\,|A(1+2 \to n+N)|^2 .$$

$$= \sum_{n=1}^{\infty} n\,\sigma^{(n)} , \qquad (4.43)$$

where $\sigma^{(n)}$ is the integrated cross section for producing $n$ particles of type 3.

Note that the total single-inclusive cross section $\sigma_{\mathrm{incl}}$ is equal to $\sum_n n\,\sigma^{(n)}$, not to $\sum_n \sigma^{(n)}$. The latter quantity, roughly speaking, measures the number of *events* with particles 3 in the final state, whereas $\sigma_{\mathrm{incl}}$ measures the number of *particles* of type 3.

The *multiplicity* of particles of type 3 is, therefore, defined as

$$\langle n \rangle = \frac{\sum_{n=0}^{\infty} n\,\sigma^{(n)}}{\sum_{n=0}^{\infty} \sigma^{(n)}} = \frac{1}{\sigma_{\text{tot}}} \sum_{n=0}^{\infty} n\,\sigma^{(n)} . \tag{4.44}$$

In view of (4.42) one has the following sum rule

$$\int \frac{\mathrm{d}^3\boldsymbol{p}}{(2\pi)^3\, 2E}\, f(\boldsymbol{p}, s) = \langle n \rangle\, \sigma_{\text{tot}} . \tag{4.45}$$

## 4.3 Unitarity

The unitarity of the $S$-matrix

$$S^\dagger S = S S^\dagger = \mathbb{1} , \tag{4.46}$$

follows directly from the conservation of probability. In fact, if we take an orthonormal basis of vectors $|k\rangle$, the probabilities to go from an arbitrary initial state $|i\rangle$ to any of the states $|k\rangle$ must sum up to unity

$$\begin{aligned} \sum_k P_{i\to k} &= \sum_k |\langle k|S|i\rangle|^2 \\ &= \sum_k \langle i|S^\dagger|k\rangle\langle k|S|i\rangle \\ &= \langle i|S^\dagger S|i\rangle = 1 , \end{aligned} \tag{4.47}$$

from which (4.46) follows, due to the arbitrariness of $|i\rangle$.

Let us now rewrite the unitarity condition (4.46) in terms of the transition matrix defined in (4.5). We have

$$(\mathbb{1} - \mathrm{i}\,T^\dagger)(\mathbb{1} + \mathrm{i}\,T) = \mathbb{1} , \tag{4.48}$$

that is

$$\mathrm{i}(T^\dagger - T) = T^\dagger T . \tag{4.49}$$

Taking now the matrix elements of the l.h.s. and r.h.s. of Eq. (4.49) between the initial state $|i\rangle$ and the final state $|f\rangle$ of the scattering process, and inserting a complete set of intermediate states $|n\rangle$ we get

$$\mathrm{i}\,\langle f|T^\dagger - T|i\rangle = \sum_{\{n\}} \langle f|T^\dagger|n\rangle\langle n|T|i\rangle , \tag{4.50}$$

that is

$$2\,\mathrm{Im}\, T_{if} = \sum_{\{n\}} T^*_{fn} T_{in} , \tag{4.51}$$

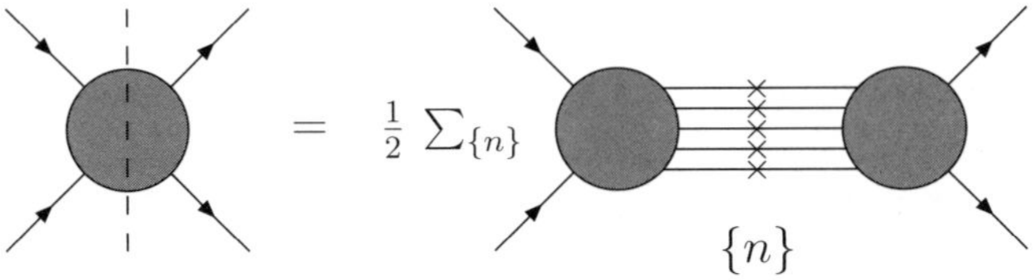


**Fig. 4.1.** Graphical representation of the unitarity equations. The cut on the left-hand side denotes the discontinuity of the amplitude. The crosses on the right-hand side denote particles on shell.

where $\sum_{\{n\}}$ contains the integration over all continuous variables (the momenta of the various particles) and the sum over all discrete quantum numbers. If the states $|n\rangle$ represent systems of $n$ spinless particles with three-momenta $\boldsymbol{q}_1, \ldots, \boldsymbol{q}_n$, and energies $E_1, \ldots, E_n$, then $\sum_{\{n\}}$ is

$$\sum_{\{n\}} = \sum_n \int \prod_{j=1}^{n} \frac{\mathrm{d}^3\boldsymbol{q}_j}{(2\pi)^3\, 2E_j} \,. \tag{4.52}$$

Equations (4.51) form a set of nonlinear coupled integral equations which incorporate the whole information about unitarity.

If we extract from the $T$-matrix elements the delta functions of energy-momentum conservation (see (4.7)), thus getting the scattering amplitudes $A$, Eq. (4.52) becomes,

$$2\,\mathrm{Im}\, A(i \to f) = \sum_n \int \mathrm{d}\Pi_n\, A^*(f \to n)\, A(i \to n) \,, \tag{4.53}$$

where $\mathrm{d}\Pi_n$ is the $n$-particle phase space measure (4.13).

The unitarity equations (4.53), or (4.51), are graphically represented in Fig. 4.1.

While it is not possible to deal in full generality with the set (4.53) of nonlinear coupled integral equations, nonetheless it is rather easy to derive from it some important consequences, as we shall see in the next sections. Here we recall that the unitarity relation (4.51) is very helpful in perturbative field theory. It tells us in fact that, in order to calculate the imaginary part of the scattering amplitude of some process at the $n$-th perturbative order, we just need to compute the amplitudes up to the $(n-1)$-th order. Then, the real part can be obtained from the imaginary part by using dispersion relations (see below, Sect. 4.6). This combination of unitarity and dispersive techniques is a common and very convenient procedure and we shall adopt it in computing the pomeron structure in perturbative QCD (Sect. 8.2).

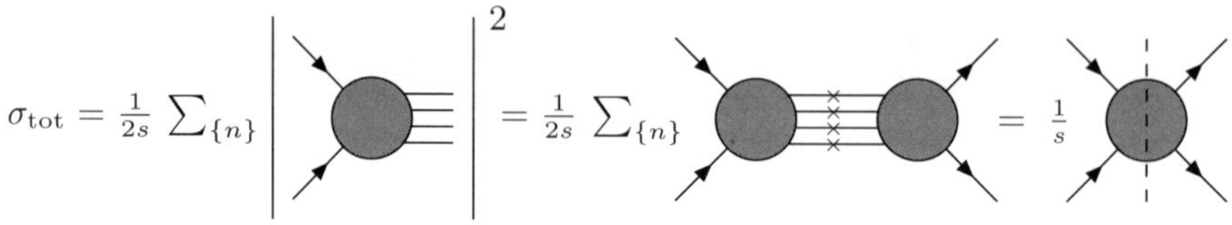

**Fig. 4.2.** The optical theorem.

### 4.3.1 The Optical Theorem

A result of great relevance following in a straighforward way from unitarity is the *optical theorem*, which is contained as a special case in Eqs. (4.53)[1]. If the initial and final states are identical (in the sense also that the individual momenta of particles remain unmodified after the collision)

$$|i\rangle = |f\rangle \ , \tag{4.54}$$

which, for a $2 \to 2$ particle collision, is the case of *forward elastic scattering* ($t = 0$), one gets from (4.53)

$$2 \,\mathrm{Im}\, A_{\mathrm{el}}(s, t = 0) = \sum_n \int \mathrm{d}\Pi_n \, |A(i \to n)|^2 \ . \tag{4.55}$$

Now, comparing (4.55) with (4.12), we see that the r.h.s. of Eq. (4.55) is the total cross section $\sigma_{\mathrm{tot}}$ times the flux factor $\Phi$, which at high energy equals $2s$. Thus we can rewrite (4.55) as

$$\begin{aligned} \sigma_{\mathrm{tot}} &= \frac{2}{\Phi} \,\mathrm{Im}\, A_{\mathrm{el}}(s, t = 0) \\ &\underset{s \to \infty}{\simeq} \frac{1}{s} \,\mathrm{Im}\, A_{\mathrm{el}}(s, t = 0) \ , \end{aligned} \tag{4.56}$$

which is the formulation of the optical theorem for relativistic scattering.

The importance of this theorem is evident and can hardly be overemphasized. It states that the total cross section (i.e. the cross section for the reaction $1 + 2 \to$ anything) is given by the imaginary part of *only one* matrix element of $S$, namely the amplitude for elastic scattering $1 + 2 \to 1 + 2$ in the forward direction. In a sense we can say that it is the sum over the (nearly) infinite number of inelastic open channels that builds up the "shadow" which we call elastic scattering: a clear bridge towards diffraction.

### 4.3.2 Other Consequences of Unitarity

From the unitarity equations (4.51) it is possible to extract some other qualitative results in the limit when $s$ is very high and consequently the number of intermediate states $|n\rangle$ tends to infinity.

[1] We have already derived the optical theorem for the diffraction of light in Sect. (2.1.3) and for nonrelativistic potential scattering in Sect. (2.2.5).

First of all, in the case of forward elastic scattering, which, as we saw, leads to the optical theorem, the fact that $\sigma_{\rm tot}$ is the sum over an increasingly larger number of positive terms (as $s \to \infty$ more and more production channels contribute), suggests that the imaginary part of the scattering amplitude should remain sizeably different from zero in the asymptotic region, and actually may increase with energy. More properly, given that no such constraint exists for the real part of the scattering amplitude, we can reasonably expect that the scattering amplitude should be *predominantly imaginary*, as the energy increases. Looking at the data we shall see that this is indeed the case.

Next, we will now show that high-energy unitarity demands elastic (or diffractive) reactions to exhibit a rapid fall as we move away from the forward direction (although we have no way to establish the analytic form of this fall).

Suppose that, still considering elastic scattering, we move slightly away from the forward direction so that

$$|i\rangle \simeq |f\rangle \, . \tag{4.57}$$

The sum in (4.51) is then a sum of almost positive-definite terms where the various phases of each term are such that the complete sum remains real. If we assume (which is probably not exactly the case) that at high energy, when the number of terms is extremely large, these phases are randomly distributed, the only way the cancellation of the imaginary contributions on the r.h.s. of (4.51) can occur is a rapid drop of the cross section. In other words, we can expect that the more we move away from the forward direction, the more random cancellations will occur to keep the sum real so that the coherent sum we had at $t = 0$ becomes rapidly totally disordered and gets drastically suppressed.

Notice that the same mechanism, outlined above for quasi-forward elastic scattering, works for *diffractive* scattering: in this case also, since no quantum numbers are exchanged, the final state is almost identical to the initial state. The highly energetic initial hadron is left almost undeflected in the final state and flies off leaving behind a stream of soft hadrons. This is known in high-energy physics as the *leading-particle effect* (Basile et al. 1980; Giffon and Predazzi 1980).

While it is essentially impossible to make more quantitative the above heuristic arguments without resorting to particular models or to other assumptions, the suggestion that comes from unitarity is that: *i)* the imaginary part of the scattering amplitude dominates over the real one in the small-angle region; *ii)* there exists a forward diffraction peak that decreases very rapidly as we move away from zero scattering angles, i.e. as $|t|$ increases.

To the extent that, in the language of Regge theory, the pomeron gives rise to a (special form of) diffraction peak, we feel authorized to conjecture that the origin of the pomeron resides precisely in the unitarity constraint or, at least, that there may be a close relationship between them. The fact that unitarity affects not only elastic amplitudes but also amplitudes involving an arbitrary number of particles in the initial and the final state, tells us that what we call pomeron may well be a very complicated object or that this object may refer to a large variety of analogous situations in which the same physical constraint, unitarity, is at work. Thus it is likely that in order to understand what the pomeron really is we will have to learn how to handle unitarity in a completely general way.

### 4.3.3 Elastic Unitarity

We discuss now a two-body elastic scattering $1(p_1)+2(p_2) \to 1'(p_3)+2'(p_4)$, in the special case when only the contribution of one intermediate state is retained on the r.h.s. of the unitarity equations (4.53), and this intermediate state is also a two-body state. Such a situation takes place when the CM energy $s$ is larger than the threshold value $(2m)^2$ for the production of *two* particles, but remains below the threshold for the production of *three* particles, $(3m)^2$.

We denote by $k_1 = (\varepsilon_1, \boldsymbol{k}_1)$ and $k_2 = (\varepsilon_2, \boldsymbol{k}_2)$ the four-momenta of the intermediate particles (see Fig. 4.3), with $k_1^2 = k_2^2 = m^2$, and define, besides the usual Mandelstam variables $s$ and $t$, two more invariants $t_1$ and $t_2$

$$t_1 = (p_1 - k_1)^2 = -2\boldsymbol{p}^2\,(1 - \cos\vartheta_1)\,, \tag{4.58}$$

$$t_2 = (p_1' - k_1)^2 = -2\boldsymbol{p}^2\,(1 - \cos\vartheta_2)\,. \tag{4.59}$$

Here $\boldsymbol{p}$ and $\vartheta$ are the CM total momentum and scattering angle, respectively, whereas $\vartheta_1$ ($\vartheta_2$) is the angle between $\boldsymbol{k}_1$ and $\boldsymbol{p}_1$ ($\boldsymbol{p}_2$).

The set of unitarity equations (4.53) reduces to one equation only, that is

$$\begin{aligned} \operatorname{Im} A_{\rm el}(s,t) = \frac{1}{2} \int \frac{\mathrm{d}^3\boldsymbol{k}_1}{(2\pi)^3\, 2\varepsilon_1} \int \frac{\mathrm{d}^3\boldsymbol{k}_2}{(2\pi)^3\, 2\varepsilon_2} \\ \times (2\pi)^4\, \delta^4(p_1 + p_2 - k_1 - k_2)\, A_{\rm el}(s,t_1)\, A^*_{\rm el}(s,t_2)\,. \end{aligned} \tag{4.60}$$

We can use $\delta^3(\boldsymbol{p}_1 + \boldsymbol{p}_2 - \boldsymbol{k}_1 - \boldsymbol{k}_2)$ to perform one of the three-dimensional integrations, thus obtaining

$$\begin{aligned} \operatorname{Im} A_{\rm el}(s,t) = \frac{1}{2\,(2\pi)^2} \int \frac{|\boldsymbol{k}_1|^2\, \mathrm{d}|\boldsymbol{k}_1|\, \mathrm{d}\Omega_1}{2\varepsilon_1\, 2\varepsilon_2} \\ \times\, \delta(E_1 + E_2 - \varepsilon_1 - \varepsilon_2)\, A_{\rm el}(s,t_1)\, A^*_{\rm el}(s,t_2)\,, \end{aligned} \tag{4.61}$$

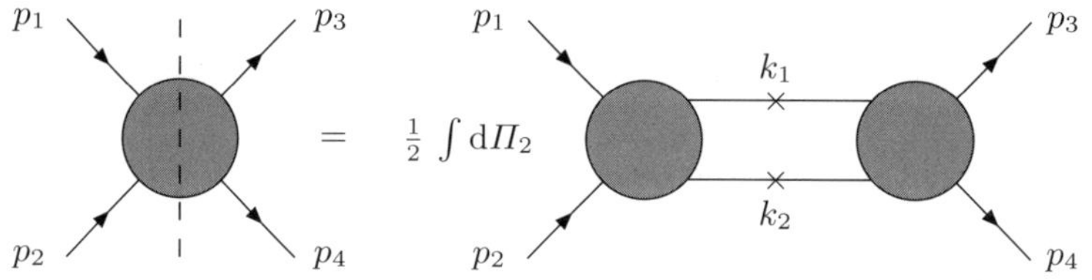


**Fig. 4.3.** Elastic unitarity.

where $\Omega_1$ is the solid angle identifying the direction of $\boldsymbol{k}_1$. In the CM frame we have

$$E_1 + E_2 = \sqrt{s}\,, \tag{4.62}$$

$$\varepsilon_1 = \varepsilon_2 = \sqrt{\boldsymbol{k}_1^2 + m^2}\,, \tag{4.63}$$

and performing the integration over $|\boldsymbol{k}_1|$ we finally get

$$\operatorname{Im} A_{\mathrm{el}}(s,t) = \frac{1}{32\pi^2\sqrt{s}}\sqrt{\frac{s}{4} - m^2}\int \mathrm{d}\Omega_1\, A_{\mathrm{el}}(s,t_1)\, A^*_{\mathrm{el}}(s,t_2)\,. \tag{4.64}$$

This is the *elastic unitarity relation*, where *elastic* means not only that the reaction that we are considering is elastic, but also that the transition from the initial state to the intermediate state is elastic.

Rewritten in terms of $s, \vartheta, \vartheta_1$ and $\vartheta_2$ Eq. (4.64) reads

$$\begin{aligned}\operatorname{Im} A_{\mathrm{el}}(s,\cos\vartheta) &= \frac{1}{32\pi^2\sqrt{s}}\sqrt{\frac{s}{4} - m^2} \\ &\times \int \mathrm{d}\Omega_1\, A_{\mathrm{el}}(s,\cos\vartheta_1)\, A^*_{\mathrm{el}}(s,\cos\vartheta_2)\,. \end{aligned} \tag{4.65}$$

If we expand $A_{\mathrm{el}}(s,\cos\vartheta)$ in partial waves as in (4.34), we can calculate the imaginary part of the partial-wave amplitude $A_\ell(s)$ by inserting Eq. (4.65) into (4.35). We thus obtain[2]

$$\begin{aligned}\operatorname{Im} A_\ell(s) &= \frac{1}{64\pi^2\sqrt{s}}\sqrt{\frac{s}{4} - m^2}\int_{-1}^{1} \mathrm{d}\cos\vartheta \\ &\times \int \mathrm{d}\Omega_1\, A_{\mathrm{el}}(s,\cos\vartheta_1)\, A^*_{\mathrm{el}}(s,\cos\vartheta_2)\,, \end{aligned} \tag{4.66}$$

and expanding again $A_{\mathrm{el}}(s,\vartheta_1)$ and $A_{\mathrm{el}}(s,\vartheta_2)$ in partial waves leads to

$$\begin{aligned}\operatorname{Im} A_\ell(s) &= \frac{1}{32\pi^2\sqrt{s}}\sqrt{\frac{s}{4} - m^2}\sum_{\ell_1}\sum_{\ell_2}(2\ell_1+1)(2\ell_2+1)\, A_{\ell_1}(s)\, A^*_{\ell_2}(s) \\ &\times \int_{-1}^{1} \mathrm{d}\cos\vartheta \int \mathrm{d}\Omega_1\, P_\ell(\cos\vartheta)\, P_{\ell_1}(\cos\vartheta_1)\, P_{\ell_2}(\cos\vartheta_2)\,. \end{aligned} \tag{4.67}$$

The angular integration over $\Omega_1$ can be performed using the relation (remember the definitions (4.58, 4.59))

$$\int \mathrm{d}\Omega_{k_1}\, P_{\ell_1}(\hat{\boldsymbol{p}}_1 \cdot \hat{\boldsymbol{k}}_1)\, P_{\ell_2}(\hat{\boldsymbol{p}}_2 \cdot \hat{\boldsymbol{k}}_1) = \frac{4\pi}{2\ell_1 + 1}\,\delta_{\ell_1\ell_2}\, P_{\ell_1}(\hat{\boldsymbol{p}}_1 \cdot \hat{\boldsymbol{p}}_2)\,. \tag{4.68}$$

---

[2] For simplicity of writing, we omit the subscript ‘el’ in the partial-wave amplitudes.

We find

$$\operatorname{Im} A_\ell(s) = \frac{1}{16\pi\sqrt{s}}\sqrt{\frac{s}{4}-m^2}\sum_{\ell_1}(2\ell_1+1)\,|A_{\ell_1}(s)|^2 \times \int_{-1}^{1}\mathrm{d}\cos\vartheta\, P_\ell(\cos\vartheta)\, P_{\ell_1}(\cos\vartheta)\,, \tag{4.69}$$

and integrating over $\cos\vartheta$ by means of the orthogonality relation of Legendre polynomials (2.71) we finally get

$$\begin{aligned}\operatorname{Im} A_\ell(s) &= \frac{1}{16\pi}\sqrt{\frac{s-4m^2}{s}}\,|A_\ell(s)|^2 \\ &\underset{s\to\infty}{\simeq} \frac{1}{16\pi}\,|A_\ell(s)|^2\,,\end{aligned} \tag{4.70}$$

which is the elastic unitarity equation for the partial wave amplitudes.

In view of the correspondence (4.29) between non relativistic and relativistic scattering amplitudes, we see that Eq. (4.70) is the analogue of the quantum mechanical unitarity condition (2.104).

The *complete* two-body unitarity equation (i.e., not restricted to a two-particle intermediate contribution) for the elastic amplitude $A_{\mathrm{el}}(s,t)$ can be formally written as

$$\operatorname{Im} A_{\mathrm{el}}(s,t) = E(s,t) + I(s,t)\,. \tag{4.71}$$

Here $E(s,t)$ (the *elastic term*) is the two-particle intermediate state contribution which we have evaluated previously (Eq. (4.64)). The *inelastic term* $I(s,t)$ takes into account the inelastic transitions to all possible intermediate states. About $I(s,t)$ we know basically nothing (unless we resort to models), except that it cannot be negative when $t\to 0$, as we learn from (4.55) (the physical reason is the same that led to the conclusion that the r.h.s. of (4.51) had to be non negative, see the discussion in Sect. 4.3.2). The unitarity condition on the partial-wave amplitudes, in its fullest generality, can be derived from (2.111) by using the relation (4.38) and reads in the large-$s$ limit

$$0 \le \frac{1}{(16\pi)^2}\,|A_\ell(s)|^2 \le \frac{1}{16\pi}\operatorname{Im} A_\ell(s) \le 1\,. \tag{4.72}$$

This inequality will be used in Sect. 4.9.1 to derive an asymptotic upper bound on the total cross sections (the Froissart-Martin bound).

## 4.4 Analyticity

The second pillar of the $S$-matrix program, as it was originally conceived, is the postulate of analyticity. It states that the $S$-matrix elements, that is the scattering amplitudes, are analytic functions of the kinematic variables, when these are continued to complex values. The *physical* amplitudes are the real-boundary values of these analytic functions.

Let us consider for definiteness a two-body process. The analyticity postulate tells us that the scattering amplitude $A(s,t)$ is an analytic function of $s$ and $t$ regarded as complex variables. We recover the physical amplitude in the limit $s,t \to$ real (and obeying the limitations of Sect. 3.2.4).

The assumption is that the only singularities of the scattering amplitudes are those dictated by unitarity. As we shall shortly see, these singularities are simple poles and branch points.

In quantum mechanics it is possible to relate analyticity to causality (see, for instance, Goldberger and Watson 1964, Chap. 10). Although it is generally believed that a similar connection should hold in quantum field theory and in relativistic $S$-matrix theory as well, no proof of this assertion is available[3]. There is (so far, at least) no general and rigorous way to derive from first principles the analyticity properties of the scattering amplitudes, without the use of perturbation theory (for some partial attempts see Eden et al. 1966). Thus the study of these properties order by order in the perturbative expansion of the $S$-matrix offers the only secure support to the $S$-matrix axioms. It turns out, in particular, that the singularity structure of Feynman's diagrams is the same as that resulting from the postulates of the theory. Let us then see in some detail this structure.

As already mentioned, among the singularities of the scattering amplitudes, we find, first of all, *simple poles* on the real axis corresponding to the exchange of physical particles. Sticking for simplicity to the case of two-body equal-mass scattering and taking $s$ and $t$ as the independent variables, this means that the $s$-channel amplitude, that we call $A_{\mathrm{I}}(s,t)$ is singular at

$$s = m^2 \,, \qquad [s - \text{channel pole}] \tag{4.73}$$

due to exchange of a particle of mass $m$ in the $s$-channel. This is most easily seen in perturbation theory, from a Feynman diagram like the one of Fig. 4.4, where the exchanged particle has a propagator

$$\frac{1}{s - m^2 + \mathrm{i}\epsilon} \,. \tag{4.74}$$

One can also infer the existence of these poles directly from the unitarity equations, but the procedure is less straightforward (we refer the reader to Eden et al. 1966; Collins 1977).

[3] One should also remember that the so-called "microcausality" condition in quantum field theory, that is the commutativity of field operators for space-like separations, is only indirectly related to the intuitive notion of causality. Thus the dispersion relations that we shall derive for the relativistic scattering amplitudes are physically less sound and less transparent than the formally analogous dispersion relations used in other fields of physics, like quantum mechanics, optics, elasticity, etc.

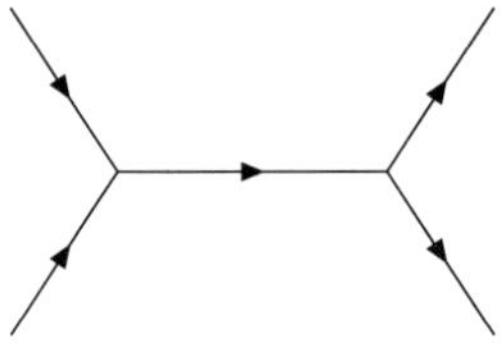

**Fig. 4.4.** Feynman diagram for the exchange of a scalar particle in the $s$ channel in a $\varphi^3$ theory.

Besides simple poles, the scattering amplitudes exhibit *branch point* singularities on the real axis corresponding to the exchange of *two or more* physical particles. Thus, $A_{\rm I}(s,t)$ has branch points at the thresholds

$$s = (2m)^2, \quad s = (3m)^2, \quad \text{etc.} \qquad [s - \text{channel branch points}] \tag{4.75}$$

This can be read out, for instance, from Eq. (4.64), where the square root at the numerator gives rise to a branch cut at $s = 4m^2$. Hence, also branch points are demanded by the unitarity equations. The singularities of $A_{\rm I}(s,t)$ in the $s$ plane are shown in Fig. 4.5.

For $s < 4m^2$ it is not possible to produce two particles and hence $A_{\rm I}(s,t)$ is real. Its only singularity is a pole at $m^2$.

So far we have considered the $s$-channel singularities. A similar discussion applies to the $t$-channel and the $u$-channel.

Since $t$ is the CM energy squared of the $t$-channel process the amplitude for this process, that we call $A_{\rm II}(s,t)$, is singular at

$$\begin{aligned} t &= m^2\,, \qquad [t - \text{channel pole}] \\ t &= (2m)^2, \;\; (3m)^2, \;\; \text{etc.} \qquad [t - \text{channel branch points}] \end{aligned} \tag{4.76}$$

Analogously, the $u$-channel amplitude $A_{\rm III}(s,t)$ presents singularities at

$$\begin{aligned} u &= m^2\,, \qquad [u - \text{channel pole}] \\ u &= (2m)^2, \;\; (3m)^2, \;\; \text{etc.} \qquad [u - \text{channel branch points}] \end{aligned} \tag{4.77}$$

In principle, the amplitudes of the $s$, $t$ and $u$ channels might be different. However, we shall see in the next section that this is not the case: the three amplitudes are given by the same function of the Mandelstam invariants analytically continued to different domains of the variables.

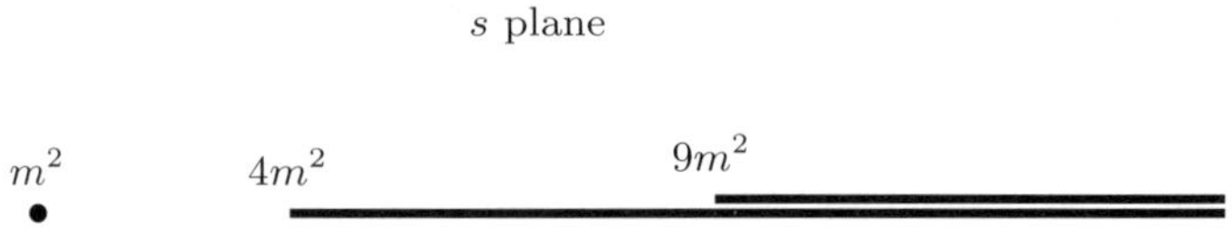


**Fig. 4.5.** Singularities of the scattering amplitude in the $s$-channel.

## 4.5 Crossing

It is an important property of *relativistic field theory* that in a given reaction, an *incoming particle* of momentum $p$ can be viewed as an *outgoing antiparticle* of momentum $-p$. This *crossing* operation relates the three channels defined in Sect. 3.2.1

$$\mathrm{I}: \quad 1+2 \quad \to 3+4 \quad s-\text{channel} \tag{4.78}$$

$$\mathrm{II}: \quad 1+\bar{3} \quad \to \bar{2}+4 \quad t-\text{channel} \tag{4.79}$$

$$\mathrm{III}: \quad 1+\bar{4} \to \bar{2}+3 \quad u-\text{channel}\,. \tag{4.80}$$

The corresponding $CPT$ (charge conjugation + parity + time reversal) transformed reactions are

$$\bar{3}+\bar{4} \to \bar{1}+\bar{2} \quad s-\text{channel} \tag{4.81}$$

$$2+\bar{4} \to \bar{1}+3 \quad t-\text{channel} \tag{4.82}$$

$$2+\bar{3} \to \bar{1}+4 \quad u-\text{channel}\,. \tag{4.83}$$

The $S$-matrix theory postulate of *crossing symmetry* states that the *same amplitude* (or appropriate combinations of amplitudes, in the case of real particles endowed with various quantum numbers like spin, isospin etc.) describes the three *different processes* (4.78–4.80), which are obtained one from the other by crossing. By virtue of the $CPT$ symmetry, the same amplitude describes also the processes (4.81–4.83), together with reactions such as 3 + 4 → 1 + 2 etc., if the validity of time reversal symmetry alone is assumed.

The key point, however, is that, as we have seen earlier (Sect. 3.2.4), the kinematical regions of the three channels are different and non overlapping. This means that, according to the crossing symmetry postulate, the same function of the Mandelstam variables $s, t, u$ describes the six physical reactions (4.78–4.83) in *different domains of these variables*. If, therefore, we know the analytic properties of the scattering amplitude in any given channel, we can, in principle, analytically continue it to the other channels. This possibility is *postulated* in $S$-matrix theory, and is *known* to be true in perturbative field theory.

Taking again $s$ and $t$ as the independent variables, $A(s,t)$ is the scattering amplitude for all processes (4.78–4.83). If $t$ is held fixed, the $u$-channel singularities listed in (4.77) appear as singularities of $A(s,t)$ in the $s$ plane at

$$s = 3m^2 - t \qquad [\text{pole}] \tag{4.84}$$

and

$$s=-t, \quad s=-t-5m^2, \quad \text{etc.} \qquad [\text{branch points}] \tag{4.85}$$

where we have used the relation $s+t+u=4m^2$. The complete singularity structure of $A(s,t)$ in the $s$ plane is shown in Fig. 4.6.

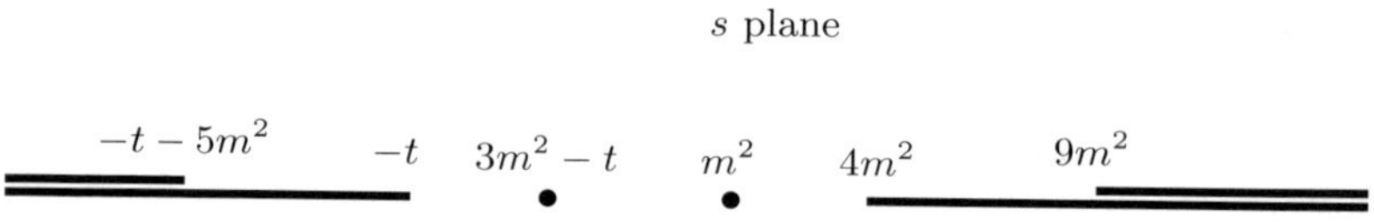


**Fig. 4.6.** Singularities of the scattering amplitude in the $s$ plane.

Note that in the massless case the amplitude $A(s,t)$ has a branch point at $s = -t$.

We assume conventionally that the $s$-channel physical amplitude $A_{\mathrm{I}}(s,t)$ is obtained by approaching the real $s$ axis from above, i.e.

$$[s - \text{channel}]: \quad \text{physical } A_{\mathrm{I}}(s,t) = \lim_{\epsilon\to 0^+} A(s+\mathrm{i}\epsilon, t) \;. \tag{4.86}$$

Due to the Schwarz reflection principle, the reality of $A(s,t)$ for $-t < s < 4m^2$ implies

$$A(s^*,t) = A^*(s,t) \;, \tag{4.87}$$

and hence one finds that the discontinuity of $A(s,t)$ across the cut associated with the $s$-channel thresholds, that we call $D_s(s,t)$ and define as

$$D_s(s,t) \equiv \mathrm{Disc}_s A(s,t,u) = \frac{1}{2\mathrm{i}} \lim_{\varepsilon\to 0^+} [A(s+\mathrm{i}\epsilon,t) - A(s-\mathrm{i}\epsilon,t)] \;, \tag{4.88}$$

coincides with the imaginary part of $A(s,t)$ (taken on the upper edge of the cut)

$$D_s(s,t) \equiv \mathrm{Im}\, A(s,t) \;. \tag{4.89}$$

The physical amplitude $A_{\mathrm{III}}(s,t)$ for the $u$-channel reaction is obtained by analytically continuing the $s$-channel physical amplitude:

$$\begin{aligned}[u - \text{channel}]: \quad \text{physical } A_{\mathrm{III}}(s,t) &= \lim_{\epsilon\to 0^+} A(s(u+\mathrm{i}\epsilon,t),t) \\ &= \lim_{\epsilon\to 0^+} A(4m^2 - u - t - \mathrm{i}\epsilon, t) \\ &= \lim_{\epsilon\to 0^+} A(s(u,t) - \mathrm{i}\epsilon, t) \end{aligned} \tag{4.90}$$

that is approaching the real $s$ axis from below.

The discontinuity of the amplitude across the $u$ branch cuts, denoted by $D_u(u,t)$ is given by

$$\begin{aligned} &D_u(u,t) \equiv \mathrm{Disc}_u A(s(u,t),t) \\ &= \frac{1}{2\mathrm{i}} \lim_{\epsilon\to 0^+} [A(s(u+\mathrm{i}\epsilon,t),t) - A(s(u-\mathrm{i}\epsilon,t),t)] \\ &= \frac{1}{2\mathrm{i}} \lim_{\epsilon\to 0^+} [A(4m^2 - u - t - \mathrm{i}\epsilon,t) - A(4m^2 - u - t + \mathrm{i}\epsilon,t)] \;. \end{aligned} \tag{4.91}$$

The crossing symmetry condition states that for identical scalar particles one has $A_{\mathrm{I}}(s,t) = A_{\mathrm{III}}(s,t) = A(s,t)$, hence the following constraint on $A(s,t)$ holds

$$A(s + \mathrm{i}\epsilon, t) = A(4m^2 - s - t - \mathrm{i}\epsilon, t)\ . \tag{4.92}$$

In the high-energy limit this becomes

$$A(s,t) = A(-s,t)\ . \tag{4.93}$$

The simple symmetry condition (4.93) is valid only for identical bosons. Everything becomes more complex when we go beyond the ideal scalar particles we are considering here, and discuss real particles, with various quantum numbers. In this case, there are a number of different amplitudes in each channel and they are related by appropriate crossing matrices obtained from Clebsch-Gordan coefficients. For example, if the reaction in the $s$-channel is

$$\mathrm{I}: \quad \pi^+ + p \to \pi^+ + p\ , \tag{4.94}$$

the amplitudes for the $s$-channel and the $u$-channel (we write here the $CPT$ transformed reaction)

$$\mathrm{II}: \quad \pi^- + p \to \pi^- + p\ , \tag{4.95}$$

are a mixture of $I = 1/2$ and $I = 3/2$ amplitudes ($I$ being the isospin), while the amplitude for the $t$-channel (again we write the $CPT$ transformed process)

$$\mathrm{III}: \quad p + \bar{p} \to \pi^- + \pi^+\ , \tag{4.96}$$

is a mixture of $I = 0$ and $I = 1$ amplitudes. Since the $s$-channel reaction differs from the $u$-channel reaction, the crossing symmetry condition (4.93) becomes now

$$A_{\pi^+p}(s,t) = A_{\pi^-p}(-s,t)\ . \tag{4.97}$$

One can still introduce a symmetric and an antisymmetric scattering amplitudes defined as

$$A^+(s,t) = A_{\pi^+p}(s,t) + A_{\pi^-p}(-s,t) \tag{4.98}$$

$$A^-(s,t) = A_{\pi^+p}(s,t) - A_{\pi^-p}(-s,t)\ . \tag{4.99}$$

whose crossing property (in the large $s$ limit) is

$$A^{\pm}(s,t) = \pm\, A^{\pm}(-s,t)\ . \tag{4.100}$$

The technical discussion of the situations involving spin and isospin is rather complicated and we will not tackle this issue here any further. For details the reader may consult, for instance, Eden (1967), Collins (1977), Pilkuhn (1979).

We shall now use the analyticity and crossing properties to derive dispersion relations for the scattering amplitude.

## 4.6 Dispersion Relations

Let us start from Cauchy's integral representation for $A(s,t)$ at fixed $t$

$$A(s,t) = \frac{1}{2\pi \mathrm{i}} \oint_{\Gamma} \frac{A(s',t)}{s'-s} \, \mathrm{d}s' \,, \qquad (4.101)$$

where $\Gamma$ is any counterclockwise closed contour in the $s$ plane, such that $A(s,t)$ is holomorphic inside it.

If we take $\Gamma$ to be the contour avoiding the cuts and encircling the two poles of $A(s,t)$ shown in Fig. 4.7, that is $\Gamma = \gamma_1 + \gamma_2 + \Gamma_1 + \Gamma_2 + C$ (see the figure), the integral (4.101) is the sum of five terms

$$A(s,t) = \frac{1}{2\pi \mathrm{i}} \left( \oint_{\gamma_1} + \oint_{\gamma_2} + \int_{\Gamma_1} + \int_{\Gamma_2} + \int_{C} \right) \frac{A(s',t)}{s'-s} \, \mathrm{d}s' \,. \qquad (4.102)$$

The first two terms in (4.102) are pole contributions

$$\frac{1}{2\pi \mathrm{i}} \left( \oint_{\gamma_1} + \oint_{\gamma_2} \right) \frac{A(s',t)}{s'-s} \, \mathrm{d}s' = \frac{f_1(t)}{s-m^2} + \frac{f_2(t)}{u-m^2} \,, \qquad (4.103)$$

where the residues $f_1$ and $f_2$ are unspecified functions of $t$. The next two terms in (4.102) can be written as

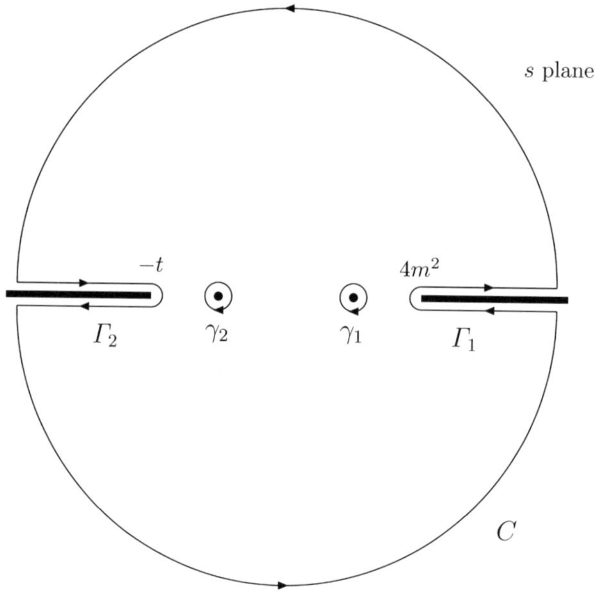


**Fig. 4.7.** The Cauchy contour.

$$\frac{1}{2\pi \mathrm{i}} \left( \int_{\Gamma_1} + \int_{\Gamma_2} \right) \frac{A(s',t)}{s'-s} \, \mathrm{d}s' = \frac{1}{\pi} \int_{4m^2}^{\infty} \frac{D_s(s',t)}{s'-s} \, \mathrm{d}s' + \frac{1}{\pi} \int_{-\infty}^{-t} \frac{D_s(s',t)}{s'-s} \, \mathrm{d}s' \,, \quad (4.104)$$

where $D_s(s',t)$ is the discontinuity defined in (4.88), or equivalently the imaginary part of $A(s,t)$ on the upper edge of the cut. Finally, if we assume that the amplitude vanishes uniformly when $|s|$ tends to infinity,

$$A(s,t) \underset{|s|\to\infty}{\to} 0 \,, \quad (4.105)$$

and let the radius $R$ of the circle $C$ go to $\infty$, the last integral in (4.102) tends to zero

$$\int_C \frac{A(s',t)}{s'-s} \, \mathrm{d}s' \underset{R\to\infty}{\to} 0 \,. \quad (4.106)$$

Putting all terms together we finally find the *single-variable dispersion relation* for the scattering amplitude at fixed $t$

$$A(s,t) = \frac{1}{\pi} \int_{4m^2}^{\infty} \frac{D_s(s',t)}{s'-s} \, \mathrm{d}s' + \frac{1}{\pi} \int_{-\infty}^{-t} \frac{D_s(s',t)}{s'-s} \, \mathrm{d}s' + \text{pole terms} \,. \quad (4.107)$$

or, making in the second integral a change of variable $s' \to u' = 4m^2 - s' - t$ to obtain a more symmetric expression

$$A(s,t) = \frac{1}{\pi} \int_{4m^2}^{\infty} \frac{D_s(s',t)}{s'-s} \, \mathrm{d}s' + \frac{1}{\pi} \int_{4m^2}^{\infty} \frac{D_u(u',t)}{u'-u} \, \mathrm{d}u' + \text{pole terms} \,. \quad (4.108)$$

'Single-variable' means that $t$ is held fixed. Analogous single-variable dispersion relations can be derived keeping one of the other Mandelstam variables ($s$ and $u$) fixed. One can also write *double dispersion relations* (Mandelstam 1958), regarding $A(s,t)$ as an analytical function of both variables.

We stress that the single dispersion relation (4.108) is valid under the convergence hypothesis (4.105). If this hypothesis is verified, dispersion relations are indeed a powerful tool, in that they allow to reconstruct the whole scattering amplitude once we know its discontinuity, that is its imaginary part. The latter can be obtained, via the unitarity equations, from quantities (cross sections) which are directly measurable (see Eq. (4.56)). Thus, in principle, analyticity and unitarity determine the amplitudes completely.

However, it turns out that the assumption (4.105) is usually not true. In this case, if the scattering amplitude behaves, as $s \to \infty$, as

$$A(s,t) \underset{s\to\infty}{\sim} s^{\lambda} \,, \quad (4.109)$$

where $\lambda$ is a real positive number, a dispersion relation can still be written for the quantity

$$\frac{A(s,t)}{(s-s_1)(s-s_2)\dots(s-s_N)}\,, \tag{4.110}$$

where $s_1, \dots s_N$ are arbitrary constants, and $N$ is the smallest integer greater than $\lambda$. Proceeding as we did above we get (the pole terms are omitted)

$$\begin{aligned} A(s,t) &= \sum_{n=0}^{N-1} c_n(t)\, s^n \\ &+ \frac{1}{\pi}(s-s_1)\dots(s-s_N)\int_{4m^2}^{\infty} \frac{D_s(s',t)}{(s'-s_1)\dots(s'-s_N)(s'-s)} \mathrm{d}s' \\ &+ \frac{1}{\pi}(u-u_1)\dots(u-u_N)\int_{4m^2}^{\infty} \frac{D_u(u',t)}{(u'-u_1)\dots(u'-u_N)(u'-u)} \mathrm{d}u'\,, \end{aligned} \tag{4.111}$$

where $u_i = 4m^2 - s_i - t$. Note that this dispersion relation (called '$N$-times subtracted dispersion relation) contains an arbitrary polynomial of degree $N-1$ which cannot be determined a priori.

In particular, if the total cross section is asymptotically constant,

$$\sigma_{\mathrm{tot}} \underset{s\to\infty}{\sim} \mathrm{const}\,, \tag{4.112}$$

as it is expected in Regge theory (see Chap. 5), then the imaginary part of the forward elastic amplitude, due to the optical theorem (4.56), increases as $s$

$$\mathrm{Im}\, A_{\mathrm{el}}(s,0) \underset{s\to\infty}{\sim} s\,. \tag{4.113}$$

Hence two subtractions are needed in the dispersion relation for $A_{\mathrm{el}}(s,0)$. This means that the forward elastic amplitude can be fully reconstructed in terms of the total cross section and two unknowns, $c_0(0)$ and $c_1(0)$ in Eq. (4.111).

## 4.7 The Froissart–Gribov Representation

If the scattering amplitude admits a subtracted or unsubtracted dispersion relation we can derive a useful integral representation for the partial-wave amplitudes, known as the *Froissart–Gribov formula* (Froissart 1961; Gribov 1961).

Let us work in the $t$ channel and assume that for fixed $t$ the asymptotic behavior of $A(s,t)$ is

$$A(s,t) \underset{s\to\infty}{\sim} s^{\lambda}\,. \tag{4.114}$$

We rewrite the scattering amplitude as a function of $t$ and of the scattering angle

$$z_t \equiv \cos\vartheta_t = 1 + \frac{2s}{t-4m^2} = -\left(1 + \frac{2u}{t-4m^2}\right), \tag{4.115}$$

so that (4.114) implies

$$A(z_t, t) \underset{z_t \to \infty}{\sim} {z_t}^{\lambda} \,. \tag{4.116}$$

Consider now the $N$-times subtracted dispersion relation for $A(z_t, t)$ at fixed $t$, where $N$ is the smallest integer greater than $\lambda$ (all subtractions are made at $z_t = 0$ for simplicity)

$$\begin{aligned} A(z_t, t) &= \sum_{n=0}^{N-1} c_n(t)\, {z_t}^n \\ &+ \frac{{z_t}^N}{\pi} \int_{z_0}^{\infty} \frac{D_s(s'(z'_t, t), t)}{{z'_t}^N (z'_t - z_t)}\, \mathrm{d}z'_t \\ &+ \frac{z_t^N}{\pi} \int_{-z_0}^{-\infty} \frac{D_u(u'(z'_t, t), t)}{{z'_t}^N (z'_t - z_t)}\, \mathrm{d}z'_t \,. \end{aligned} \tag{4.117}$$

Here $z_0$ is the branch point in the $z_t$ variable. If we substitute $A(z_t, t)$, given by (4.117), in (4.37) we get

$$\begin{aligned} A_\ell(t) &= \frac{1}{2} \sum_{n=0}^{N-1} c_n(t) \int_{-1}^{1} \mathrm{d}z_t\, P_\ell(z_t)\, z_t^n \\ &+ \frac{1}{2\pi} \int_{-1}^{1} \mathrm{d}z_t\, z_t^N\, P_\ell(z_t) \int_{z_0}^{\infty} \frac{D_s(s'(z'_t, t), t)}{{z'_t}^N (z'_t - z_t)}\, \mathrm{d}z'_t \\ &+ \frac{1}{2\pi} \int_{-1}^{1} \mathrm{d}z_t\, z_t^N\, P_\ell(z_t) \int_{-z_0}^{-\infty} \frac{D_u(u'(z'_t, t), t)}{{z'_t}^N (z'_t - z_t)}\, \mathrm{d}z'_t \,. \end{aligned} \tag{4.118}$$

For $\ell \geq N$ the first term is zero. Changing the order of integration in the other two terms and using

$$\frac{1}{2} \int_{-1}^{1} \mathrm{d}z_t\, P_\ell(z_t)\, \frac{z_t^N}{{z'_t}^N (z'_t - z_t)} = Q_\ell(z'_t) \,, \quad \ell \geq N \,, \tag{4.119}$$

where $Q_\ell(z'_t)$ is the Legendre function of the second kind, we finally obtain the Froissart-Gribov expression for the partial wave amplitude

$$\begin{aligned} A_\ell(t) &= \frac{1}{\pi} \int_{z_0}^{\infty} \mathrm{d}z_t\, D_s(s(z_t, t), t)\, Q_\ell(z_t) \\ &+ \frac{1}{\pi} \int_{-z_0}^{-\infty} \mathrm{d}z_t\, D_u(u(z_t, t), t)\, Q_\ell(z_t) \,. \end{aligned} \tag{4.120}$$

This representation for $A_\ell(t)$ is valid for $\ell \geq N$. A similar representation can be derived for the $s$-channel partial-wave amplitudes $A_\ell(s)$.

The Froissart-Gribov projection will be used in Sect. 4.9.1 to derive the Froissart bound, and in Chap. 5 in the framework of Regge theory.

## 4.8 Mueller's Generalized Optical Theorem

Combining unitarity, analyticity and crossing it is possible to extend the optical theorem to relate inclusive cross sections to discontinuities of three-body forward elastic amplitudes. This generalized optical theorem was formulated by Mueller (1970) and marked in the early Seventies a rise of interest towards inclusive and multiparticle soft processes.

Let us consider the inclusive differential cross section (4.41) that we rewrite here as

$$f(\boldsymbol{p}, s) = (2\pi)^3\, 2E\,\frac{\mathrm{d}^3\sigma}{\mathrm{d}^3\boldsymbol{p}} = \frac{1}{2s}\sum_X |\langle 3\,X|T|1\,2\rangle|^2\,, \tag{4.121}$$

Here we have used the flux factor in the large $s$ limit and embodied all the phase space sums and integrations into $\sum_X$.

We now express the squared amplitude in (4.121) as

$$\sum_X |\langle 3\,X|T|1\,2\rangle|^2 = \sum_X \langle 1\,2|T^\dagger|3\,X\rangle\,\langle 3\,X|T|1\,2\rangle\,. \tag{4.122}$$

The crossing postulate states that

$$\langle 3\,X|T|1\,2\rangle = \langle X|T|1\,2\,\bar{3}\rangle\,, \tag{4.123}$$

where $\bar{3}$ means, as usual, that momentum and quantum numbers of particle 3 have been reversed. The next step is to substitute (4.123) in (4.122) and use the unitarity relation (4.50), thus obtaining

$$\begin{aligned}\sum_X |\langle 3\,X|T|1\,2\rangle|^2 &= \sum_X \langle 1\,2\,\bar{3}|T^\dagger|X\rangle\langle X|T|1\,2\,\bar{3}\rangle\\ &= \langle 1\,2\,\bar{3}|T^\dagger T|1\,2\,\bar{3}\rangle\\ &= \mathrm{i}\left(\langle 1\,2\,\bar{3}|T^\dagger|1\,2\,\bar{3}\rangle - \langle 1\,2\,\bar{3}|T|1\,2\,\bar{3}\rangle\right)\,. \end{aligned}\tag{4.124}$$

This is the imaginary part of the elastic amplitude for the three-body reaction $1+2+\bar{3} \to 1+2+\bar{3}$. Introducing the discontinuity (remember that $M^2 = (p_1+p_2-p_3)^2$)

$$\mathrm{Disc}_{M^2}\, A_{12\bar{3}}(s,t,M^2) = \frac{1}{2\mathrm{i}}\left[A_{12\bar{3}}(s,t,M^2+\mathrm{i}\varepsilon) - A_{12\bar{3}}(s,t,M^2-\mathrm{i}\varepsilon)\right]\,, \tag{4.125}$$

we finally find Mueller's theorem

$$(2\pi)^3\, 2E\,\frac{\mathrm{d}^3\sigma}{\mathrm{d}^3\boldsymbol{p}} = \frac{1}{s}\,\mathrm{Disc}_{M^2}\, A_{12\bar{3}}(s,t,M^2)\,. \tag{4.126}$$

The derivation presented above is shown pictorially in Fig. 4.8.

Note that in (4.123) we performed an analytical continuation to the unphysical amplitude $A_{12\bar{3}}$. We simply assume that this is a legitimate procedure. There are also some other subtleties related to the definition of the discontinuity (4.125), but we do not consider them here (the reader may consult, for instance, Brower, DeTar and Weis 1974).

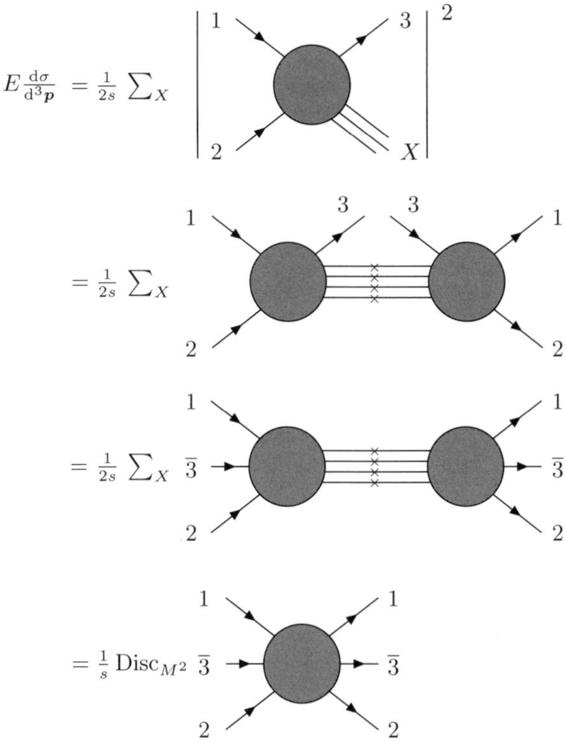


**Fig. 4.8.** Derivation of Mueller's theorem.

## 4.9 Rigorous Theorems

In this section we shall present a few of the rigorous theorems which have left a mark in the development of particle physics. We shall focus on the most prominent among them, namely the Froissart bound and the Pomeranchuk theorem on the total cross sections. Extensive discussions of other general results (including upper and lower bounds on various physical observables and a wealth of asymptotic relations) can be found in Eden (1967) and Roy (1972).

### 4.9.1 The Froissart-Martin Bound

The Froissart – or, perhaps more correctly, Froissart-Martin – theorem (Froissart 1961; Martin 1963) states that total cross sections cannot grow faster than $\ln^2 s$, that is

$$\sigma_{\mathrm{tot}} \leq C \ln^2 s\,, \quad \text{as } s \to \infty\,, \tag{4.127}$$

where $C$ is a constant.

This bound puts a strict limit to the rate of growth with energy of *any* total cross section.

There are various degrees of sophistication at which the result (4.127) can be proved[4], and follows from the positivity of the imaginary part of the scattering amplitude (therefore, in essence, from unitarity), and from the existence of an $N$-times subtracted dispersion relation at fixed $s$ (therefore, from analyticity).

Before presenting a rigorous proof, we discuss a very simple quantum mechanical model in which the bound (4.127) holds (Eden 1967). Assume that the target acts as an absorptive Yukawa-like potential, i.e., neglecting an unessential $1/r$ factor, as $\mathrm{e}^{-\chi r}$, where $r$ is the distance from its mean position. If we suppose now that for an incoming particle of energy $E$ the probability of interaction is bounded by some power $E^N$, where $N$ is fixed, the probability of an interaction at some distance $r$ from the center of the target will be

$$P(E,r) \lesssim E^N \,\mathrm{e}^{-\chi r} \,. \tag{4.128}$$

This quantity will be negligible for $r > r_0$ where

$$r_0 \sim \frac{N}{\chi} \ln E \,. \tag{4.129}$$

We thus have the following bound over $\sigma_{\text{tot}}$

$$\sigma_{\text{tot}} \lesssim (\text{const}) \cdot \ln^2 E \,, \tag{4.130}$$

which is nothing but (4.127), since $E$ is proportional to $s$.

**Proof.** The proof of the Froissart bound goes as follows. Let us start from the Froissart–Gribov representation of the $s$-channel partial-wave amplitudes

$$\begin{aligned} A_\ell(s) &= \frac{1}{\pi} \int_{z_0}^{\infty} \mathrm{d}z \, D_s(s, t(s,z))\, Q_\ell(z) \\ &+ \frac{1}{\pi} \int_{-z_0}^{-\infty} \mathrm{d}z \, D_u(s, u(s,z))\, Q_\ell(z) \,. \end{aligned} \tag{4.131}$$

For large $\ell$, the Legendre functions of the second kind behave as (Bateman 1953)

$$Q_\ell(z) \underset{\ell\to\infty}{\sim} \ell^{-\frac{1}{2}} \exp\left[ - \left( \ell + \frac{1}{2} \right) \zeta(z) \right] \,, \tag{4.132}$$

with

$$\zeta(z) = \ln \left[ z + (z^2 - 1)^{\frac{1}{2}} \right] \,. \tag{4.133}$$

Thus from (4.131) we get

[4] In its fullest generality it has been proved in axiomatic field theory (Martin 1963).

$$A_\ell(s) \underset{\ell, s \to \infty}{\sim} f(s) \exp[-\ell\, \zeta(z_0)] \,, \tag{4.134}$$

where $f(s)$ is some function of $s$ which behaves at most as a power of $s$ due to the assumptions underlying the Froissart-Gribov projection (in the $s$ channel) ·

$$f(s) \underset{s \to \infty}{\sim} s^\delta \,, \tag{4.135}$$

and $z_0$ is the lowest $t$-singularity. For $t = m^2$ (the pole singularity) we have

$$z_0 \underset{s \to \infty}{\simeq} 1 + \frac{2m^2}{s} \,, \tag{4.136}$$

and (4.134) becomes

$$A_\ell(s) \underset{\ell, s \to \infty}{\sim} f(s) \exp\left[ -\left( \frac{2m}{\sqrt{s}} \right) \ell \right] \sim \exp\left[ -\left( \frac{2m}{\sqrt{s}} \right) \ell + \delta \ln s \right] . \tag{4.137}$$

Therefore, at large $s$, the partial waves with

$$\ell \gtrsim c \sqrt{s} \ln s \,, \tag{4.138}$$

where $c$ is some constant, are negligible and the partial wave series can be truncated approximately as

$$A(s,t) \underset{s \to \infty}{\simeq} \sum_{\ell=0}^{c\sqrt{s}\,\ln s} (2\ell + 1)\, A_\ell(s)\, P_\ell(z) \,. \tag{4.139}$$

Now, making use of the unitarity bound (4.72) and recalling that $|P_\ell(z)| \le 1$ for $-1 \le z \le 1$, we find

$$|A(s,t)| \lesssim 16\pi \sum_{\ell=0}^{c\sqrt{s}\,\ln s} (2\ell + 1) \simeq (\text{const}) \cdot s \ln^2 s \,, \quad \text{as } s \to \infty \,. \tag{4.140}$$

Finally, the optical theorem leads to (4.127)

$$\sigma_{\text{tot}}(s) \le C \cdot \ln^2 s \,, \quad \text{as } s \to \infty \,. \tag{4.141}$$

Going a little more into the details, one finds that the constant $C$ is bounded as

$$C \ge \frac{\pi}{m_\pi^2} \,, \tag{4.142}$$

where $m_\pi$ is the pion mass. Note that the quantity on the r.h.s. of (4.142) is rather large, about 60 mb.

### 4.9.2 The Pomeranchuk Theorems

Back in 1956 it was shown (Pomeranchuk 1956; Okun and Pomeranchuk 1956) that the smallness and apparent vanishing of charge exchange reactions at high energies, together with isospin conservation, implies quite straightforwardly a variety of simple relationships between the cross sections for different processes. Two years later, Pomeranchuk (1958) proved that asymptotically (i.e. as $s \to \infty$) the total cross sections for particle-particle and particle-antiparticle collisions tend to become equal. Before offering the proof of the latter result (known as the Pomeranchuk theorem on the total cross sections), we shall derive the above mentioned Okun-Pomeranchuk relations.

**Okun-Pomeranchuk Relations.** Consider, for instance, the antinucleon-nucleon ($\overline{N}N$) reactions

$$1) \quad \overline{p} + p \to \overline{p} + p \tag{4.143}$$

$$2) \quad \overline{p} + n \to \overline{p} + n \tag{4.144}$$

$$3) \quad \overline{p} + p \to \overline{n} + n \tag{4.145}$$

Isospin symmetry implies that the three reactions

$$1) \quad \overline{n} + n \to \overline{n} + n \tag{4.146}$$

$$2) \quad \overline{n} + p \to \overline{n} + p \tag{4.147}$$

$$3) \quad \overline{n} + n \to \overline{p} + p \tag{4.148}$$

are equivalent to (4.143–4.145).

All the $\overline{N}N$ processes can be described by two amplitudes $A_0$ and $A_1$, corresponding to total isospin $I = 0$ and $I = 1$ respectively. The differential cross sections for the reactions (4.143–4.145) are then (we omit irrelevant kinematic factors)

$$\frac{\mathrm{d}\sigma_1}{\mathrm{d}t} \sim \frac{1}{4} |A_0 + A_1|^2 , \tag{4.149}$$

$$\frac{\mathrm{d}\sigma_2}{\mathrm{d}t} \sim |A_1|^2 , \tag{4.150}$$

$$\frac{\mathrm{d}\sigma_3}{\mathrm{d}t} \sim \frac{1}{4} |A_0 - A_1|^2 . \tag{4.151}$$

Notice now that $\overline{p}p \to \overline{p}p$ is an elastic reaction to which *all* annihilation channels contribute. Among these, $\overline{p}p \to \overline{n}n$ is only one of the many that contribute progressively as the energy increases. As a consequence, we expect that the charge exchange cross section $\sigma_3$ should asymptotically become small compared to the elastic cross sections, namely

$$|A_0 - A_1| \ll |A_0 + A_1| , \quad \text{as } s \to \infty , \tag{4.152}$$

and hence

$$A_0(s,t) \underset{s\to\infty}{\simeq} A_1(s,t) . \qquad (4.153)$$

We thus have

$$\frac{d\sigma_1}{dt} \underset{s\to\infty}{\simeq} \frac{d\sigma_2}{dt} . \qquad (4.154)$$

Taking (4.153) at $t = 0$ and using the optical theorem we finally come to the conclusion that all total cross sections for $\overline{N}N$ scattering are asymptotically equal

$$\sigma_{\text{tot}}(\overline{p}p) \underset{s\to\infty}{\simeq} \sigma_{\text{tot}}(\overline{n}p) \underset{s\to\infty}{\simeq} \sigma_{\text{tot}}(\overline{n}n) , \qquad (4.155)$$

where the last equality follows from isospin symmetry.

Eq. (4.155) is the Okun-Pomeranchuk relation between nucleon-antinucleon cross sections. As we have seen, it is based on the (plausible) assumption that charge-exchange reactions are negligible at high energies.

Had we considered nucleon-nucleon reactions, we would have obtained another Okun-Pomeranchuk relation

$$\sigma_{\text{tot}}(pp) \underset{s\to\infty}{\simeq} \sigma_{\text{tot}}(np) \underset{s\to\infty}{\simeq} \sigma_{\text{tot}}(nn) . \qquad (4.156)$$

The Okun-Pomeranchuk rules do not relate in general[5] particle-particle to particle-antiparticle cross sections, e.g. $\sigma_{\text{tot}}(pp)$ to $\sigma_{\text{tot}}(\overline{p}p)$. This relation is instead the content of the Pomeranchuk theorem for total cross sections, that we are now going to prove.

**Pomeranchuk Theorem for Total Cross-sections.** The Pomeranchuk theorem states that at high energy the total cross sections for particle-particle and particle-antiparticle scattering become equal, that is

$$\sigma_{\text{tot}}(ab) \underset{s\to\infty}{\simeq} \sigma_{\text{tot}}(a\bar{b}) . \qquad (4.157)$$

This theorem relies essentially on the possibility of writing dispersion relations for the scattering amplitudes, and on some (not so restrictive) assumptions on the energy dependence of the amplitudes.

**Proof.** Many proofs of the Pomeranchuk theorem have been offered, based on slightly different hypotheses. Here we shall follow the presentation of Eden (1967).

Consider the generic particle-particle and particle-antiparticle reactions

$$1) \quad a + b \to a + b , \qquad (4.158)$$

$$2) \quad a + \bar{b} \to a + \bar{b} . \qquad (4.159)$$

We call $A_1(s,t)$ and $A_2(s,t)$ the scattering amplitudes of reaction 1 and 2 respectively. At large $s$ crossing symmetry implies

[5] But they do constrain $\sigma_{\text{tot}}(\pi^+ p)$ to be asymptotically the same as $\sigma_{\text{tot}}(\pi^- p)$ by requiring the charge-exchange cross section $\sigma(\pi^- p \to \pi^0 n)$ to become negligible in the limit $s \to \infty$.

$$A_1(s,t) = A_2(-s,t) . \tag{4.160}$$

Hence we can introduce two amplitudes with definite parity

$$A^{\pm} \equiv A_1 \pm A_2 , \tag{4.161}$$

which satisfy

$$A^{\pm}(s,t) = \pm A^{\pm}(-s,t) . \tag{4.162}$$

Let us assume now that the total cross sections for the processes (4.158, 4.159) tend asymptotically to a constant, i.e.

$$\sigma_{\text{tot}}(ab) \underset{s\to\infty}{\to} \kappa_1 , \tag{4.163}$$

$$\sigma_{\text{tot}}(a\bar{b}) \underset{s\to\infty}{\to} \kappa_2 . \tag{4.164}$$

The optical theorem translates the conditions (4.163–4.164) into

$$\text{Im}\, A_1(s,0) \underset{s\to\infty}{\sim} \kappa_1 s , \tag{4.165}$$

$$\text{Im}\, A_2(s,0) \underset{s\to\infty}{\sim} \kappa_2 s , \tag{4.166}$$

or, equivalently

$$D_s^+(s,0) \equiv \text{Im}\, A^+(s,0) \underset{s\to\infty}{\sim} \kappa_+ s , \tag{4.167}$$

$$D_s^-(s,0) \equiv \text{Im}\, A^-(s,0) \underset{s\to\infty}{\sim} \kappa_- s , \tag{4.168}$$

where $\kappa_{\pm} = \kappa_1 \pm \kappa_2$.

We can then write a twice-subtracted dispersion relations for $A_1$ and $A_2$, or equivalently for $A^+$ and $A^-$. For the even amplitude $A^+(s,t)$ at $t=0$ we have

$$\begin{aligned} A^+(s,0) &= c_0 + \frac{s^2}{\pi}\int_{s_0}^{\infty} \frac{D_s^+(s',0)}{s'^2(s'-s)}\, \mathrm{d}s' + \frac{s^2}{\pi}\int_{-\infty}^{-s_0} \frac{D_s^+(s',0)}{s'^2(s'-s)}\, \mathrm{d}s' , \\ &= c_0 + \frac{2s^2}{\pi}\int_{s_0}^{\infty} \frac{D_s^+(s',0)}{s'(s'^2-s^2)}\, \mathrm{d}s' , \end{aligned} \tag{4.169}$$

where there is no term linear in $s$ due to the parity of $A^+$, and the second equality follows from changing the sign of the integration variable in the second integral and using (4.162). Note also that, since we are interested in the high-energy reaction, the actual value of the branch points is irrelevant and we chose them to be symmetric: $s_0$ is a small quantity (of the order of $m^2$) which simply acts as an infrared regulator of the integrals.

Similarly, we have for the odd amplitude $A^-(s,0)$

$$\begin{aligned} A^-(s,0) &= c_1 s + \frac{s^2}{\pi}\int_{s_0}^{\infty} \frac{D_s^-(s',0)}{s'^2(s'-s)}\, \mathrm{d}s' + \frac{s^2}{\pi}\int_{-\infty}^{-s_0} \frac{D_s^-(s',0)}{s'^2(s'-s)}\, \mathrm{d}s' , \\ &= c_1 s + \frac{2s^3}{\pi}\int_{s_0}^{\infty} \frac{D_s^-(s',0)}{s'^2(s'^2-s^2)}\, \mathrm{d}s' . \end{aligned} \tag{4.170}$$

Inserting the asymptotic behavior (4.167) into the dispersion relation (4.169) we find

$$A^{+}(s,0) \underset{s\to\infty}{\sim} \mathrm{i}\,\kappa_{+} s \,. \tag{4.171}$$

Thus the real part of the symmetric amplitude is asymptotically negligible.

On the other hand, inserting (4.168) into (4.170) we get

$$A^{-}(s,0) \underset{s\to\infty}{\sim} -\frac{2\kappa_{-}}{\pi}\, s\,\ln s + \mathrm{i}\,\kappa_{-} s \,, \tag{4.172}$$

and the real part of the antisymmetric amplitude turns out to be asymptotically dominant.

Now comes the assumption on the energy dependence of the amplitudes. Let us suppose that, as $s \to \infty$, the behavior of the real and of the imaginary part of the amplitudes for $ab$ and $a\bar{b}$ scattering is such that

$$\frac{\mathrm{Re}\, A_{1,2}(s,0)}{\ln s\, \mathrm{Im}\, A_{1,2}(s,0)} \underset{s\to\infty}{\to} 0 \,. \tag{4.173}$$

Combining Eqs. (4.171) and (4.172) we obtain

$$\mathrm{Re}\, A_{1}(s,0) \underset{s\to\infty}{\sim} -\frac{\kappa_{-}}{\pi}\, s\,\ln s \,. \tag{4.174}$$

The asymptotic behaviors (4.165) and (4.174) are compatible with the hypothesis (4.173) only if $\kappa_{-} \equiv \kappa_1 - \kappa_2 = 0$, that is, only if

$$\sigma_{\mathrm{tot}}(ab) - \sigma_{\mathrm{tot}}(a\bar{b}) \underset{s\to\infty}{\simeq} 0 \,. \tag{4.175}$$

This relation has been derived under the assumption of asymptotically constant cross sections, (4.163, 4.164). It has been also proven (Grunberg and Truong 1973) that, if the cross sections grow with energy, one has

$$\frac{\sigma_{\mathrm{tot}}(ab)}{\sigma_{\mathrm{tot}}(a\bar{b})} \underset{s\to\infty}{\to} 1 \,. \tag{4.176}$$

We shall see in Sect. 7 that the predictions of Pomeranchuk theorem are well verified experimentally: at the largest presently attainable energies, the total cross sections for $pp$, $\pi^{+}p$ and $K^{+}p$ collisions approach the total cross sections for $\bar{p}p$, $\pi^{-}p$ and $K^{-}p$ collisions, respectively.

A theorem similar to Pomeranchuk's was proven by Cornille and Martin (1972) for the forward slope $B$ of the elastic differential cross section, defined as

$$B(s,t=0) = \left[\frac{\mathrm{d}}{\mathrm{d}t}\left(\ln \frac{\mathrm{d}\sigma_{\mathrm{el}}}{\mathrm{d}t}\right)\right]_{t=0} \,. \tag{4.177}$$

According to the Cornille-Martin theorem one has asymptotically

$$\frac{B_{ab}(s,t=0)}{B_{a\bar{b}}(s,t=0)} \underset{s\to\infty}{\to} 1 \,. \tag{4.178}$$

This prediction has been checked for $pp$ and $\bar{p}p$ and turns out to be verified by the data (see Chap. 7).

# 5. Regge Theory

Donnachie and Landshoff (1992) conclude their analysis of total cross sections based on Regge-type fits by stating that "Regge theory remains one of the great truths of particle physics". Even taking a less optimistic attitude, one cannot dispute the phenomenological success of Regge theory in describing in a unified way a large class of reactions for which no alternative theoretical framework is – at least presently – available.

Regge theory, however, while standing on a firm ground as far as non-relativistic potential scattering is concerned, when extended to high-energy particle physics, relies on a series of assumptions. Its incorporation in a fundamental theory such as QCD would therefore be an important step towards a full understanding of strong interactions.

In this chapter we present those elements of Regge theory which are still of some utility to the modern particle physicist. For more details and information about the historical developments we refer the reader to some comprehensive treatises written in the golden years of Reggeology (Chew 1961, Squires 1963, Frautschi 1963, Eden 1967, Collins and Squires 1968, Collins 1977; see also Caneschi 1989 for a collection of commented reprints, and Levin 1997 for a modern revisitation of the subject).

## 5.1 The Regge Pole Idea

Before entering into the technical details of Regge theory, it is worth sketching the main ideas of this approach to hadronic phenomena.

Let us first consider non-relativistic quantum mechanics, which is the theoretical ground where the Regge pole idea was originally formulated.

The bound states for an attractive spherically symmetric potential fall into families with increasing angular momentum and energy. These bound states appear as poles of the partial wave amplitude[1] $a_\ell(k)$ (i.e., as poles of the $S$-matrix), for a given integer $\ell$.

Regge's idea (Regge 1959, 1960) was to continue $a_\ell(k)$ to complex values of $\ell$, thus obtaining an interpolating function $a(\ell, k)$, which reduces to $a_\ell(k)$

[1] We shall denote the partial wave amplitudes by $a_\ell(k)$ when working in quantum mechanics, and by $A_\ell(s)$, or $A_\ell(t)$, when working in relativistic $S$-matrix theory.

for $\ell = 0, 1, 2, \ldots$ For well behaved potentials (for instance, Yukawa-type potentials) the singularities of $a(\ell, k)$ turn out to be simple 'moving' poles (the *Regge poles*) located at values defined by a relation of the kind

$$\ell = \alpha(k) \,, \tag{5.1}$$

where $\alpha(k)$ is a function of the energy called *Regge trajectory*. Each family of bound states or resonances corresponds to a single Regge trajectory like (5.1). The energies of these states are obtained from Eq. (5.1), assigning physical (i.e., integer) values to the angular momentum $\ell$.

If we limit ourselves to quantum mechanics, the Regge pole idea appears to be little more than an interesting curiosity. It is in the context of $S$-matrix theory that it really proves useful. The extension of Regge's technique to high-energy particle physics is originally due to Chew and Frautschi (1961) and Gribov (1961), but many more authors contributed to the theory and its applications (too many to be mentioned here). By 1963 over 300 articles had been written on the subject (they are listed in Newton 1964). Other references can be found in Squires (1963), Frautschi (1963), Eden (1967), Collins (1977).

In the $S$-matrix framework we do not have a Schrödinger equation and therefore we are unable to construct explicitly the scattering amplitude and study directly its analyticity properties. Thus the existence of Regge poles is essentially a conjecture. However this conjecture helps solving in a very elegant way some theoretical difficulties and, most of all, is sustained by a rich phenomenological evidence.

Anticipating some results, we can say that, under some plausible hypotheses and using the general properties of the $S$-matrix, the relativistic partial wave amplitude $A_\ell(t)$ can be analytically continued to complex $\ell$ values in a unique way. The resulting function, $A(\ell, t)$, has simple poles at

$$\ell = \alpha(t) \,. \tag{5.2}$$

Each pole contributes to the scattering amplitude a term which behaves asymptotically (i.e. for $s \to \infty$ and $t$ fixed) as

$$A(s,t) \underset{s \to \infty}{\sim} s^{\alpha(t)} \,. \tag{5.3}$$

Thus the *leading* singularity (i.e. the singularity with the largest real part) in the $t$-channel determines the asymptotic behavior of the scattering amplitude in the $s$-channel.

Unfortunately, the story is not really so simple. Complicated singularities, such as cuts, may exist, and in this case Regge theory becomes definitely more intricate.

What is surprising is the wide success of Regge theory in its simplest form, namely the fact that a large class of processes is accurately described by such simple predictions as (5.3). That is why people tend to believe that Regge theory must contain at least a grain of truth.

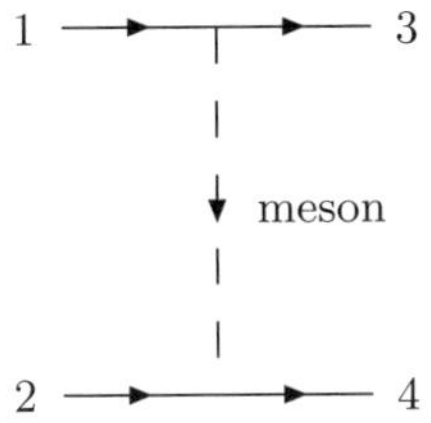


**Fig. 5.1.** Single meson exchange in the $t$-channel.

## 5.2 Meson Exchange vs. Reggeon Exchange

Regge theory belongs to the class of the so-called $t$-channel models. These models describe hadronic processes in terms of the $t$-channel exchange of 'something', in the spirit of Yukawa's original proposal (for the alternative (or complementary), $s$-channel, point of view see Chap. 6).

In the simplest version of $t$-channel models, this 'something' is a (virtual) *particle*. Thus, for instance, nuclear forces are usually attributed to the exchange of mesons ($\pi$, $\rho$, etc.), just like electromagnetic interactions arise from exchanging virtual photons between electrons.

This picture, however, becomes inapplicable at high energies. The reason is the following. Consider a generic reaction $1 + 2 \to 3 + 4$ mediated by a single-particle exchange in the $t$-channel (Fig. 5.1).

The scattering amplitude for the exchange of a meson of mass $M$ and spin $J$ goes as

$$A_{\mathrm{mes}}(s,t) \sim A_J(t)\, P_J(\cos\vartheta_t) \,, \tag{5.4}$$

where ($m$ is the mass of the interacting particles)

$$\cos\vartheta_t = 1 + \frac{2s}{t - 4m^2} \,. \tag{5.5}$$

According to the general discussion in Sect. 4.4, the partial wave amplitude $A_J(t)$ contains a pole singularity[2], thus

$$A_{\mathrm{mes}}(s,t) \sim \frac{P_J(\cos\vartheta_t)}{t - M^2} \,. \tag{5.6}$$

If we keep $t$ fixed and let $s \to \infty$, since $P_\ell(z) \underset{z\to\infty}{\sim} z^\ell$, we have

$$A_{\mathrm{mes}}(s,t) \underset{s\to\infty}{\sim} s^J \,. \tag{5.7}$$

The amplitude for the exchange of a single resonance is real and hence does not contribute to the cross section, which is proportional to the imaginary part of the scattering amplitude. To calculate $\mathrm{Im}\, A(s, t = 0)$ for the process

[2] Exchanging more than one particle instead leads to cuts.

$1 + 2 \to 3 + 4$ we can resort to the unitarity equation (4.60). We then find, using (5.7)

$$\text{Im}\, A(s, t = 0) \underset{s\to\infty}{\sim} s^{2J-1} , \tag{5.8}$$

and the optical theorem implies

$$\sigma_{\text{tot}} = \frac{1}{s} \text{Im}\, A(s, t = 0) \underset{s\to\infty}{\sim} s^{2J-2} . \tag{5.9}$$

It is clear that, if the exchanged mesons have spin greater than 1, the model leads to the violation of the Froissart–Martin bound (4.127), i.e. to the violation of unitarity.

As we will see, Regge theory overcomes this drawback preserving the idea of the $t$-channel exchange. According to Regge theory, the strong interaction is due not to the exchange of *particles* with a definite spin, but rather to the exchange of a *Regge trajectory*, i.e., of a whole family of resonances. The large $s$-limit of a hadronic process is determined by the exchange of one or more Regge trajectories in the $t$-channel. Adopting a particle physics language, Regge trajectories are often called *reggeons*.

Exchanging *reggeons* instead of *particles* leads to scattering amplitudes of the type of (5.3), which are in general less divergent. They do not violate the Froissart-Martin bound if $\alpha(0) < 1$.

## 5.3 Convergence of the Partial-Wave Expansion

We have already mentioned that the crucial tool in Regge theory is the continuation to complex angular momenta. This technique is fairly old. It was first applied to study the propagation of electromagnetic waves (for its early history, see Sommerfeld 1949), and introduced in quantum mechanics by Regge (1959, 1960) in his investigation of the analyticity properties of the scattering amplitude in potential scattering (see also Bottino, Longoni and Regge 1962). As we shall now see, it emerges rather naturally when establishing the convergence domain of the scattering amplitude to allow a correct analytic continuation to arbitrarily large energies.

Our starting point will be the partial-wave expansion of the scattering amplitude in the $s$-channel[3]

$$A(s, z) = \sum_{\ell=0}^{\infty} (2\ell + 1)\, A_\ell(s) P_\ell(z) , \tag{5.10}$$

$$A_\ell(s) = \frac{1}{2} \int_{-1}^{+1} \mathrm{d}z\, P_\ell(z) A(s, t(z, s)) , \tag{5.11}$$

[3] We limit our considerations to two-body elastic reactions. Also, to avoid unnecessary complications, we stick to the case of equal-mass spinless particles. For non-zero spin particle scattering see Collins (1977).

where

$$z \equiv \cos\vartheta = 1 + \frac{2t}{s - 4m^2} \,. \tag{5.12}$$

Clearly the representation (5.10) is well defined in the physical $s$-channel domain

$$s \geq 4m^2 \quad \text{and} \quad -1 \leq z \leq 1 \,. \tag{5.13}$$

The question arises whether it converges in a domain of the complex $s$, $t$ and $u$ variables larger than just (5.13), and, more specifically, in a domain large enough to contain at least part of the physical regions of the $t$ and $u$ channels. As we shall see, the answer to the second (crucial) question is negative. Thus the series (5.10) cannot be used as a representation of the crossing symmetric scattering amplitude.

The fact that (5.10) cannot be continued to the $t$-channel is easy to understand: the $s$ singularities of $A(s,t)$ defined by (5.10) are contained in the partial-wave amplitudes $A_\ell(s)$, but the $t$ dependence of $A(s,t)$ is embodied in the Legendre polynomials, which are entire functions. Thus, the singularities in $t$ manifest themselves in that the series (5.10) diverges, becoming senseless.

The problem of finding a representation for $A(s,t)$ connecting the various channels and hence suitable to describe all physical reactions related by crossing can be solved by introducing a seemingly unphysical concept, that of *complex angular momenta*. Before seeing how this idea works let us make a step back and determine the convergence domain of (5.10) in the complex $\vartheta$ plane. Using the asymptotic behavior of $P_\ell(\cos\vartheta)$ for $\ell$ real and tending to $\infty$,

$$P_\ell(\cos\vartheta) = \mathcal{O}(\mathrm{e}^{\ell\,|\mathrm{Im}\,\vartheta|}) \,, \tag{5.14}$$

we see that the series (5.10) converges only if, as $\ell \to \infty$

$$A_\ell(s) \sim \mathrm{e}^{-\ell\eta(s)} \,, \tag{5.15}$$

and

$$|\mathrm{Im}\,\vartheta| \leq \eta(s) \,. \tag{5.16}$$

Thus the convergence in the complex $\vartheta$ plane is insured in a horizontal strip symmetric with respect to the imaginary $\vartheta$ axis of width $\eta(s)$.

Setting $\chi = \cosh\eta(s)$ (which is always $> 1$), the corresponding convergence domain of the partial wave expansion (5.10) in the complex $\cos\vartheta$ plane ($z \equiv \cos\vartheta \equiv x + \mathrm{i}y$) plane is

$$\frac{x^2}{\chi^2} + \frac{y^2}{\chi^2 - 1} = 1 \,. \tag{5.17}$$

This domain is an ellipse with foci $z = \pm 1$ and semiaxes $\chi$ and $\sqrt{\chi^2 - 1}$, known as the *Lehmann ellipse* (see Fig. 5.2).

The above result implies that the usual partial wave expansion converges in a domain which, although larger than the simple physical domain

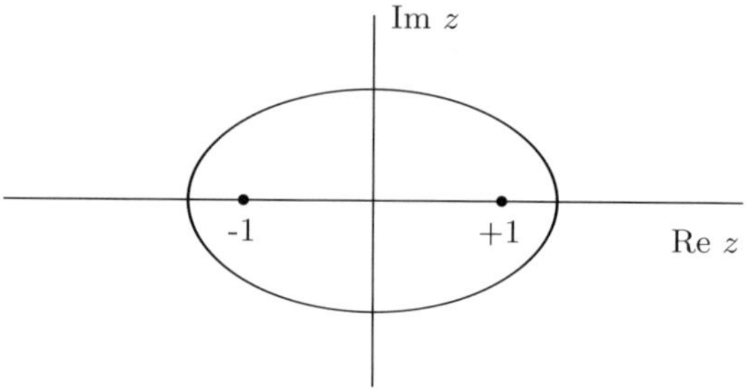


**Fig. 5.2.** Lehmann ellipse (Lehmann 1958).

$-1 \leq z \leq 1$ (to which it reduces if $\eta(s) \to 0$), never extends to arbitrarily large values of the complex variable $z$. Translated into the language of the Mandelstam variables (3.9–3.11), this means that, for any finite $s$, the expansion (5.10) converges in a finite domain of the $|t|$ and $|u|$ variables. Thus, we cannot continue the amplitude, as defined by (5.10) to regions where any of the Mandelstam variables $t$ or $u$ can become arbitrarily large[4]. Thus, if we want to construct a representation of $A(s,t)$ valid in all channels, the series (5.10) does not help us.

As we said above, the way to circumvent this difficulty is to continue the expansion (5.10) to *complex values* of $\ell$. We shall carry out this program in the next section.

It is instructive to see how the situation would change in a simple extreme case, that is if the expansion (5.10) was not over purely real but, rather, over purely imaginary values of $\ell$. In this case, in fact, provided

$$A_{|\ell|}(s) \underset{\ell \to \mathrm{i}\infty}{\sim} \mathrm{e}^{-|\ell|\,\delta(s)} , \tag{5.18}$$

convergence would be insured in a vertical strip parallel symmetric with respect to the real axis in the complex $\vartheta$ plane, i.e. in

$$|\mathrm{Re}\,\vartheta| \leq \delta(s) . \tag{5.19}$$

Setting, accordingly, $\xi = \cos\delta$ (which is always $\leq 1$), the convergence domain in the complex $z \equiv x + \mathrm{i}y$ plane would be

$$\frac{x^2}{\xi^2} - \frac{y^2}{1-\xi^2} = 1 . \tag{5.20}$$

Contrary to the previous case, (5.20) is an *open* domain (a hyperbola with foci $z = \pm 1$) and convergence is insured outside one of its halves. This hyperbola, in addition, overlaps in part with the Lehmann ellipse which guarantees that, if we can continue (5.10) to imaginary $\ell$ values, the new expansion represents the same analytic function (i.e. the same scattering amplitude) in a domain where $|t|$ (and/or $|u|$) can become arbitrarily large.

---

[4] Owing to the constraint $s + t + u = 4m^2$, if any of the Mandelstam variables becomes unboundedly large, at least another variable must also do so.

## 5.4 Complex Angular Momenta

We now show explicitly how the introduction of complex angular momenta allows getting a representation for $A(s,t)$ valid in all channels. This program requires that the partial-wave amplitudes possess a number of properties.

Let us assume, first of all, that we can continue the partial-wave amplitude $A_\ell(s)$ to complex values of $\ell$ and construct an interpolating function $A(\ell, s)$ which reduces to $A_\ell(s)$ for real integer $\ell$. Suppose also that:

I: $A(\ell, s)$ has only isolated singularities in the complex $\ell$ plane.

II: $A(\ell, s)$ is holomorphic for $\mathrm{Re}\,\ell \geq L$.

III: $A(\ell, s) \to 0$ as $|\ell| \to \infty$, for $\mathrm{Re}\,\ell > 0$.

We shall see that an amplitude $A(\ell, s)$ with these properties indeed exists in two contexts:

*i)* in non-relativistic quantum mechanics, for a certain class of potentials;
*ii)* in the relativistic case, if the scattering amplitude meets some extra requirements (essentially, it should obey appropriate dispersion relations).

If such an $A(\ell, s)$ exists, then it is possible to show that it is unique. A theorem due to Carlson (see Titchmarsh 1939, p. 185), in fact, guarantees that a function $A(\ell, s)$ satisfying conditions II and III is uniquely determined by the values it takes for integer $\ell$. (In general, Carlson's theorem applies to functions $A(\ell, s)$ bounded by an exponential $\mathrm{e}^{\pi|\ell|}$ for $|\ell| \to \infty$.)

Using the properties I and II of $A(\ell, s)$ we can now rewrite the partial wave expansion (5.10) as

$$A(s,z) = \sum_{\ell=0}^{N-1} (2\ell+1)\, A_\ell(s) P_\ell(z) - \frac{1}{2\mathrm{i}} \int_C (2\ell+1)\, A(\ell, s)\, \frac{P_\ell(-z)}{\sin \pi\ell}\, \mathrm{d}\ell \,, \tag{5.21}$$

where $N$ is the first integer greater than $L$ and $C$ is the contour shown in Fig. 5.3 (it avoids all singularities of $A(\ell, s)$). The equivalence of Eqs. (5.10) and (5.21) is ensured by the theorem of residues: the integrand $f(\ell)$ of (5.21) has simple poles at $\ell = n$ (where $n$ is any integer $\geq N$), with residues

$$\mathrm{Res}\, f(\ell)|_{\ell=n} = 2\mathrm{i}\,(2n+1)\, A_n(s)\, P_n(z)\,. \tag{5.22}$$

In (5.22) we used $P_n(-z) = (-1)^n\, P_n(z)$.

We now deform the contour $C$ into $C' = (a - \mathrm{i}\infty, a + \mathrm{i}\infty)$, a line parallel to the imaginary $\ell$ axis, to the right of all singularities of $A(\ell, s)$ (see Fig. 5.3). The property III of $A(\ell, s)$ and the asymptotic behavior for $\ell \to \infty$ of the Legendre polynomials (Bateman 1953, vol. 1, p. 162)

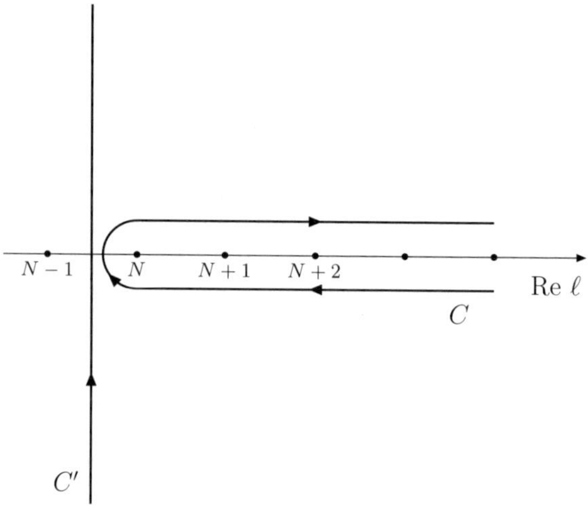


**Fig. 5.3.** The integration contour for the Watson-Sommerfeld representation of the scattering amplitude (see Eqs. (5.21) and (5.24)).

$$\left|\frac{P_\ell(-z)}{\sin \pi\ell}\right| < \ell^{-\frac{1}{2}} \times \exp\left\{|\mathrm{Im}\,\vartheta\,\mathrm{Re}\,\ell + (\pi - \mathrm{Re}\,\vartheta)\,\mathrm{Im}\,\ell| - \pi\,|\mathrm{Im}\,\ell|\right\}\, f(z)\,, \quad (5.23)$$

guarantee that the integral on the semicircle at infinity closing $C'$ vanishes. Thus we have

$$A(s,z) = \sum_{\ell=0}^{N-1} (2\ell+1)\, A_\ell(s) P_\ell(z) - \frac{1}{2\mathrm{i}} \int_{a-\mathrm{i}\infty}^{a+\mathrm{i}\infty} (2\ell+1)\, A(\ell,s)\, \frac{P_\ell(-z)}{\sin \pi\ell}\, \mathrm{d}\ell\,, \quad (5.24)$$

with $\mathrm{Re}\, a \geq L$.

We next displace $C'$ to the left. If the singularities of $A(\ell, s)$ are *poles*, then, as the contour $C'$ is moved towards smaller $\mathrm{Re}\,\ell$ values, we get contributions from the residues of the poles of $A(\ell, s)$ that are crossed over (the circles in Fig. 5.4), together with residues from the poles of $(\sin \pi\ell)^{-1}$, which cancel some of the terms of the truncated partial-wave series in (5.24). We can push $C'$ down to $-\frac{1}{2} \leq \mathrm{Re}\,\ell < 0$, thus getting (with $-\frac{1}{2} \leq \mathrm{Re}\, c < 0$, see Fig. 5.4)

$$A(s,z) = -\sum_i \pi\,(2\alpha_i(s)+1)\,\beta_i(s)\, \frac{P_{\alpha_i}(-z)}{\sin \pi\alpha_i} - \frac{1}{2\mathrm{i}} \int_{c-\mathrm{i}\infty}^{c+\mathrm{i}\infty} (2\ell+1)\, A(\ell,s)\, \frac{P_\ell(-z)}{\sin \pi\ell}\, \mathrm{d}\ell\,, \quad (5.25)$$

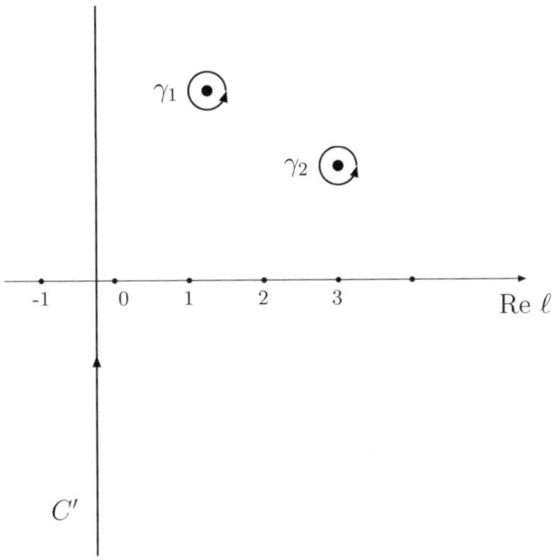


**Fig. 5.4.** The deformed integration contour for the Watson-Sommerfeld representation of the scattering amplitude (see Eq. (5.25)).

where $\alpha_i(s)$ is the location in the complex $\ell$ plane of the $i$-th pole of $A(\ell, s)$ (a *Regge pole*) and $\beta_i(s)$ is the residue at that pole.

Equation (5.25) is called *Watson-Sommerfeld representation* (Watson 1918, Sommerfeld 1949), and is the crucial mathematical tool of the complex angular momentum formalism. It is originally due to Poincaré (for its classical applications see Sommerfeld 1949) and was used by Regge (1959, 1960) in the context of potential scattering.

Because of (5.23), the representation (5.25) is well defined in a domain larger than the Lehmann ellipse, and actually is valid in a region of the $z$ plane outside one of the halves of the hyperbola (5.20).

Consider now the large-$t$ behavior of the amplitude (5.25) for fixed $s$ (i.e., the large-$|z|$ behavior). Since for $\mathrm{Re}\,\ell \geq -1/2$ one has (Bateman 1953, vol. 1, p. 164)

$$P_\ell(z) \underset{|z|\to\infty}{\sim} z^\ell \,, \tag{5.26}$$

the integral in (5.25) behaves as $|z|^{-\frac{1}{2}}$ as $|z| \to \infty$ and therefore gives an asymptotically negligible contribution. Thus, in the large $|z|$-limit, only the pole series survives

$$A(s, z) \underset{|z|\to\infty}{\simeq} -\sum_i \beta_i(s) \frac{(-z)^{\alpha_i(s)}}{\sin \pi\alpha_i(s)} \,, \tag{5.27}$$

where we have lumped into $\beta_i(s)$ all other constant and $s$-dependent factors appearing in (5.25).

The dominant term of this series is the right-most pole, i.e. the one for which $\mathrm{Re}\,\alpha_i$ is the largest. If we call this Regge trajectory simply $\alpha(s)$, the asymptotic (i.e., large-$t$) behavior of the scattering amplitude for fixed $s$ is

$$A(s,t) \underset{t\to\infty}{\sim} -\beta(s)\,\frac{t^{\alpha(s)}}{\sin\pi\alpha(s)}\;. \tag{5.28}$$

Nothing can be said about the residue function $\beta(s)$.

Equation (5.28) has been obtained by working out the partial-wave series (5.10) written for the $s$-channel. Starting instead from the $t$-channel partial-wave expansion leads to an expression where the roles of $s$ and $t$ are exchanged, that is

$$A(s,t) \underset{s\to\infty}{\sim} -\beta(t)\,\frac{s^{\alpha(t)}}{\sin\pi\alpha(t)}\;. \tag{5.29}$$

Invoking the validity of crossing, we thus find the (amazingly simple) Regge theory prediction for the large-$s$ behavior of the scattering amplitude. As anticipated, it is the leading singularity in the crossed $t$-channel that governs the asymptotic behavior of $A(s,t)$ as $s\to\infty$. We shall actually see in Sect. 5.6 that crossing slightly modifies (5.29), but leaves its $s$-dependence intact.

Equation (5.29) is valid up to subasymptotic corrections, represented by other poles with a smaller $\mathrm{Re}\,\alpha$ and by other types of singularities (cuts) which alter the simple behavior (5.29) (for instance, they can produce a $\log s$ behavior). Let us see how the Watson-Sommerfeld representation is modified by the presence of cuts. If $A(\ell,s)$ has a branch point at $\alpha_c(s)$, the integration contour of Eq. (5.21) is the one depicted in Fig. 5.5 and Eq. (5.25) acquires the additional term

$$-\frac{1}{2\mathrm{i}}\int_{\Gamma}(2\ell+1)\,\mathrm{Disc}_{\Gamma}A(\ell,s)\,\frac{P_\ell(-z)}{\sin\pi\ell}\,\mathrm{d}\ell\;, \tag{5.30}$$

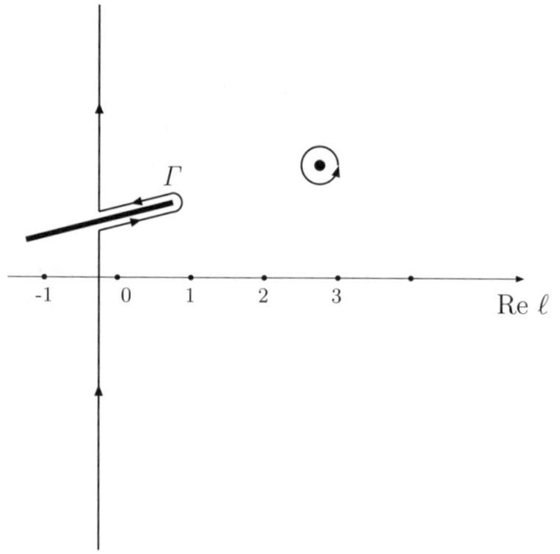


**Fig. 5.5.** The integration contour for the Watson-Sommerfeld representation of the scattering amplitude in presence of a cut (see Eq. (5.30)).

where $\Gamma$ is the contour along the cut ending at $\alpha_c(s)$ and $\text{Disc}_\Gamma A(\ell, s)$ is the discontinuity of the partial-wave amplitude across the cut.

We now proceed to show how a partial-wave amplitude $A(\ell, s)$ meeting the requirements of Sect. 5.4 can be constructed in quantum mechanics and in relativistic $S$-matrix theory.

## 5.5 Regge Poles in Quantum Mechanics

In quantum mechanics it is possible to check directly whether a partial wave amplitude $a_\ell(k)$ satisfying the requisites of Sect. 5.4 exists. What one has to do is to solve the Schrödinger equation.

For a potential $U(r)$ which is less singular than $1/r^2$ at the origin

$$r^2 U(r) \underset{r \to 0}{\to} 0 \,, \tag{5.31}$$

and goes to zero at infinity more rapidly than $1/r$

$$r\, U(r) \underset{r \to \infty}{\to} 0 \,, \tag{5.32}$$

the partial wave amplitude, analytically continued to complex $\ell$ values, can be written as

$$a(\ell, k) = \frac{\mathrm{e}^{\mathrm{i}\pi\ell}\, g(\ell, k) - g(\ell, -k)}{2\mathrm{i}k\, g(\ell, -k)} \,, \tag{5.33}$$

where $g(\ell, k)$ is the so-called *Jost function*[5] (see, e.g., de Alfaro and Regge 1965).

An important class of potentials having the properties (5.31–5.32) is given by the generalized Yukawa potentials

$$U(r) = \int_m^\infty c(\mu)\, \mathrm{e}^{-\mu r}\, \mathrm{d}\mu \,, \tag{5.34}$$

For $c(\mu) = C = \text{const}$, the integral (5.35) reproduces the original Yukawa potential

$$U(r) = C\, \frac{\mathrm{e}^{-mr}}{r} \,. \tag{5.35}$$

The corresponding Born scattering amplitude

$$f(q) \sim \frac{1}{t - m^2} \,, \tag{5.36}$$

with $t = -\boldsymbol{q}^2 = -(\boldsymbol{k}' - \boldsymbol{k})^2$, resembles a scalar meson exchange amplitude. Thus Yukawa-type potentials are reminiscent of $S$-matrix (or field theory) interactions.

[5] In the literature the Jost function is usually denoted by $f(\ell, k)$. Here we call it $g(\ell, k)$ to avoid any confusion with the quantum mechanical scattering amplitude that we called $f(k, \vartheta)$.

The analyticity properties of the Jost function for generalized Yukawa potentials have been studied in great detail by Bottino, Longoni and Regge (1962). It has been shown (for a general account, see de Alfaro and Regge 1965) that

*i)* For $\mathrm{Re}\,\ell > -1/2$, the only singularities of $a(\ell, k)$ as a function of $\ell$ are a *finite* number of *simple poles* (the Regge poles) which lie in the upper quadrant of the complex $\ell$ plane.
*ii)* The amplitude $a(\ell, k)$ tends to zero exponentially for $|\ell| \to \infty$, when $\mathrm{Re}\,\ell > -1/2$.

Thus the properties I, II and III of Sect. 5.4 are fulfilled and one can apply Carlson's theorem to prove that the analytical continuation of the partial wave amplitude to the complex $\ell$ plane is unique. Furthermore, the Watson-Sommerfeld representation holds and for $|\cos\vartheta| \to \infty$ and fixed energy $k$ the scattering amplitude behaves as

$$f(k, \vartheta) \underset{|\cos\vartheta|\to\infty}{\sim} -\beta(k) \frac{(-\cos\vartheta)^{\alpha(k)}}{\sin \pi\alpha(k)}, \tag{5.37}$$

where $\alpha(k)$ is the trajectory of the dominant Regge pole, the one with the largest real part.

It is clear that in the frame of potential scattering $|\cos\vartheta|$ cannot exceed unity so that the exercise of letting it become unrestrictedly large is rather academic. But in relativistic scattering the limit $|\cos\vartheta| \to \infty$ can be viewed as the limit $|t| \to \infty$ which is physically meaningful.

Before proceeding further, let us see what happens when there are exchange forces. In this case, a factor $(-1)^\ell$ appears in the partial wave amplitude. Thus the condition III of Sect. 5.4 is not satisfied, since $(-1)^\ell$ diverges when $|\ell| \to \infty$ along the imaginary axis. As a consequence, Carlson's theorem does not hold and no unique continuation of $A_\ell(s)$ to the complex $\ell$-plane is guaranteed. As we shall discuss in detail in the next section, this problem can be solved in a clever way, by introducing a new quantum number, the *signature*.

Whereas in quantum mechanics the factor $(-1)^\ell$ appears in special cases (i.e., in presence of exchange forces) and the need for signature is only occasional, in $S$-matrix theory, because of crossing, there is always an "exchange" potential term, and the signature, as we shall see, is always necessary.

## 5.6 Regge Poles in Relativistic Scattering

We have seen in Sect. 4.7 that under suitable conditions (essentially, the existence of an $N$-times subtracted dispersion relation) it is possible to write an integral representation – the Froissart-Gribov projection (Froissart 1961,

Gribov 1961) – for the partial wave amplitude $A_\ell(t)$, valid for $\ell \geq N$. For convenience, we rewrite here this representation

$$A_\ell(t) = \frac{1}{\pi}\int_{z_0}^{\infty} D_s(s(z_t,t),t)\, Q_\ell(z_t)\, \mathrm{d}z_t + \frac{1}{\pi}\int_{-z_0}^{-\infty} D_u(u(z_t,t),t)\, Q_\ell(z_t)\, \mathrm{d}z_t \ , \tag{5.38}$$

where

$$z_t \equiv \cos\vartheta_t = 1 + \frac{2s}{t-4m^2} \ . \tag{5.39}$$

Making the change of variable $z_t \to -z_t$ in the second integral in (5.38), and using $u(-z_t,t) = s(z_t,t)$ and

$$Q_\ell(-z_t) = -\mathrm{e}^{-\mathrm{i}\pi\ell}\, Q_\ell(z_t), \tag{5.40}$$

we get

$$A_\ell(t) = \frac{1}{\pi}\int_{z_0}^{\infty} [D_s(s(z_t,t),t) + \mathrm{e}^{-\mathrm{i}\pi\ell}\, D_u(s(z_t,t),t)]\, Q_\ell(z_t)\, \mathrm{d}z_t \ . \tag{5.41}$$

exhibiting explicitly the $(-1)^\ell$ factor whose appearance was anticipated in Sect. 5.5.

Let us see what happens when $\ell$ tends to infinity. The asymptotic behavior of $Q_\ell(z)$ is given by Eq. (4.132). Thus, the first term in (5.41) vanishes exponentially for $\ell \to \infty$, but the second term, due to the presence of the $(-1)^\ell$ factor has a bad behavior when $\mathrm{Im}\,\ell$ becomes unrestrictedly large. Hence, according to the general discussion in Sect. 5.4, the amplitude (5.41) cannot be analytically continued to complex $\ell$ values.

We can circumvent this difficulty by defining two different amplitudes $A_\ell^\pm(t)$

$$A_\ell^\pm(t) = \frac{1}{\pi}\int_{z_0}^{\infty} [D_s(s,t) \pm D_u(s,t)]\, Q_\ell(z_t)\, \mathrm{d}z_t \ , \tag{5.42}$$

which coincide with (5.41) for even and odd integer values of $\ell$, respectively

$$A_\ell^+(t) = A_\ell(t) \quad \text{for } \ell \text{ even} \ , \tag{5.43}$$

$$A_\ell^-(t) = A_\ell(t) \quad \text{for } \ell \text{ odd} \ , \tag{5.44}$$

In (5.42) we introduced a new quantum number, the *signature* $\xi$, which takes two values: $\xi = \pm 1$. We say that the amplitude (5.43) has positive signature, $\xi = +1$, whereas the amplitude (5.44) has negative signature, $\xi = -1$. We shall rewrite (5.42) as

$$A_\ell^\xi(t) = \frac{1}{\pi}\int_{z_0}^{\infty} D_s^\xi(s,t)\, Q_\ell(z_t)\, \mathrm{d}z_t \ , \tag{5.45}$$

where the signatured discontinuity $D_s^\xi(s,t)$ is defined as

$$D_s^\xi(s,t) \equiv D_s(s,t) + \xi\, D_u(s,t)\,. \tag{5.46}$$

$A_\ell(t)$ is obtained from $A_\ell^\xi(t)$ by

$$A_\ell(t) = \frac{1}{2}\sum_{\xi=\pm1}\left(1+\xi\,\mathrm{e}^{-\mathrm{i}\pi\ell}\right) A_\ell^\xi(t)\,. \tag{5.47}$$

In terms of $A_\ell^\xi(t)$, the definite-signature scattering amplitude is given by the partial-wave expansion

$$A^\xi(z_t,t) = \sum_{\ell=0}^{\infty}(2\ell+1)\,A_\ell^\xi(t)\,P_\ell(z_t)\,, \tag{5.48}$$

from which, using $P_\ell(-z_t) = (-1)^\ell P_\ell(z_t)$, the full scattering amplitude is obtained as

$$A(z_t,t) = \frac{1}{2}\left[A^+(z_t,t) + A^+(-z_t,t) + A^-(z_t,t) - A^-(-z_t,t)\right]\,. \tag{5.49}$$

Equation (5.49) shows that, since at high energies the crossing operation corresponds to $z_t \to -z_t$, only the even(odd)-signature amplitude contributes to the crossing-even(odd) part of the physical amplitude.

The definite-signature partial-wave amplitudes (5.42) have a good behavior for $\ell \to \infty$ and hence can be analytically continued in a unique way to the complex $\ell$ plane, becoming $A^\xi(\ell,t)$. We can then construct the definite-signature scattering amplitude $A^\xi(z_t,t)$ by the Watson-Sommerfeld transform

$$A^\xi(z_t,t) = -\frac{1}{2\mathrm{i}}\int_C (2\ell+1)\,A^\xi(\ell,t)\,\frac{P_\ell(-z_t)}{\sin\pi\ell}\,\mathrm{d}\ell\,. \tag{5.50}$$

The manipulations presented in Sect. 5.4, which are legitimate due to the properties of $A^\xi(\ell,t)$, lead to

$$\begin{aligned} A(z_t,t)^\xi &= -\sum_{i_\xi}\pi\,(2\alpha_{i_\xi}(s)+1)\,\beta_{i_\xi}(s)\,\frac{P_{\alpha_{i_\xi}}(-z)}{\sin\pi\alpha_{i_\xi}} \\ &\quad -\frac{1}{2\mathrm{i}}\int_{c-\mathrm{i}\infty}^{c+\mathrm{i}\infty}(2\ell+1)\,A(\ell,s)\,\frac{P_\ell(-z)}{\sin\pi\ell}\,\mathrm{d}\ell\,, \end{aligned} \tag{5.51}$$

where the sum $\sum_{i_\xi}$ is over definite-signature Regge poles. The full amplitude is

$$\begin{aligned} A(z_t,t) &= -\sum_{\xi=\pm1}\sum_{i_\xi}\frac{1+\xi\,\mathrm{e}^{-\mathrm{i}\pi\alpha_{i_\xi}(t)}}{2}\,\pi\,\left(2\alpha_{i_\xi}(t)+1\right)\,\beta_{i_\xi}(t)\,\frac{P_{\alpha_{i_\xi}}(-z_t)}{\sin\pi\alpha_{i_\xi(t)}} \\ &\quad -\frac{1}{2\mathrm{i}}\sum_{\xi=\pm1}\int_{c-\mathrm{i}\infty}^{c+\mathrm{i}\infty}\frac{1+\xi\,\mathrm{e}^{-\mathrm{i}\pi\ell}}{2}\,(2\ell+1)\,A(\ell,t)\,\frac{P_\ell(-z_t)}{\sin\pi\ell}\,\mathrm{d}\ell\,, \end{aligned} \tag{5.52}$$

In the large-$|z_t|$ the pole series gives the dominant contribution and, with the asymptotic behavior (5.26) of the Legendre polynomials, we have

$$A(z_t, t) \underset{|z_t| \to \infty}{\simeq} - \sum_{\xi = \pm 1} \sum_{i_\xi} \beta_{i_\xi}(t) \, \frac{1 + \xi \, \mathrm{e}^{-i\pi\alpha_{i_\xi}(t)}}{\sin \pi \alpha_{i_\xi}(t)} \, (-z_t)^{\alpha_{i_\xi}(t)} \,, \tag{5.53}$$

where some irrelevant factors have been incorporated into $\beta_{i_\xi}(t)$. Note that the only difference between (5.53) and (5.29) is the factor $(1 + \xi \, \mathrm{e}^{-i\pi\alpha_{i_\xi}})$ appearing in front of each pole contribution to the scattering amplitude. Finally, if we keep only the leading pole, which is the one with the largest $\mathrm{Re}\, \alpha_{i_\xi}$ (we call $\alpha(t)$ its trajectory and $\beta(t)$ its residue), we get

$$A(s, t) \underset{s \to \infty}{\sim} -\beta(t) \, \frac{1 + \xi \, \mathrm{e}^{-i\pi\alpha(t)}}{\sin \pi \alpha(t)} \, s^{\alpha(t)} \,. \tag{5.54}$$

Proceeding along similar lines – but starting from the $s$-channel partial wave expansion for the signatured amplitude –, one can derive the large-$t$ asymptotic behavior of $A(s, t)$

$$A(s, t) \underset{t \to \infty}{\sim} -\beta(s) \, \frac{1 + \xi \, \mathrm{e}^{-i\pi\alpha(s)}}{\sin \pi \alpha(s)} \, t^{\alpha(s)} \,, \tag{5.55}$$

which is the relativistic analog of (5.37).

Equations (5.54) and (5.55) represent a fundamental result which can be interpreted as follows: the leading complex angular momentum singularity (i.e., the singularity with the largest real part) of the partial wave amplitude in a given channel determines the asymptotic behavior in the crossed channels. In particular, Eq. (5.54) shows that the behavior of the scattering amplitude for $s \to \infty$ and fixed $t$ is governed by the rightmost complex angular momentum singularity in the $t$-channel.

Note that the result (5.54) can be formally obtained from the "naïve" expression (5.29) by the replacement

$$s^{\alpha(t)} \to s^{\alpha(t)} + \xi \, u^{\alpha(t)} = s^{\alpha(t)} + \xi \, (-s)^{\alpha(t)} = (1 + \xi \, \mathrm{e}^{-i\pi\alpha(t)}) \, s^{\alpha(t)} \,, \tag{5.56}$$

which sums the $s$ and $u$-channel contributions.

**The High-Energy Limit.** In deriving the result (5.54) we performed the *Regge limit*

$$s \to \infty \quad \text{with fixed } t \,, \tag{5.57}$$

only at the end of the calculation. If we are interested in the high-energy domain, it is convenient to work from the beginning in the limit (5.57). Using the asymptotic behaviors of the Legendre functions of the second kind

$$Q_\ell(z) \underset{|z| \to \infty}{\sim} \sqrt{\pi} \, \frac{\Gamma(\ell + 1)}{\Gamma\left(\ell + \frac{3}{2}\right)} \, (2z)^{-\ell - 1} \,, \tag{5.58}$$

we can introduce the Froissart-Gribov projection of the definite-signature partial-wave amplitudes in the Regge limit (cfr. (5.45) with $z_t \simeq -2s/|t|$)

$$a^\xi(\ell,t) = \int_1^\infty D_s^\xi(s,t) \left(\frac{s}{|t|}\right)^{-\ell-1} \mathrm{d}\left(\frac{s}{|t|}\right) . \tag{5.59}$$

Thus $a^\xi(\ell,t)$ is the *Mellin transform* of the discontinuity $D_s^\xi(s,t)$ (for details on Mellin transforms see Bateman 1954, vol. 1, p. 305; some of their properties are reviewed in App. B). The inverse Mellin transform of (5.59) is

$$D_s^\xi(s,t) = \frac{1}{2\pi\mathrm{i}} \int_{c-\mathrm{i}\infty}^{c+\mathrm{i}\infty} a^\xi(\ell,t) \left(\frac{s}{|t|}\right)^\ell \mathrm{d}\ell , \tag{5.60}$$

where the integration contour is a line parallel to the imaginary axis and to the right of all the singularities in $a^\xi(\ell,t)$ in the complex $\ell$ plane. By exploiting the asymptotic behavior of the Legendre polynomial,

$$P_\ell(z) \underset{|z|\to\infty}{\sim} \frac{1}{\sqrt{\pi}} \frac{\Gamma\left(\ell+\frac{1}{2}\right)}{\Gamma(\ell+1)} (2z)^\ell , \tag{5.61}$$

we can obtain from (5.52) the Watson-Sommerfeld expression of the asymptotic scattering amplitude in terms of $a^\xi(\ell,t)$, that is

$$A(s,t) = -\frac{1}{4\pi\mathrm{i}} \sum_{\xi=\pm1} \int_{c-\mathrm{i}\infty}^{c+\mathrm{i}\infty} \frac{\mathrm{e}^{-\mathrm{i}\pi\ell}+\xi}{\sin\pi\ell} \left(\frac{s}{|t|}\right)^\ell a^\xi(\ell,t)\,\mathrm{d}\ell . \tag{5.62}$$

In conclusion let us see some relevant examples. If $D_s^\xi(s,t)$ behaves as a power of $s$, the amplitude in the $\ell$ plane has a simple pole, namely

$$D_s^\xi(s,t) \sim s^\alpha \quad \Rightarrow \quad a(\ell,t) \sim \frac{1}{\ell-\alpha} , \tag{5.63}$$

where one recognizes the Regge pole behavior. If there is an additional $\ln^{-1} s$ dependence, the situation changes as follows

$$D_s^\xi(s,t) \sim s^\alpha (\ln s)^{-1} \quad \Rightarrow \quad a(\ell,t) \sim \ln(\ell-\alpha) , \tag{5.64}$$

and we have a cut at $\ell = \alpha$, instead of a pole. We shall encounter the latter type of singularity in Sect. 5.10.

## 5.7 Regge Trajectories

From the discussion in (5.6) it should be clear that, in presence of a Regge pole, the partial-wave amplitude $A(\ell,t)$ behaves for $\ell \to \alpha(t)$ as

$$A(\ell, t) \underset{\ell \to \alpha(t)}{\sim} \frac{\beta(t)}{\ell - \alpha(t)} . \tag{5.65}$$

For $t$ physical in the $s$-channel (i.e. $t \leq 0$), the $\ell$-plane singularities of the partial wave amplitude (5.65) are in general complex. The trajectory $\alpha(t)$ takes integer values of $\ell$ at some non-physical value of $t$ (i.e., for $t > 0$). These Regge poles correspond to resonances or bound states. Let us see concretely what happens.

Suppose that for some real $t_0$, we have

$$\alpha(t_0) = \ell + \mathrm{i}\epsilon , \tag{5.66}$$

where $\ell$ is an integer and $\epsilon$ is a real number small compared to unity. Expanding $\alpha(t)$ around $t_0$

$$\alpha(t) = \ell + \mathrm{i}\epsilon + \alpha'(t_0)\,(t - t_0) + \dots \tag{5.67}$$

we find that the denominator of (5.65) behaves as

$$\frac{1}{\ell - \alpha(t)} \propto \frac{1}{t - t_0 + \mathrm{i}\Gamma} , \tag{5.68}$$

where

$$\Gamma = \frac{\operatorname{Im} \alpha(t_0)}{\alpha'(t_0)} = \frac{\epsilon}{\alpha'(t_0)} \tag{5.69}$$

This is the typical structure of a Breit-Wigner term for a resonance of mass $M = \sqrt{t_0}$. In order for $\Gamma$ to be real, we have to assume

$$\left.\frac{\mathrm{d} \operatorname{Im} \alpha(t)}{\mathrm{d}t}\right|_{t_0} \ll \left.\frac{\mathrm{d} \operatorname{Re} \alpha(t)}{\mathrm{d}t}\right|_{t_0} . \tag{5.70}$$

Below the lowest threshold in $t$ (which for equal-mass particles is $4m^2$), we have $\operatorname{Im} \alpha(t) = 0$, and the Regge poles correspond to bound states.

Thus, for real and positive values of $t$ (i.e., when $t$ is a mass squared), Regge poles represent resonances and bound states of increasing angular momentum (i.e., spin) $\ell$. We say that $\alpha(t)$ is a *Regge trajectory* (or *reggeon*) interpolating such resonances (or bound states). Reggeons are often collectively denoted by the symbol $I\!R$.

Now consider the scattering amplitude (5.54). Its denominator vanishes whenever $\alpha(t)$ crosses an integer. Owing to the signature factor $(1 + \xi\, \mathrm{e}^{-\mathrm{i}\pi\alpha(t)})$, the numerator of (5.54) also vanishes, but at *every other integer* value of $\ell$. As a consequence, a trajectory with *positive signature* ($\xi = 1$) interpolates between *even angular momentum* resonances whereas a *negative signature* ($\xi = -1$) trajectory interpolates between *odd angular momentum* resonances.

The important message to learn is that the $s$-channel asymptotic behavior (5.54) is due to the *exchange of a family of resonances in the crossed channel.* Understanding this point is fundamental since it clarifies the nature and the role of a singularity in the complex angular momentum. Notice, incidentally, that this is in line with the original message from the Yukawa conjecture about the existence of the meson. The novelty is that this role of dominance of the exchange in the crossed channels is now transferred to the asymptotic behavior.

Different processes will, in general, receive contribution from different trajectories. The contributions to a given reaction depend on the quantum numbers that this reaction exhibits in the crossed channels.

A simple way to visualize the Regge trajectories, is to expand $\alpha(t)$ in power series around $t = 0$. In this case, for $t$ small enough, we can write

$$\alpha(t) = \alpha(0) + \alpha' t , \tag{5.71}$$

where $\alpha(0)$ and $\alpha'$ are known as the *intercept* and the *slope*, respectively, of the trajectory.

Quite unexpectedly, when interpolating resonances with the same quantum numbers (other than the spin), one finds that the expansion (5.71), which was a priori justified only for small $t$, holds actually for rather large values of $t$ (up to several units of GeV$^2$). In addition, this is true for both mesonic and baryonic trajectories (i.e., for trajectories which interpolate between integer and half-integer spin, respectively).

The situation is exemplified in Fig. 5.6 where the leading mesonic trajectories (the ones with largest $\alpha(0)$, i.e. the $\rho$, the $f_2$, the $a_2$ and the $\omega$, etc.) are shown (they are all superimposed).

Each trajectory has the quantum numbers (parity, charge conjugation, $G$-parity, isospin, strangeness, etc.) of the first recurrence of which takes the name. More specifically, we have for $f_2, \rho, \omega, a_2$

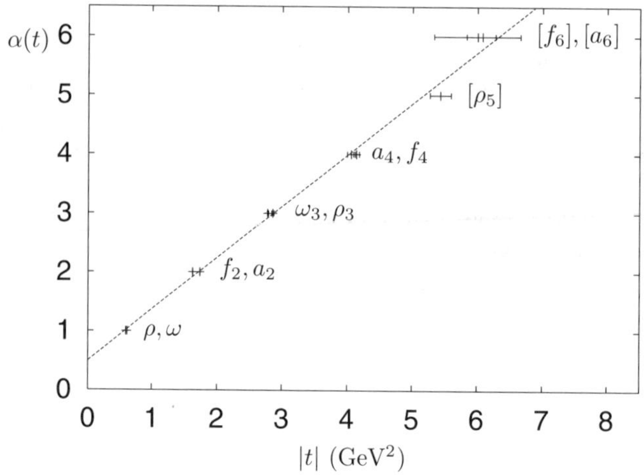


**Fig. 5.6.** The leading mesonic trajectories, the $\rho$, the $f_2$, the $a_2$, the $\omega$, etc. (all superimposed).

$$f_2: \quad P=+1,\ C=+1,\ G=+1,\ I=0,\ \xi=+1\,, \tag{5.72}$$
$$\rho: \quad P=-1,\ C=-1,\ G=+1,\ I=1,\ \xi=-1\,, \tag{5.73}$$
$$\omega: \quad P=-1,\ C=-1,\ G=-1,\ I=0,\ \xi=-1\,, \tag{5.74}$$
$$a_2: \quad P=+1,\ C=+1,\ G=-1,\ I=1,\ \xi=+1\,. \tag{5.75}$$

Note that, among these trajectories, the $f_2$ trajectory has the quantum numbers of the vacuum. We shall see in Sect. 5.8.3 that there exists another trajectory with vacuum quantum numbers, the *pomeron*. Its existence has been postulated in order to explain the total cross section data at high energies, but it does not correspond to any meson (see Sect. 5.8.3).

From Fig. 5.6 one sees that all the leading mesonic trajectories have $\alpha(0) \simeq 0.5$ (by contrast, the pomeron intercept is $\alpha(0) = 1$). The other mesonic trajectories not shown in Fig. 5.6 (for instance those interpolating strange particles) have lower intercepts. Equation (5.54) shows clearly that the larger the intercept, the more important the contribution will be as $s$ increases. The slope $\alpha'$ of all the mesonic trajectories in Fig. 5.6 is of the order of 1 GeV$^{-2}$.

The main baryonic trajectories have a very similar slope but a considerably lower intercept. For some of these trajectories $\alpha(0)$ is negative.

Many comments are in order. First, notice the unexpectedly large interval of masses for which the trajectories are basically linear. Second, the slope $\alpha'$ is essentially universal (in the language of string theory, it is called the *string tension*). The third point is that the leading mesonic trajectories are essentially *degenerate*, in the sense that they all lie one on top of the other, as one can see from Fig. 5.6. This property is called in the literature *exchange degeneracy*[6]. Thus, even though, in principle, each trajectory interpolates among even or odd angular momenta according to whether it has positive or negative signature, in practice, $\xi = +1$ trajectories (the $f_2$ and the $a_2$) are undistinguishable from $\xi = -1$ trajectories (the $\rho$ and the $\omega$).

## 5.8 Regge Phenomenology

We have seen that Regge theory describes a two-body scattering process $1 + 2 \to 3 + 4$, in the large $s$ limit, in terms of the exchange of Regge trajectories. We rewrite here for convenience the amplitude of this process, in the simple case of a single reggeon exchange (see Fig. 5.7)

$$A(s,t) = \beta(t)\,\eta(t)\,s^{\alpha(t)}\,, \tag{5.76}$$

where $\beta(t)$ is the *residue* and

[6] A careful analysis reveals that the exchange degeneracy is only approximate. The $\rho$ intercept, for instance, is slightly higher than the $\omega$ intercept (and similarly for $f_2$ and $a_2$). As a consequence, one often prefers to talk of $\rho - f_2$ and $\omega - a_2$ degeneracies.

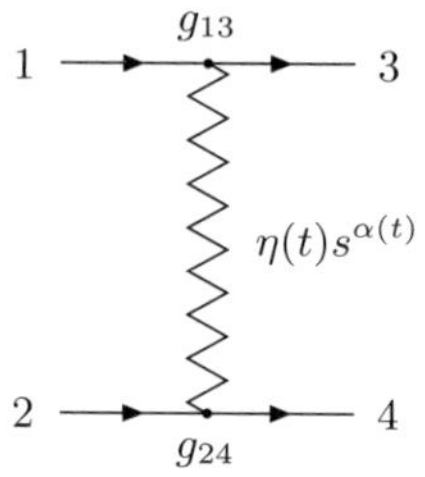


**Fig. 5.7.** Reggeon exchange.

$$\eta(t) = -\frac{1+\xi\,\mathrm{e}^{-\mathrm{i}\,\pi\,\alpha(t)}}{\sin\pi\alpha(t)} \,. \tag{5.77}$$

is the *signature factor*. This can be also expressed as

$$\eta(t) = -\frac{\mathrm{e}^{-\mathrm{i}\,\frac{\pi}{2}\,\alpha(t)}}{\sin\frac{\pi}{2}\alpha(t)} \quad (\xi=+1)\,, \tag{5.78}$$

$$\eta(t) = -\mathrm{i}\,\frac{\mathrm{e}^{-\mathrm{i}\,\frac{\pi}{2}\,\alpha(t)}}{\cos\frac{\pi}{2}\alpha(t)} \quad (\xi=-1)\,. \tag{5.79}$$

For linear trajectories $\alpha(t) \simeq \alpha(0) + \alpha'\,t$, we can write $\eta(t)$ as

$$\eta(t) \simeq \eta(0)\,\mathrm{e}^{-\mathrm{i}\,\frac{\pi}{2}\,\alpha'\,t}\,, \tag{5.80}$$

where we neglected the $t$-dependence of the denominators in (5.79), to the extent that $|t|$ is small. Assuming for the residue an exponential behavior in $t$, i.e.

$$\beta(t) = \beta(0)\,\mathrm{e}^{B_0\,t/2}\,, \tag{5.81}$$

the Regge amplitude at large $s$ and small $|t|$ reads

$$A(s,t) = \beta(0)\,\eta(0)\,s^{\alpha(0)}\,\exp\left[\frac{B_0}{2} + \alpha'\left(\ln s - \mathrm{i}\,\frac{\pi}{2}\right)\right]\,. \tag{5.82}$$

Since the Regge trajectories and the residue functions are expected to be real below threshold, the signature factor $\eta(t)$ completely determines the phase of the Regge pole amplitude. The ratio of the real to imaginary part of $A(s,t)$, if a single Regge pole is exchanged, is given by

$$\frac{\mathrm{Re}\,A(s,t)}{\mathrm{Im}\,A(s,t)} = -\frac{\xi+\cos\pi\alpha(t)}{\sin\pi\alpha(t)}\,. \tag{5.83}$$

The value of this ratio in the forward direction ($t=0$), the so-called $\rho$ parameter, is determined in Regge theory by the intercept of the exchanged trajectory.

Equation (5.76), or (5.82), gives rise to a very rich phenomenology, that we are now going to review (for more comprehensive accounts see Collins 1977, Irving and Worden 1977, Kaidalov 1979, Ganguli and Roy 1980). The comparison with some experimental findings is postponed to Chap. 7.

### 5.8.1 Factorization

General properties of the $S$-matrix lead to the expectation that the Regge pole residue $\beta(t)$ factorizes as

$$\beta(t) = g_{13}(t)\, g_{24}(t) \,, \tag{5.84}$$

where $g_{13}(t)$ and $g_{24}(t)$ are the couplings at each vertex of the reggeon exchange diagram (Fig. 5.7). This factorization property allows one to relate cross sections of different processes. For instance, it gives

$$\mathrm{d}\sigma^2(1\,2 \to 3\,4) = \mathrm{d}\sigma(1\,1 \to 3\,3)\, \mathrm{d}\sigma(2\,2 \to 4\,4) \,, \tag{5.85}$$

which is valid when a single Regge trajectory dominates. However, a relation such has (5.85) is difficult to test, as it would require different targets. In practice, only nucleon targets are used in experiments. So, in order to test Regge factorization in two-body processes, one needs to compare elastic scattering off nucleons with the production of nucleon resonances. In this case the factorizability of Regge amplitudes implies

$$\frac{\mathrm{d}\sigma(a\,N \to a\,N)}{\mathrm{d}\sigma(b\,N \to b\,N)} = \frac{\mathrm{d}\sigma(a\,N \to a\,N^*)}{\mathrm{d}\sigma(b\,N \to b\,N^*)} \,, \tag{5.86}$$

where $N^*$ is a nucleon resonance. More complete tests of factorization can be performed in inclusive reactions.

An important consequence of Regge factorization is the property known as *line reversal symmetry*. The two processes $1+2 \to 3+4$ and $1+\bar{4} \to 3+\bar{2}$ have the same quantum numbers in the $t$-channel, so they exchange the same Regge poles with the same couplings. There is only a change of sign for negative signature trajectories, which are odd under $s \leftrightarrow u$. We shall see later how this property leads to very useful relations between cross sections of different processes.

Finally, it should be stressed that Regge factorization breaks down when more than one Regge trajectory contibutes to the process, or in presence of cuts.

### 5.8.2 Total and Elastic Cross-sections

From (5.76) it is immediate, via the optical theorem, to obtain the Regge theory prediction for total cross sections. The single-pole contribution is[7]

[7] Hereafter the powers of $s$ should be normalized to some fixed value $s_0$ setting the scale beyond which we are authorized to use the asymptotic expressions for the scattering amplitudes and the cross sections. This scale is typically of the order of 1 $\mathrm{GeV}^2$. For simplicity we shall omit $s_0$ but it should be borne in mind that any asymptotic formula is intended to be a function of $s/s_0$.

$$\sigma_{\text{tot}} \underset{s\to\infty}{\simeq} \frac{1}{s} \operatorname{Im} A(s,t=0) \underset{s\to\infty}{\sim} s^{\alpha(0)-1} . \tag{5.87}$$

If more than one pole contributes (this is the case when $s$ is not so large, see below), the total cross section is given by a sum of terms of the form (5.87)

$$\sigma_{\text{tot}} \sim \sum_i A_i \, s^{\alpha_i(0)-1} . \tag{5.88}$$

The elastic cross section is given by Eq. (4.33) and its expression in Regge theory is

$$\frac{d\sigma_{\text{el}}}{dt} = F(t)\, s^{2\alpha(t)-2} , \tag{5.89}$$

where $F(t)$ is a function of $t$ incorporating the residue function and the signature factor. If many poles contribute, interference terms in general appear.

Let us consider one single reggeon with a linear trajectory

$$\alpha(t) = \alpha(0) + \alpha' t . \tag{5.90}$$

Inserting (5.90) into (5.89) gives

$$\frac{d\sigma_{\text{el}}}{dt} = F(t)\, s^{2\alpha(0)-2}\, \mathrm{e}^{-2\,\alpha'\,|t|\,\ln s} . \tag{5.91}$$

If we suppose, for simplicity, that the colliding particles are alike and assume the simple exponential parametrization (5.81) for the residue function, the elastic cross section becomes

$$\frac{d\sigma_{\text{el}}}{dt} \sim s^{2\alpha(0)-2}\, \mathrm{e}^{-B\,|t|} , \tag{5.92}$$

with

$$B = B_0 + 2\alpha' \ln s . \tag{5.93}$$

We see that the width of the forward peak $\Delta|t| = (B_0 + 2\alpha' \ln s)^{-1}$ decreases as the energy increases. This is the phenomenon known as the *shrinkage* of the diffraction peak, which can be interpreted as an increase of the interaction radius $R_{\text{int}} \underset{s\to\infty}{\sim} \sqrt{\alpha' \ln s}$. The shrinkage, which is indeed observed, is not suggested by the optical analogy. The prediction of it is therefore a non trivial achievement of Regge theory.

### 5.8.3 The Pomeron

The Regge trajectories discussed in Sect. 5.7 have intercepts which do not exceed 0.5. Their exchange, according to (5.87), leads to total cross sections decreasing with energy. However, it is experimentally known (see Chap. 7) that hadronic total cross sections, as a function of $s$, are rather flat around $\sqrt{s} \sim (10-20)$ GeV$^2$ and increase at higher energies.

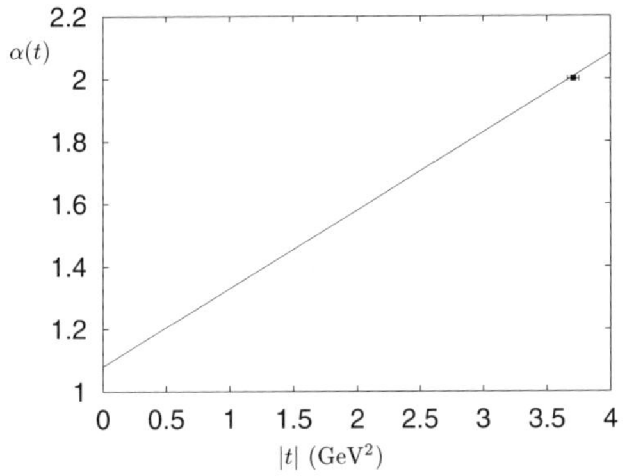


**Fig. 5.8.** The pomeron trajectory with a glueball candidate (Abatzis et al. 1994). The line corresponds to $\alpha_{I\!P}(t) = 1.08 + 0.25\,t/\text{GeV}^2$ (Donnachie and Landshoff 1992).

In order to account for asymptotically constant total cross sections (this was the expectation in the early 60's), Chew and Frautschi (1961) and Gribov (1961) introduced a Regge trajectory with intercept 1. This reggeon was named *pomeron*, after I.Ya. Pomeranchuk, and is denoted by $I\!P$.

The pomeron trajectory (Fig. 5.8) does not correspond to any known particle. Its recurrences are expected to be *glueballs*, rather than conventional resonances. It results from a complicated exchange of gluons (at least two), as we shall see in Chap. 8. A $2^{++}$ glueball candidate (Abatzis et al. 1994) lying on the $I\!P$ trajectory is shown in Fig. 5.8.

From fitting elastic scattering data, one finds that the $I\!P$ trajectory is much flatter than the others ($\alpha'_{I\!P} \simeq 0.25$ GeV$^{-2}$). As for the intercept, we already recalled that requiring the constancy of total cross sections implies $\alpha_{I\!P}(0) \simeq 1$.

The pomeron is the dominant trajectory in the elastic and diffractive processes, which are known to proceed via the exchange of vacuum quantum numbers in the $t$-channel. Thus we have

$$I\!P: \quad P = +1,\ C = +1,\ G = +1,\ I = 0,\ \xi = +1\,. \tag{5.94}$$

Since $\xi = +1$, and

$$\lim_{x\to 1} \frac{1 + \mathrm{e}^{-\mathrm{i}\pi x}}{\sin \pi x} = -\mathrm{i} \tag{5.95}$$

the pomeron-dominated scattering amplitude has $\eta(0) = \mathrm{i}$, and at $t = 0$ behaves asymptotically

$$A_{I\!P}(s, t = 0) \underset{s\to\infty}{\simeq} \mathrm{i}\,\beta_{I\!P}(0)\, s^{\alpha_{I\!P}(0)} \tag{5.96}$$

and is purely imaginary. Note that, with an intercept equal to 1, a negative signature would give rise to a diverging forward amplitude.

The pomeron intercept $\alpha_{I\!P}(0) = 1$ saturates the unitarity bound: an intercept larger than 1 would violate Froissart-Martin constraint. However, as we shall see in Chap. 7, total cross sections are known to rise with energy. This would imply a pomeron intercept $\alpha_{I\!P}(0) > 1$. There are some possible ways out of this dilemma. One can claim that at the presently achieved energies we are still far from asymptopia, and therefore a (presently unknown) mechanism should eventually set in and unitarize cross sections. We must also recall that poles are not the only singularities in the complex angular momentum plane. As shown by Amati, Fubini and Stanghellini (1962), Mandelstam (1963), Gribov, Pomeranchuk and Ter-Martirosyan (1965) cut singularities occur as well. These contribute to tame the growth of cross sections, as we shall see in Sect. 5.10.

### 5.8.4 The Odderon

Another hypothetical Regge trajectory which may play some rôle in high-energy scattering is the so-called *odderon*, introduced by Łukaszuk and Nicolescu (1973) (see also Joynson et al. 1975 and for an early discussion Gribov et al. 1971). The odderon is conceived as the $C = P = -1$ partner of the pomeron. It is a $j$-plane singularity of the crossing-odd amplitude $A_-$ near $j = 1$. The existence of the odderon (for which there is presently no evidence from the data – at least at low $|t|$) would entail differences between the asymptotic scattering amplitudes and cross sections of $pp$ and $p\bar{p}$ (see Sect. 7.4.1).

### 5.8.5 Regge Trajectories and Hadron Scattering

Having completed, with the introduction of the pomeron (and of the odderon), the list of the main Regge trajectories, we can now see how they contribute to some hadronic processes. Let us consider, for example, the most important elastic reactions. To each of them we associate the corresponding exchanged reggeons, using line reversal symmetry (couplings and other factors are omitted; no odderon contribution is assumed)

$$\begin{aligned}
\pi^- p &\sim I\!P + f_2 + \rho \, , \\
\pi^+ p &\sim I\!P + f_2 - \rho \, , \\
K^- p &\sim I\!P + f_2 + \rho + a_2 + \omega \, , \\
K^+ p &\sim I\!P + f_2 - \rho + a_2 - \omega \, , \\
pp &\sim I\!P + f_2 - \rho + a_2 - \omega \, , \\
p\bar{p} &\sim I\!P + f_2 + \rho + a_2 + \omega \, , \\
pn &\sim I\!P + f_2 + \rho - a_2 - \omega \, .
\end{aligned} \tag{5.97}$$

In the particle-antiparticle total cross section differences the pomeron cancels out, so we have

$$\begin{aligned} \sigma(K^-p) - \sigma(K^+p) &\sim 2\,(\omega + \rho)\,, \\ \sigma(p\bar{p}) - \sigma(pp) &\sim 2\,(\omega + \rho)\,, \\ \sigma(pn) - \sigma(pp) &\sim 2\,(\rho - a_2)\,, \\ \sigma(\pi^-p) - \sigma(\pi^+p) &\sim 2\,\rho\,. \end{aligned} \tag{5.98}$$

Since these differences are determined by subleading trajectories, they tend to vanish as $s \to \infty$, in agreement with the Pomeranchuk theorems.

More relations between cross sections (and residue functions) can be obtained if other theoretical inputs (such as isospin symmetry, $SU(3)$, etc.) are used. For this, we refer the reader to Eden (1967) and Collins (1977).

### 5.8.6 Diffractive Dissociation

Let us consider the single-inclusive reaction $1 + 2 \to 3 + X$ in the limit $s \gg M^2 \gg t$ (the so-called *triple-Regge limit*), where $M^2$ is the invariant mass of the hadronic system $X$ (see Fig. 5.9a). The particle 3 is produced in the fragmentation region of particle 1. If 3 has the same quantum numbers as 1, we have single diffractive dissociation.

In the triple-Regge limit the scattering amplitude of the process is

$$A(12 \to 3X) \underset{s\to\infty}{\sim} \sum_i g^i_{13}(t) g^i_{2X}(t)\, \eta_i(t) \left(\frac{s}{M^2}\right)^{\alpha_i(t)}\,, \tag{5.99}$$

where the sum is over the contributing reggeons, $\eta_i(t)$ is the signature factor, and we have used (5.53) and $z_t \equiv \cos\vartheta_t \sim (s/M^2)$, as $s \to \infty$, see (5.39).

According to Mueller's theorem (Sect. 4.8) the cross section is (we omit the subscript 3 in the kinematic variables of particle 3; $SD$ means single diffraction)

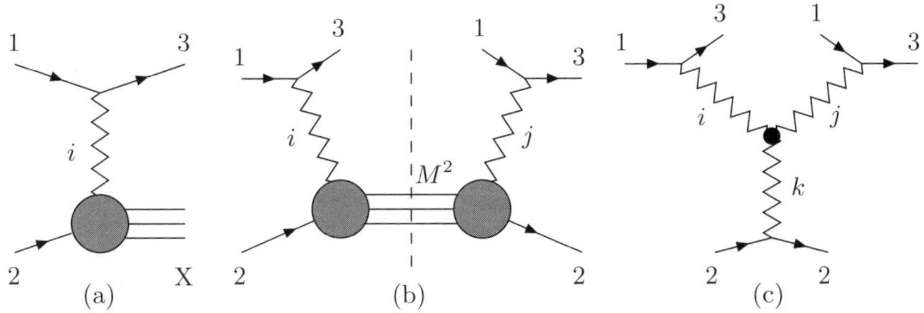


**Fig. 5.9.** (**a**) Single-inclusive reaction $1 + 2 \to 3 + X$ mediated by the exchange of a reggeon $i$. (**b**) The discontinuity across $M^2$ of the scattering amplitude. (**c**) The triple-reggeon diagram.

$$\begin{aligned}
f(s,t,M^2) &\equiv 16\pi^2 s\,\frac{\mathrm{d}^2\sigma^{SD}}{\mathrm{d}M^2\,\mathrm{d}t} \\
&= \frac{1}{s}\,\mathrm{Disc}_{M^2}\,A(12\bar{3}\to 12\bar{3}) \\
&\underset{s\to\infty}{\sim} \frac{1}{s}\sum_{ij} g_{13}^{i}(t) g_{13}^{j*}(t)\,\eta_i(t)\,\eta_j^*(t)\,\left(\frac{s}{M^2}\right)^{\alpha_i(t)+\alpha_j(t)} \\
&\quad \times \mathrm{Disc}_{M^2}\,A(i\,2\to j\,2)\,. 
\end{aligned} \tag{5.100}$$

Here $A(i\,2 \to j\,2)$ is the reggeon-particle scattering amplitude (see Fig. 5.9b). Its discontinuity, in the limit $M^2 \to \infty$, is predicted by Regge theory to be (see Fig. 5.9c)

$$\mathrm{Disc}_{M^2}\,A(i\,2\to j\,2) \underset{M^2\to\infty}{\sim} \sum_k g_{22}^{k}(0)\,g_{ijk}(t)\,(M^2)^{\alpha_k(0)}\,. \tag{5.101}$$

Note that reggeons $i$ and $j$ carry the momentum squared $t$, whereas $k$ carries no momentum. In (5.101) $g_{ijk}$ is the triple-reggeon coupling and $s_0$ is an arbitrary reference scale.

Putting (5.101) into (5.100) gives, in the limit $s \gg M^2 \gg t$

$$\begin{aligned}
16\pi^2 s\,\frac{\mathrm{d}^2\sigma^{SD}}{\mathrm{d}M^2\,\mathrm{d}t} &= \frac{1}{s}\sum_{ijk} g_{13}^{i}(t) g_{13}^{j*}(t)\,\eta_i(t)\,\eta_j^*(t)\,\left(\frac{s}{M^2}\right)^{\alpha_i(t)+\alpha_j(t)} \\
&\quad \times g_{22}^{k}(0)\,g_{ijk}(t)\,(M^2)^{\alpha_k(0)} \\
&= \sum_{ijk} G_{ijk}(t)\,s^{\alpha_i(t)+\alpha_j(t)-1}\,(M^2)^{\alpha_k(0)-\alpha_i(t)-\alpha_j(t)}\,,
\end{aligned} \tag{5.102}$$

where, in the last line, all couplings and signatures have been incorporated into the functions $G_{ijk}(t)$.

In terms of the Feynman variable $x_F \simeq 1 - M^2/s$ the cross section reads (in the limit $x_F \to 1$)

$$\begin{aligned}
f(s,t,x_F) &= 16\pi^2\,\frac{\mathrm{d}^2\sigma^{SD}}{\mathrm{d}x_F\,\mathrm{d}t} \\
&= \sum_{ijk} G_{ijk}(t)\,(1-x_F)^{\alpha_k(0)-\alpha_i(t)-\alpha_j(t)}\,s^{\alpha_k(0)-1}\,.
\end{aligned} \tag{5.103}$$

Let us focus now on a specific single-inclusive reaction, $1\,2 \to 1'\,X_2$. If the outgoing particle is equal to one of the incoming particles and carries most of its momentum, and the hadronic state $X_2$ has the same quantum numbers as the other incoming particle, we have a diffractive process. In this case the two trajectories that we previously called $i$ and $j$ (i.e., those exchanged between particles 1 and 2) are the pomeron trajectory: $\alpha_i(t) = \alpha_j(t) = \alpha_{I\!P}(t)$. The trajectory $k$ can be either a pomeron or another reggeon $I\!R$ (the former dominates when $M^2$ is very large). Thus we have

$$16\pi^2 s \frac{\mathrm{d}^2\sigma^{SD}}{\mathrm{d}M^2\,\mathrm{d}t} = G_{I\!PI\!PI\!P}(t) s^{2\,\alpha_{I\!P}(t)-1}\,(M^2)^{\alpha_{I\!P}(0)-2\,\alpha_{I\!P}(t)} + G_{I\!PI\!PI\!R}(t) s^{2\,\alpha_{I\!P}(t)-1}\,(M^2)^{\alpha_{I\!R}(0)-2\,\alpha_{I\!P}(t)}\,. \quad (5.104)$$

In particular, the triple-pomeron term reads explicitly

$$16\pi^2 s\,\frac{\mathrm{d}^2\sigma^{SD}}{\mathrm{d}M^2\,\mathrm{d}t} = |g_{I\!P}(t)|^2\, g_{I\!P}(0)\, g_{I\!PI\!PI\!P}(0)\, \left(\frac{s}{M^2}\right)^{2\,\alpha_{I\!P}(t)-1} (M^2)^{\alpha_{I\!P}(0)-1} = |g_{I\!P}(t)|^2\, g_{I\!P}(0)\, g_{I\!PI\!PI\!P}(0)\, s^{2\,\alpha_{I\!P}(t)-1}\,(M^2)^{\alpha_{I\!P}(0)-2\,\alpha_{I\!P}(t)}\,. \quad (5.105)$$

If we take $\alpha_{I\!P}(0) = 1$ and $\alpha_{I\!R}(0) = 1/2$, we get the following contributions to $f(s, t = 0, M^2)$

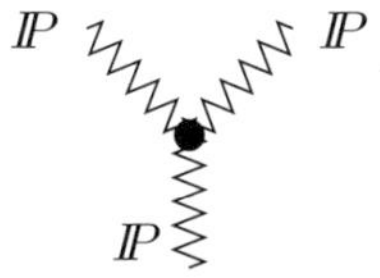


$$f(s, t = 0, M^2) \sim (M^2)^{-1}\, s^1 \sim (1 - x_F)^{-1}\, s^0$$

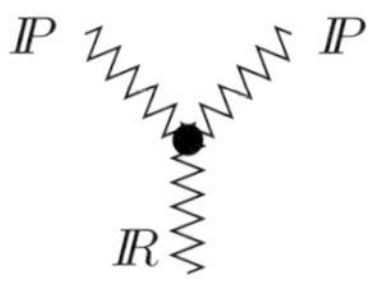


$$f(s, t = 0, M^2) \sim (M^2)^{-\frac{3}{2}}\, s^1 \sim (1 - x_F)^{-\frac{3}{2}}\, s^{-\frac{1}{2}}$$

As anticipated, in the limit of very large $M^2$, the $I\!PI\!PI\!R$ term dies off more rapidly and the triple pomeron term dominates, with the behavior (if $\alpha_{I\!P}(0) = 1$)

$$\left.\frac{\mathrm{d}^2\sigma^{SD}}{\mathrm{d}M^2\,\mathrm{d}t}\right|_{t=0} \sim \frac{1}{M^2}\,. \quad (5.106)$$

Let us now introduce the total cross section for the interaction between the pomeron and particle 2 ($M^2$ is the energy squared in the c.m. frame of the pomeron and particle 2)

$$\sigma_{I\!P}(M^2) = \frac{1}{M^2}\,\mathrm{Disc}_{M^2}\, A(I\!P\,2 \to I\!P\,2) \underset{M^2\to\infty}{\sim} g_{I\!P}(0)\, g_{I\!PI\!PI\!P}(0)\, (M^2)^{\alpha_{I\!P}(0)-1} + g_{I\!R}(0)\, g_{I\!PI\!PI\!R}(0)\, (M^2)^{\alpha_{I\!R}(0)-1}. \quad (5.107)$$

Here we have used the experimentally known fact (Cool et al. 1981) that the triple-pomeron coupling is nearly independent of $t$, so that $g_{I\!PI\!PI\!P}(t) \simeq g_{I\!PI\!PI\!P}(0)$. Thus we can put (5.104) in the factorized form

$$s\,\frac{\mathrm{d}^2\sigma^{SD}}{\mathrm{d}M^2\,\mathrm{d}t} = \frac{1}{16\pi^2}\,|g_{I\!P}(t)|^2 \left(\frac{s}{M^2}\right)^{2\,\alpha_{I\!P}(t)-1} \sigma_{I\!P}(M^2)\,. \quad (5.108)$$

This factorization property of the diffractive cross section is a remarkable consequence of Regge theory. However, one should keep in mind that the pomeron is not a particle and hence the cross section $\sigma_{\mathbb{P}}$ is not a physical quantity. Its normalization is given by (5.108) and there is no way to measure it or determine it independently of (5.108).

If $s$ is large, but not really asymptotic, non-diffractive contributions to the process $1\,2 \to 1\,X$ may become important. In this case, at least one of the trajectories $i$ and $j$ is not a pomeron, but rather a subleading reggeon with a smaller intercept. So we have ($ND$ means non diffractive)

$$\begin{aligned} 16\pi^2 s \frac{\mathrm{d}^2\sigma^{ND}}{\mathrm{d}M^2\,\mathrm{d}t} &= G_{\mathbb{R}\mathbb{R}\mathbb{P}}(t) s^{2\,\alpha_{\mathbb{R}}(t)-1}\,(M^2)^{\alpha_{\mathbb{P}}(0)-2\,\alpha_{\mathbb{R}}(t)} \\ &+ G_{\mathbb{R}\mathbb{R}\mathbb{R}}(t) s^{2\,\alpha_{\mathbb{P}}(t)-1}\,(M^2)^{\alpha_{\mathbb{R}}(0)-2\,\alpha_{\mathbb{R}}(t)} \\ &+ \text{interference terms}\,. \end{aligned} \tag{5.109}$$

and, with the classical intercepts, we find

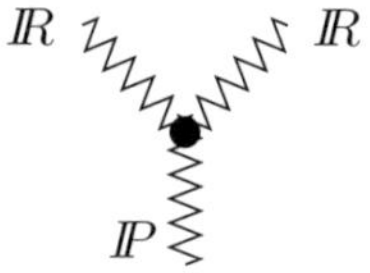


$$f(s, t=0, M^2) \sim (M^2)^0\, s^0 \sim (1-x_F)^0\, s^0$$

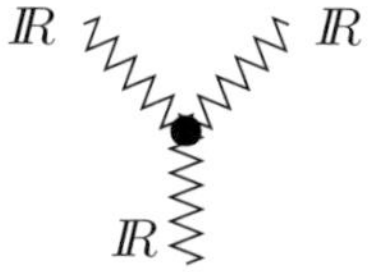


$$f(s, t=0, M^2) \sim (M^2)^{-\frac{1}{2}}\, s^0 \sim (1-x_F)^{-\frac{1}{2}}\, s^{-\frac{1}{2}}$$

For completeness, we give also the interference terms $\mathbb{R}\mathbb{P}\mathbb{P}$ and $\mathbb{R}\mathbb{P}\mathbb{R}$

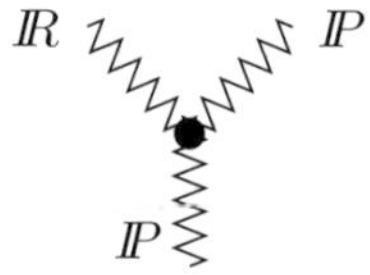


$$f(s, t=0, M^2) \sim (M^2)^{-\frac{1}{2}}\, s^{\frac{1}{2}} \sim (1-x_F)^{-\frac{1}{2}}\, s^0$$

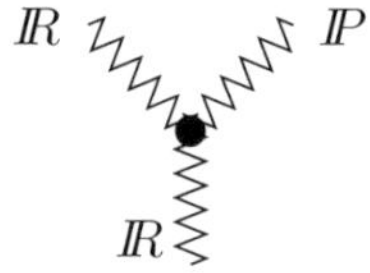


$$f(s, t=0, M^2) \sim (M^2)^{-1}\, s^{\frac{1}{2}} \sim (1-x_F)^{-1}\, s^{-\frac{1}{2}}$$

Another inclusive process that deserves at least a mention is double diffractive dissociation (Fig. 5.10a): $1\,2 \to X_1\,X_2$, where $X_1$ and $X_2$ carry

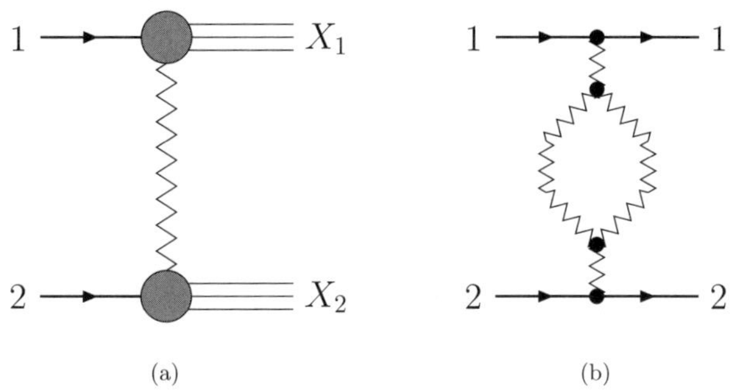


**Fig. 5.10.** (**a**) Double diffractive dissociation. (**b**) The pomeron loop (a discontinuity is to be taken across the loop).

the same quantum numbers of particles 1 and 2, respectively (from the experimental point of view, the reaction is characterized by a large rapidity gap between $X_1$ and $X_2$). If the masses $M_1^2$ and $M_2^2$ of the produced states are large, we can proceed as we did above for single diffraction, and we find that the process is dominated by a pomeron loop, which arises from gluing together two triple pomeron diagrams (Fig. 5.10b). Factorization relates the cross section of double diffraction dissociation ($DD$) to the cross sections of single diffractive dissociation ($SD$) and elastic scattering

$$\frac{\mathrm{d}\sigma^{DD}(12 \to X_1 X_2)}{\mathrm{d}M_1^2\,\mathrm{d}M_2^2\,\mathrm{d}t} = \frac{\mathrm{d}\sigma^{SD}(12 \to X_1 2)}{\mathrm{d}M_1^2\,\mathrm{d}t}\,\frac{\mathrm{d}\sigma^{SD}(12 \to 1 X_2)}{\mathrm{d}M_2^2\,\mathrm{d}t} \Big/ \frac{\mathrm{d}\sigma_{\mathrm{el}}(12 \to 12)}{\mathrm{d}t}\,. \tag{5.110}$$

Using (5.105) and (5.89) we obtain

$$16\pi^3\, s\, \frac{\mathrm{d}\sigma^{DD}}{\mathrm{d}M_1^2\,\mathrm{d}M_2^2\,\mathrm{d}t} = g_{I\!P}^2(0)\, g_{I\!P I\!P I\!P}^2(0) \left(\frac{s}{M_1^2\, M_2^2}\right)^{2\,\alpha_{I\!P}(t)-1} \times (M_1^2)^{\alpha_{I\!P}(0)-1}\,(M_2^2)^{\alpha_{I\!P}(0)-1}\,. \tag{5.111}$$

We shall make large use of the Regge theory predictions for diffraction dissociation presented above, when discussing diffractive deep inelastic scattering (Sect. 10.1).

## 5.9 Regge Poles in Field Theory

In the late Sixties many efforts were made to incorporate the Regge pole idea into field theory. We shall see in Chap. 8 that this program has been actually accomplished only in more recent years, after the advent of quantum

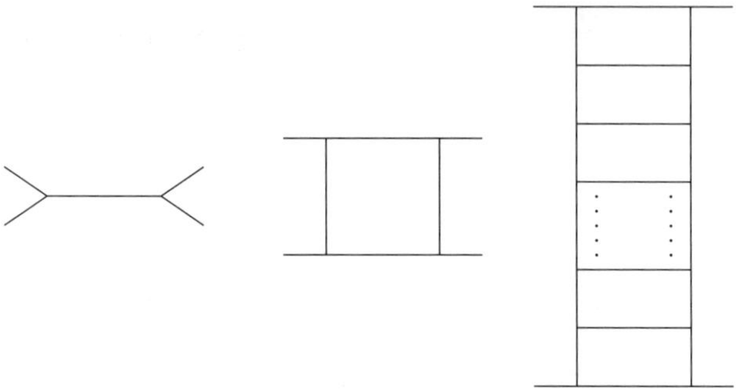

(a) (b) (c)

**Fig. 5.11.** Ladder diagrams in $\varphi^3$ field theory: (**a**) one rung; (**b**) two rungs; (**c**) $n$ rungs.

chromodynamics. Thus, it is mainly for historical and pedagogical reasons that we give here a brief account of the early, pre-QCD, attempts.

Let us consider the simplest field theoretical model of Regge poles, based on a massive $\varphi^3$ theory with coupling constant $g$ (Eden et al. 1966). The tree diagram for the two-body scattering (Fig. 5.11a) gives the amplitude (the superscript denotes the power of $g^2$, i.e. the perturbative order)

$$A^{(1)} \underset{s\to\infty}{\sim} \frac{g^2}{s} . \tag{5.112}$$

At next order we have the box diagram (Fig. 5.11b), which is easily calculated (see for instance Eden et al. 1966, Collins 1977) and gives, as $s \to \infty$ with $t$ fixed

$$A^{(2)} \underset{s\to\infty}{\sim} \frac{g^2}{s} K(t) \ln s , \tag{5.113}$$

where $K(t)$ is given by ($t \equiv q^2$, and $q^2 \simeq -\boldsymbol{q}_\perp^2$ in the large-$s$ limit)

$$K(t) \sim g^2 \int \frac{\mathrm{d}^2\boldsymbol{k}_\perp}{(\boldsymbol{k}_\perp^2 + m^2)\,[(\boldsymbol{k}_\perp + \boldsymbol{q}_\perp)^2 + m^2]} . \tag{5.114}$$

It turns out that the ladder diagram with $n$ rungs (Fig. 5.11c) gives, in the $s \to \infty$ limit with $t$ fixed, the amplitude

$$A^{(n)} \underset{s\to\infty}{\sim} \frac{g^2}{s} \frac{(K(t) \ln s)^{n-1}}{(n-1)\,!} . \tag{5.115}$$

Thus, if we sum an infinite series of ladder diagrams the asymptotic behavior of the resulting amplitude is

$$\begin{aligned} A(s,t) = \sum_{n=1}^{\infty} A^{(n)} &\underset{s\to\infty}{\sim} \sum_{n=1}^{\infty} \frac{g^2}{s} \frac{(K(t) \ln s)^{n-1}}{(n-1)\,!} \\ &\sim \frac{g^2}{s} \mathrm{e}^{K(t) \ln s} \sim g^2 \, s^{\alpha(t)} , \end{aligned} \tag{5.116}$$

where

$$\alpha(t) = -1 + K(t) \, . \tag{5.117}$$

We have found that summing an infinite number of diagrams, each contributing a power of $\ln s$ in the large-$s$ limit, ultimately builds up the Regge-type behavior

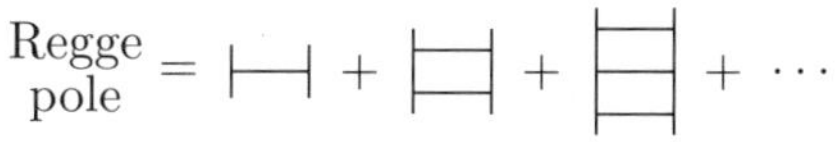


Clearly, what has been presented is simply a *model* of Regge poles. A scalar field theory has little to do with strong interactions – as we can see from the unrealistic exponent (5.117). However, the idea that Regge behavior is the result of a sum of leading $\ln s$ terms proves essentially correct. This idea is inherited by QCD, where we will see that reggeization arises from summing a series of gluon ladders in the high-energy limit.

It is instructive to see the source of the $\ln s$ powers associated to ladder diagrams. Unitarity relates the imaginary part of such diagrams to the cross section for multiparticle production.

Consider the production of $n+2$ particles (Fig. 5.12). In the large-$s$ limit, the cross section of this process is

$$\sigma^{(n+2)} \sim \frac{1}{s^2} \int \prod_{i=1}^{n} \left( \frac{\mathrm{d}x_i}{x_i} \, \mathrm{d}^2\boldsymbol{\kappa}_{i\perp} \right) \mathrm{d}^2\boldsymbol{\kappa}_{n+1\perp} \, |\mathcal{M}(x_j, \boldsymbol{\kappa}_{j\perp}^2)|^2 \, , \tag{5.118}$$

where $x_i$ and $\boldsymbol{\kappa}_{\perp i}$ are the longitudinal momentum fractions and the transverse momenta, respectively, of the produced particles. In (5.118) one of the factors $1/s$ is the flux factor, the other one comes from phase space integration.

At leading $\ln s$, the main contribution to multiparticle production comes from a region of the phase space where the longitudinal momentum fractions $x_i$ are strongly ordered

$$1 \gg x_0 \gg x_1 \gg x_2 \gg \ldots \gg x_n \gg x_{n+1} \sim \frac{\boldsymbol{\kappa}_\perp^2}{s} \, , \tag{5.119}$$

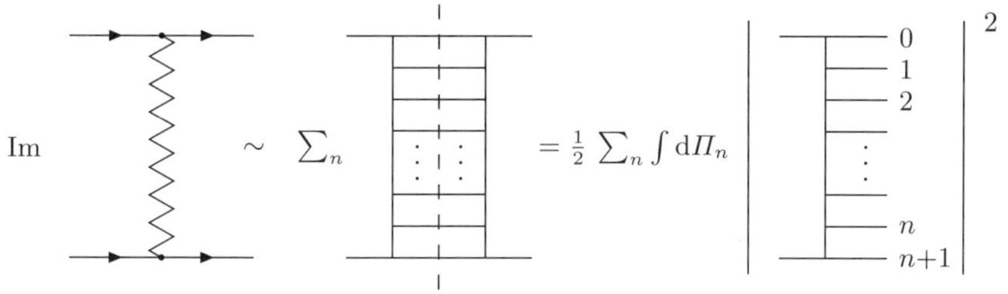


**Fig. 5.12.** From reggeon exchange to multiparticle production, via unitarity.

whereas the transverse momenta $\boldsymbol{\kappa}^2_{\perp i}$ are of comparable size

$$\boldsymbol{\kappa}^2_{0\perp} \simeq \boldsymbol{\kappa}^2_{1\perp} \simeq \boldsymbol{\kappa}^2_{2\perp} \simeq \ldots \simeq \boldsymbol{\kappa}^2_{n\perp} \simeq \boldsymbol{\kappa}^2_{n+1\perp} \simeq \boldsymbol{\kappa}^2_{\perp} \ll s \,. \tag{5.120}$$

The conditions (5.119) and (5.120) define the so-called *multi-Regge kinematics*. Setting $x_i = 0$ in the amplitude $\mathcal{M}$, due to the condition (5.119), we get for the total cross section

$$\begin{aligned}\sigma^{(n+2)} &\sim \frac{1}{s^2} \int \prod_{i=1}^{n+1} \mathrm{d}^2\boldsymbol{\kappa}_{i\perp} \, |\mathcal{M}(\boldsymbol{\kappa}^2_{j\perp})|^2 \\ &\quad \times \int_{\boldsymbol{\kappa}^2_\perp/s}^{1} \frac{\mathrm{d}x_1}{x_1} \int_{\boldsymbol{\kappa}^2_\perp/s}^{x_1} \frac{\mathrm{d}x_2}{x_2} \cdots \int_{\boldsymbol{k}^2_\perp/s}^{x_{n-1}} \frac{\mathrm{d}x_n}{x_n} \\ &\sim \int \prod_{i=1}^{n+1} \mathrm{d}^2\boldsymbol{\kappa}_{i\perp} \, |\mathcal{M}(\boldsymbol{\kappa}^2_{j\perp})|^2 \, \frac{1}{n!} \ln^n s \,. \end{aligned} \tag{5.121}$$

This is consistent with the result we derived previously, Eq. (5.115). Using dispersion relations, in fact, one obtains from (5.115) the imaginary part of the amplitude for production of $n+2$ particles

$$\mathrm{Im}\, A^{(n+2)} \sim \frac{1}{s} \frac{(K \ln s)^n}{n!} \,, \tag{5.122}$$

and the optical theorem gives for the cross section $\sigma^{(n+2)}$

$$\sigma^{(n+2)} \sim \frac{1}{s^2} \frac{(K \ln s)^n}{n!} \,, \tag{5.123}$$

in agreement with (5.121).

From (5.121), we see that the powers $\ln^n s$ arise from the integration over the longitudinal momenta, under the requirement of strong ordering. Something very similar happens in QCD, as we shall show in Chap. 8.

In terms of the rapidities $y_i$ of the produced particles, the condition (5.119) traslates into the strong ordering

$$y_0 \gg y_1 \gg y_2 \gg \ldots \gg y_n \gg y_{n+1} \,. \tag{5.124}$$

A diagram for the production of $n$ particles strongly ordered in rapidity (see Fig. 5.12c) is called *multiperipheral diagram* (Bertocchi, Fubini and Tonin 1962; Amati, Fubini and Stanghellini 1962). The multiplicity $\langle n \rangle$ of the particles is given by

$$\langle n \rangle = \frac{\sum_n n\, \sigma^{(n)}}{\sum_n \sigma^{(n)}} \sim K \ln s \,, \tag{5.125}$$

and their distribution is

$$P(n) = \frac{\sigma^{(n)}}{\sum_n \sigma^{(n)}} = \frac{\langle n \rangle^n \, \mathrm{e}^{-\langle n \rangle}}{n!} \,, \tag{5.126}$$

that is a Poisson distribution, signaling that the produced particles in a multiperipheral diagram are not correlated.

## 5.10 Regge Cuts

Regge poles are only a part of the whole story. In the $\ell$-plane there are other singularities, the *Regge cuts* (Gribov and Pomeranchuk 1962; Amati, Fubini and Stanghellini 1962; Mandelstam 1963; Gribov, Pomeranchuk and Ter-Martirosyan 1965), corresponding to the exchange of two or more reggeons (Fig. 5.13). The existence of cuts is also a phenomenological need, as there are observed failures of factorization which cannot be explained by poles alone. Moreover, being related to the iteration of reggeon exchanges, Regge cuts are important to understand $s$-channel unitarity in the framework of Regge theory.

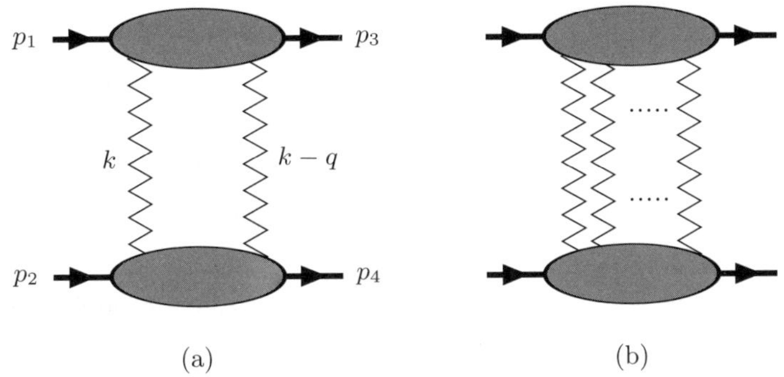


**Fig. 5.13.** (**a**) Exchange of two reggeons. (**b**) Exchange of $n$ reggeons.

Unfortunately, in the case of Regge cuts, we do not get much insight from potential scattering models, since they admit no branch points if the potentials are well behaved. In a scalar field theory, the simplest diagram that might be expected to give rise to a Regge cut is a double ladder like the one in Fig. 5.14a (Amati, Fubini and Stanghellini 1962). However it turns out that the two-ladder amplitude is

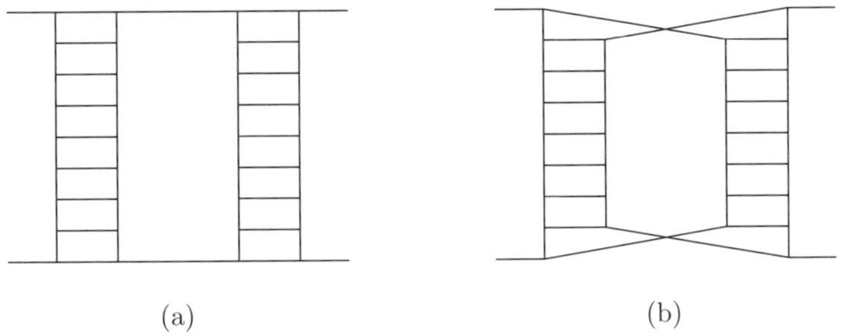


**Fig. 5.14.** (**a**) Two-ladder diagram. (**b**) Double-cross diagram (Mandelstam diagram) giving rise to a Regge cut.

$$A(s,t) \underset{s\to\infty}{\sim} s^{-3} \ln s \,, \tag{5.127}$$

independent of the number of rungs (the technical explanation of this behavior can be found in Eden et al. 1966).

So there is no way to obtain a Regge cut by summing this kind of *planar* diagrams.

In order to get a branch point singularity one must consider a slightly complicated diagram, the double-cross graph of Fig. 5.14b (Mandelstam 1963). If we sum over all possible numbers of rungs we find (see, e.g. Collins 1977)

$$A(s,t) \underset{s\to\infty}{\sim} \frac{s^{\alpha_c(t)}}{\ln s} \,, \tag{5.128}$$

with

$$\alpha_c(t) = 2\,\alpha(t/4) - 1 \,, \tag{5.129}$$

where $\alpha(t)$ is the trajectory corresponding to the ladder. In the $\ell$-plane this amplitude exhibits a cut at $\ell = \alpha_c(t)$. In particular, $A(\ell,t) \sim \ln(\ell - \alpha_c)$.

A convenient theoretical tool to calculate multi-reggeon diagrams is the reggeon calculus invented by Gribov (1968). This method consists of using, in Feynman diagrams, reggeon lines with propagators $\eta(k^2)\, s^{\alpha(k^2)}$. The two–reggeon diagram of Fig. 5.13a, for instance, is computed as follows (see, e.g., Baker and Ter-Martirosyan 1976; Collins 1977, Sect. 8.3; Kaidalov 1979). Its amplitude is obtained by applying ordinary Feynman rules and reads (the superscript is the number of reggeons exchanged)

$$\begin{aligned} A^{(2)}(s,t) = \frac{-\mathrm{i}}{2!} \int \frac{\mathrm{d}^4 k}{(2\pi)^4}\, \eta(k^2)\, \eta((q-k)^2)\, s^{\alpha(k^2)+\alpha((q-k)^2)} \\ \times T(p_1,k,q)\, T(p_2,k,q) \,, \end{aligned} \tag{5.130}$$

where $t \equiv q^2$, and $T(p_1,k,q)$ and $T(p_2,k,q)$ are the reggeon-particle scattering amplitudes representing the upper and lower blobs in Fig. 5.13a. Each of these amplitudes is the sum of many terms, some of which are shown in Fig. 5.15: single-particle intermediate state, multiparticle intermediate states, triple-Regge diagram, etc.

Let us introduce the c.m. energies of the $I\!R$-particle scattering

$$s_1 = (p_1 + k)^2 \,, \quad s_2 = (p_2 - k)^2 \,, \tag{5.131}$$

in terms of which, using

$$\int \mathrm{d}^4 k = \frac{1}{2s} \int \mathrm{d}^2 \boldsymbol{k}_\perp \, \mathrm{d}s_1 \, \mathrm{d}s_2 \,, \tag{5.132}$$

Eq. 5.130 becomes

$$\begin{aligned} A^{(2)}(s,t) = \frac{-\mathrm{i}}{2!\,2s} \frac{1}{(2\pi)^2} \int \frac{\mathrm{d}^2 \boldsymbol{k}_\perp}{(2\pi)^2} \mathrm{d}s_1\, \mathrm{d}s_2\, \eta(k^2)\, \eta((q-k)^2)\, s^{\alpha(k^2)+\alpha((q-k)^2)} \\ \times T(s_1,t,k^2,(q-k)^2)\, T(s_2,t,k^2,(q-k)^2) \,, \end{aligned} \tag{5.133}$$

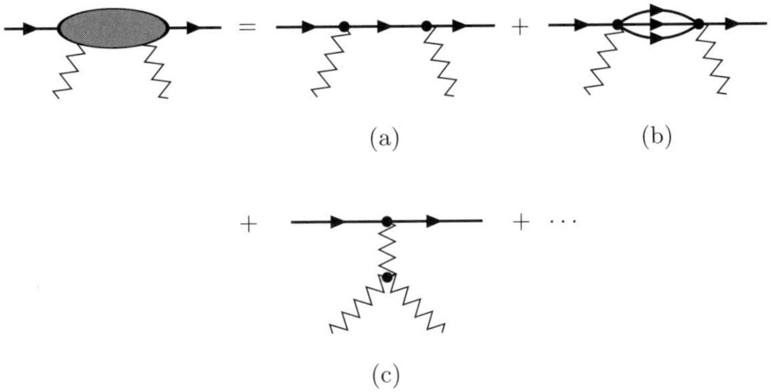


**Fig. 5.15.** The reggeon-particle scattering amplitude as a sum of many diagrams: single-particle rescattering (**a**), multiparticle rescattering (**b**), triple-Regge contribution (**c**), etc.

The integral of $T(s_1, t, k^2, (q-k)^2)$ over $s_1$ is along the contour $C$ shown in Fig. 5.16a. If the amplitude $T$ decreases sufficiently fast at large $s_1$ we can deform the contour $C$ into $C'$ (Fig. 5.16b) and the $s_1$ integral gives

$$\begin{aligned}\int_C \mathrm{d}s_1\, T(s_1, t, k^2, (q-k)^2) &= \int_{C'} \mathrm{d}s_1\, T(s_1, t, k^2, (q-k)^2) \\ &= 2\mathrm{i} \int_0^\infty \mathrm{d}s_1\, \mathrm{Im}\, T(s_1, t, k^2, (q-k)^2) \,. \end{aligned} \tag{5.134}$$

The integration over $s_2$ is analogous.

Thus, what we need in order to compute (5.133) is only the discontinuity of the amplitudes $T$ in the $s_1(s_2)$ plane. In other terms, the intermediate states in the $I\!R$-particle scattering are to be taken on shell. Considering only elastic rescattering in the $s$-channel (Fig. 5.15a) we have

$$\mathrm{Im}\, T(s_1, t, k^2, (q-k)^2) = \frac{1}{2}\, g(k^2)\, g((q-k)^2)\, 2\pi\, \delta(s_1 - m^2) \,, \tag{5.135}$$

so that

$$\int \mathrm{d}s_1\, T(s_1, t, k^2, (q-k)^2) = 2\pi \mathrm{i}\, g(k^2)\, g((q-k)^2) \,, \tag{5.136}$$

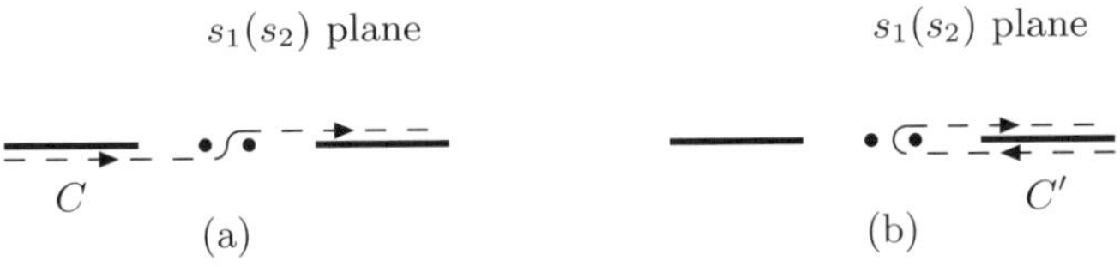


**Fig. 5.16.** (**a**) Integration contour $C$ in the $s_1(s_2)$ plane. (**b**) The deformed contour $C'$.

where $g$ is the $I\!R$-particle coupling. The on-shellness of the intermediate particles imply, in the large-$s$ limit, that the exchanged momenta are predominantly transverse, that is

$$k^2 \simeq -\boldsymbol{k}_\perp^2 , \quad (q-k)^2 \simeq -(\boldsymbol{q}_\perp - \boldsymbol{k}_\perp)^2 . \tag{5.137}$$

Putting everything together, the amplitude for two-reggeon exchange, in the hypothesis of elastic rescattering (Fig. 5.17a) becomes

$$A^{(2)}(s,t) = \frac{\mathrm{i}}{2!\,2s} \int \frac{\mathrm{d}^2\boldsymbol{k}_\perp}{(2\pi)^2}\, \beta(\boldsymbol{k}_\perp^2)\, \beta((\boldsymbol{q}_\perp - \boldsymbol{k}_\perp)^2) \\ \times \eta(\boldsymbol{k}_\perp^2)\, \eta((\boldsymbol{q}_\perp - \boldsymbol{k}_\perp)^2)\, s^{\alpha(\boldsymbol{k}_\perp^2)+\alpha((\boldsymbol{q}_\perp-\boldsymbol{k}_\perp)^2)} , \tag{5.138}$$

where $\beta(\boldsymbol{k}_\perp^2) = g^2(\boldsymbol{k}_\perp^2)$ is the residue function. Note that (5.139) corresponds to

$$A^{(2)}(s,t) = \frac{\mathrm{i}}{2!\,2s} \int \frac{\mathrm{d}^2\boldsymbol{k}_\perp}{(2\pi)^2}\, A^{(1)}(s,\boldsymbol{k}_\perp^2)\, A^{(1)}((\boldsymbol{q}_\perp - \boldsymbol{k}_\perp)^2) , \tag{5.139}$$

where $A^{(1)}$ is the Regge pole amplitude. Recalling that, for small $|t|$ and linear trajectories, this can be put in the form

$$A^{(1)}(s,\boldsymbol{k}_\perp^2) = \beta(0)\,\eta(0)\,s^{\alpha(0)}\,\mathrm{e}^{-\Lambda(s)\,\boldsymbol{k}_\perp^2} , \tag{5.140}$$

with $\Lambda(s) = \frac{B_0}{2} + \alpha' (\ln s - \mathrm{i}\pi/2)$, the integration over the transverse momentum in (5.139) is easily performed and gives

$$A^{(2)}(s,t) = \frac{\mathrm{i}}{16\pi}\,\beta^2(0)\,\eta^2(0)\,\frac{s^{2\alpha(0)-1}}{2\,\Lambda(s)}\,\mathrm{e}^{\Lambda(s)\,t/2} . \tag{5.141}$$

Asymptotically this behaves as

$$A^{(2)}(s,t) \underset{s\to\infty}{\sim} \frac{s^{\alpha_c(t)}}{\ln s} , \tag{5.142}$$

which shows the existence of a cut at

$$\alpha_c(t) = 2\,\alpha(0) - 1 + \frac{\alpha'}{2}\,t , \tag{5.143}$$

in agreement with (5.129).

Let us focus on the two–pomeron exchange. In this important case one finds that the cut amplitude has an opposite phase with respect to the pomeron pole amplitude. This can be seen from (5.141) recalling that $\eta_{I\!P}(0) = \mathrm{i}$. An important consequence of this is that the contribution of the two-pomeron cut to the total cross section is negative. Using the optical theorem and taking into account the $I\!P$ pole and the $I\!P \otimes I\!P$ cut, we get in fact

$$\sigma_{\mathrm{tot}} \sim A\,s^{\alpha_{I\!P}(0)-1} - B\,\frac{s^{2(\alpha_{I\!P}(0)-1)}}{\ln s} , \tag{5.144}$$

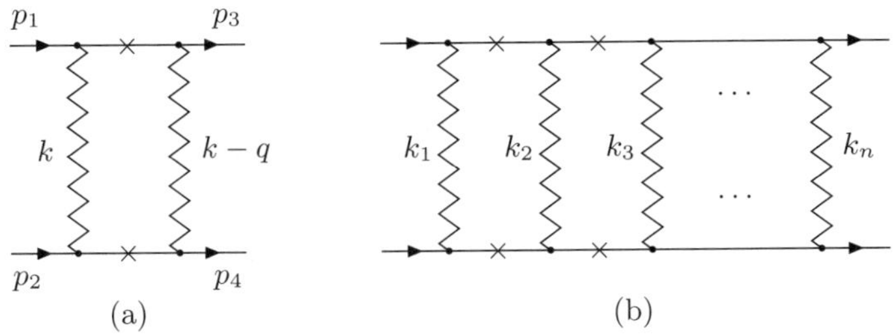


**Fig. 5.17.** Two–reggeon (**a**) and multireggeon (**b**) exchange with elastic rescattering in the $s$-channel.

where $A$ and $B$ are positive constants. A recent fit of total cross sections (Cudell, Kang and Kim 1996) with a form of the type (5.144) (plus other subdominant contributions) has shown that the two–pomeron cut contribution is rather small and has the effect of raising the extracted pomeron pole intercept.

If $n$ identical reggeons are exchanged (Fig. 5.17b), proceeding as done above for the two–reggeon case, one finds (again, in the case of elastic rescatterings)

$$A^{(n)}(s,t) = \frac{\mathrm{i}^{n-1}}{n!}\,\frac{1}{(2s)^{n-1}}\,(2\pi)^2 \int \frac{\mathrm{d}^2\boldsymbol{k}_{1\perp}}{(2\pi)^2} \cdots \int \frac{\mathrm{d}^2\boldsymbol{k}_{n\perp}}{(2\pi)^2} \times A^{(1)}(s,\boldsymbol{k}_{1\perp}^2)\,\ldots\,A^{(1)}(s,\boldsymbol{k}_{n\perp}^2)\,\delta^2\left(\boldsymbol{q}_\perp - \sum_{i=1}^{n} \boldsymbol{k}_{i\perp}\right) . \quad (5.145)$$

which gives (omitting irrelevant constants)

$$A^{(n)}(s,t) \sim \mathrm{i}^{n-1}\,\beta^n(0)\,\eta^n(0)\,\frac{s^{n(\alpha(0)-1)+1}}{\Lambda^{n-1}(s)}\,\mathrm{e}^{\Lambda(s)\,t/n} . \quad (5.146)$$

Asymptotically this amplitude behaves as

$$A(s,t) \underset{s\to\infty}{\sim} \frac{s^{\alpha_c(t)}}{\ln^{n-1} s} , \quad (5.147)$$

with

$$\alpha_c(t) = n\,(\alpha(0)-1)+1+\frac{\alpha'}{n}\,t . \quad (5.148)$$

More generally, the position of the cut is

$$\alpha_c(t) = n\,\alpha(t/n^2) - n + 1 . \quad (5.149)$$

Let us now discuss the $t$-dependence of the branch point singularities, focusing on pomeron cuts. The exchange of $n$ pomerons, with $\alpha_{I\!P} = 1+\alpha'_{I\!P}\,t$, gives a branch point at

$$\alpha_c(t) = 1 + \frac{\alpha'_{I\!P}}{n}\, t\,. \tag{5.150}$$

All these cuts coincide at $t = 0$, and become flatter as $n$ increases. Thus, higher order cuts lie above lower order cuts in the region $t < 0$. This means that, while the pomeron pole dominates at small $|t|$, at larger $|t|$ the two-pomeron cut, with $\alpha_c(t) = 1 - \alpha'_{I\!P}\,|t|/2$, becomes important and the destructive interference between the two terms, which have opposite phases, leads to a depletion of the cross section. The observed dip in the $pp$ elastic cross section can be attributed to this effect. The fact that, as the number of exchanged reggeons grows, the trajectory of the resulting cut gets flatter, means that, at large $|t|$, scattering processes are governed by the exchange of many reggeons, and hence their description in Regge theory becomes extremely complicate. On the other hand, when $|t|$ becomes large, one enters the realm of perturbative QCD and hence predictions are still possible in a different (and more fundamental) theoretical framework.

In Sect. 5.9 we showed that the unitarity cut of a Regge pole diagram corresponds to multiperipheral production (Fig. 5.12). Let us consider now the unitarity cuts of a two-reggeon exchange diagram. There are three possible cuttings: *i)* cutting between the reggeon lines corresponds to double (diffractive) dissociation (Fig. 5.18a and cross section $\sigma_a$); it ii) cutting one of the reggeons corresponds to absorption corrections to multiperipheral production (Fig. 5.18b and cross section $\sigma_b$); *iii)* cutting simultaneously both reggeons corresponds to the production of two multiperipheral showers (Fig. 5.18c and cross section $\sigma_c$).

Using Reggeon calculus, Abramovsky, Gribov and Kanchely (1974) found an important relation between the contributions of the three cuttings to the total cross section. For two-pomeron exchange the AGK cutting rules state

$$\sigma_a = \sigma_{DD} = -\sigma_{\rm tot}^{I\!P\otimes I\!P}\,, \quad \sigma_b = 4\,\sigma_{\rm tot}^{I\!P\otimes I\!P}\,, \quad \sigma_c = -2\,\sigma_{\rm tot}^{I\!P\otimes I\!P}\,, \tag{5.151}$$

where $\sigma_{DD}$ is the double-diffractive cross section and $\sigma_{\rm tot}^{I\!P\otimes I\!P}$ is the two-pomeron contribution to the total cross section. Note that the latter is a

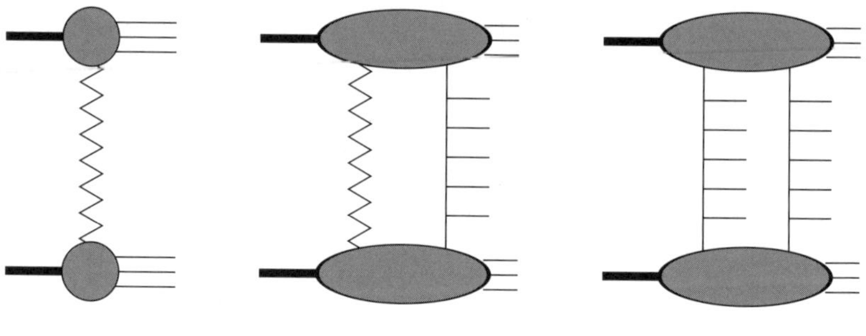

(a) (b) (c)

**Fig. 5.18a–c.** The unitarity cuts of a two-reggeon diagram.

negative quantity, so that $\sigma_a$ and $\sigma_c$, which measure physical processes, are positive.

The AGK cutting rules can be extended to the case of $n$-pomeron exchange. Thus, they establish a connection between Reggeon calculus (and, more generally, a field theory like QCD provided a factorization theorem holds) and $s$-channel unitarity.

# 6. Geometrical and $s$-Channel Models

The complex angular momentum, or Regge, approach teaches us that the exchange of crossed (i.e. $t$-) channel reggeons (singularities in the complex angular momentum) determines the asymptotic behavior in the direct (i.e. $s$-) channel. In spite of the complications already discussed (such as the appearance of cuts induced by unitarity), and of the persistent ambiguities (such as the nature of the leading singularity, the pomeron, which will be the main subject of Chap. 8), the above mentioned property makes the Regge-pole approach the prototype and the most powerful of the so-called $t$-channel models.

By contrast, the optical or eikonal approaches (of which we have briefly discussed an example, the eikonal approximation, in Sect. 2.3, starting from the non-relativistic Schrödinger equation) belongs to the class of the so-called $s$-channel models.

Both approaches are variously used in phenomenological descriptions of the data and both have merits and shortcomings. It is quite puzzling that in each channel it is the dependence on the complementary variable which appears somewhat constrained. In $t$-channel models the form of the energy dependence is almost unambiguous (Regge behaved) and the conservation of $t$-channel unitarity can be enforced in various ways, but the $t$-dependence of the residue functions is not prescribed by the theory. In $s$-channel models the $t$-dependence is strongly suggested and non-violation of unitarity in the $s$-channel can be guaranteed, but the energy dependence is arbitrary and must be inferred from some educated guess. No one, so far – and to the best of our knowledge –, has been able to combine in a satisfactory way and unify in a single form the various qualities of the different approaches.

To the extent that relativistic and crossing invariances are properties of the $S$-matrix (Chap. 4), the real theory must be one in which no channel is privileged. Although many attempts have been made to construct a model where no specific channel is chosen, none has been successful in the sense of being also realistic. The only attempt which guarantees crossing symmetry and the proper asymptotic behavior is the Veneziano model (Veneziano, 1968) whose unitarity problems appear, alas, incurable.

In recent years the $s$-channel approach has regained some popularity thanks to the work of Nikolaev and Zakharov (1991b, 1992, 1994b) and

Mueller (1994), who have shown that hard diffractive physics turns out to be more transparent when expressed in the impact parameter representation (see Sects. 9.9, 9.10, 11.2, 11.10).

In what follows, we shall provide a sketchy account of the traditional $s$-channel models. Reviews on this subject are Halzen (1993) and Matthiae (1994).

## 6.1 The Eikonal Picture

The prototype of $s$-channel models is the eikonal picture (see, for instance, Fried 1990, Wu 2000). We already summarized the eikonal representation and its derivations in Sect. 2.3. Here we reconsider it in the context of high-energy particle diffraction.

In the limit of large $s$ and small scattering angles the partial wave expansion (4.34) for the relativistic amplitude $A(s,t)$ can be rewritten as an integral over the impact parameter $b$, as we did in Sect. 2.3 for the quantum mechanical amplitude $f(k,\vartheta)$ (azimuthal symmetry will be assumed hereafter)

$$\begin{aligned} A(s,t) &= \frac{s}{4\pi}\int \mathrm{d}^2\boldsymbol{b}\, \mathrm{e}^{-i\boldsymbol{q}\cdot\boldsymbol{b}}\, A(s,b) \\ &= 2\mathrm{i}s\int \mathrm{d}^2\boldsymbol{b}\, \mathrm{e}^{-i\boldsymbol{q}\cdot\boldsymbol{b}}\, \Gamma(s,b) \\ &= 4\pi\mathrm{i}s\int \mathrm{d}b\, b\, J_0(qb)\, \Gamma(s,b)\;, \end{aligned} \tag{6.1}$$

where $t \simeq -\boldsymbol{q}^2$ and $A(s,b)$ is the correspondent of $A_\ell(s)$ in the impact-parameter space. The profile function is related to the phase shifts $\delta(s,b)$ in the $b$-space by

$$\Gamma(s,b) = -\frac{\mathrm{i}}{8\pi}\, A(s,b) = 1 - \mathrm{e}^{2\,\mathrm{i}\,\delta(s,b)}\;. \tag{6.2}$$

The definitions of the cross sections in terms of the profile function are essentially the same as given earlier (Eqs. (2.140)–(2.142)). The integrated elastic cross section reads

$$\sigma_{\mathrm{el}} = \int \mathrm{d}^2\boldsymbol{b}\, |\Gamma(s,b)|^2\;. \tag{6.3}$$

From the optical theorem we get the total cross section

$$\sigma_{\mathrm{tot}} = 2\int \mathrm{d}^2\boldsymbol{b}\, \mathrm{Re}\,\Gamma(s,b)\;. \tag{6.4}$$

We write the inelastic cross section as

$$\sigma_{\mathrm{in}} = \int \mathrm{d}^2\boldsymbol{b}\, G_{\mathrm{in}}(s,b)\;, \tag{6.5}$$

where $G_{\text{in}}$ is known as the *shadow profile function* or *inelastic overlap function* (Van Hove 1964), and represents the probability of absorption associated to each value of the impact parameter $b$. From the unitarity relation

$$2\,\mathrm{Re}\,\Gamma(s,b) - |\Gamma(s,b)|^2 = G_{\text{in}}(s,b)\;, \tag{6.6}$$

which is equivalent to $\sigma_{\text{tot}} = \sigma_{\text{el}} + \sigma_{\text{in}}$, we get $\sigma_{\text{in}}$ in terms of $\Gamma(s,b)$

$$\sigma_{\text{in}} = \int \mathrm{d}^2\boldsymbol{b}\,\left[2\,\mathrm{Re}\,\Gamma(s,b) - |\Gamma(s,b)|^2\right]\;. \tag{6.7}$$

It is customary to introduce the *opacity* $\Omega(s,b)$ defined as

$$\Omega(s,b) = -2\mathrm{i}\delta(s,b)\;, \tag{6.8}$$

so that the profile function is written as

$$\Gamma(s,b) = 1 - \mathrm{e}^{-\Omega(s,b)}\;, \tag{6.9}$$

and the scattering amplitude reads

$$A(s,t) = 2\mathrm{i}s \int \mathrm{d}^2\boldsymbol{b}\;\mathrm{e}^{-i\boldsymbol{q}\cdot\boldsymbol{b}}\left[1 - \mathrm{e}^{-\Omega(s,b)}\right]\;. \tag{6.10}$$

In terms of the opacity the cross sections (6.3, 6.4, 6.7) read

$$\sigma_{\text{el}} = \int \mathrm{d}^2\boldsymbol{b}\,|1 - \mathrm{e}^{-\Omega(s,b)}|^2\;, \tag{6.11}$$

$$\sigma_{\text{tot}} = 2\int \mathrm{d}^2\boldsymbol{b}\,\mathrm{Re}\left[1 - \mathrm{e}^{-\Omega(s,b)}\right]\;, \tag{6.12}$$

$$\sigma_{\text{in}} = \int \mathrm{d}^2\boldsymbol{b}\,\left[1 - |\mathrm{e}^{-\Omega(s,b)}|^2\right]\;. \tag{6.13}$$

The limit $\Omega \to \infty$ corresponds to absolute blackness. In this case, setting $\Gamma = 1$ for $b \leq R$ (black disk) one gets the relation

$$\sigma_{\text{el}} = \sigma_{\text{in}} = \frac{1}{2}\sigma_{\text{tot}} = \pi R^2 \qquad \text{(black disk)}\;. \tag{6.14}$$

Note that if $\Omega$ is real the amplitude is purely imaginary.

Let us come now to the consequences of unitarity. This demands

$$\mathrm{Im}\,\delta(s,b) \geq 0\;, \tag{6.15}$$

namely

$$\mathrm{Re}\,\Omega(s,b) \geq 0\;, \qquad G_{\text{in}}(s,b) \leq 1\;. \tag{6.16}$$

The relation between ($s$-channel) unitarity and eikonalization has been studied by several authors but has never been clarified in a completely satisfactory way. In the impossibility to quote all authors, we refer the interested reader to a recent paper (Desgrolard et al. 2000) where, besides some new results, quite a large coverage of the previous literature is provided.

Concerning the angular (or $t$) distribution, we recall that the $t$-variation of peripheral collisions is concentrated at $|t| = 0$ (the limiting case $\Gamma(s,b) \propto \delta(b-R)$ gives $J_0(R\sqrt{-t})$). A central collision will give either a $J_1(R\sqrt{-t})$ (in this case $\Gamma(s,b)$ has a sharp edge $\propto \theta(R-b)$) or an exponential variation $e^{tR^2/4}$ (corresponding to a Gaussian $b$-distribution $e^{-b^2/R^2}$). These qualitative behaviors are illustrated in Fig. 6.1. Some of these examples were discussed earlier in Sect. 2.1.4.

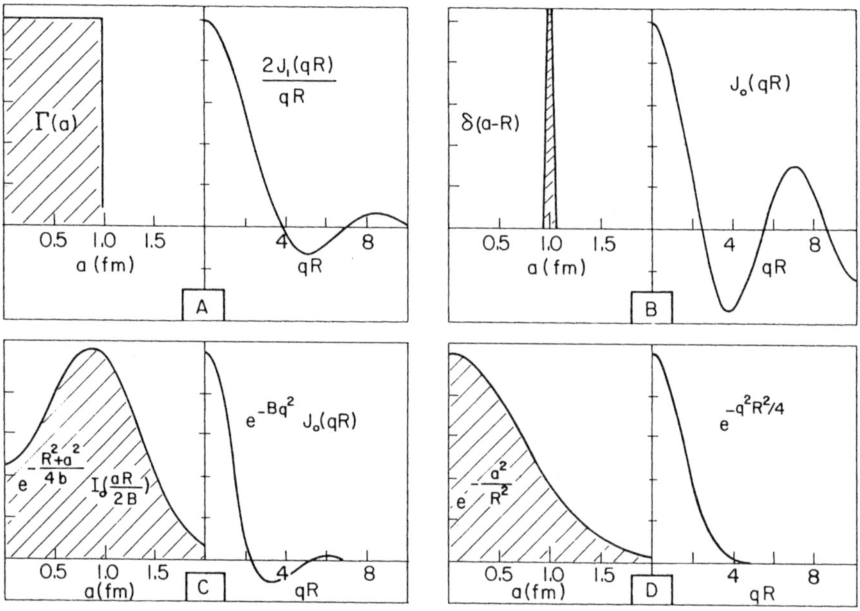


**Fig. 6.1.** Various examples of $t$-distributions (*right*) corresponding to different profile functions (*left*). Figure from Amaldi, Jacob and Matthiae (1976).

## 6.2 A Model for the Eikonal Amplitude

A field-theoretical basis for the eikonal amplitude (6.10) is provided, to some extent, by the model that we are now going to describe (see, e.g., Wu 2000). Consider two massive fermions (the colliding particles) interacting via meson-exchange. With the following Feynman rules fermion   $\frac{\mathrm{i}}{\not{p}-m+\mathrm{i}\epsilon}$

meson   $\frac{-\mathrm{i}g^{\mu\nu}}{p^2-\mu^2+\mathrm{i}\epsilon}$

vertex 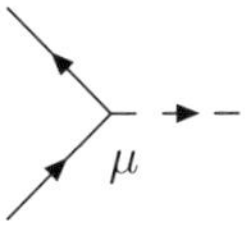 $-\mathrm{i}g\gamma^{\mu}$

the amplitude for one-meson exchange (Fig. 6.2a) in the high-energy limit ($s \to \infty$), with $t$ fixed, is

$$A^{(1)}(s,t) \sim \frac{2g^2 s}{t-\mu^2} \,. \tag{6.17}$$

We can rewrite this expression as

$$A^{(1)}(s,t) \sim -2s \int \mathrm{d}^2\boldsymbol{b}\, \mathrm{e}^{-\mathrm{i}\boldsymbol{q}\cdot\boldsymbol{b}} \left[\frac{g^2\, K_0(\mu b)}{2\pi}\right] , \tag{6.18}$$

where $K_0$ is the Neumann function. At order $g^4$ (two-meson exchange, Fig. 6.2b), one easily finds asymptotically ($s \to \infty$, with fixed $t$)

$$A^{(2)}(s,t) \sim \mathrm{i}sg^4 \int \frac{\mathrm{d}^2\boldsymbol{k}}{(2\pi)^2} \frac{1}{(\boldsymbol{k}^2+\mu^2)[(\boldsymbol{q}-\boldsymbol{k})^2+\mu^2]} , \tag{6.19}$$

which becomes, in the impact-parameter representation

$$A^{(2)}(s,t) \sim 2\mathrm{i}s \int \mathrm{d}^2\boldsymbol{b}\, \mathrm{e}^{-\mathrm{i}\boldsymbol{q}\cdot\boldsymbol{b}} \frac{1}{2} \left[\frac{g^2 K_0(\mu b)}{2\pi}\right]^2 . \tag{6.20}$$

One can see that the bracketed term in the $\mathcal{O}(g^2)$ amplitude $A^{(1)}(s,t)$, Eq. (6.18), is squared in the $\mathcal{O}(g^4)$ amplitude $A^{(2)}(s,t)$, Eq. (6.20). This trend continues at higher orders, and the asymptotic amplitude for the exchange of $n$ mesons (Fig. 6.2c) is

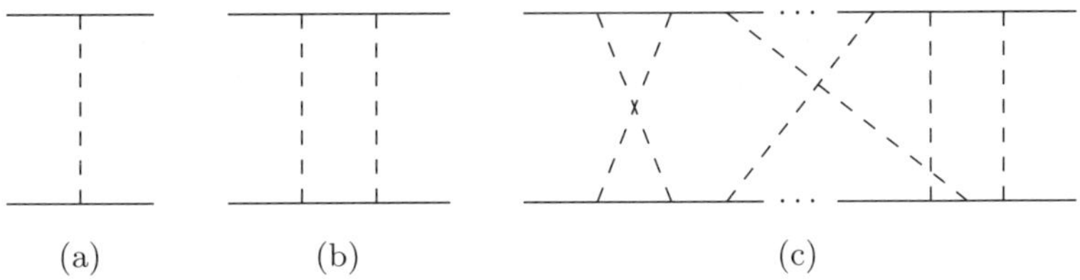


**Fig. 6.2.** Fermion–fermion scattering via meson exchange. (**a**) one-meson exchange; (**b**) two-meson exchange; (**c**) multi-meson exchange.

$$A^{(n)}(s,t) \sim -2\mathrm{i}s \int \mathrm{d}^2\boldsymbol{b}\, \mathrm{e}^{-\mathrm{i}\boldsymbol{q}\cdot\boldsymbol{b}}\, \frac{1}{n!} \left[\frac{-\mathrm{i}g^2 K_0(\mu b)}{2\pi}\right]^n . \tag{6.21}$$

Summing over $n$ gives the multi-meson exchange amplitude for fermion-fermion scattering as

$$\begin{aligned} A(s,t) &= \sum_{n=1}^{\infty} A^{(n)}(s,t) \\ &\sim 2\mathrm{i}s \int \mathrm{d}^2\boldsymbol{b}\, \mathrm{e}^{-\mathrm{i}\boldsymbol{q}\cdot\boldsymbol{b}} \left\{1 - \exp\left[-\mathrm{i}\,\frac{g^2\, K_0(\mu b)}{2\pi}\right]\right\} , \end{aligned} \tag{6.22}$$

valid for $s \to \infty$ and fixed $t$.

Note the similarity between (6.22) and the eikonal amplitude (6.10). There is an important difference, however, between these two expressions. While the exponent in (6.22) is purely imaginary, the opacity $\Omega(s,b)$ in (6.10) is in general complex. A non-zero real part of $\Omega(s,b)$, which must be positive (see (6.16)), means attenuation. Because of unitarity, the attenuation can only come from particle production diagrams, like the one of Fig. 6.3, which were not taken into account in the eikonal series. Another feature of (6.22) is that it leads, via optical theorem, to a total cross section constant with energy. The rise of $\sigma_{\mathrm{tot}}$ with $s$ is indeed a consequence of particle production intermediate processes in high-energy scattering (Cheng and Wu 1970a; Frolov, Gribov and Lipatov 1970a).

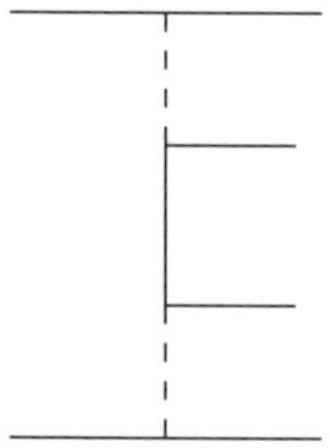

**Fig. 6.3.** A particle production diagram not included in the eikonal series.

## 6.3 The Structure of Hadrons in the Impact-Parameter Space

In a very phenomenological way the data in the small-$t$ range can be represented by an exponential

$$A(s,t) \sim \mathrm{i}\, s\, \sigma_{\mathrm{tot}}\, \mathrm{e}^{B(s)\, t/2} , \tag{6.23}$$

which, transformed to the $b$-representation (6.1) gives, as a profile function, a Gaussian[1]

$$\Gamma(s,b) = \frac{\sigma_{\text{tot}}}{4\pi\, B(s)} \exp\left[-\frac{b^2}{2\,B(s)}\right] , \tag{6.24}$$

corresponding to an interaction radius $R(s) \sim \sqrt{2B(s)}$.

The analysis of the ISR data at $\sqrt{s} = 53$ GeV (see Fig. 6.4) shows that $R$ is somewhat smaller than 1 fm and the shadow profile function reaches about 94% of the black disk limit at $b = 0$.

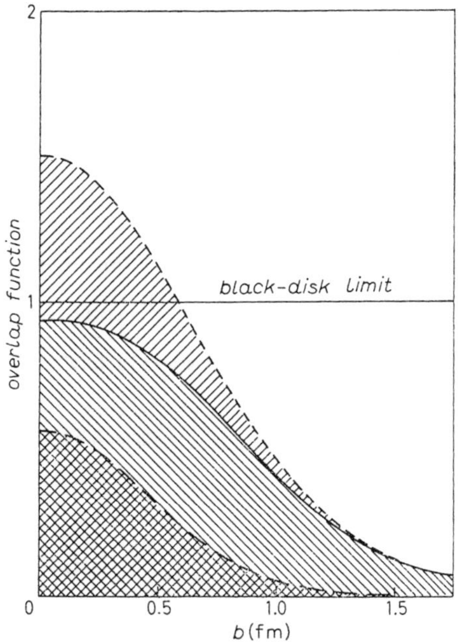


**Fig. 6.4.** The impact parameter structure at $\sqrt{s} = 53$ GeV. From bottom to top: the elastic contribution $|\Gamma(s,b)|^2$, the inelastic contribution $G_{\text{in}}(s,b)$, elastic + inelastic $2\,\text{Re}\,\Gamma(s,b)$. From Predazzi (1976).

Other analyses (Amaldi and Schubert 1980, Henzi and Valin 1983, 1985) lead to the conclusion that both the interaction radius and the central opacity grow mildly with energy. In particular, $G_{\text{in}}(s,b)$ grows when going from ISR to SPS energies (see Fig. 6.5): the proton becomes (slightly) bigger and blacker. Another interesting feature is the rapid increase of $G_{\text{in}}$ at the periphery (around $b \sim 1$ fm) (see inset of Fig. 6.5). A related effect is present in diffractive inclusive data (Kopeliovich, Povh and Predazzi 1997). An extensive eikonal-type study of $pp$ and $p\bar{p}$ cross sections has been performed

[1] Analyticity teaches us that as $b \to \infty$ a Gaussian decrease is not allowed but we shall not worry about this here.

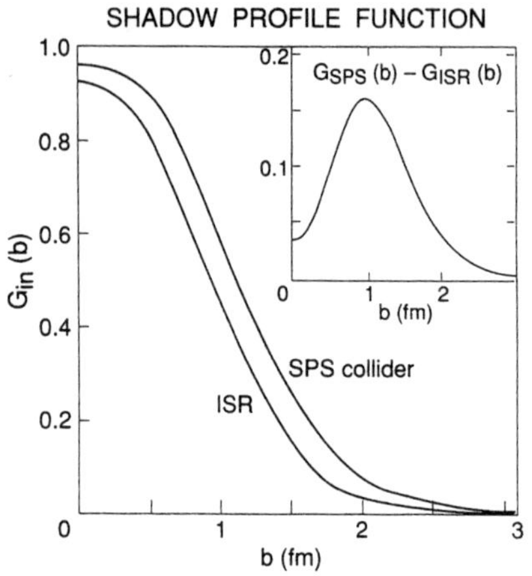


**Fig. 6.5.** Variation of the shadow profile function at the center and at the periphery, going from $\sqrt{s} = 53$ GeV (ISR) to $\sqrt{s} = 546$ GeV (SPS). From Matthiae (1994).

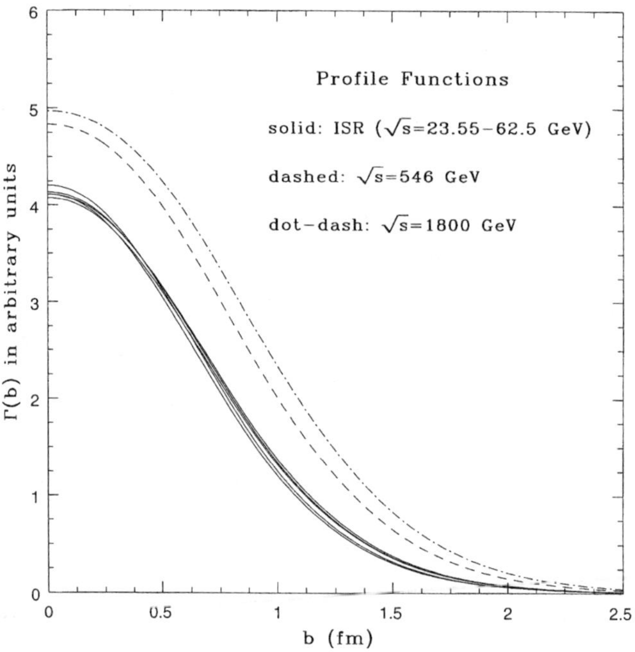


**Fig. 6.6.** Profile functions obtained from an analysis of various $pp$ and $p\bar{p}$ elastic cross section data (adapted from Ferreira and Pereira 1999).

by Ferreira and Pereira (1999). The resulting profile functions at different energies are plotted in Fig. 6.6.

We now proceed to a brief review of some models (for a wider and more detailed survey the reader may consult Matthiae 1994).

## 6.4 s-Channel Models

### 6.4.1 The Geometrical Model

Historically, the first approach to take to the extreme consequences the idea of diffraction as outlined in Sect. 2.1 was the *geometrical model* proposed by Yang and collaborators (Wu and Yang 1965, Byers and Yang 1966, Chou and Yang 1968, 1970). The starting point is basically the remark that high-energy scattering is the shadow of absorption. Accordingly, the interacting hadrons are viewed as extended objects made of some hadronic matter (partons, we would say nowadays) flying through each other (Fig. 6.7). At each point, the interaction is proportional to the local density of hadronic matter which is assumed to have the same shape as the electric charge distribution. The opacity, which is taken to be real, so that the amplitude is purely imaginary, is factorized as

$$\Omega(s,b) = K(s)\,D(b)\,, \tag{6.25}$$

where $D(b)$ is related to the electric form factors of the colliding particles, while the energy-dependent quantity $K(s)$ is a free parameter of the model, which is fitted to the $\sigma_{\rm tot}$ data. In detail, $D(b)$ is obtained as follows ($A$ and $B$ are the two hadrons). We start from

$$D(\boldsymbol{b}) = \int \mathrm{d}^2\boldsymbol{b}'\, T_A(\boldsymbol{b}-\boldsymbol{b}')\, T_B(\boldsymbol{b}')\,, \tag{6.26}$$

where $T(\boldsymbol{b})$ is related to the charge density $\rho(\boldsymbol{b},z)$ of the hadron by $T(\boldsymbol{b}) = \int_{-\infty}^{+\infty} \mathrm{d}z\, \rho(\boldsymbol{b},z)$. Introducing the form factors

$$G_{A,B}(\boldsymbol{q}^2) = \int \mathrm{d}^2\boldsymbol{b}\, \mathrm{e}^{-\mathrm{i}\boldsymbol{q}\cdot\boldsymbol{b}}\, T_{A,B}(\boldsymbol{b})\,, \tag{6.27}$$

$D(\boldsymbol{b})$, which depends only on $b \equiv |\boldsymbol{b}|$, turns out to be given by

$$D(b) = \int \frac{\mathrm{d}^2\boldsymbol{q}}{(2\pi)^2}\, \mathrm{e}^{\mathrm{i}\boldsymbol{q}\cdot\boldsymbol{b}}\, G_A(\boldsymbol{q}^2) G_B(\boldsymbol{q}^2)\,. \tag{6.28}$$

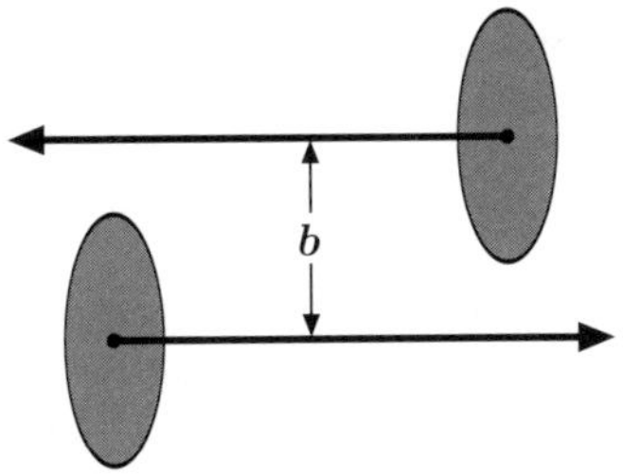


**Fig. 6.7.** Schematic representation of the collisions between two hadrons.

A clear indication of the geometrical picture (Chou and Yang 1968, Durand and Lipes 1968) is the appearance of a diffraction pattern in the elastic cross section, with a sharp minimum and a secondary maximum. The forward slope $B(s, t = 0)$, see (4.177), and the $|t|$ value of the dip are expected to be proportional to $\sigma_{\rm tot}$ and $1/\sigma_{\rm tot}$, respectively. All these predictions are indeed borne out by the data (see Chap. 7). The Chou-Yang model also predicts a link between $\sigma_{\rm tot}$, the ratio $\sigma_{\rm el}/\sigma_{\rm tot}$ and the value of $d\sigma/dt$ at the second maximum (Chou and Yang 1979), which seems to be supported, at least qualitatively, by the observations at the SPS Collider and at the Tevatron (Fig. 6.8). A basic question, which is left completely unanswered by the geometrical approach, is the $s$-dependence of the observables.

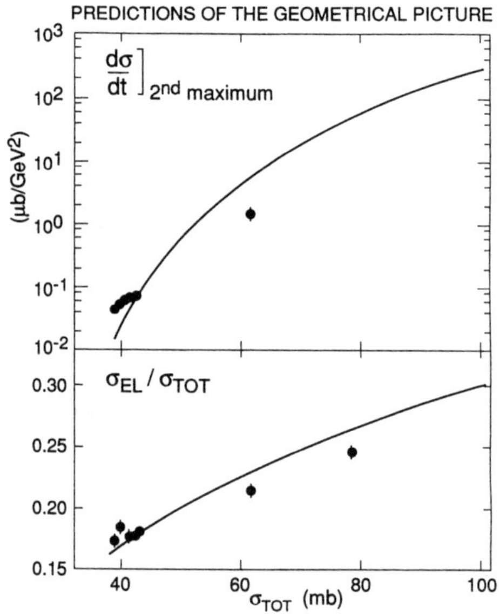


**Fig. 6.8.** Comparison with the data of some predictions of the geometrical model of Chou and Yang (1979). Figure from Matthiae (1994).

### 6.4.2 The Impact Picture

An attempt to incorporate an energy dependence derived from a perturbative field-theoretical calculation into the geometrical model was made by Bourrely, Soffer and Wu (1984, 1988, 1990). In their *impact picture* the opacity is still of the form (6.25), apart from an additive subleading term, and the function $K(s)$ is taken from the Cheng-Wu classical study of ladder graphs in QED (Cheng and Wu 1969, 1970a, 1970b), which gave the first theoretical indication of a growth with energy of total cross sections. Working in massive quantum electrodynamics, Cheng and Wu found that the asymptotic behavior of the scattering amplitude is

$$s^{1+\epsilon}(\ln s)^{-3/2}\,, \tag{6.29}$$

where $\epsilon$ is a positive quantity which depends on the coupling constant of the theory. Motivated by this result, Bourrely, Soffer and Wu use for $K(s)$ the crossing symmetric form

$$K(s) = \frac{s^a}{(\ln s)^b} + \frac{u^a}{(\ln u)^b}\,, \tag{6.30}$$

where $a$ and $b$ are constants. $D(b)$ is still given by (6.28). The impact picture predicts that asymptotically $\sigma_{\rm tot}$, $\sigma_{\rm el}$ and $B(s,t=0)$ should all increase as $\ln^2 s$, and that the ratio $\sigma_{\rm el}/\sigma_{\rm tot}$ should approach $1/2$. These predictions are in good agreement with the data (see Chap. 7). A schematic representation of the expanding proton in the impact picture is shown in Fig. 6.9: the proton core, almost completely absorbing (i.e. black), has a radius growing as $\ln s$, whereas the peripheral region, only partially absorbing (i.e. gray), has a width independent of $s$. A complete account of this theory can be found in Cheng and Wu (1987).

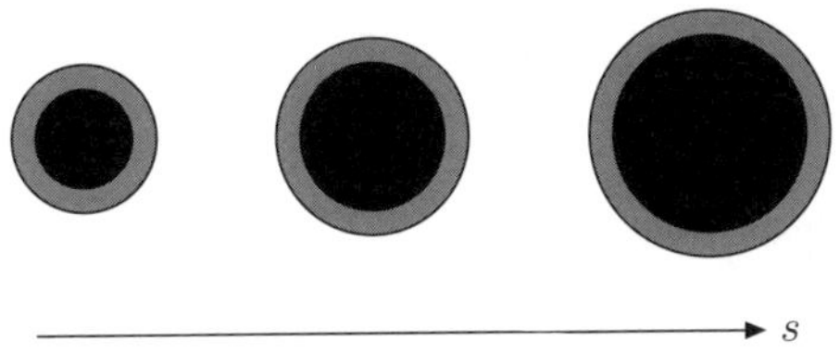


**Fig. 6.9.** Schematic representation of the expanding proton according to the impact picture.

### 6.4.3 Geometrical Scaling

Let us now focus on the conjecture that asymptotically $\sigma_{\rm el}$, $\sigma_{\rm tot}$ and the forward slope $B(s,t=0)$ grow like $\ln^2 s$,

$$\sigma_{\rm el} \sim \sigma_{\rm tot} \sim B(s,0) \sim R^2(s) \sim \ln^2 s\,. \tag{6.31}$$

As we have just seen, this is a prediction of the impact picture. We also obtain the behavior (6.31) if we assume that the Froissart-Martin bound is saturated, $\sigma_{\rm tot} \sim \ln^2 s$, and use two other general inequalities: a bound on the integrated elastic cross section (Eden 1966; Singh and Roy 1970)

$$\sigma_{\rm el} \geq {\rm const}\,\frac{\sigma_{\rm tot}^2}{\ln^2 s}\,, \qquad \text{as } s \to \infty\,, \tag{6.32}$$

and a bound on the slope $B(s,0)$ (MacDowell and Martin 1964)

$$B(s,0) \geq \frac{1}{18\pi} \frac{\sigma_{\rm tot}^2}{\sigma_{\rm el}} , \qquad \text{as } s \to \infty , \tag{6.33}$$

which is almost saturated already by the present data.

If (6.31) holds, an important scaling property (Auberson, Kinoshita and Martin 1971) states that the scattering amplitude becomes a function of the scaling variable $\tau = -t \ln^2 s \sim -t\, \sigma_{\rm tot}$, namely,

$$A(s,t) = A(s,0)\, \Theta(\tau) . \tag{6.34}$$

Such a behavior is at the basis of the so-called *geometrical scaling model* (Dias de Deus and Kroll 1978, 1983) which takes for the opacity the form $\Omega(s,b) = \Omega(b/R(s))$ with $R(s) \sim \ln s$.

This model enjoyed some popularity since one of its main predictions, the receding to smaller $|t|$ values of the dip in the amplitude with increasing energy, was indeed met by the ISR data (Fig. 6.10). The trouble with geometrical scaling, however, is that it predicts also a constant value for the ratio $\sigma_{\rm el}/\sigma_{\rm tot}$. As we shall see, the experimental evidence is that $\sigma_{\rm el}/\sigma_{\rm tot}$, while nearly constant around ISR energies, grows when we go to SPS and to Tevatron energies (see Fig. 7.7).

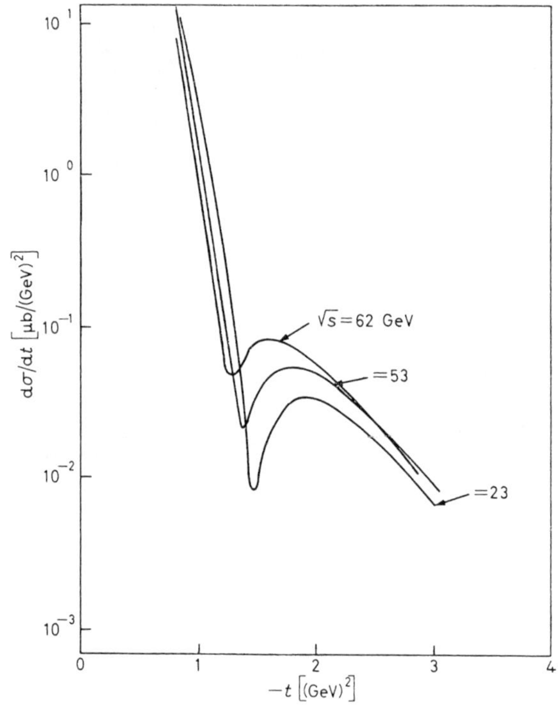


**Fig. 6.10.** Trend of the $t$-distributions of $pp$ elastic scattering at various ISR energies (from Predazzi 1976).

### 6.4.4 Other Models

Other applications of the $s$-channel approach take explicitly into account the partonic structure of hadrons.

In the 80's, Glauber applied to hadron–hadron scattering (Glauber and Velasco 1984) his multiple-diffraction theory, originally developed for hadron-nucleus scattering (Glauber 1958). Hadrons are viewed as clusters of interacting partons and the opacity is consequently written as

$$\Omega(s,b) = N_A\, N_B \int \mathrm{d}^2\boldsymbol{q}\; \mathrm{e}^{\mathrm{i}\boldsymbol{q}\cdot\boldsymbol{b}}\, G_A(\boldsymbol{q}^2)\, f(s,\boldsymbol{q}^2)\, G_B(\boldsymbol{q}^2)\, f(s,\boldsymbol{q}^2)\,, \qquad (6.35)$$

where $N_A$ ($N_B$) is the number of partons in hadron $A$ ($B$), $G_{A,B}$ are form factors and $f(s,\boldsymbol{q}^2)$ is a parton–parton scattering amplitude. Inserting (6.35) in (6.10) and expanding the exponential leads to a multiple-scattering series, which allows computing the various observables. The predictions of this model are rather similar to those of the impact picture and reproduce quite well the general features of the small-$t$ phenomenology.

Many attempts have been also made to provide a picture of soft diffraction somehow based on QCD. Some authors (Margolis et al. 1988; Block et al. 1990; Jenkovszky, Paccanoni and Predazzi 1992), for example, have proposed a link between the growth of total cross sections with energy and the rise of the gluon distribution at low $x$ (for the latter see Chap. 9). The main difficulty of this kind of approach lies in the fact that perturbative QCD is completely inadequate to treat small-$t$ processes and hence model assumptions are needed in order to obtain quantitative results. The contact with the fundamental theory is therefore lost.

It is worth recalling, in conclusion, that an ambitious and promising program for describing soft hadronic processes by means of non-perturbative QCD and semiclassical tools has been recently undertaken by the Heidelberg group (Nachtmann 1991; Dosch, Ferreira and Krämer 1992, 1994; Dosch and Rueter 1996; Ferreira and Pereira 1997, 1999).

## 6.5 Duality

The notion of duality in high-energy physics was introduced by Dolen, Horn and Schmid (1968) who showed how, in some simple cases at least, the superposition of direct-channel resonances is averaged by the corresponding reggeon(s) exchanged in the crossed channel, that is symbolically

$$\left\langle \sum_{\mathrm{res}} A^{(s)}_{\mathrm{res}}(s,t) \right\rangle = \left\langle \sum_{I\!R} A^{(t)}_{I\!R}(s,t) \right\rangle \qquad (6.36)$$

where the average is performed over the energy.

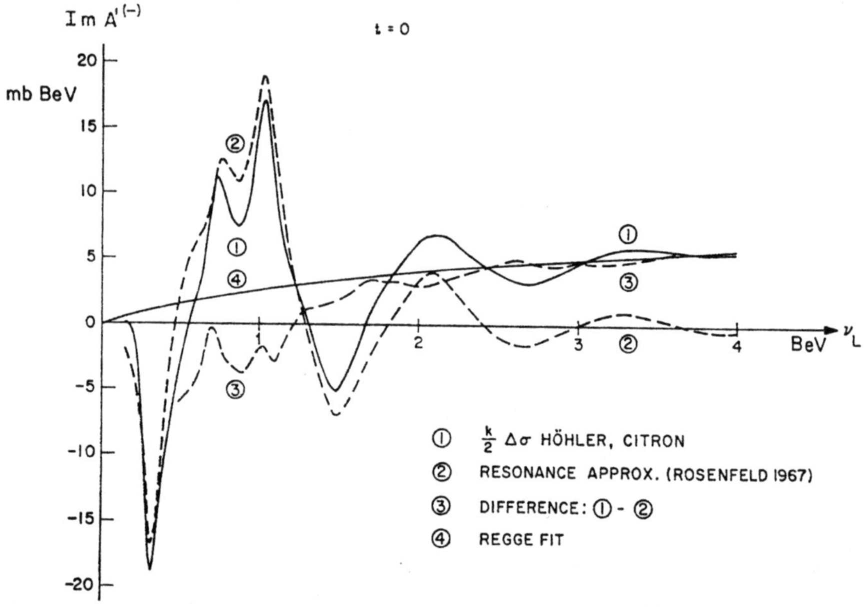


**Fig. 6.11.** The difference $\sigma_{\text{tot}}^{\pi^- p} - \sigma_{\text{tot}}^{\pi^+ p}$ vs. various models. Curve 1 is fit to the data. Curve 2 is the resonance contribution. Curve 4 is the $\rho$-pole Regge fit. From Dolen, Horn and Schmid (1968).

The most remarkable (but not the only) example of duality, shown in Fig. 6.11, occurs in pion–nucleon scattering. The bumpy dashed line in Fig. 6.11 is the sum of the resonances contributing to the total cross section difference $\sigma_{\text{tot}}^{\pi^- p} - \sigma_{\text{tot}}^{\pi^+ p}$; the smooth line (labeled 4) represents the contribution of admissible Regge trajectories (in this reaction only the $\rho$ trajectory contributes). As one can see, the asymptotic form seems to interpolate nicely between the resonant contributions.

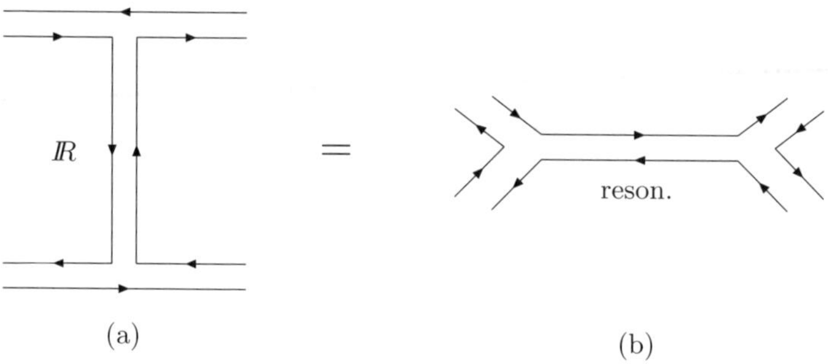


**Fig. 6.12.** Duality diagrams for meson-meson scattering: (**a**) reggeon exchange in $t$-channel; (**b**) resonance exchange in $s$-channel.

In the language of quark model, the property of duality can be reinterpreted by saying that dual diagrams, such as those shown in Fig. 6.12a,b for the case of meson-meson scattering, are equivalent to each other, and represent the same contribution to the scattering amplitude.

## 6.6 The Veneziano Model

The main shortcoming of $t$- and $s$-channel models, as already mentioned, lies in their lack of crossing symmetry (which is a property of relativistic amplitudes) and in the asymmetric way unitarity is usually enforced.

A remarkable generalization of a Regge-behaved crossing symmetric amplitude was proposed long ago by Veneziano (1968) who noticed that the combination of Euler Gamma functions

$$V(s,t) = \frac{\Gamma(1-\alpha(s))\,\Gamma(1-\alpha(t))}{\Gamma(2-\alpha(s)-\alpha(t))} \tag{6.37}$$

has the following properties:

- it is $t \leftrightarrow s$ crossing symmetric;
- owing to Stirling's formula,

$$\lim_{x\to\infty} \frac{\Gamma(a+x)}{\Gamma(b+x)} = x^{(a-b)}\,, \tag{6.38}$$

  the following behavior

$$\lim_{\alpha(s)\to\infty} V(s,t) = \Gamma(1-\alpha(t))[-\alpha(s)]^{\alpha(t)} \tag{6.39}$$

  holds and this, in turn, implies that
- $V(s,t)$ is Regge behaved, i.e. $\propto s^{\alpha(t)}$, so long as $\alpha(s) \propto s$ as $s \to \infty$ or, equivalently, so long as $\alpha(s)$ can be approximated by a linear function of $s$, as discussed in Sect. 5.7. In addition,
- $V(s,t)$ has a (simple) pole whenever either $\Gamma(1-\alpha(t))$ blows up (i.e., its argument vanishes or becomes negative)

$$\alpha(t) = n \qquad (n \text{ integer}) \tag{6.40}$$

  or $\Gamma(1-\alpha(s))$ blows up, that is when

$$\alpha(s) = m \qquad (m \text{ integer}) \tag{6.41}$$

  but, in spite of its superficial appearance,
- $V(s,t)$ does not have any double pole, since when both $\alpha(s)$ and $\alpha(t)$ become integer and, therefore, both $\Gamma(1-\alpha(t))$ and $\Gamma(1-\alpha(s))$ diverge, a zero is provided by the denominator $\Gamma(2-\alpha(s)-\alpha(t))$ whose argument also vanishes or becomes negative.

Therefore, if $\alpha(s)$ is identified with the (linearly growing) Regge trajectory introduced in Sect.5.7, $V(s,t)$ has (almost) all the properties one would like to have for it to fulfill the requirements discussed in Chap. 5 and, in addition, the form (6.37) is crossing symmetric.

As a consequence, $V(s,t)$ can be viewed either as a sum of infinitely many poles (resonances) in the $t$-channel

$$V(s,t) = \sum_{m=0}^{\infty} \frac{\Gamma(m+\alpha(s))}{\Gamma(\alpha(s))\, m!} \frac{1}{m+1-\alpha(t)} \tag{6.42}$$

or as a sum of infinitely many poles (resonances) in the $s$-channel

$$V(s,t) = \sum_{n=0}^{\infty} \frac{\Gamma(n+\alpha(t))}{\Gamma(\alpha(t))\, n!} \frac{1}{n+1-\alpha(s)} . \tag{6.43}$$

The fact that $V(s,t)$ is the sum of infinitely many resonances in either channel and, at the same time, is asymptotically Regge behaved, shows that it satisfies the property we named "duality" (see Sect. 6.5).

From the phenomenological viewpoint, however, the Veneziano model is affected by an incurable disease which has slowly led to abandoning it as an amplitude appropriate for describing the properties of hadronic reactions. This resides in the fact that the resonances whose infinite sum gives the Regge behavior are in fact zero width, i.e. they are poles on the real axis (that is for either $s$ or $t$ real). As a consequence, unitarity is violated and attempts to construct *simple* realistic extensions of it have been so far unsuccessful.

The Veneziano amplitude has also enjoyed some popularity as a model for particle production but, most important, its developments have led to string models (the pathway from dual models to strings is outlined in Green, Schwarz and Witten 1987, Chap. 1).

# 7. Soft Diffraction: a Phenomenological Survey

The high-energy data on total, elastic and diffractive cross sections accumulated in the last decades concern mostly $\bar{p}p$ (starting from the ISR, which explored c.m. energies ranging from $\sqrt{s} \simeq 20$ GeV up to $\sqrt{s} \simeq 62$ GeV, and going up to the CERN SPS at $\sqrt{s} = 546$ GeV and to the FNAL Tevatron at $\sqrt{s} = 1800$ GeV). By contrast, $pp$ has (so far) been explored only up to ISR energies, but both RHIC and LHC should, hopefully, fill the gap in the near future bringing information on the several TeV domain. All these data come from colliders. Other reactions, such as $\pi^{\pm}p$, $K^{\pm}p$, etc., have been studied only at fixed target accelerators, that is at much lower energies.

In this chapter we shall give a brief account of the soft diffraction data. For greater details, the reader is referred to the many reviews existing on the subject (e.g., Amaldi, Jacob and Matthiae 1976; Predazzi 1976; Kaidalov 1979; Alberi and Goggi 1981; Goulianos 1983; Matthiae 1994; Goulianos 2001a).

## 7.1 Total Cross-sections

It is now nearly forty years since total cross sections have been found to grow with energy after it was believed for long time that they would become asymptotically constant. The very first signal of this growth came from preliminary results of the Serpukhov machine on $\pi^{\pm}p$ and $K^{\pm}p$ scattering at $p_L \sim 60$ GeV. A clear evidence of the rise of both $pp$ and $\bar{p}p$ total cross sections was then provided by the ISR and FNAL experiments (for a review of these measurements and a list of references see Amaldi, Jacob and Matthiae 1976; Predazzi 1976). These data were compatible with the asymptotic equality of $\sigma_{\mathrm{tot}}(pp)$ and $\sigma_{\mathrm{tot}}(\bar{p}p)$ predicted by the Pomeranchuk theorem (see below). The growth of $\sigma_{\mathrm{tot}}(\bar{p}p)$ with $s$ became macroscopically visible with the SPS data at $\sqrt{s} = 0.546$ TeV (UA4 Collaboration: Bozzo et al. 1984) and $\sqrt{s} = 0.90$ TeV (UA5 Collaboration: Alner et al. 1986), and with the Tevatron data at $\sqrt{s} = 1.8$ TeV (E710 Collaboration: Amos et al. 1990, 1992; CDF Collaboration: Abe et al. 1994). This growth has been discussed on a theoretical ground in Chap. 6 as the evidence that the proton becomes larger and blacker as seen by an incoming hadron of increasing energy. Recently, a semi-quantitative explanation has been offered in the realm of QCD. As a rough statement, we can say that the actual energy growth is induced by soft gluon radiation superimposed to a rather large, non-perturbative, energy independent term (Kopeliovich et al. 2000).

On the experimental side, the measurement of total cross sections is performed in the following way. At fixed-target accelerators, total cross sections are measured with the transmission technique, i.e. they are determined from the observed attenuation of the beam after it strikes a liquid hydrogen target. At colliders, one has to resort to a different method. The total cross section is related to the observed number of elastic ($N_{\rm el}$) and inelastic ($N_{\rm in}$) events via

$$N_{\rm el} + N_{\rm in} = \mathcal{L}\, \sigma_{\rm tot} \,, \tag{7.1}$$

where $\mathcal{L}$ is the machine luminosity. Since this quantity is not very accurately known, one cannot directly use (7.1) to extract $\sigma_{\rm tot}$. A "luminosity-independent method" has to be adopted. We can relate $\sigma_{\rm tot}$ to the elastic scattering rate at $t = 0$ by

$$\left.\frac{\mathrm{d}N_{\rm el}}{\mathrm{d}t}\right|_{t=0} = \mathcal{L}\left.\frac{\mathrm{d}\sigma_{\rm el}}{\mathrm{d}t}\right|_{t=0} = \mathcal{L}\,\frac{1+\rho^2}{16\pi}\,\sigma_{\rm tot}^2 \,, \tag{7.2}$$

where $\rho$ is the ratio of the real to the imaginary forward amplitude

$$\rho = \frac{\mathrm{Re}\, A(s, t=0)}{\mathrm{Im}\, A(s, t=0)} \,. \tag{7.3}$$

This quantity is rather small at high energy ($\rho \sim 0.15$) and so it does not have to be known with very great accuracy. Eliminating $\mathcal{L}$ between (7.1) and (7.2) one gets

$$\sigma_{\rm tot} = \frac{16\pi}{1+\rho^2}\,\frac{(\mathrm{d}N_{\rm el}/\mathrm{d}t)|_{t=0}}{N_{\rm el}+N_{\rm in}} \,. \tag{7.4}$$

This formula is at the basis of the luminosity-independent approach adopted by the most recent experiments for measuring $\sigma_{\rm tot}$. The errors are typically of few percent, larger than for fixed-target experiments (which are very accurate, at a level of 0.2-0.3 %).

The $pp$ and $\bar{p}p$ total cross section data are shown in Fig. 7.1, together with a fit by Augier et al. (1993) to a $\ln^\gamma s$ behavior. An approximate $\ln^2 s$ growth ($\gamma = 2.2 \pm 0.3$) is suggested, which seems to saturate the energy growth permitted by the Froissart-Martin bound (4.141) (even though the multiplicative constant in the Froissart-Martin formula, $\pi/m_\pi^2 \simeq 60$ mb, is so large that numerically the data lie much below the bound).

The issue of the *exact* growth with energy of the total cross sections is both delicate and unresolved; the mild growth of total cross sections could be simulated by essentially any form and logarithmic physics is exceedingly difficult to resolve in a clear cut way.

From a phenomenological point of view, the uncertainties of the data do not allow to rule out the possibility of a $\ln s$ growth advocated by some

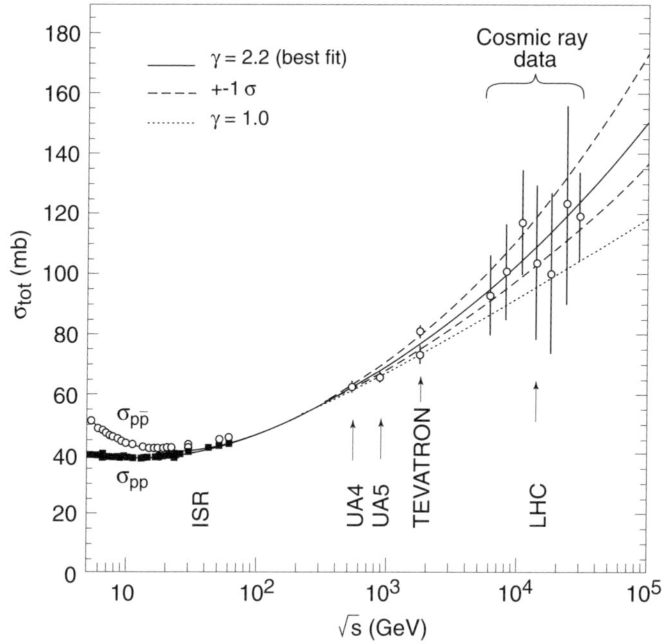


**Fig. 7.1.** Total $pp$ and $\bar{p}p$ cross sections fitted to a $(\ln s)^\gamma$ behavior (from TOTEM 1997).

authors. Note that there is a discrepancy between the two Tevatron determinations at $\sqrt{s} = 1.8$ TeV. The E710 result (Amos et al. 1990, 1992) tends to favor a $\ln s$ increase, while the CDF result (Abe et al. 1994) favors the $\ln^2 s$ dependence. Cosmic ray data (Baltrusaitis et al. 1984; Honda et al. 1993) are too uncertain to be really conclusive (for a critical discussion of these data see Kopeliovich, Nikolaev and Potashnikova 1989; Nikolaev 1993; Engel, Gaisser and Lipari 1998). At $\sqrt{s} = 14$ TeV (the energy to be explored at LHC) the difference between the $\ln s$ and the $\ln^2 s$ fits amounts to about 15 mb. The measurement of the TOTEM experiment (TOTEM 1997) is expected to have an accuracy of $\sim 1$ mb and will certainly put a much stronger constraint on the fits.

The available total cross section data can be also fitted quite well by a mild power dependence, as shown by Donnachie and Landshoff (1992) in their analysis based on Regge theory. The total cross sections for $pp$, $\bar{p}p$, $K^{\pm}p$, $\pi^{\pm}p$ and $\gamma p$ scattering are fitted (see Fig. 7.2) with the simple Regge-inspired expression

$$\sigma_{\text{tot}} = X\, s^{0.0808} + Y\, s^{-0.4525} \,, \tag{7.5}$$

where $X$ and $Y$ are reaction-dependent free parameters. Recalling that Regge theory predicts

$$\sigma_{\text{tot}} \sim \sum_i A_i\, s^{\alpha_i(0)-1} \,, \tag{7.6}$$

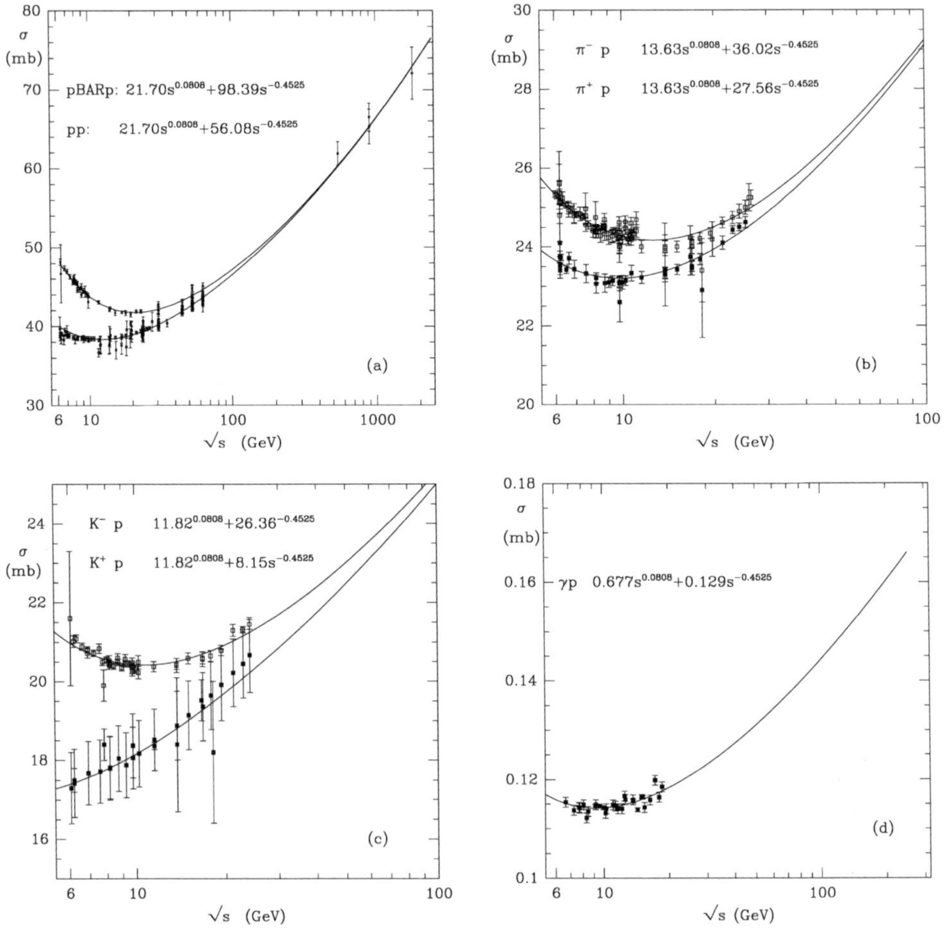


**Fig. 7.2a–d.** Total cross sections fitted versus $s^{0.08}$, plus a subdominant reggeon contribution (from Donnachie and Landshoff 1992).

where the sum is over the exchanged trajectories, we can interpret the first term in (7.5) as due to the pomeron exchange, with $\alpha_{I\!P}(0) = 1.0808$, and the second term as due to a reggeon exchange (a mesonic trajectory), with $\alpha_{I\!R}(0) = 0.5475$.

The pomeron, as it emerges from the fits to total cross sections, with an intercept $\alpha_{I\!P}(0) \simeq 1.08$, is often called *soft pomeron*. More recent analyses of total cross sections (Cudell, Kang and Kim 1997), still based on simple-pole exchanges, give slightly larger values for $\alpha_{I\!P}(0)$. In particular, Covolan, Montanha and Goulianos (1996) obtain $\alpha_{I\!P}(0) = 1.104 \pm 0.002$, whereas Cudell, Kang and Kim (1997) find $\alpha_{I\!P}(0) = 1.0964^{+0.0115}_{-0.0091}$. It has been observed (Covolan, Montanha and Goulianos 1996) that using eikonalized Regge amplitudes in the fits, the soft pomeron intercept slightly increases (from 1.104 to 1.122).

From a strictly mathematical point of view, the power growth à la Donnachie and Landshoff may not be distinguishable from a logarithmic growth unless one goes to fantastically higher energies (which will never be attained). Similarly, the combination of a $\ln s$ plus a constant term is essentially undistinguishable from a combination containing also a $\ln^2 s$ term.

From the physics point of view, on the other hand, any $s^\lambda$ power behavior (with positive $\lambda$), taken at face value, would violate unitarity and, therefore, should at some point be modified by unitarity corrections, whereas no arguments exist against any $\ln^\gamma s$ behavior unless $\gamma$ exceeds 2. The empirical argument put forward by some authors that a rate of growth $\sim s^{0.08}$ would violate unitarity only at astronomically large energies, on the other hand, is challenged by the fact that actually unitarity in the impact parameter would be violated already just above Tevatron energies (Kopeliovich, Povh and Predazzi 1997).

We should also recall that not only there exists a power growth of the form $s^\lambda$ with the value $\lambda \approx 0.08 - 0.10$, as suggested by Donnachie and Landshoff, and other authors (the *soft pomeron*), but the perturbative QCD approach by Lipatov and coworkers (Kuraev, Lipatov and Fadin 1976, 1977; Balitsky and Lipatov 1978; Lipatov 1986) predicts a higher value of $\lambda \approx 0.15$, corresponding to the so-called *BFKL* or *hard pomeron* which is expected to manifest itself in hard diffractive processes. We shall devote Chap. 8 to this subject.

According to the Pomeranchuk theorem (see Sect. 4.9.2), the $\bar{p}p$ and $pp$ total cross sections should become equal at asymptotic energies. Aside from the ambiguity inherent in the very term "asymptotic energies", and to the extent that the present data can be used as if they were already asymptotic, this prediction seems to be supported by the experimental results, as shown in Fig. 7.3. The continuous line in Fig. 7.3 represents the result of a power-

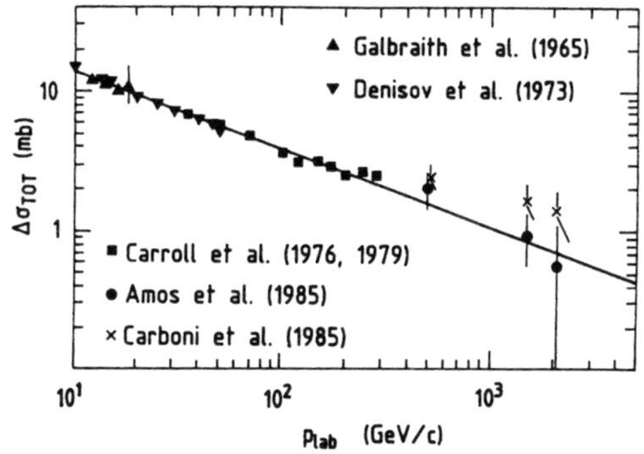


**Fig. 7.3.** The difference $\Delta\sigma_{\rm tot} = \sigma_{\rm tot}(\bar{p}p) - \sigma_{\rm tot}(pp)$ as a function of the energy (from Matthiae 1994).

law fit to the data which gives $\Delta\sigma_{\text{tot}} \sim s^{-0.56}$, in agreement with the Regge theory expectations. The pomeron contribution, in fact, cancels out in the $\bar{p}p - pp$ difference, and the behavior of $\Delta\sigma_{\text{tot}}$ is dominated by a secondary reggeon trajectory with an intercept close to 1/2.

Unfortunately, the highest energies for which we have both $\bar{p}p$ and $pp$ are in the ISR range ($\sqrt{s} \simeq 62$ GeV) and this makes the evidence in support of Pomeranchuk theorem not quite so conclusive as one would have liked. It is a great pity that, at least for the time being, there are no options to extend LHC to cover also the $\bar{p}p$ channel. This extension would allow a detailed comparison of $pp$ and $\bar{p}p$ data at very high energies essentially free of normalization ambiguities. It is also a great pity that there are no plans of extracting a secondary meson beam from either RHIC (the Relativistic Heavy Ion Collider operating at Brookhaven) or LHC. Used in a fixed-target configuration against liquid hydrogen this could allow measuring $\pi^{\pm}p$ and $K^{\pm}p$ cross sections up to $p_L \sim 800 - 900$ GeV; a gigantic jump from the maximum energy values at which meson-nucleon scattering has been measured at the Serpukhov machine.

## 7.2 The Real Part of the Forward Elastic Amplitude

We have already commented on the importance of the constraint represented by the optical theorem on the data. The optical theorem (Sect. 4.3.1) can be viewed as a sum rule telling us us that the imaginary part of the forward amplitude (which is proportional to the total cross section) grows with energy, while no such constraint exists over the real part. On the other hand, general analyticity and crossing properties relate the real and the imaginary parts of the amplitude (Khuri and Kinoshita 1965) so that the measurement of the $\rho$ parameter, as defined in (7.3), is an important complement to the measurement of $\sigma_{\text{tot}}$. In addition, $\rho$ is a very sensitive indicator of several theoretical properties.

To see how the behavior of $\rho$ is related to the energy dependence of $\sigma_{\text{tot}}$, let us consider an extremely simplified model. Suppose that the total cross section grows like $\ln^{\nu} s$. A very simple amplitude compatible with this growth is

$$A(s,0) = \mathrm{i}\, C\, s\, \ln^{\nu} s\,, \tag{7.7}$$

where $C$ is a constant. To get a crossing-even amplitude $A^{+}$ (which must remain the same when $s \to s\,\mathrm{e}^{i\pi}$) we modify (7.7) as

$$A^{+}(s,0) = \mathrm{i}\, C\, s\, \ln^{\nu}(s\,\mathrm{e}^{-\mathrm{i}\pi/2}) = \mathrm{i}\, C\, s\, \left(\ln s - i\,\frac{\pi}{2}\right)^{\nu}\,. \tag{7.8}$$

In this case, if $\nu$ is $> 0$ (i.e. $\sigma_{\text{tot}}$ grows with $s$), we would find, when $s \to \infty$

$$\rho^{+}(s) \underset{s\to\infty}{\sim} \frac{\pi\,\nu}{2}\,\frac{1}{\ln s} \tag{7.9}$$

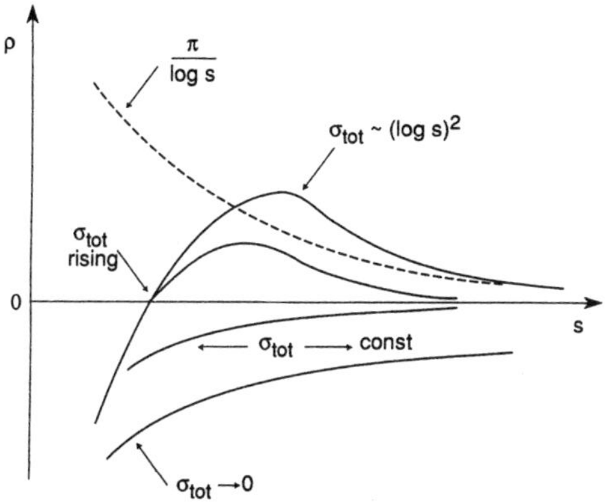


**Fig. 7.4.** Behavior of $\rho$ for different assumptions on the energy dependence of the total cross section (from Matthiae 1994).

implying that $\rho(s)$ should approach asymptotically zero from above. (A variety of other behaviors is shown in Fig. 7.4.) The importance of this prediction (Khuri and Kinoshita 1965) is that it was made when the $\rho$ data for both $\bar{p}p$ and $pp$ were negative. Soon after, it was indeed found that at ISR energies $\rho(s)$ becomes positive. More generally, one can show by a simplified form of dispersion relations (Bronzan, Kane and Sukhatme 1974) that the even-signature real-to-imaginary amplitude ratio $\rho^+$ is related to the corresponding cross section $\sigma^+_{\rm tot}$ by

$$\rho^+ \simeq \frac{\pi}{2\sigma^+_{\rm tot}} \frac{{\rm d}\sigma^+_{\rm tot}}{{\rm d}\ln s} \,. \tag{7.10}$$

This implies that in the region where the total cross section is first decreasing with energy and then rising, $\rho$, which is initially negative, will rise, going through zero when the cross section has a minimum and becoming positive at high energy. This trend is indeed visible in the data (Fig. 7.5). If asymptotically $\sigma_{\rm tot} \sim \ln^\nu s$, then (7.10) reduces to (7.9). The prediction that $\rho(s)$ approaches zero from above as $s \to \infty$, however, is far from being a property already exhibited by the data even at the highest energies so far explored, as one can see from Fig. 7.5. Many would agree that these data imply that we have not yet reached the asymptotic regime and that $\rho$ is a sensitive indicator of how *Asymtopia* is being approached.

In practice, the measurement of $\rho(s)$ is performed by observing the interference of the hadronic amplitude $A_h$ parametrized as

$$A_h(s,t) = s\,({\rm i}+\rho)\,\sigma_{\rm tot}\,{\rm e}^{Bt/2}\,, \tag{7.11}$$

in the small $|t|$ region, with the (known) Coulomb amplitude given by (with our normalization)

$$A_C = \pm\, 8\pi\,\alpha_{\rm em}\, s\, \frac{G^2(t)}{|t|} \tag{7.12}$$

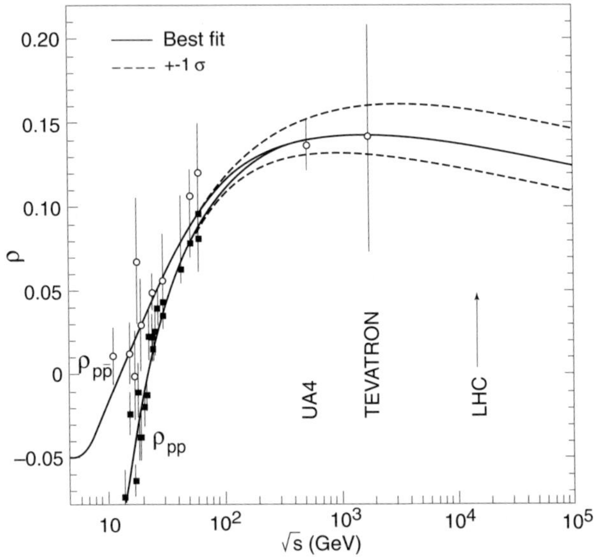


**Fig. 7.5.** The $\rho$ parameter for $\bar{p}p$ (open circles) and $pp$ scattering (black circles), as a function of the energy. The solid line is a dispersion relation fit with the $1\sigma$ uncertainty region delimited by the dashed lines (from TOTEM 1997).

where $G(t)$ is the proton electromagnetic form factor and $+/-$ refer to $\bar{p}p$ and $pp$ respectively. Coulomb scattering dominates at small $|t|$, and the two amplitudes become comparable when the the momentum transfer is of order

$$|t_0| \sim \frac{8\pi\alpha_{\rm em}}{\sigma_{\rm tot}} \tag{7.13}$$

where $\alpha_{\rm em}$ is the fine structure constant. At the present energies one has typically $|t_0| \sim 10^{-3}$ GeV$^2$. This is the $|t|$ region where the measurement of the real part of the amplitude is possible (hence $\mathrm{Re}\, A$ is known only in the forward direction).

The small-$|t|$ differential cross section is then written as

$$\frac{d\sigma_{\rm el}}{dt} = \frac{1}{16\pi s^2} \, |A_h + A_C \, e^{\pm i\, \alpha_{\rm em}\, \Phi}|^2 \, , \tag{7.14}$$

where $\alpha_{\rm em}\Phi$ is the relative Coulomb-hadronic phase. As one can see, the interference term (where $|t|$ is small enough that we can neglect $|A_h|^2$) is proportional to $(\rho \pm \alpha_{\rm em}\Phi)$. The relative phase $\alpha_{\rm em}\Phi$, first calculated by Bethe (1958) in a potential scattering model, has been investigated by several authors (West and Yennie 1968; Buttimore, Gotsman and Leader 1978; Cahn 1982; Kopeliovich and Tarasov 2001). Numerically one finds $\alpha_{\rm em}\Phi \approx 0.027$ at $t = t_0$, with $B = 15$ GeV$^{-2}$.

## 7.3 Elastic Cross-sections

At hadron colliders elastic events are identified by detecting two, and only two, back-to-back particles in the final state. The difficulty is that the higher the energy, the smaller the scattering angles (typically fractions of mrad), so that the detectors must be placed very close to the beam pipe. To this aim, an ingenious device known as the "Roman pot" is used (Amaldi et al. 1973). The detectors are placed into movable sections of the vacuum chamber of the accelerator which normally are left in a retreated position so that the beam, when injected, circulates freely along the vacuum chamber pipe. When the wanted energy has been reached and the beam is stable, the Roman pots are slid into their operational position until the inner part of the detector is just a few millimeters from the beam (Fig. 7.6). (Roman pots are also extensively used in diffractive deep inelastic scattering experiments.) The detectors inserted in the Roman pots are usually designed to accept high rate and have good spatial resolution ($\sim 100\ \mu$m). They consist of combinations of drift chambers, hodoscopes of scintillating fibers and silicon detectors.

Normally, hadron colliders are used in search of rare events for which the highest luminosity is desirable. For this, the transverse size of the beam is reduced as much as possible at the crossing point by means of focusing quadrupoles. The angular divergence of the beam is then increased and a

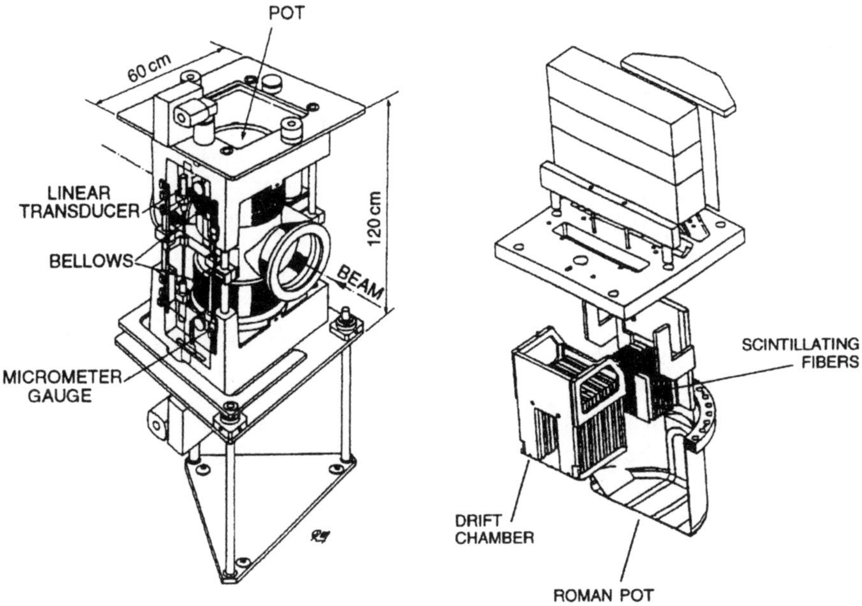


**Fig. 7.6.** A hero of experimental diffractive physics: the Roman pot. The system shown here is used by the UA4 experiment at SPS. On the right: an exploded view of the detectors. From Matthiae (1994).

large fraction of scattered particles are not accessible to detection. When measuring elastic scattering, one needs the opposite. The beam size at the crossing point is made relatively large, and consequently the luminosity is reduced (which is not a problem since the differential cross section is quite large at small $|t|$).

The ratio of the (integrated) elastic to the total cross section is known to decrease at low energies and to reach a constant value at ISR energies. For $pp$, the measurements are rather precise and give an average value $\sigma_{el}/\sigma_{tot} = 0.175$ between $\sqrt{s} = 23 - 62$ GeV. The data for $\bar{p}p$ are not so accurate but the result is compatible with the same value within the errors. The constancy of this ratio, by the way, was one of the predictions of the geometrical model mentioned in Sect. 6.4.1. The fact that at higher energies (SPS and Tevatron), the ratio $\sigma_{el}/\sigma_{tot}$ increases with energy (see Fig. 7.7) is not only a strong argument against that model (at least at subasymptotic energies), but can also be taken as an evidence that hadrons become more and more opaque as the energy increases.

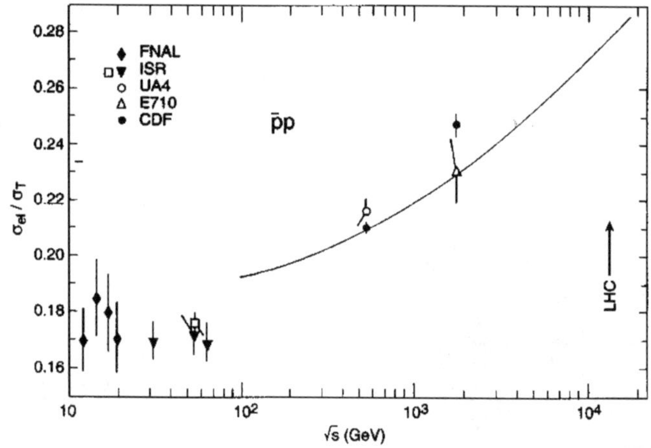


**Fig. 7.7.** The ratio $\sigma_{el}/\sigma_{tot}$ as a function of the energy (from TOTEM 1997).

The growth with energy of $\sigma_{el}/\sigma_{tot}$ is actually compatible with various models. In particular, let us recall that the model of Cheng and Wu (1970a, 1970b), and Bourrely, Soffer and Wu (1984) – see Sect. 6.4.2 –, predicts that this ratio should approach the limit $\sigma_{el}/\sigma_{tot} = 1/2$. This limit is actually very remote. If the prediction of an increase with energy of $\sigma_{el}/\sigma_{tot}$ is correct and if we extrapolate the trend of the present data, at LHC one should expect a value $\sigma_{el}/\sigma_{tot} \approx 0.28$.

### 7.3.1 The Forward Peak

The high energy $t$-distribution exhibits a pronounced forward peak (the *diffraction peak*) whose origin we have already commented upon previously. For simplicity, we will assume, as it is usual, that it can be represented by an exponential of the form $e^{B(s)t}$. Actually, there are good reasons, both theoretical and phenomenological, which would recommend different, more appropriate (but also more complicated), forms. The exponential, however, is certainly the simplest form, and this we shall use. Theory and data show that the slope of the diffraction peak depends on $s$. Naively, the fact that in natural units $B(s)$ has dimensions $(length)^2$ suggests that there should be a relationship between this quantity and the only length in the game (the hadron's size or, equivalently, the total cross section) so that one expects it to grow as some power of $\ln s$ with increasing energies. Regge pole phenomenology predicts, as we have seen, a straight $\ln s$ behavior, whereas a rough unitarity motivated approach (Giffon, Nahabetian and Predazzi 1987, 1988) predicts a growth with energy whose rate is proportional to the average multiplicity.

Whatever the exact form, indeed, $B(s)$ grows with energy as the data closest to $t = 0$ ($|t| \simeq 0.02$ GeV$^2$) show (see Fig. 7.8). In particular the solid line in Fig. 7.8 represents the Regge prediction for the growth of $B(s)$. At large $s$, this is a straight line with a slope given by the pomeron slope: $B(s) = B_0 + 2\,\alpha'_{I\!P} \ln s$. From Fig. 7.8 one derives $\alpha'_{I\!P} \approx 0.25$ GeV$^{-2}$ in good agreement with other estimates.

The only direct high-energy comparison between $pp$ and $\bar{p}p$ slopes in the diffraction region is at the ISR energies and was performed by Breakstone et al. (1984) by measuring with the same experimental apparatus both reactions. The ratio $B(\bar{p}p)/B(pp)$ decreases towards 1 as the energy increases

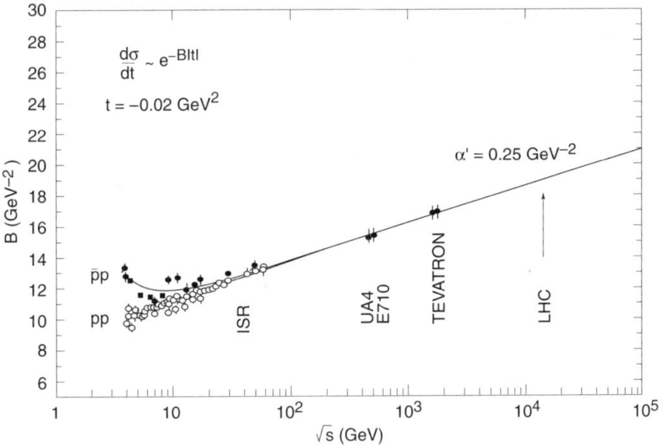


**Fig. 7.8.** The forward slope parameter as a function of $\sqrt{s}$ (from TOTEM 1997).

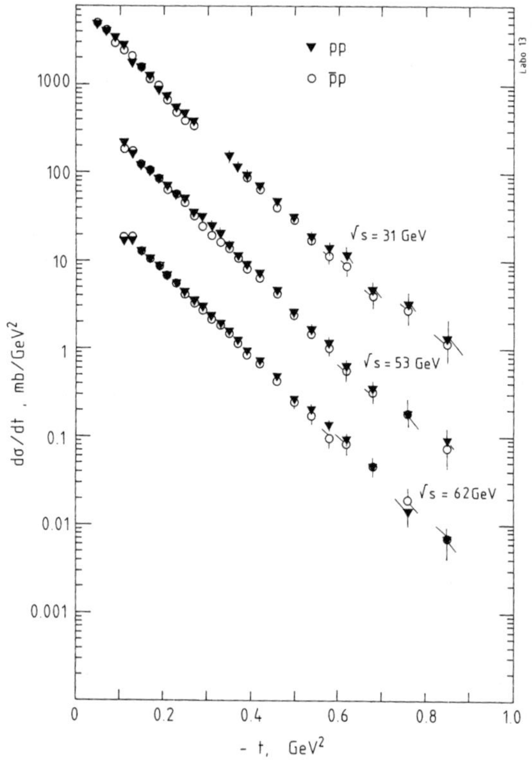


**Fig. 7.9.** The diffraction cone for $pp$ and $\bar{p}p$ in the ISR energy region (Breakstone et al. 1984).

and reaches unity (within the experimental errors) at $\sqrt{s} = 62$ GeV (see Fig. 7.9), as expected from the Cornille-Martin theorem (Cornille and Martin 1972) presented in Sect. 4.9.2.

Phenomenologically, it turns out that the overall diffraction peak for $|t| < 0.5$ GeV$^2$ is not described by a simple exponential. The slope $B(s)$ is found to decrease when measured at $|t|$ values larger than 0.02 GeV$^2$ (to which Fig. 7.8 refers). This feature is visible both in the ISR data (Barbiellini et al. 1972) and in the SPS data (Bozzo et al. 1984) – for the latter see Fig. 7.10. At $|t| \simeq 0.2{-}0.3$ GeV$^2$ the slope is about 2 units of GeV$^{-2}$ smaller than at $|t| \simeq 0$. On the contrary, the Tevatron data (Amos et al. 1990) show no evidence for a change of slope as $|t|$ increases and are fitted by a simple exponential $e^{16.3\,t}$ up to $|t| \approx 0.5$ GeV$^2$. The situation is summarized in Fig. 7.11. The decrease of the slope is accounted for by various models (for instance, Bourrely, Soffer and Wu 1988; Giffon, Nahabetian and Predazzi 1987, 1988).

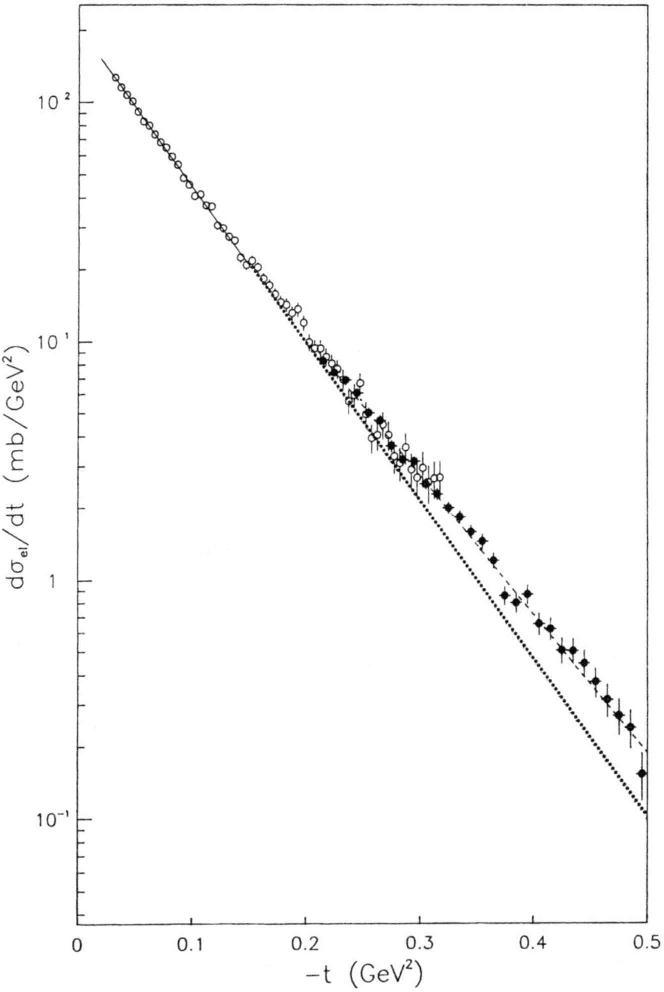


**Fig. 7.10.** The forward $\bar{p}p$ peak at the SPS (Bozzo et al. 1984). The thin dashed curve is a $e^{13.4\,t}$ fit. The dotted curve is a $e^{15.3\,t}$ fit.

## 7.4 The Dip-Shoulder Region

As one moves to larger $|t|$ values, a dip is encountered in $d\sigma_{el}/dt$, followed by a broad maximum, according to typical diffraction patterns. Such a behavior was anticipated (Chou and Yang 1968) long before the data were collected. The dip appears to recede towards zero with increasing energy, the $|t|$ value at its position being roughly proportional to $1/\sigma_{tot}$, as expected in the geometrical picture (Sect. 6.4.1). There is an interesting difference between the $pp$ and $p\bar{p}$ data: while $pp$ scattering shows a pronounced dip, $p\bar{p}$ exhibits no dip but only a shoulder (Fig. 7.12).

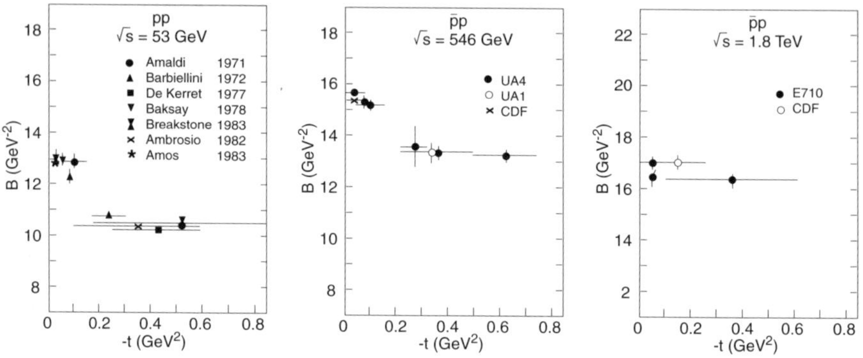


**Fig. 7.11.** The slope $B$ as a function of $t$ (from TOTEM 1997).

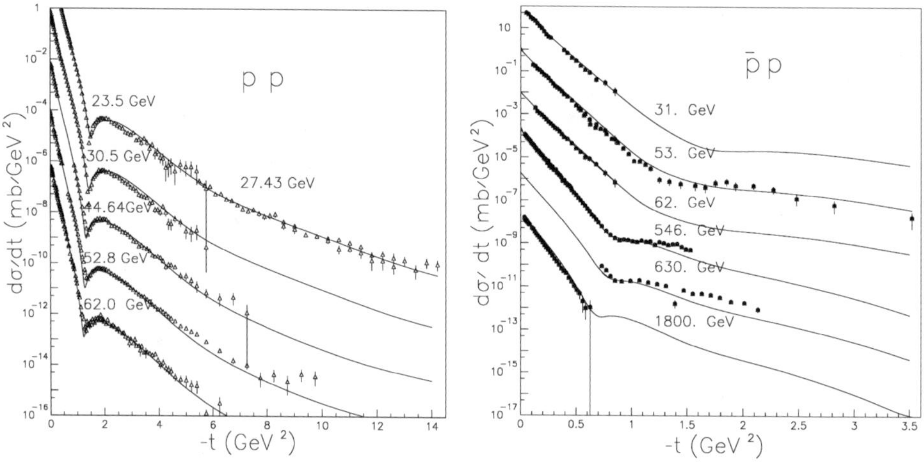


**Fig. 7.12.** The $pp$ and $\bar{p}p$ angular distributions at high energies. The curves are the predictions of Desgrolard et al. (2000).

Various models are able to reproduce the experimental findings. The appearance of secondary diffractive maxima and minima has been also predicted by several authors, and future experiments at LHC and at RHIC have been planned to verify this prediction. In Fig. 7.12 we give the angular distributions as reproduced by one of the fits on the market (Desgrolard et al. 2000), and in Fig.7.13 we give their predictions for how the secondary dips are expected to develop with increasing energy.

### 7.4.1 Odderon Phenomenology

The different behavior of $\mathrm{d}\sigma_{\mathrm{el}}^{pp}/\mathrm{d}t$ and $\mathrm{d}\sigma_{\mathrm{el}}^{p\bar{p}}/\mathrm{d}t$ can be interpreted, in the framework of Regge theory, as a support for the existence of the odderon (see Sect. 5.8.4). Data at $t = 0$, however, seem to give an opposite indication.

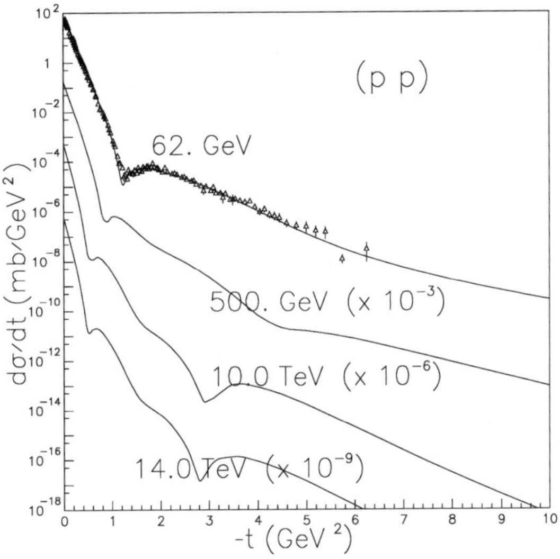


**Fig. 7.13.** Evolution expected for the secondary diffraction minima in the $pp$ and $\bar{p}p$ angular distributions according to the model of Desgrolard et al. (2000).

In particular, the dispersion relation fit of $\rho_{pp}$ and $\rho_{p\bar{p}}$ shown in Fig. 7.5 (Augier et al. 1993) provides a tight bound on the $pp - p\bar{p}$ difference

$$|\rho_{pp}(s) - \rho_{p\bar{p}}| \lesssim 0.05 \,, \quad \text{for } \sqrt{s} \gtrsim 0.5\,\text{GeV} \,, \tag{7.15}$$

which seems to leave very small room (if any) to a hypothetical odderon contribution. Unfortunately, the scarcity of combined $pp$ and $p\bar{p}$ data makes the whole matter very uncertain.

At high $|t|$ values, i.e. in the perturbative region, the odderon is expected to correspond to the exchange of three gluons in a $C = P = -1$ configuration. As we shall see below, three-gluon exchange indeed dominates large-angle $pp$ scattering.

The situation is therefore the following: while there is some evidence for the odderon (intended as three-gluon exchange) at (relatively) high $|t|$, the "soft" odderon has not been observed so far at $t = 0$. Various explanations for the elusiveness of the soft odderon have been advanced (e.g., Donnachie and Landshoff 1991; Dosch and Rueter 1996). In the Stochastic Vacuum Model of the Heidelberg group (Nachtmann 1991; Dosch, Ferreira and Krämer 1992, 1994; Ferreira and Pereira 1997) the odderon is expected to give a vanishing contribution to $pp$ and $p\bar{p}$ scattering if the proton and the antiproton have a quark-diquark structure (Dosch and Rueter 1996), as first pointed out by Zakharov (1989). Other authors (Giffon, Predazzi and Samokhin 1996) have suggested that $t = 0$ data at RHIC and LHC energies may already be sufficient to detect possible odderon effects. This can be done by looking at some particular correlations between $t = 0$ observables, whose sign is quite sensitive to the existence of $C$-odd contribution.

A process which can directly probe the odderon is diffractive electroproduction of $C = +1$ mesons, with or without proton breakup (Dosch and Rueter 1996; Rueter, Dosch and Nachtmann 1999; Czyżewski et al. 1997; Engel et al. 1998), $\gamma^* + p \to M + p(N^*)$. The advantage of this reaction is that the odderon is in competition with the photon only and is not obscured by the large pomeron contribution as in hadron–hadron scattering. In particular, if $M$ is a light meson ($\pi^0$, $\eta$, $\eta'$, ...) and $Q^2 \lesssim 0.5$ GeV$^2$, the soft odderon is involved; if $M$ is a heavy meson ($\eta_c$, $\eta_b$) and/or $Q^2$ is large, the process is expected to be dominated by three-gluon exchange (the "hard" odderon). The interference between $\gamma$ and odderon exchange amplitudes gives also information on the phase of the odderon trajectory. Diffractive electroproduction of $C = +1$ mesons is currently under investigation at HERA. For more information on the odderon phenomenology the reader may consult Landshoff and Nachtmann (1998) and Nicolescu (1999).

### 7.4.2 The Large-$|t|$ Region

At large $|t|$ values ($|t| \gtrsim 3$ GeV$^2$) the elastic cross section $\mathrm{d}\sigma_{\mathrm{el}}/\mathrm{d}t$ decreases not exponentially, but as a power of $t$. In particular it is found (see Fig. 7.14)

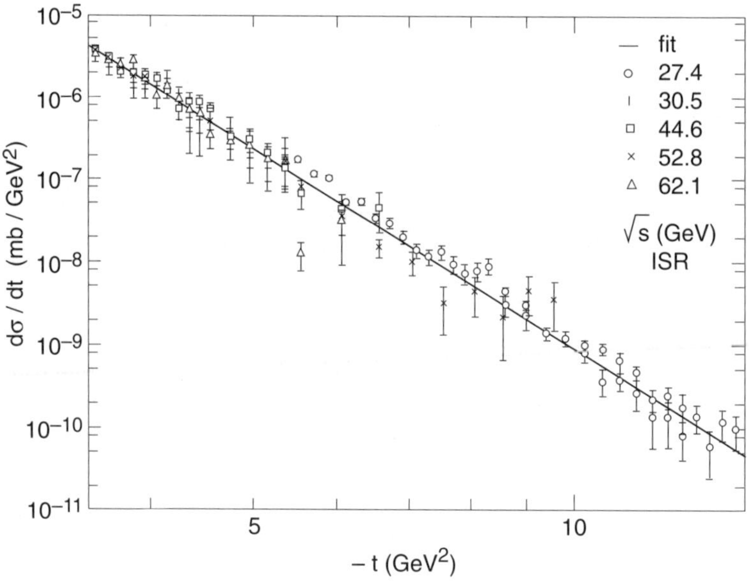


**Fig. 7.14.** *pp* elastic cross section at large $|t|$, showing a $1/t^8$ behavior (the solid line).

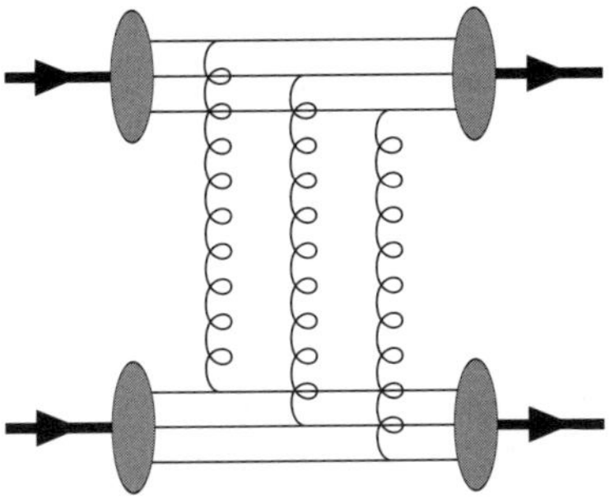

**Fig. 7.15.** Three-gluon exchange.

$$\frac{d\sigma_{el}}{dt} \sim \frac{1}{t^8} . \tag{7.16}$$

This behavior is accounted for by the three-gluon exchange model (Donnachie and Landshoff 1979). The idea is that $pp$ and $p\bar{p}$ scattering at high $|t|$ proceeds via the exchange of three gluons which couple to the valence quarks of the proton, or antiproton (Fig. 7.15). The amplitude of this process has opposite sign for $pp$ and $p\bar{p}$, and this would explain also the difference between the $pp$ and $p\bar{p}$ data in the dip-shoulder region. It should be mentioned that the Donnachie-Landshoff model predicts no secondary dips in $pp$ at high $|t|$. The investigation of the region of $|t| \approx 10 - 15\ \mathrm{GeV}^2$ at LHC will shed light on this matter. As mentioned above, three-gluon exchange can be seen as a QCD model for the odderon.

## 7.5 Diffractive Dissociation

At high energies it becomes very difficult to detect specific exclusive channels and most of the time at colliders the experimental activity is limited to measuring the mass and the fragmentation properties of some excited hadronic state which it is customary to denote by $X$ (inclusive production). As already mentioned (see Sect. 3.3.3), diffractive dissociation is a special case of inclusive production in a quasi two-body process, wherein all quantum numbers of the final states (charge, isospin, strangeness etc.) are the same as for the initial hadrons. Spin and parity can, of course, be different since orbital angular momentum can be transferred in the collision. Diffractive dissociation, therefore, is closely connected with elastic scattering, as it may be visualized by the quasi two-body reaction (*single diffraction*) $1 + 2 \to 1' + X_2$. *Double diffraction* occurs when both 1 and 2 are excited to systems with the same initial quantum numbers: $1+2 \to X_1+X_2$. Whenever the basic conditions for single or double diffraction are satisfied, the differential cross sections exhibit a sharp forward peak and the production cross sections depend on energy not unlike the elastic cross sections (for a review see Goulianos 1983).

A point which should be briefly recalled concerns the leading-particle effect (Basile et al. 1980; Giffon and Predazzi 1980). As seen in Sect. 4.3.2, by this term one denotes a well known property typical of high-energy hadronic interactions. When a highly energetic hadron strikes a target and we do not look for the complete final state (i.e., we look at inclusive processes) a non negligible portion of the events (say $\approx 10\%$ of the total) has a rather peculiar configuration which is especially simple in the laboratory system: the same incident particle flies off essentially unscathed in the (near) forward direction leaving behind a stream of slow particles produced. Experimentally, the resulting inclusive cross section has most of the features of elastic distributions. The situation is clearly understood in terms of diffraction. The configuration we have just depicted is a special case meeting the requirement of our general criterion: between the initial colliding particle and the final fast emerging one, there is no exchange of quantum numbers (the particle is the same) and therefore the reaction is diffractive and has all the properties we have seen for these processes. It should be stressed that this leading-particle effect requires the fast particle to be exactly the same as the initial one. For instance, if a proton strikes, a proton must come out, not a neutron. If a neutron comes out as the fast particle, charge is being exchanged. The process, accordingly, cannot be attributed to the exchange of zero quantum numbers (i.e. to diffraction) but, for instance, to pion exchange.

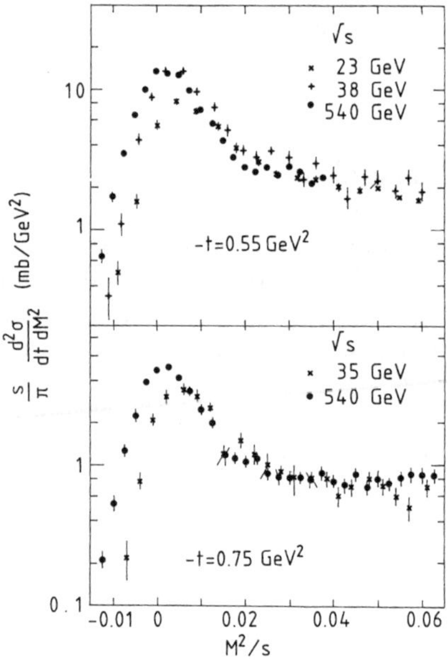


**Fig. 7.16.** The diffractive cross section measured at ISR (Albrow et al. 1976) and SPS (Bozzo et al. 1984) for two values of the momentum transfer $t$.

Let us now come to the data. Considering the single-inclusive process $p(\bar{p}) + p \to p(\bar{p}) + X$, clear evidence of diffraction is seen up to about $M^2/s \sim 0.05$, where $M^2$ is the mass squared of the unresolved hadronic system $X$. The differential cross section $\mathrm{d}\sigma^{SD}/\mathrm{d}M^2\,\mathrm{d}t$ as measured at ISR (Albrow et al. 1976) and SPS (Bozzo et al. 1984) is shown in Fig. 7.16. Notice that it scales with $\xi \equiv 1 - x_F = M^2/s$.

In the framework of Regge theory, diffractive dissociation, when $s \gg M^2 \gg |t|$, is dominated by the so-called triple-pomeron vertex (Sect. 5.8.6). The predicted cross section is (we call $g_{3I\!P}$ the triple-pomeron coupling)

$$\frac{\mathrm{d}\sigma^{SD}}{\mathrm{d}\xi\mathrm{d}t} = f_{I\!P}(\xi, t)\,\sigma_{I\!P}(M^2)\,, \tag{7.17}$$

where the pomeron flux factor $f_{I\!P}(\xi, t)$ and the pomeron cross section $\sigma_{I\!P}(M^2)$ are given by

$$f_{I\!P}(\xi, t) = \frac{1}{16\pi^2}\,|g_{I\!P}(t)|^2 \xi^{1-2\alpha_{I\!P}(t)}\,, \tag{7.18}$$

$$\sigma_{I\!P}(M^2) = g_{I\!P}(0) g_{3I\!P}(0)\,(M^2)^{\alpha_{I\!P}(0)-1}\,. \tag{7.19}$$

The mass spectrum is

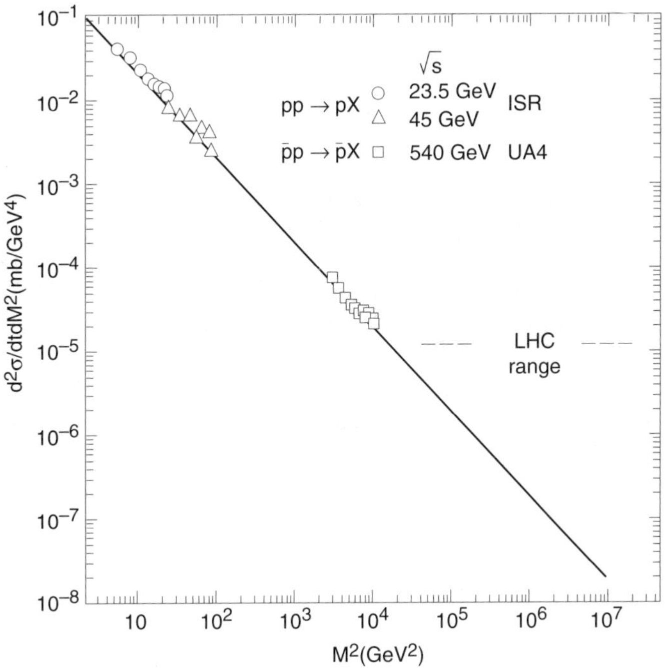


**Fig. 7.17.** $1/M^2$ behavior of the inclusive $\bar{p}p$ data (Tevatron excluded). From TOTEM (1997).

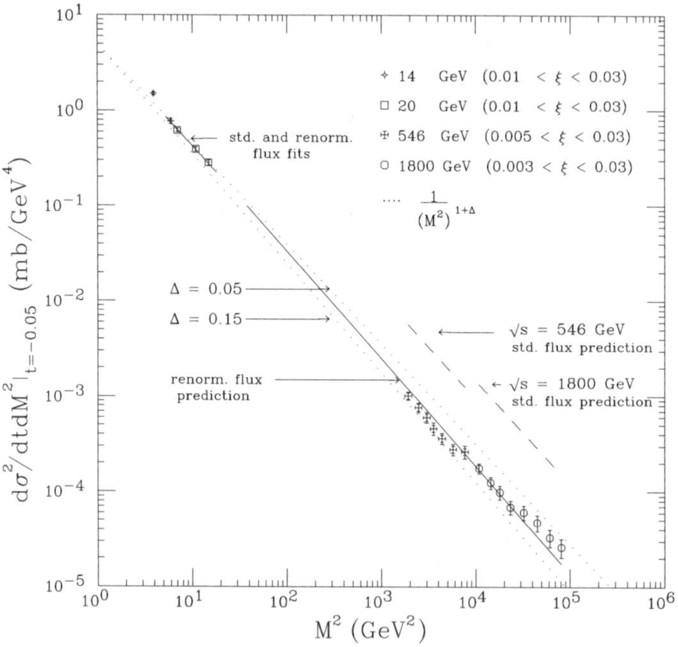


**Fig. 7.18.** $M^2$ dependence of the single diffractive Tevatron data (from Goulianos and Montanha 1999).

$$\left.\frac{\mathrm{d}\sigma^{SD}}{\mathrm{d}M^2\mathrm{d}t}\right|_{t=0} \sim \frac{1}{(M^2)^{\alpha_{I\!P}(0)}} \sim \begin{cases} \frac{1}{M^2} & \text{if } \alpha_{I\!P}(0) = 1 \,, \\ \frac{1}{(M^2)^{1+\epsilon}} & \text{if } \alpha_{I\!P}(0) = 1+\epsilon \,. \end{cases} \tag{7.20}$$

The pre-Tevatron data, shown in Fig. 7.17 follow a $1/M^2$ behavior. At Tevatron (Abe et al. 1994) a slightly steeper spectrum is found (Fig. 7.18), with $\epsilon = 0.121 \pm 0.011$ at $\sqrt{s} = 546$ GeV and $\epsilon = 0.103 \pm 0.017$ at $\sqrt{s} = 1800$ GeV. We stress that the asymptotic behavior (7.20) follows only if the triple-pomeron vertex dominates and all subasymptotic contributions can be neglected.

Results on the integrated cross section of single diffraction dissociation ($\sigma_{\mathrm{SD}}$) are shown in Fig. 7.19. Note that $\sigma_{\mathrm{SD}}$ is a sizeable fraction of the total cross section. As for the energy dependence, the ratio $\sigma^{SD}/\sigma_{\mathrm{tot}}$ is found to decrease with $s$. This is in contrast with the Regge theory expectation, which gives for $\alpha_{I\!P}(0) = 1 + \epsilon$

$$\sigma_{\mathrm{tot}} \sim s^{\epsilon} \,, \quad \sigma^{SD} \sim s^{2\epsilon} \,, \quad \frac{\sigma^{SD}}{\sigma_{\mathrm{tot}}} \sim s^{\epsilon} \,. \tag{7.21}$$

(According to Regge theory the integrated elastic cross section $\sigma_{\mathrm{el}}$ behaves as $s^{2\epsilon}/\ln s$, having assumed a $\ln s$ dependence of the $t$ slope, and consequently $\sigma_{\mathrm{el}}/\sigma_{\mathrm{tot}} \sim s^{\epsilon}/\ln s$).

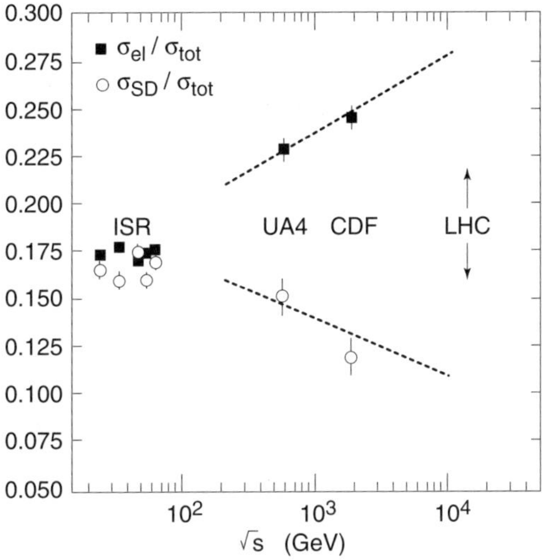


**Fig. 7.19.** A compilation of integrated single-diffractive cross section data (from TOTEM 1997).

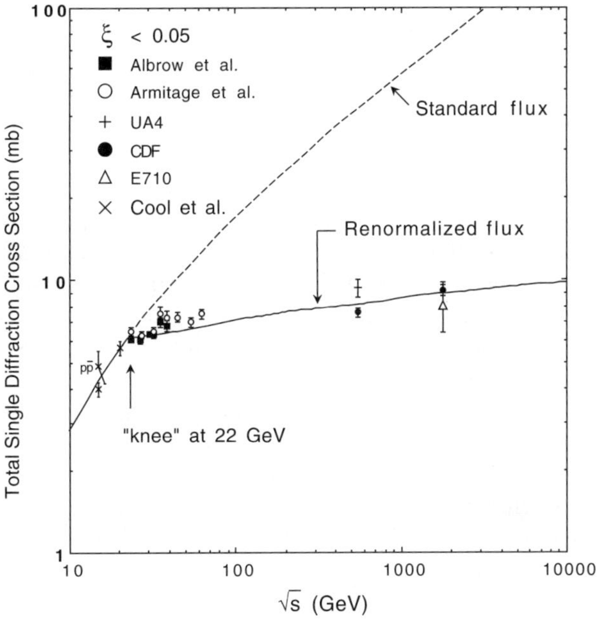


**Fig. 7.20.** Energy dependence of the integrated single-diffraction cross section (from Goulianos 1995).

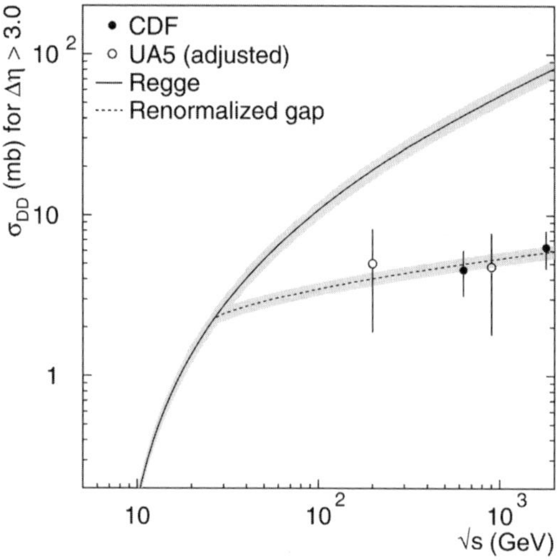


**Fig. 7.21.** Energy dependence of the integrated double-diffraction cross section (Affolder et al. 2001a).

Fig. 7.20, taken from Goulianos (1995), collects the existing measurements of $\sigma^{SD}$ and shows that $\sigma^{SD}$ does not continue to increase with energy following the triple-Regge behavior (which would eventually violate unitarity). The observed flattening of the integrated single-diffraction cross section has been attributed by Goulianos (1995) (see also Goulianos and Montanha 1999) to the saturation of the pomeron flux factor. In other terms, it is suggested that $f_{I\!P}(\xi, t)$ integrated over $\xi$ and $t$ should not exceed unity. Therefore, above some energy value ($\sqrt{s} = 22$ GeV) $f_{I\!P}(\xi, t)$ is renormalized, and this gives the solid curve in Fig. 7.20. If one reinterprets the pomeron flux as a rapidity gap probability (recall that $\Delta\eta = \ln\frac{1}{\xi}$), renormalizing the pomeron flux is equivalent to demanding that the integrated gap probability be always $\leq 1$. As shown in Fig. 7.21, also the integrated double-diffractive cross section seems to scale with $s$ (Affolder et al. 2001a).

A detailed study of single diffractive dissociation in the medium-$|t|$ range ($(0.8 < |t| < 2.0)$ GeV$^2$) has been recently presented by the UA8 Collaboration (Brandt et al. 1986). An interesting result of the combined analysis of these data and of the old ISR data is that the pomeron trajectory requires a term quadratic in $t$. The best fit is achieved with $\alpha_{I\!P}(t) = 1.10 + 0.25\,t + \alpha''\,t^2$ and $\alpha'' = (0.079 \pm 0.012)$ GeV$^{-4}$.

## 7.6 Soft Diffraction at RHIC and LHC

We conclude this chapter by a brief overview of the perspectives of soft diffraction at RHIC and LHC.

The Relativistic Heavy Ion Collider (RHIC) at Brookhaven National Laboratory operates with proton beams in the $\sqrt{s}$ range from 60 to 500 GeV and with a luminosity $\mathcal{L} \approx 10^{32}$ cm$^{-2}$ s$^{-1}$. Initially the diffraction studies at RHIC will center around the PP2PP experiment (Guryn 2001). Until 2003 PP2PP will measure elastic cross sections, single diffraction dissociation and double diffraction dissociation in the medium $|t|$ range, $0.006 < |t| < 1.3$ GeV$^2$. After 2003, an unpgrade of the RHIC machine will allow PP2PP to reach very small $|t|$ values, down to $4 \cdot 10^{-4}$ GeV$^2$, with the possibility of measuring the $\rho$ parameter and total cross sections. Note that the highest c.m. energy at RHIC ($\sqrt{s} = 500$ GeV) is about the same as at SPS. The comparison of the $pp$ and $p\bar{p}$ data from these two colliders will certainly be very interesting. An added value of RHIC, which has the unique capability of accelerating polarized protons with a high average polarization (70% for each beam), is the investigation of spin effects in elastic $pp$ scattering at large $\sqrt{s}$ (for an overview of spin physics see Leader 2001).

At the CERN Large Hadron Collider (LHC) protons and antiprotons will collide at the c.m. energy $\sqrt{s} = 14$ TeV. The physics of diffraction will be explored by the TOTEM experiment (TOTEM 1997; Matthiae 2001). As mentioned in Sect. 7.1, the measurement of the $pp$ total cross section at $\sqrt{s} = 14$ TeV is extremely important to discriminate between different asymptotic behaviors of $\sigma_{\text{tot}}$. Also, the large design luminosity of the LHC allows elastic scattering to be measured up to high values of $|t|$, in the range 10-15 GeV$^2$. TOTEM plans to study in detail the first dip (expected around $|t| \sim 0.6$ GeV$^2$ at the LHC energy) and the high-$|t|$ tail, with a search for possible secondary dips. Due to the very large c.m. energy, the mass spectrum of diffractive dissociation will be explored up to very large $M$ (when $x_F = 0.95$ the mass $M$ is as large as 3 TeV). On the other hand, the measurement of the real part of the amplitude near the forward direction is not so easy at the LHC, since it requires the study of the Coulomb-nuclear interference region around $|t| \approx 10^{-3}$ (corresponding to a scattering angle of few μrad). From an experimental point of view detecting events at such small angles is a serious challenge (TOTEM 1997).

We conclude this section on the experimental perspectives in the domain of soft diffraction by recalling that the Tevatron machine at Fermilab has recently undergone some upgrades which have brought its c.m. energy to $\sqrt{s} = 2$ TeV and increased its luminosity. Run II at the Tevatron will provide new data on elastic scattering, total cross sections and single diffraction, besides a wealth of other measurements of hard diffraction observables (see e.g., Santoro 2001).

# 8. The Pomeron in QCD

In Regge theory the pomeron is a trajectory with vacuum quantum numbers and intercept $\alpha_{I\!P}(0) = 1$, which was originally introduced to explain the constancy with energy of the total cross sections.

Although quite successful, Regge theory is nonetheless only a phenomenological picture of hadronic phenomena. The fundamental theory of strong interactions is *quantum chromodynamics*. One may then wonder if there is any room for the pomeron in QCD. We shall see in this chapter that the answer is positive. As shown by Lipatov and collaborators (Lipatov 1976; Kuraev, Lipatov and Fadin 1976, 1977; Balitsky and Lipatov 1978; Lipatov 1986), the perturbative QCD pomeron is essentially a gluon ladder (with some subtleties to be explained below).

In the following we shall explore the QCD nature of the pomeron and the BFKL theory (after Balitsky, Fadin, Kuraev, Lipatov). Detailed treatments of the subject can be found in Lipatov (1989), Del Duca (1995), Lipatov (1997), Forshaw and Ross (1997).

## 8.1 Early Approaches

In the early 70's some attempts were made to incorporate the pomeron in the quark picture of hadrons. A model in this direction was proposed by Landshoff and Polkinghorne and predated QCD (Landshoff and Polkinghorne 1971; Jaroszkiewicz and Landshoff 1974). The main assumption of the model is that the pomeron couples to quarks as a $C = +1$ photon. Thus, elastic scattering is described by the diagram in Fig. 8.1a, where the elementary process, i.e. quark–quark scattering (Fig. 8.1b), has the following amplitude

$$-g^2 \left[ \frac{1 + \mathrm{e}^{-i\pi\alpha_{I\!P}(t)}}{\sin \pi\alpha_{I\!P}(t)} \left( \frac{s}{s_0} \right)^{\alpha_{I\!P}(t)-1} \right] (\overline{u}\gamma^\mu u)\,(\overline{u}\gamma_\mu u)\ . \tag{8.1}$$

Here $g$ is the $I\!P$-quark coupling, $u$ is the quark spinor and the quantity in square brackets is the pomeron propagator.

The main support to a pointlike $I\!P$-quark coupling comes from the success of the additive quark model (Levin and Frankfurt 1965; Lipkin and Scheck 1966; Kokkedee and Van Hove 1966) which, among other things, correctly

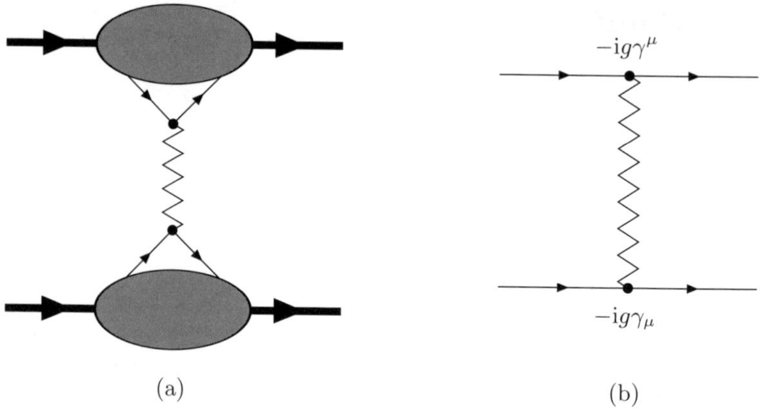


**Fig. 8.1.** (**a**) Elastic scattering in the Landshoff–Polkinghorne model. (**b**) The elementary process: quark–quark scattering mediated by the pomeron.

predicts some ratios of total cross sections (for instance, $\sigma_{\rm tot}^{\pi p}/\sigma_{\rm tot}^{pp} = 2/3$). Another element in favor of a photon-like pomeron is the (quasi) absence of helicity flip in pomeron exchange.

Using the amplitude (8.1) one obtains for the elastic $pp$ differential cross section

$$\frac{d\sigma_{\rm el}}{dt} = \frac{g^4\,[3\,F_1(t)]^4}{4\pi\,\sin^2\left(\frac{\pi\alpha_{I\!P}(t)}{2}\right)}\left(\frac{s}{s_0}\right)^{2\alpha_{I\!P}(t)-2}, \qquad (8.2)$$

where $F_1(t)$ is the charge form factor of the proton. Donnachie and Landshoff (1979, 1984) fitted Eq. (8.2) to the data, thus extracting $\alpha_{I\!P}(t)$ and $g$. The values they obtained are

$$\alpha_{I\!P}(0) = 1.08, \quad \alpha'_{I\!P} = 0.25\ \mathrm{GeV}^{-2}, \quad g^4 = 3.21\,\mathrm{GeV}^{-2}\,. \qquad (8.3)$$

Here we find two figures which have been extensively used and quoted in the modern literature on the pomeron: the intercept $\alpha_{I\!P}(0) = 1.08$, traditionally associated to the *soft pomeron*[1], and the slope $\alpha'_{I\!P} = 0.25$ GeV$^{-}$2. As discussed in Sect. 7.1, a fit of a large set of total cross sections (Donnachie and Landshoff 1992) has confirmed these values.

When it was realized that QCD was more than a simple candidate to the theory of hadronic interactions, the problem arised as to interpret the pomeron in terms of the QCD degrees of freedom, quarks and gluons. Low and Nussinov (Low 1975; Nussinov 1975, 1976) proposed to picture pomeron exchange as a *two-gluon exchange.* Two is the minimal number of gluons needed to reproduce the $I\!P$ quantum numbers. In the Low-Nussinov model

[1] The same value of $\alpha_{I\!P}(0)$ had already been found by Collins, Gault and Martin (1974). Values of $\alpha_{I\!P}(0)$ greater than unity were also considered by Capella and Kaplan (1974), and by Dubovikov et al. (1977).

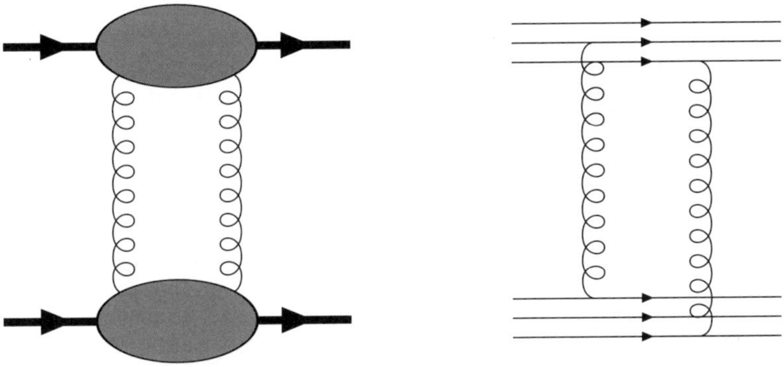

(a) (b)

**Fig. 8.2.** (**a**) Elastic scattering via two–gluon exchange in the Low–Nussinov model. (**b**) The elementary process: quark–quark scattering mediated by two gluons (only one of the possible diagrams is drawn).

elastic scattering is described as in Fig. 8.2a and the elementary process is quark–quark scattering via two-gluon exchange (Fig. 8.2b). This general picture represents the Born approximation to the pomeron in QCD.

It is worth mentioning that an attempt was made (Landshoff and Nachtmann 1987) to reconciliate the pomeron model of Landshoff and Polkinghorne with the two-gluon QCD picture. In general, the two exchanged gluons may couple either to the same quark or to different quarks in the colliding hadrons. In the latter case, it is not possible to describe the process in terms of a point-like quark-pomeron coupling. However, if the gluon correlation length in the QCD vacuum is much smaller than the hadron radius, the diagrams where the gluons do not couple to the same quark are strongly suppressed and the additive quark rule holds. Thus, the coupling of the gluon pair turns out to be similar to that of a $C = +1$ photon (Landshoff and Nachtmann 1987). The conclusion is that non-perturbative effects establish a bridge between the two models discussed above.

## 8.2 The Perturbative QCD Pomeron

The aim of the present chapter is to give a perturbative QCD content to the notion of *pomeron exchange* (Fig. 8.3). The basic process we shall study is quark–quark scattering at high energy (i.e., in the limit $s \gg |t|$). The amplitude of this process will be computed in the *leading* $\ln(s/|t|)$ *approximation* $\ln s$ approximation (LLA).

The resummation procedures of large logarithms of the type $\ln(s/|t|)$ date back to the studies of the high-energy limit of quantum electrodynamics by Cheng and Wu (1969, 1970a, 1970b) and Frolov, Gribov and Lipatov (1970a,

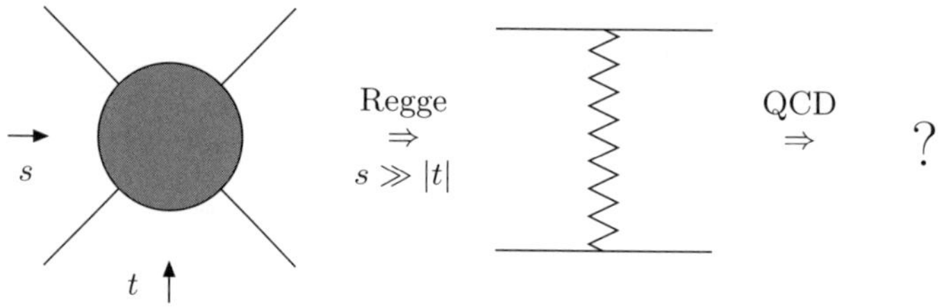


**Fig. 8.3.** Hadron–hadron scattering at high energies.

1970b). It is easy to see, using power-counting arguments, that at large $s$ and fixed $t$ the leading contribution to a given scattering amplitude comes from photon exchange in the crossed channel. This contribution leads to a total cross section approaching asymptotically a constant value. Let us then suppose that along the photon exchanged in the $t$-channel a fermion pair is produced. This $\mathcal{O}(\alpha_{\rm em}^2)$ radiative correction contains a large logarithm, $\ln(s/|t|)$. Resumming the leading powers of $\alpha_{\rm em}^2 \ln(s/|t|)$ in the perturbative expansion one gets a total cross section growing with $s$. This technique was extended to non-abelian gauge theories, and in particular to QCD, by Lipatov and coworkers (Lipatov 1976; Kuraev, Lipatov and Fadin 1976, 1977; Balitsky and Lipatov 1978). An important difference emerges with respect to QED: due to the gluon self-coupling, radiative corrections containing a large logarithm $\ln(s/t)$ appear already at $\mathcal{O}(\alpha_s)$. Therefore, they are a factor $\alpha_s/\alpha_{\rm em}^2$ stronger than the corresponding corrections in QED, and may be relevant in high-energy scattering processes.

The important achievement of Kuraev, Lipatov and Fadin (1976, 1977) (see also Lipatov 1976; Tyburski 1976; Cheng and Lo 1976, 1977; Frankfurt and Sherman 1977) is the proof of the *gluon reggeization* to all orders in perturbation theory (keeping the leading $\ln s$ terms at each order). By "reggeized gluon" we mean a gluon with a modified propagator of the form (in Feynman gauge)

$$D_{\mu\nu}(s,q^2) = -{\rm i}\,\frac{g_{\mu\nu}}{q^2}\left(\frac{s}{s_0}\right)^{\alpha_g(q^2)-1}, \tag{8.4}$$

where $\alpha_g(q^2) = 1 + \epsilon(q^2)$ is the (perturbatively calculable) Regge trajectory of the gluon. In this context, the pomeron emerges as a gluon ladder in a color-singlet configuration, with reggeized gluons on the vertical lines and non-local effective triple-gluon vertices (the so-called Lipatov vertices).

## 8.3 Quark–Quark Scattering in Leading $\ln s$ Approximation

We shall now compute the quark–quark scattering amplitude in the leading $\ln s$ approximation (LLA). This means that, at each order in $\alpha_s$, only the leading powers of ln $s$ will be retained. The calculation will be carried out in parallel for color-singlet and color-octet exchange.

### 8.3.1 One-Gluon Exchange

At first order in the strong coupling constant, quark–quark scattering proceeds via one-gluon exchange (Fig. 8.4). This process does not contribute to the pomeron structure since a single gluon is a colored object (whereas the pomeron is colorless). As already mentioned, the lowest-order diagrams contributing to the pomeron are the two-gluon exchange diagrams. Our discussion, however, will start from one-gluon exchange because this is the first step in the construction of the reggeized gluon.

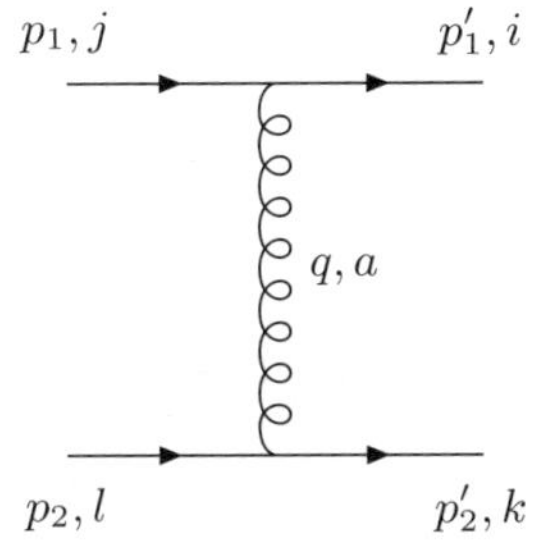


**Fig. 8.4.** Quark–quark scattering via one-gluon exchange.

Consider the $\mathcal{O}(\alpha_s)$ process of Fig. 8.4. The quarks are assumed to have different flavors so that there is no $u$-channel diagram with the outgoing legs interchanged. In order to simplify our formulas we shall omit the helicity indices: one should just recall that quark helicities are conserved. We introduce the Sudakov parametrization of the gluon momentum (see App. A.2)

$$q = \alpha\, p_1 + \beta\, p_2 + q_\perp \,, \tag{8.5}$$

where $q_\perp = (0, \boldsymbol{q}, 0)$ is a four-vector with only transverse components (with respect to the axis of the incoming quarks)[2]. The c.m. energy squared is

$$s = 2\, p_1 \cdot p_2 \,, \tag{8.6}$$

whereas the momentum transfer squared is

$$t = q^2 = 2\,\alpha\beta\, p_1 \cdot p_2 - \boldsymbol{q}^2 = \alpha\beta\, s - \boldsymbol{q}^2 \,. \tag{8.7}$$

The mass-shell conditions for the outgoing quarks read

[2] Throughout this chapter, boldface vectors are transverse two-vectors.

$$(p_1 - q)^2 = -(1-\alpha)\beta\, s - \boldsymbol{q}^2 = 0\,, \tag{8.8}$$
$$(p_2 + q)^2 = \alpha(1+\beta)\, s - \boldsymbol{q}^2 = 0\,. \tag{8.9}$$

Equations (8.8, 8.9) imply $\alpha = -\beta$ and when $s \gg -q^2$ one finds

$$\alpha = |\beta| \simeq \frac{\boldsymbol{q}^2}{s} \ll 1 \quad \text{and} \quad q^2 \simeq -\boldsymbol{q}^2\,. \tag{8.10}$$

In the following we shall use the fact that in the large-$s$ limit all components of the exchanged momentum are much smaller than $p_1$ and $p_2$.

The one-gluon exchange amplitude, in Feynman gauge, is

$$\mathrm{i}A^{(0)}(s,t) = \mathrm{i}\, g_s^2\, t^a_{ij} t^a_{kl}\, \bar{u}(p_1')\gamma^\mu u(p_1)\, \frac{g_{\mu\nu}}{q^2}\, \bar{u}(p_2')\gamma^\nu u(p_2)\,. \tag{8.11}$$

From now on, $i, j, k, l, \ldots$ are quark color indices, whereas $a, b, c, \ldots$ are gluon color indices. Note that $A^{(0)}$ should read $A^{(0)}_{ijkl}$, but for simplicity we shall systematically omit the color indices, unless otherwise stated.

The square of (8.11), averaged and summed over colors, is

$$\overline{|A^{(0)}|^2} = g_s^4\, \frac{N_c^2-1}{4\,N_c^2}\, 2\left(\frac{s^2+u^2}{t^2}\right)\,, \tag{8.12}$$

where the color factor is computed as follows (repeated indices are summed)

$$\begin{aligned} \text{color factor} &= \frac{1}{N_c^2}\,(t^a_{ij}t^a_{kl})(t^b_{ij}t^b_{kl})^* = \frac{1}{N_c^2}\, t^a_{ij}t^a_{kl}\, t^b_{ji}t^b_{lk} \\ &= \frac{1}{N_c^2}\,\mathrm{Tr}\,(t^a t^b)\,\mathrm{Tr}\,(t^a t^b) = \frac{N_c^2-1}{4\,N_c^2} = \frac{2}{9}\,. \end{aligned} \tag{8.13}$$

In the limit of large $s$ one has $u \simeq -s$ and (8.12) becomes

$$\overline{|A^{(0)}|^2} = \frac{8}{9}\, g_s^4\, \frac{s^2}{t^2}\,. \tag{8.14}$$

**Eikonal Vertices.** There is another, more convenient, way to derive the high-energy behavior (8.14). It consists in performing the large-$s$ limit already at the amplitude level by means of the so-called *eikonal vertices*. All the subsequent calculations will be based on this procedure, that we now outline.

Consider the upper vertex in the diagram of Fig. 8.4

$$-ig_s\, \bar{u}(p_1+q)\gamma^\mu u(p_1)\,. \tag{8.15}$$

We have said above that when $s \gg |t|$ all $q^\mu$ components are small with respect to the components of $p_1^\mu$ and $p_2^\mu$. Thus we can approximate (8.15) as

$$\begin{aligned} -\mathrm{i}\, g_s\, \bar{u}(p_1+q)\gamma^\mu u(p_1) &\simeq -\mathrm{i}\, g_s\, \bar{u}(p_1)\gamma^\mu u(p_1) \\ &= -2\mathrm{i}\, g_s\, p_1^\mu\,. \end{aligned} \tag{8.16}$$

This is the *quark-gluon eikonal vertex.*

Approximating the lower vertex in a similar way, we get for the amplitude at large-$s$

$$A^{(0)} = g_s^2\, t^a_{ij} t^a_{kl}\, \frac{4\, p_1 \cdot p_2}{q^2} = 8\pi\alpha_s\, t^a_{ij} t^a_{kl}\, \frac{s}{t}\,. \tag{8.17}$$

Squaring (8.17) we obtain

$$\overline{|A^{(0)}|^2} = \frac{8}{9}\, g_s^4\, \frac{s^2}{t^2}\,. \tag{8.18}$$

which coincides with (8.14).

### 8.3.2 Digression: Gluon–Gluon Scattering

Let us now make a digression from $qq$ scattering to show how eikonal vertices similar to (8.16) can be introduced for the three-gluon coupling as well. This has the important consequence that the high-energy limits of $gg$ scattering and $qq$ scattering differ only by color factors.

Consider the diagram of Fig. 8.5 (the $t$-channel contribution to gluon–gluon scattering). Its amplitude is

$$\begin{aligned} \mathrm{i}\, A^{(0)} &= -\mathrm{i}\, g_s^2\, f^{aa'c} f^{bb'c} \left[ g_{\mu\mu'}(p_1 + p_1')_\rho + g_{\rho\mu'}(-p_1' + q)_\mu - g_{\mu\rho}(p_1 + q)_{\mu'} \right] \\ &\quad \times \frac{g^{\rho\sigma}}{q^2} \left[ g_{\nu\nu'}(p_2 + p_2')_\sigma + g_{\sigma\nu'}(-p_2' - q)_\nu - g_{\nu\rho}(p_2 - q)_{\nu'} \right] \\ &\quad \times \epsilon^{\mu}_{\lambda_1}(p_1)\, \epsilon^{\mu'}_{\lambda_1'}(p_1')\, \epsilon^{\nu}_{\lambda_2}(p_2)\, \epsilon^{\nu'}_{\lambda_2'}(p_2')\,. \end{aligned} \tag{8.19}$$

Neglecting $q$ with respect to the four-momenta of the external gluons we can approximate the upper vertex as

$$\begin{aligned} g_s f^{aa'c} & \left[ g_{\mu\mu'}(p_1 + p_1')_\rho + g_{\rho\mu'}(-p_1' + q)_\mu - g_{\mu\rho}(p_1 + q)_{\mu'} \right] \\ & \simeq g_s f^{aa'c} \left[ 2 g_{\mu\mu'} p_{1\rho} - g_{\rho\mu'} p_{1\mu} - g_{\mu\rho} p'_{1\mu'} \right]\,. \end{aligned} \tag{8.20}$$

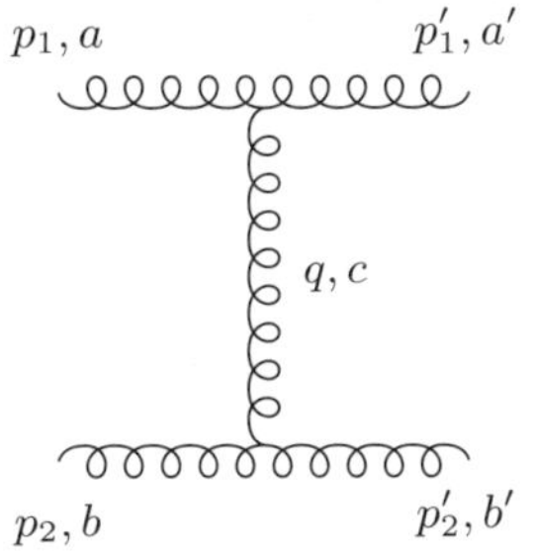


**Fig. 8.5.** Gluon–gluon scattering via one–gluon exchange.

Observing that the last two terms vanish if the external gluons have physical polarizations, because $\epsilon(p_1)\cdot p_1 = \epsilon(p_1')\cdot p_1' = 0$, we get the *three-gluon eikonal vertex*

$$2g_s f^{aa'c}\, g_{\mu\mu'} p_{1\rho} \,. \tag{8.21}$$

The similarity with the quark-gluon eikonal vertex (8.16) is evident. Both vertices are proportional to the momentum of the incoming particle, which has large components.

We can now rewrite the amplitude (8.20) in the high-energy limit as

$$A^{(0)} = -\mathrm{i}\, g_s^2\, f^{aa'c} f^{bb'c}\, g_{\mu\mu'} \frac{4\, p_1 \cdot p_2}{q^2}\, g_{\nu\nu'} \\ \times \epsilon^{\mu}_{\lambda_1}(p_1)\, \epsilon^{\mu'}_{\lambda_1'}(p_1')\, \epsilon^{\nu}_{\lambda_2}(p_2)\, \epsilon^{\nu'}_{\lambda_2'}(p_2') \,. \tag{8.22}$$

Squaring (8.22), we perform the sum over the gluon helicities by

$$\sum_{\lambda} \epsilon^{\mu}_{\lambda}(p) \epsilon^{\nu *}_{\lambda}(p) = -\left(g^{\mu\nu} - \frac{p^{\mu} n^{\nu} + p^{\nu} n^{\mu}}{p \cdot n}\right) , \tag{8.23}$$

where $n$ is a null vector, $n^2 = 0$. In particular, we choose $n = p_2$ when summing over $\epsilon^{\mu}_{\lambda_1}(p_1)$, $n = p_1$ when summing over $\epsilon^{\nu}_{\lambda_2}(p_1)$, and so on. Equation (8.23) guarantees that only physical polarization states are taken into account (remember that this is the crucial assumption for the legitimate use of the eikonal vertex (8.21)). In fact, writing the metric tensor in terms of Sudakov vectors as

$$g^{\mu\nu} = g^{\mu\nu}_{\perp} + \frac{2}{s}(p_1^{\mu} p_2^{\nu} + p_1^{\nu} p_2^{\mu}) \,, \tag{8.24}$$

the sum (8.23) for the upper incoming gluon, with the choice $n = p_2$, becomes

$$\sum_{\lambda_1} \epsilon^{\mu}_{\lambda_1}(p_1) \epsilon^{\mu' *}_{\lambda_1}(p_1) = -\left[g^{\mu\nu} - \frac{2}{s}\,(p_1^{\mu} p_2^{\nu} + p_1^{\nu} p_2^{\mu})\right] = -g^{\mu\nu}_{\perp} \,, \tag{8.25}$$

which shows explicitly that only physical polarizations are summed.

In conclusion, we obtain for $\overline{|A^{(0)}_{gg}|^2}$

$$\overline{|A^{(0)}|^2} = g_s^4\, \frac{C_A^2}{N_c^2 - 1}\, 4\, \frac{s^2}{t^2} = \frac{9}{2}\, g_s^4\, \frac{s^2}{t^2} \,, \tag{8.26}$$

where the color factor has been computed as follows

$$\text{color factor} = \frac{1}{(N_c^2 - 1)^2}\, f^{aa'c} f^{bb'c} f^{a'ad} f^{b'bd} \\ = \frac{1}{(N_c^2 - 1)^2}\, C_A^2 (N_c^2 - 1) = \frac{C_A^2}{N_c^2 - 1} = \frac{9}{8} \,. \tag{8.27}$$

Here $C_A = N_c$ is the Casimir factor of the adjoint representation of $SU(N_c)$, under which the gluons transform.

Equation (8.26) has been obtained considering only the $t$-channel diagram of Fig. 8.5. There are other diagrams contributing to $gg$ scattering at tree level, with one gluon exchange in the $s$-channel and in the $u$-channel. It turns out however that, if one uses (8.23), these are subleading contributions in $s/|t|$. Thus (8.26) is actually the end of the story as far as the large-$s$ limit is taken.

Comparing (8.14) with (8.26), one sees that at large $s$ the cross section for $qq$ scattering is the same as for $gg$ scattering, except a different color factor (Combridge, Kripfgang and Ranft 1977; Cutler and Sivers 1977, 1978)

$$\overline{|A_{gg}^{(0)}|^2} = \left(\frac{9}{4}\right)^2 \overline{|A_{qq}^{(0)}|^2} \, . \tag{8.28}$$

Here $9/4 = C_A/C_F$ is the relative color strength of gluons and quarks. $C_F = (N_c^2 - 1)/2N_c$ is the Casimir factor of the fundamental representation of $SU(N_c)$, under which the quarks transform. For completeness we also give the relation between the quark-gluon scattering amplitude and the quark–quark amplitude, namely

$$\overline{|A_{qg}^{(0)}|^2} = \frac{9}{4} \overline{|A_{qq}^{(0)}|^2} \, . \tag{8.29}$$

The different powers of 9/4 in (8.28) and (8.29) are evidently related to the different number of interacting gluons in the two cases.

### 8.3.3 Two-Gluon Exchange

Let us return to $qq$ scattering and consider one-loop (i.e., $\mathcal{O}(\alpha_s)$) corrections.

The relevant diagrams are the two-gluon exchange diagrams (Fig. 8.6). We can easily convince ourselves that diagrams with self-energy insertions

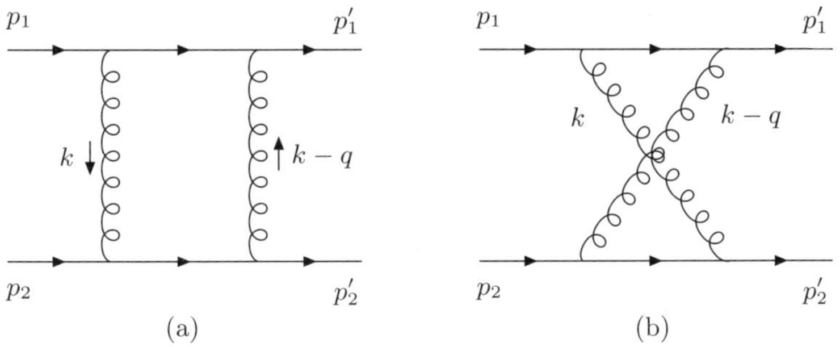


**Fig. 8.6.** Quark–quark scattering via two–gluon exchange: (**a**) the box diagram; (**b**) the crossed diagram.

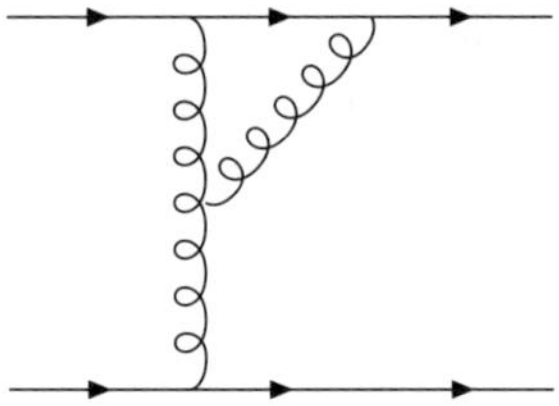


**Fig. 8.7.** Example of a diagram subleading in $\ln s$ (vertex correction).

and vertex corrections (Fig. 8.7) are subleading in $\ln s$ and can be neglected in LLA (for a justification of this statement, see, e.g., Forshaw and Ross 1997). Since these diagrams determine the running of the strong coupling, in a leading logarithmic calculation $\alpha_s$ must be regarded as a *fixed constant.*

The most convenient way to calculate the scattering amplitudes, or, to be precise, the imaginary part of them, is to use unitarity in the form of cutting rules (Cutkosky 1960). The real part of the amplitudes can then be obtained by dispersive techniques.

In the case under discussion we have

$$\mathrm{Im}\, A^{(1)}(s,t) = \frac{1}{2} \int \mathrm{d}\Pi_2\, A^{(0)}(s,k^2)\, A^{(0)\dagger}(s,(k-q)^2)\,, \tag{8.30}$$

where $t \equiv q^2$. For $|t| \ll s$ the momentum transferred $q$ is dominated by its transverse components, as can be seen by imposing the on-shellness of the outgoing quarks. Thus we have $t \simeq -\boldsymbol{q}^2$.

The amplitudes on the right-hand side of (8.30) are the one-gluon exchange amplitudes computed in Sect. 8.3.1. The two-body phase space is

$$\begin{aligned}\int \mathrm{d}\Pi_2 &= \int \frac{\mathrm{d}^4\kappa_1}{(2\pi)^3}\frac{\mathrm{d}^4\kappa_2}{(2\pi)^3}\,\delta(\kappa_1^2)\,\delta(\kappa_2^2)\,(2\pi)^4\,\delta(p_1+p_2-\kappa_1-\kappa_2)\\ &= \int \frac{\mathrm{d}^4 k}{(2\pi)^2}\,\delta((p_1-k)^2)\,\delta((p_2+k)^2)\end{aligned} \tag{8.31}$$

Introducing the Sudakov decomposition of $k^\mu$

$$k = \alpha\, p_1 + \beta\, p_2 + k_\perp\,, \tag{8.32}$$

and using

$$\mathrm{d}^4 k = \frac{s}{2}\,\mathrm{d}\alpha\,\mathrm{d}\beta\,\mathrm{d}^2\boldsymbol{k}\,, \tag{8.33}$$

we can rewrite (8.31) as

$$\int \mathrm{d}\Pi_2 = \frac{s}{8\pi^2}\int \mathrm{d}\alpha\,\mathrm{d}\beta\,\mathrm{d}^2\boldsymbol{k}\,\delta(-\beta(1-\alpha)s+\boldsymbol{k}^2)\,\delta(\alpha(1+\beta)s-\boldsymbol{k}^2)\,. \tag{8.34}$$

In the large-$s$ limit the delta functions of (8.34) imply

$$\alpha = |\beta| \simeq \frac{\boldsymbol{k}^2}{s} \ll 1 , \tag{8.35}$$

$$k^2 \simeq -\boldsymbol{k}^2 , \quad (k-q)^2 \simeq -(\boldsymbol{k}-\boldsymbol{q})^2 , \tag{8.36}$$

with

$$\boldsymbol{k}^2 \simeq (\boldsymbol{k}-\boldsymbol{q})^2 \simeq \boldsymbol{q}^2 . \tag{8.37}$$

Thus Eq. (8.34) simplifies to

$$\int \mathrm{d}\Pi_2 = \frac{1}{8\pi^2 s} \int \mathrm{d}\alpha\, \mathrm{d}\beta\, \mathrm{d}^2\boldsymbol{k}\, \delta\left(\beta + \frac{\boldsymbol{k}^2}{s}\right) \delta\left(\alpha - \frac{\boldsymbol{k}^2}{s}\right) = \frac{1}{8\pi^2 s} \int \mathrm{d}^2\boldsymbol{k} . \tag{8.38}$$

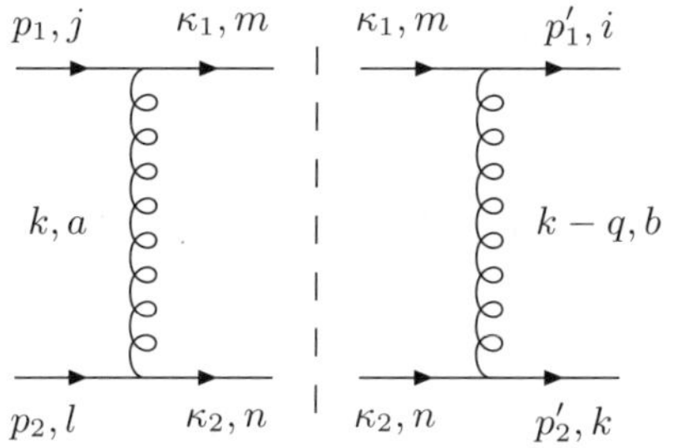


**Fig. 8.8.** The imaginary part of the box diagram.

Let us now consider the box diagram (Fig. 8.8). Using (8.38) and the results of Sect. 8.3.1 for $A^{(0)}$, that is – see (8.17)

$$A^{(0)}(s, k^2) = -8\pi\alpha_s\, (t^a_{mj} t^a_{nl})\, \frac{s}{\boldsymbol{k}^2} , \tag{8.39}$$

$$A^{(0)\dagger}(s, (k-q)^2) = -8\pi\alpha_s\, (t^b_{mi} t^b_{nk})^*\, \frac{s}{(\boldsymbol{k}-\boldsymbol{q})^2} , \tag{8.40}$$

we get from (8.30) the imaginary part of the box amplitude $\mathrm{Im}\, A^{(1)}_{\mathrm{a}}$:

$$\mathrm{Im}\, A^{(1)}_{\mathrm{a}}(s,t) = 4\alpha_s^2\, (t^a t^b)_{ij} (t^a t^b)_{kl}\, s \int \frac{\mathrm{d}^2\boldsymbol{k}}{\boldsymbol{k}^2\, (\boldsymbol{k}-\boldsymbol{q})^2} . \tag{8.41}$$

The full amplitude is reconstructed by means of a dispersion relation (see Sect. 4.6). In leading ln $s$ one finds

$$A^{(1)}_{\mathrm{a}}(s,t) = -\frac{4\alpha_s^2}{\pi}\, (t^a t^b)_{ij} (t^a t^b)_{kl}\, s \ln\left(\frac{s}{t}\right) \int \frac{\mathrm{d}^2\boldsymbol{k}}{\boldsymbol{k}^2\, (\boldsymbol{k}-\boldsymbol{q})^2} . \tag{8.42}$$

Equation (8.42) yields immediately the real and imaginary parts if one uses $\ln(-s) = \ln s - i\pi$,

$$A_{\rm a}^{(1)}(s,t) = -\frac{4\alpha_s^2}{\pi}\,(t^a t^b)_{ij}(t^a t^b)_{kl}\, s\left[\ln\left(\frac{s}{|t|}\right) - \mathrm{i}\pi\right]\int \frac{\mathrm{d}^2\boldsymbol{k}}{\boldsymbol{k}^2\,(\boldsymbol{k}-\boldsymbol{q})^2}\,. \quad (8.43)$$

Note that this amplitude is predominantly real. In view of future developments we rewrite the amplitude (8.42) as

$$A_{\rm a}^{(1)}(s,t) = -\frac{16\pi\alpha_s}{N_c}\,(t^a t^b)_{ij}(t^a t^b)_{kl}\,\frac{s}{t}\ln\left(\frac{s}{t}\right)\epsilon(t)\,, \quad (8.44)$$

where the dimensionless function $\epsilon(t)$ incorporates the transverse-momentum integration

$$\epsilon(t) = \frac{N_c\alpha_s}{4\pi^2}\int \mathrm{d}^2\boldsymbol{k}\,\frac{-\boldsymbol{q}^2}{\boldsymbol{k}^2\,(\boldsymbol{k}-\boldsymbol{q})^2}\,. \quad (8.45)$$

The integral in (8.45) is infrared divergent. This singularity arises because the external quarks are on mass-shell. In a real situation, however, quarks are confined inside hadrons and their off-shellness provides an infrared cutoff $\mu^2$. Equation (8.45) then becomes

$$\epsilon(t) = \frac{N_c\alpha_s}{4\pi^2}\int \mathrm{d}^2\boldsymbol{k}\,\frac{-\boldsymbol{q}^2}{(\boldsymbol{k}^2+\mu^2)\,[(\boldsymbol{k}-\boldsymbol{q})^2+\mu^2]}\,, \quad (8.46)$$

and the evaluation of the integral gives

$$\epsilon(t) = -\frac{N_c\alpha_s}{2\pi}\ln\left(\frac{\boldsymbol{q}^2}{\mu^2}\right)\,. \quad (8.47)$$

Consider now the contribution of the crossed diagram (Fig. 8.9). Its amplitude $A_{\rm b}^{(1)}$ is obtained from the box contribution $A_{\rm a}^{(1)}$ by replacing $s$ with $u$ and properly changing the color factor, that is

$$A_{\rm b}^{(1)}(s,t) = -\frac{16\pi\alpha_s}{N_c}\,(t^a t^b)_{ij}(t^b t^a)_{kl}\,\frac{u}{t}\ln\left(\frac{u}{t}\right)\epsilon(t)\,, \quad (8.48)$$

and, since $u \simeq -s$, this becomes

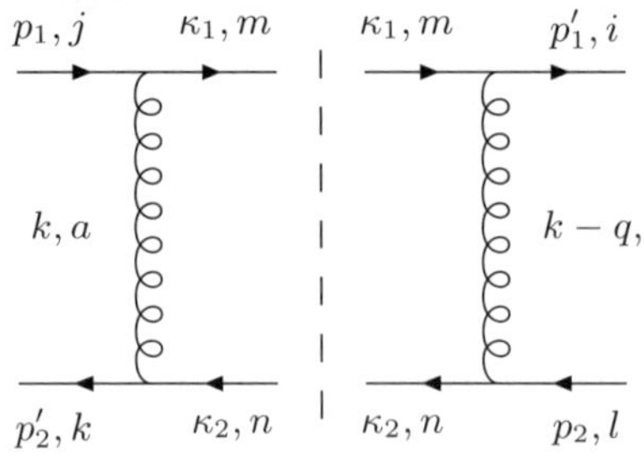


**Fig. 8.9.** The imaginary part of the crossed diagram.

$$A_{\rm b}^{(1)}(s,t) = \frac{16\pi\alpha_s}{N_c}\,(t^a t^b)_{ij}(t^b t^a)_{kl}\,\frac{s}{t}\ln\left(\frac{s}{|t|}\right)\,\epsilon(t)\;. \tag{8.49}$$

Summing the box and the crossed amplitudes we finally obtain

$$\begin{aligned} A^{(1)}(s,t) &= A_{\rm a}^{(1)}(s,t) + A_{\rm b}^{(1)}(s,t) \\ &= -\frac{16\pi\alpha_s}{N_c}\,(t^a t^b)_{ij}\,\frac{s}{t} \\ &\times \left\{[t^a,t^b]_{kl}\ln\left(\frac{s}{|t|}\right) - \mathrm{i}\pi\,(t^a t^b)_{kl}\right\}\epsilon(t)\;. \end{aligned} \tag{8.50}$$

The reason why we retained the imaginary term, although it is subleading in ln $s$, is that the color-singlet component of (8.50) receives contribution only from that term, as we shall shortly see.

**Color Projectors.** To determine the scattering amplitudes for color-singlet and color-octet exchange, we proceed as follows. The quark–quark scattering amplitude $A_{kl}^{ij}$ (we restore temporarily the color indices) can be decomposed as a sum over the $SU(3)$ representations in the $t$-channel

$$A_{kl}^{ij}(s,t) = \sum_R \mathcal{P}_{lk}^{ji}(R)\,\mathcal{A}_R(s,t)\;, \tag{8.51}$$

where the color dependence on the right-hand side is contained in $\mathcal{P}_{lk}^{ji}(R)$, the color projectors for quark–quark scattering (see, e.g., Lipatov 1989). The amplitudes for color-singlet and color-octet exchange are

$$A_{\underline{1},kl}^{ij}(s,t) = \mathcal{P}_{lk}^{ji}(\underline{1})\,\mathcal{A}_{\underline{1}}(s,t)\;, \tag{8.52}$$

$$A_{\underline{8},kl}^{ij}(s,t) = \mathcal{P}_{lk}^{ji}(\underline{8})\,\mathcal{A}_{\underline{8}}(s,t)\;, \tag{8.53}$$

with

$$\mathcal{P}_{kl}^{ij}(\underline{1}) = \frac{1}{N_c}\,\delta_{ij}\delta_{kl}\;, \tag{8.54}$$

$$\mathcal{P}_{kl}^{ij}(\underline{8}) = 2\,t_{ij}^a t_{kl}^a\;, \tag{8.55}$$

The normalization of these projectors is

$$\mathcal{P}_{kl}^{ij}(R)\mathcal{P}_{mn}^{lk}(R') = \mathcal{P}_{mn}^{ij}(R)\,\delta_{RR'}\;. \tag{8.56}$$

Using (8.56) we get from (8.51)

$$\mathcal{A}_{\underline{1}}(s,t) = \mathcal{P}_{lk}^{ji}(\underline{1})A_{kl}^{ij}(s,t)\;, \tag{8.57}$$

$$\mathcal{A}_{\underline{8}}(s,t) = \frac{1}{N_c^2-1}\,\mathcal{P}_{lk}^{ji}(\underline{8})A_{kl}^{ij}(s,t)\;. \tag{8.58}$$

**Color-Octet Exchange.** Making use of (8.58) and (8.55), we extract from (8.50) the $\mathcal{O}(\alpha_s^2)$ amplitude for quark–quark scattering via color-octet exchange in leading $\ln s$ approximation

$$\mathcal{A}_{\underline{8}}^{(1)}(s,t) = -\frac{16\pi\alpha_s}{N_c}\,\mathcal{C}_{\underline{8}}^{(1)}\,\frac{s}{t}\,\ln\left(\frac{s}{|t|}\right)\,\epsilon(t)\,, \tag{8.59}$$

where the color factor is (we use the formulas of Appendix C)

$$\begin{aligned}
\mathcal{C}_{\underline{8}}^{(1)} &= \frac{1}{N_c^2-1}\,\mathcal{P}_{lk}^{ji}(\underline{8})\,(t^a t^b)_{ij}\,[t^a,t^b]_{kl} \\
&= \frac{2}{N_c^2-1}\,t_{lk}^c t_{ji}^c\,(t^a t^b)_{ij}\,\mathrm{i}f^{abd}t_{kl}^d \\
&= \frac{2}{N_c^2-1}\,\mathrm{Tr}\,(t^a t^b t^c)\,\mathrm{i}f^{abd}\,\mathrm{Tr}\,(t^c t^d) \\
&= \frac{1}{N_c^2-1}\,\mathrm{Tr}\,(t^a t^b t^c)\,\mathrm{i}f^{abc} \\
&= \frac{1}{N_c^2-1}\,\frac{1}{4}\,(d^{abc}+\mathrm{i}f^{abc})\,\mathrm{i}f^{abc} \\
&= \frac{1}{N_c^2-1}\left(-\frac{1}{4}\right)f^{abc}f^{abc} \\
&= \frac{1}{N_c^2-1}\left(-\frac{1}{4}\right)N_c(N_c^2-1) = -\frac{N_c}{4}\,.
\end{aligned} \tag{8.60}$$

Therefore we have, using (8.53) and (8.55)

$$A_{\underline{8}}^{(1)}(s,t) = 8\pi\alpha_s\,t_{ij}^a t_{kl}^a\,\frac{s}{t}\,\ln\left(\frac{s}{|t|}\right)\,\epsilon(t)\,. \tag{8.61}$$

Note that the color-octet amplitude is real and $\mathcal{O}(\ln s)$ at one-loop level.

**Color-Singlet Exchange.** For the color-singlet amplitude (which contributes to the pomeron) we proceed analogously. From (8.50) and (8.54) one can see that, since $\delta_{kl}\,[t^a,t^b]_{kl}=0$, the terms proportional to $\ln(s/|t|)$ coming from the box and the crossed diagrams cancel each other, and the surviving (box) contribution is subdominant in $\ln s$. It reads

$$\mathcal{A}_{\underline{1}}^{(1)}(s,t) = \frac{16\mathrm{i}\pi^2\,\alpha_s}{N_c}\,\mathcal{C}_{\underline{1}}^{(1)}\,\frac{s}{t}\,\epsilon(t)\,, \tag{8.62}$$

with

$$\begin{aligned}
\mathcal{C}_{\underline{1}}^{(1)} &= \mathcal{P}_{lk}^{ji}(\underline{1})\,(t^a t^b)_{ij}(t^a t^b)_{kl} \\
&= \frac{1}{N_c}\,\mathrm{Tr}\,(t^a t^b)\,\mathrm{Tr}\,(t^a t^b) \\
&= \frac{1}{4N_c}\,\delta_{ab}\delta_{ab} = \frac{N_c^2-1}{4N_c}\,.
\end{aligned} \tag{8.63}$$

Hence, using (8.52) and (8.54), we get

$$A_{\underline{1}}^{(1)}(s,t) = 4\mathrm{i}\pi^2\,\alpha_s\,\delta_{ij}\delta_{kl}\,\frac{N_c^2-1}{N_c^2}\,\frac{s}{t}\,\epsilon(t)\ , \tag{8.64}$$

Note that color-singlet gluon exchange, which starts to develop at order $\alpha_s^2$, is suppressed by a factor $\ln s$ with respect to the color-octet exchange at the same order in $\alpha_s$.

Let us summarize the one-loop results. We found that combining direct and crossed channel contributions yields a *real* color-octet amplitude in leading $\ln s$ approximation and an *imaginary* color-singlet amplitude. The former, which starts already at $\mathcal{O}(\alpha_s)$, is enhanced by a factor $\ln s$ compared to the latter, which starts at $\mathcal{O}(\alpha_s^2)$.

In the language of Regge theory (see Sect. 5.6) this behavior can be rephrased by saying that color-octet and color-singlet amplitudes have opposite signature, $\xi_{\underline{1}} = +1$ and $\xi_{\underline{8}} = -1$, and contain a factor

$$\left(1 + \xi_R\,\mathrm{e}^{-\mathrm{i}\pi\alpha_R(t)}\right)\ , \tag{8.65}$$

where $\alpha_R(t) = 1 + \mathcal{O}(\alpha_s)$ (see below, Sect. 8.3.5). If we expand the exponential as a power series in $\alpha_s$ and retain the dominant term, we find that the factor (8.65) is real and $\mathcal{O}(1)$ for color octet, and imaginary and $\mathcal{O}(\alpha_s)$ for color singlet. As it will be confirmed by the subsequent calculations at higher orders, in leading logarithmic approximation the color-octet amplitude is real, whereas the color-singlet amplitude is imaginary.

We now go further, to $\mathcal{O}(\alpha_s^2)$ corrections.

### 8.3.4 The Simplest Gluon Ladder

The two-loop graphs contributing to the quark–quark scattering amplitude are shown in Fig. 8.10. In particular, the diagram of Fig. 8.10a represents the simplest gluon ladder, with one rung. We shall now see, by an explicit calculation, that, in LLA, the whole $\mathcal{O}(\alpha_s^2)$ correction can be incorporated into a ladder diagram like that of Fig. 8.10a, but with a non-local three-gluon vertex.

We use again the unitarity relations. Hence we have to cut the diagrams of Fig. 8.10 in all possible ways, with the simplification that one-loop graphs with self-energy or vertex corrections to the left or to the right of the cut are subdominant in $\ln s$ and therefore can be ignored.

Let us start from the *real-gluon emission* diagrams. Three of them diagrams are shown in Fig. 8.11. The momenta of the exchanged gluons are parametrized as

$$k_1 = \alpha_1\,p_1 + \beta_1\,p_2 + k_{1\perp}\ , \tag{8.66}$$

$$k_2 = \alpha_1\,p_2 + \beta_2\,p_2 + k_{2\perp}\ . \tag{8.67}$$

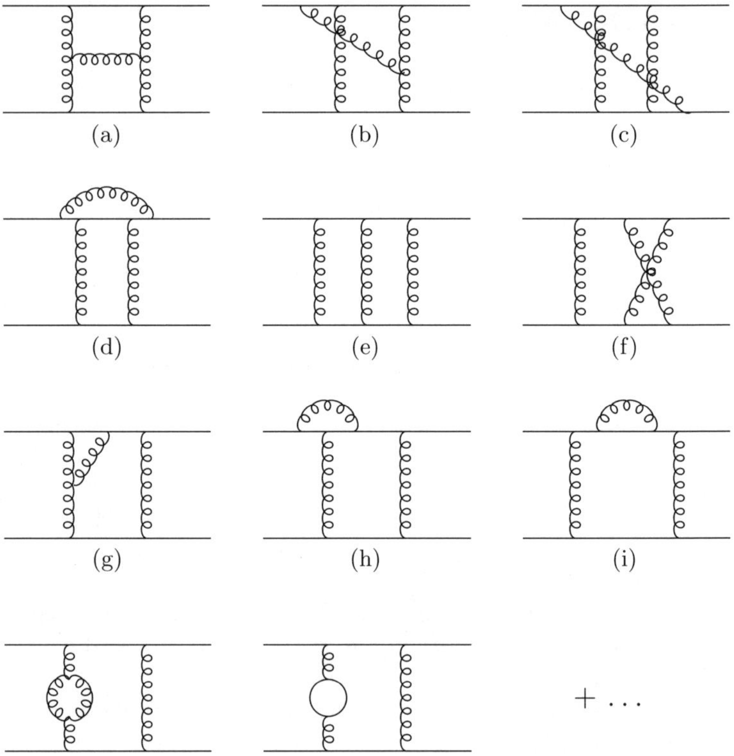


**Fig. 8.10a–k.** A sample of two-loop diagrams contributing to $qq$ scattering.

The leading $\ln s$ contributions come from the kinematical regime

$$1 \gg \alpha_1 \gg \alpha_2 \,, \tag{8.68}$$

$$1 \gg |\beta_2| \gg |\beta_1| \,. \tag{8.69}$$

The on-shellness of the outgoing gluon implies

$$\alpha_1 \beta_2 s = -(\boldsymbol{k}_1 - \boldsymbol{k}_2)^2 \,, \tag{8.70}$$

and therefore

$$k_1^2 \simeq -\boldsymbol{k}_1^2 \,, \quad k_2^2 \simeq -\boldsymbol{k}_2^2 \,. \tag{8.71}$$

All transverse momenta are of the same order of magnitude

$$\boldsymbol{k}_1^2 \simeq \boldsymbol{k}_2^2 \simeq \boldsymbol{q}^2 \,. \tag{8.72}$$

Using eikonal couplings and working in Feynman gauge, the amplitude corresponding to the diagram of Fig. 8.11a is

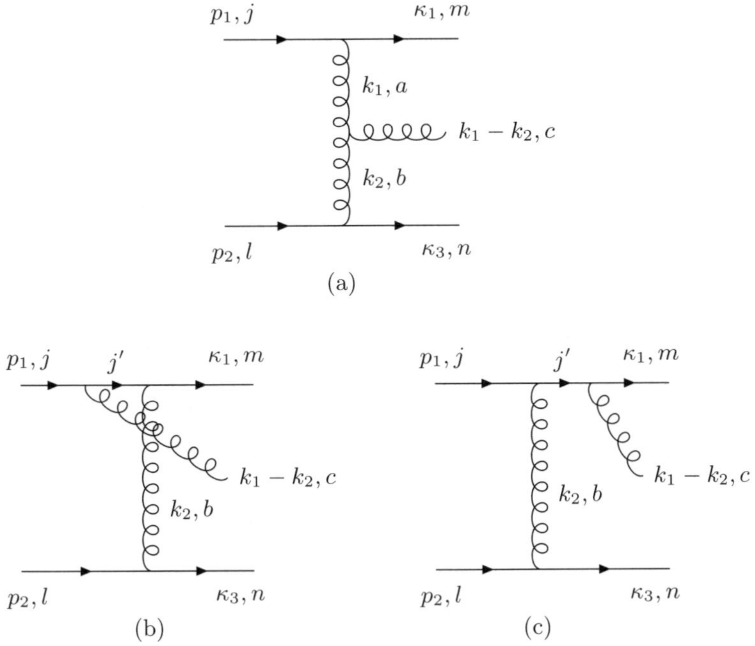


**Fig. 8.11a–c.** Diagrams for the process $qq \to qqg$ (real gluon emission).

$$\begin{aligned} \mathrm{i}A^{\rho}_{2\to3,\mathrm{a}} = {} & (-2\mathrm{i}g_s\, p_1^{\mu})\, t^a_{mj} \left(-\frac{\mathrm{i}}{k_1^2}\right) \\ & \times g_s f_{abc} \left[(k_1+k_2)^{\rho} g^{\mu\nu} + (k_1 - 2k_2)^{\mu} g^{\nu\rho} + (k_2 - 2k_1)^{\nu} g^{\rho\mu}\right] \\ & \times \left(-\frac{\mathrm{i}}{k_2^2}\right) (-2\mathrm{i}g_s\, p_2^{\nu})\, t^b_{nl} \,. \end{aligned} \tag{8.73}$$

With the kinematics (8.68–8.71) this becomes

$$A^{\rho}_{2\to3,\mathrm{a}} = -2\mathrm{i}g_s^3\, f_{abc}\, t^a_{mj} t^b_{nl}\, \frac{1}{\boldsymbol{k}_1^2 \boldsymbol{k}_2^2} \left[\alpha_1 p_1^{\rho} + \beta_2 p_2^{\rho} - (k_1^{\rho} + k_2^{\rho})\right] . \tag{8.74}$$

Consider now the diagrams with gluon emission from the quark lines (Fig. 8.11b,c). We have

$$\begin{aligned} \mathrm{i}A^{\rho}_{2\to3,\mathrm{b}} = {} & (-2\mathrm{i}g_s\, p_1^{\rho})\, t^c_{j'j}\, \frac{\mathrm{i}}{(p_1 - k_1 + k_2)^2}\, (-2\mathrm{i}g_s)(p_1^{\mu} - k_1^{\mu} + k_2^{\mu}) t^b_{mj'} \\ & \times \left(-\frac{\mathrm{i}}{k_2^2}\right) (-2\mathrm{i}g_s p_{2\mu})\, t^b_{nl} \,, \end{aligned} \tag{8.75}$$

for the diagram of Fig. 8.11b, and

$$\mathrm{i}A^{\rho}_{2\to3,\mathrm{c}} = (-2\mathrm{i}g_s\, p_{1\mu})\, t^b_{j'j} \left(-\frac{\mathrm{i}}{k_2^2}\right) (-2\mathrm{i}g_s p_2^{\mu}) t^b_{nl}$$

$$\times \frac{\mathrm{i}}{(p_1 - k_2)^2} (-2\mathrm{i}g_s)(p_1^\rho - k_2^\rho)\, t^c_{mj'} \,, \tag{8.76}$$

for the diagram of Fig. 8.11c. With the simplifications arising from the kinematics (8.68–8.71) we get

$$A^\rho_{2\to 3,\mathrm{b}} = -4g_s^3 s\, (t^b t^c)_{mj} t^b_{nl} \frac{1}{\beta_2 s \boldsymbol{k}_2^2}\, p_1^\rho \,, \tag{8.77}$$

$$A^\rho_{2\to 3,\mathrm{c}} = 4g_s^3 s\, (t^c t^b)_{mj} t^b_{nl} \frac{1}{\beta_2 s \boldsymbol{k}_2^2}\, p_1^\rho \,. \tag{8.78}$$

Summing these two contributions, we find, using $[t^b, t^c] = \mathrm{i} f_{abc} t^a$,

$$A^\rho_{2\to 3,\mathrm{b+c}} = -4\mathrm{i}g_s^3 s\, f_{abc} t^a_{mj} t^b_{nl} \frac{1}{\beta_2 s \boldsymbol{k}_2^2}\, p_1^\rho \,. \tag{8.79}$$

The two similar diagrams with the real gluon attached to the lower quark line (Fig. 8.12) give

$$-4\mathrm{i}g_s^3 s\, f_{abc} t^a_{mj} t^b_{nl} \frac{1}{\alpha_1 s \boldsymbol{k}_1^2}\, p_2^\rho \,. \tag{8.80}$$

Putting (8.74), (8.79) and (8.80) together, we obtain the full $\mathcal{O}(g_s^3)$ contribution to the amplitude for $qq$ scattering with emission of a real gluon, namely

$$A^\rho_{2\to 3} = -4\mathrm{i}g_s^3 \frac{p_1^\mu p_2^\nu}{\boldsymbol{k}_1^2 \boldsymbol{k}_2^2}\, t^a_{mj} t^b_{nl}\, f_{abc}\, \Gamma^\rho_{\mu\nu} \,, \tag{8.81}$$

where $\Gamma^\rho_{\mu\nu}$ is the *Lipatov effective vertex*

$$\Gamma^\rho_{\mu\nu}(k_1, k_2) = \frac{2\, p_{2\mu} p_{1\nu}}{s} \Bigg[ \left( \alpha_1 + \frac{2\, \boldsymbol{k}_1^2}{\beta_2 s} \right) p_1^\rho + \left( \beta_2 + \frac{2\, \boldsymbol{k}_2^2}{\alpha_1 s} \right) p_2^\rho - (k^\rho_{1\perp} + k^\rho_{2\perp}) \Bigg] \,. \tag{8.82}$$

This vertex, which is non-local as it incorporates the propagators of the emitted gluon in the diagrams of Fig. 8.11b,c and Fig. 8.12, has the important

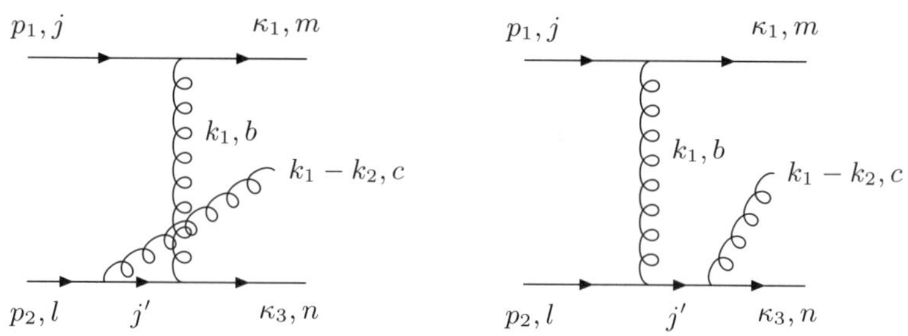


**Fig. 8.12.** Diagrams for the process $qq \to qqg$ (real gluon emission).

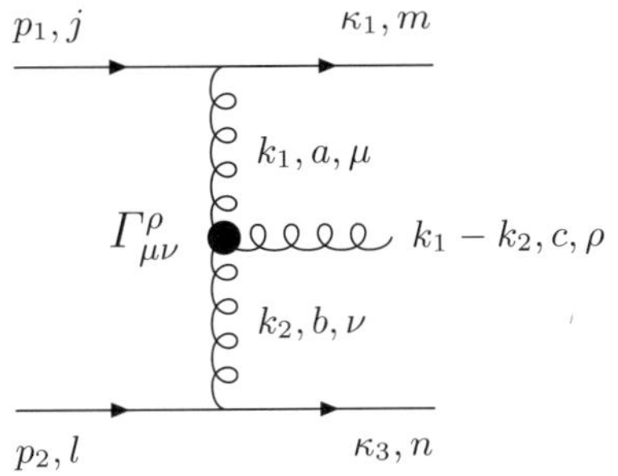


**Fig. 8.13.** The Lipatov effective vertex.

property of being gauge-invariant. It is easy to verify in fact that it satisfies the Ward identity

$$(k_{1\rho} - k_{2\rho})\, \Gamma^\rho_{\mu\nu}(k_1, k_2) = 0\,. \tag{8.83}$$

The lesson we learned is that all graphs with one gluon in the final state are summed up by the effective diagram of Fig. 8.13, where the Lipatov three–gluon vertex is represented by the fat dot. The Feynman rule for this vertex is $g_s f_{abc} \Gamma^\rho_{\mu\nu}$ and the amplitude for the diagram of Fig. 8.13 is consequently written as

$$\begin{aligned} \mathrm{i}A^\rho_{2\to 3} &= (-2\mathrm{i}g_s\, p_1^\mu)\, t^a_{mj} \left(-\frac{\mathrm{i}}{k_1^2}\right) f_{abc}\, g_s\, \Gamma^\rho_{\mu\nu}(k_1, k_2) \\ &\times \left(-\frac{\mathrm{i}}{k_2^2}\right) (-2\mathrm{i}g_s\, p_2^\nu)\, t^b_{nl}\,, \end{aligned} \tag{8.84}$$

which coincides with (8.81).

For future convenience we introduce the quantity

$$C^\rho(k_1, k_2) = \left(\alpha_1 + \frac{2\,\boldsymbol{k}_1^2}{\beta_2 s}\right) p_1^\rho + \left(\beta_2 + \frac{2\,\boldsymbol{k}_2^2}{\alpha_1 s}\right) p_2^\rho - (k^\rho_{1\perp} + k^\rho_{2\perp})\,, \tag{8.85}$$

so that

$$\Gamma^\rho_{\mu\nu} = \frac{2p_{2\mu}p_{1\nu}}{s}\, C^\rho\,, \tag{8.86}$$

and

$$p_{1\mu}p_{2\nu}\, \Gamma^\rho_{\mu\nu} = \frac{s}{2}\, C^\rho\,. \tag{8.87}$$

The amplitude (8.84 then reads

$$A^\rho_{2\to 3} = 2\mathrm{i}g_s\, t^a_{mj} \left(\frac{\mathrm{i}}{\boldsymbol{k}_1^2}\right) f_{abc}\, g_s\, C^\rho(k_1, k_2) \left(\frac{\mathrm{i}}{\boldsymbol{k}_2^2}\right) g_s\, t^b_{nl}\,. \tag{8.88}$$

Using the unitarity relations, the contribution of the diagrams of Fig. 8.11 and 8.12 to the imaginary part of the $qq$ scattering amplitude – that we call *real-gluon contribution* – is given by (see Fig. 8.14)

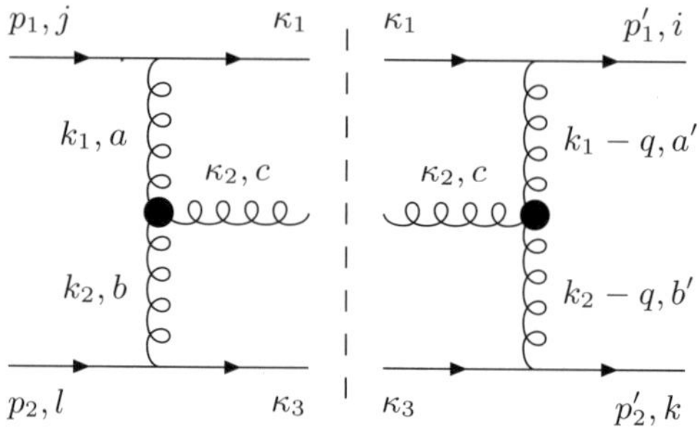


**Fig. 8.14.** The real-gluon contribution to $qq$ scattering at order $\alpha_s^3$.

$$\operatorname{Im} A^{(2)}_{\text{real}}(s,t) = \frac{-g_{\rho\sigma}}{2} \int \mathrm{d}\Pi_3 \, A^{\rho}_{2\to 3}(k_1, k_2)\, A^{\sigma\dagger}_{2\to 3}(k_1 - q, k_2 - q) \,. \tag{8.89}$$

The sum over the helicity of the intermediate gluon has been done by means of

$$\sum_{\lambda} \varepsilon^{\mu}_{\lambda}(p) \varepsilon^{\nu *}_{\lambda}(p) = -g^{\mu\nu} \,, \tag{8.90}$$

since the Lipatov vertex is gauge invariant.

The three-body phase space in (8.89) is (see Fig. 8.14 for notations)

$$\int \mathrm{d}\Pi_3 = \frac{1}{(2\pi)5} \int \mathrm{d}^4 k_1 \, \mathrm{d}^4 k_2 \, \delta((p_1 - k_1)^2)\, \delta((p_2 + k_2)^2)\, \delta((k_1 - k_2)^2) \,, \tag{8.91}$$

and, with the parametrizations (8.66) and (8.67), we have

$$\begin{aligned} \int \mathrm{d}\Pi_3 = \frac{s^2}{4(2\pi)^5} & \int \mathrm{d}\alpha_1 \, \mathrm{d}\beta_1 \, \mathrm{d}^2 \boldsymbol{k}_1 \int \mathrm{d}\alpha_2 \, \mathrm{d}\beta_2 \, \mathrm{d}^2 \boldsymbol{k}_2 \\ & \times \delta(-\beta_1(1-\alpha_1)s - \boldsymbol{k}_1^2)\, \delta(\alpha_2(1+\beta_2)s - \boldsymbol{k}_2^2) \\ & \times \delta((\alpha_1 - \alpha_2)(\beta_1 - \beta_2)s - (\boldsymbol{k}_1 - \boldsymbol{k}_2)^2) \,. \end{aligned} \tag{8.92}$$

Using (8.68, 8.69, 8.71) we can approximate (8.92) by

$$\begin{aligned} \int \mathrm{d}\Pi_3 &= \frac{s^2}{4(2\pi)^5} \int \mathrm{d}\alpha_1 \, \mathrm{d}\beta_1 \, \mathrm{d}^2 \boldsymbol{k}_1 \int \mathrm{d}\alpha_2 \, \mathrm{d}\beta_2 \, \mathrm{d}^2 \boldsymbol{k}_2 \\ & \quad \times \delta(-\beta_1 s - \boldsymbol{k}_1^2)\, \delta(\alpha_2 s - \boldsymbol{k}_2^2)\, \delta(-\alpha_1 \beta_2 s - (\boldsymbol{k}_1 - \boldsymbol{k}_2)^2) \\ &= \frac{1}{4(2\pi)^5} \int_{\alpha_2}^{1} \frac{\mathrm{d}\alpha_1}{\alpha_1} \int_0^1 \mathrm{d}\alpha_2 \int \mathrm{d}^2 \boldsymbol{k}_1 \int \mathrm{d}^2 \boldsymbol{k}_2 \, \delta(\alpha_2 s - \boldsymbol{k}_2^2) \\ &= \frac{1}{4(2\pi)^5 s} \int_{\boldsymbol{q}^2/s}^{1} \frac{\mathrm{d}\alpha_1}{\alpha_1} \int \mathrm{d}^2 \boldsymbol{k}_1 \int \mathrm{d}^2 \boldsymbol{k}_2 \,. \end{aligned} \tag{8.93}$$

The amplitude to the right of the cut in the diagram of Fig. 8.14 is

$$A^{\rho\dagger}_{2\to 3} = -2\mathrm{i}g_s\, t^{a'}_{im} \left(\frac{-\mathrm{i}}{(k_1-q)^2}\right)$$
$$\times (-f_{a'bc'}g_s)\, C^\rho(-(k_1-q), -(k_2-q)) \left(\frac{-\mathrm{i}}{\boldsymbol{k}_2^2}\right) g_s\, t^{b'}_{kn}\,. \quad (8.94)$$

From (8.87) and (8.94) we get

$$A^\rho_{2\to 3}(k_1,k_2)\, A^\dagger_{2\to 3,\rho}(k_1-q, k_2-q)$$
$$= 4g_s^6 s^2\, \mathcal{G}_{\mathrm{real}}\, \frac{C^\rho(k_1,k_2) C_\rho(-k_1+q, -k_2+q)}{\boldsymbol{k}_1^2\boldsymbol{k}_2^2\,(\boldsymbol{k}_1-\boldsymbol{q})^2\,(\boldsymbol{k}_1-\boldsymbol{q})^2}\,, \quad (8.95)$$

where $\mathcal{G}_{\mathrm{real}}$ is the color factor

$$\mathcal{G}_{\mathrm{real}} = -(t^{a'}t^a)_{ij}(t^{b'}t^b)_{kl}\, f_{abc}f_{a'b'c}\,. \quad (8.96)$$

The contraction of the Lipatov vertices yields

$$C^\rho(k_1,k_2)\, C_\rho(-(k_1-q), -(k_2-q))$$
$$= -2\left[\boldsymbol{q}^2 - \frac{\boldsymbol{k}_1^2\,(\boldsymbol{k}_2-\boldsymbol{q})^2}{(\boldsymbol{k}_1-\boldsymbol{k}_2)^2} - \frac{\boldsymbol{k}_2^2\,(\boldsymbol{k}_1-\boldsymbol{q})^2}{(\boldsymbol{k}_1-\boldsymbol{k}_2)^2}\right]. \quad (8.97)$$

Using (8.97) in (8.95), and replacing (8.95) into (8.89) with the phase space (8.93), we obtain the imaginary part of the $\mathcal{O}(\alpha_s^2)$ real radiative correction to the $qq$ scattering amplitude

$$\mathrm{Im}\, A^{(2)}_{\mathrm{real}}(s,t) = \frac{2\alpha_s^3}{\pi^2}\, \mathcal{G}_{\mathrm{real}}\, s\, \ln\left(\frac{s}{|t|}\right) \int \mathrm{d}^2\boldsymbol{k}_1^2 \int \mathrm{d}^2\boldsymbol{k}_2^2$$
$$\times \left[\frac{\boldsymbol{q}^2}{\boldsymbol{k}_1^2\boldsymbol{k}_2^2(\boldsymbol{k}_1-\boldsymbol{q})^2(\boldsymbol{k}_2-\boldsymbol{q})^2} - \frac{1}{\boldsymbol{k}_2^2(\boldsymbol{k}_1-\boldsymbol{q})^2(\boldsymbol{k}_1-\boldsymbol{k}_2)^2}\right.$$
$$\left. - \frac{1}{\boldsymbol{k}_1^2(\boldsymbol{k}_2-\boldsymbol{q})^2(\boldsymbol{k}_1-\boldsymbol{k}_2)^2}\right]. \quad (8.98)$$

There is another class of two-loop contributions to be taken into account: the virtual radiative corrections shown in Fig. 8.15 (there are two extra diagrams, not displayed, with one gluon exchanged to the left of the cut). The imaginary part of the corresponding amplitude is given by

$$\mathrm{Im}\, A^{(2)}_{\mathrm{virt}}(s,t) = \frac{1}{2}\int \mathrm{d}\Pi_2\, A^{(1)}(s,k_2^2)\, A^{(0)\dagger}(s,(k_2-q)^2)$$
$$+\frac{1}{2}\int \mathrm{d}\Pi_2\, A^{(0)}(s,k_1^2)\, A^{(1)\dagger}(s,(k_1-q)^2)\,. \quad (8.99)$$

Let us consider the first term of (8.99), that we call $\mathrm{Im}\, A^{(2)}_{\mathrm{virt,a}}$. Making use of results previously obtained (cfr. Eqs. (8.17) and (8.61)),

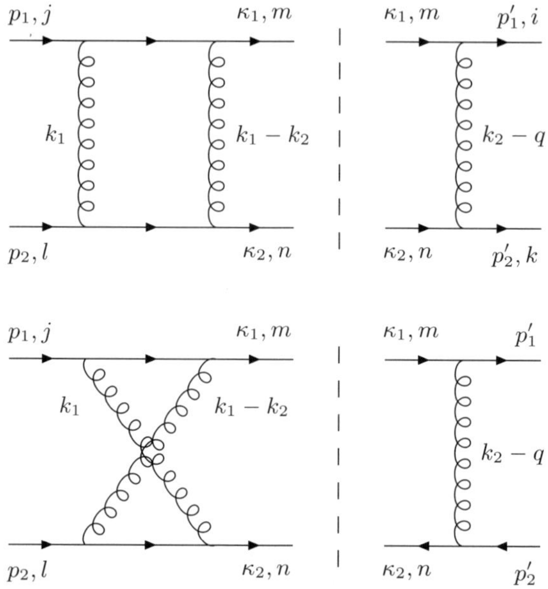


**Fig. 8.15.** Virtual-gluon contributions to $qq$ scattering at order $\alpha_s^3$.

$$A^{(1)}(s, k_2^2) = 8\pi\alpha_s \, t^b_{mj} t^b_{nl} \frac{s}{k_2^2} \ln\left(\frac{s}{\boldsymbol{k}_2^2}\right) \epsilon(t) \,, \tag{8.100}$$

$$A^{(0)\dagger}(s, (k_2 - q)^2) = 8\pi\alpha_s \, t^{a*}_{mi} t^{a*}_{nk} \frac{s}{(k_2 - q)^2} \,, \tag{8.101}$$

and recalling (8.38) and the definition (8.45), we find

$$\begin{aligned} \mathrm{Im}\, A^{(2)}_{\mathrm{virt,a}}(s,t) &= -\frac{N_c \alpha_s^3}{\pi^2} \, \mathcal{G}_{\mathrm{virt}} \, s \ln\left(\frac{s}{|t|}\right) \\ &\quad \times \int \mathrm{d}^2\boldsymbol{k}_1 \int \mathrm{d}^2\boldsymbol{k}_2 \, \frac{1}{\boldsymbol{k}_1^2 (\boldsymbol{k}_2 - \boldsymbol{q})^2 (\boldsymbol{k}_1 - \boldsymbol{k}_2)^2} \,, \end{aligned} \tag{8.102}$$

where we used $\ln(s/\boldsymbol{k}_2^2) \simeq \ln(s/|t|)$ and the color factor is

$$\mathcal{G}_{\mathrm{virt}} = (t^a t^b)_{ij} \, (t^a t^b)_{kl} \,. \tag{8.103}$$

The second term in (8.99) – that we call $\mathrm{Im}\, A^{(2)}_{\mathrm{virt,b}}$ – is computed analogously and reads

$$\begin{aligned} \mathrm{Im}\, A^{(2)}_{\mathrm{virt,b}}(s,t) &= -\frac{N_c \alpha_s^3}{\pi^2} \, \mathcal{G}_{\mathrm{virt}} \, s \ln\left(\frac{s}{|t|}\right) \\ &\quad \times \int \mathrm{d}^2\boldsymbol{k}_1 \int \mathrm{d}^2\boldsymbol{k}_2 \, \frac{1}{\boldsymbol{k}_2^2 (\boldsymbol{k}_1 - \boldsymbol{q})^2 (\boldsymbol{k}_1 - \boldsymbol{k}_2)^2} \,. \end{aligned} \tag{8.104}$$

The full virtual-gluon discontinuity is obtained by summing (8.102) and (8.104) and is given by

$$\mathrm{Im}\, A^{(2)}_{\mathrm{virt}}(s,t) = -\frac{N_c \alpha_s^3}{\pi^2}\, \mathcal{G}_{\mathrm{virt}}\, s \ln\left(\frac{s}{|t|}\right) \int \mathrm{d}^2\boldsymbol{k}_1 \int \mathrm{d}^2\boldsymbol{k}_2 \times \left[\frac{1}{\boldsymbol{k}_1^2(\boldsymbol{k}_2-\boldsymbol{q})^2(\boldsymbol{k}_1-\boldsymbol{k}_2)^2} + \frac{1}{\boldsymbol{k}_2^2(\boldsymbol{k}_1-\boldsymbol{q})^2(\boldsymbol{k}_1-\boldsymbol{k}_2)^2}\right]. \quad (8.105)$$

We are now going to treat separately the color-octet and color-singlet exchange.

**Color-Octet Exchange.** In the color-octet case, the $u$-channel contribution is equal and opposite to the $s$-channel contribution, but with the interchange $t^b \leftrightarrow t^{b'}$ in (8.96) and $(t^a t^b)_{kl} \leftrightarrow (t^b t^a)_{kl}$ in (8.103). Thus, the $u$-channel terms are automatically accounted for by the replacements

$$\mathcal{G}_{\mathrm{real}} \to \mathcal{G}'_{\mathrm{real}} = -(t^{a'} t^a)_{ij}\, [t^{b'}, t^b]_{kl}\, f_{abc} f_{a'b'c}\,, \quad (8.106)$$

$$\mathcal{G}_{\mathrm{virt}} \to \mathcal{G}'_{\mathrm{virt}} = (t^a t^b)_{ij}\, [t^a, t^b]_{kl}\,. \quad (8.107)$$

It proves convenient to rewrite $\mathcal{G}'_{\mathrm{real}}$ as

$$\begin{aligned}\mathcal{G}'_{\mathrm{real}} &= -(t^{a'} t^a)_{ij}\, [t^{b'}, t^b]_{kl}\, \frac{1}{2}\left(f_{abc} f_{a'b'c} - f_{ab'c} f_{a'bc}\right)\\ &= -(t^{a'} t^a)_{ij}\, [t^{b'}, t^b]_{kl}\, \frac{1}{2}\, f_{aa'c} f_{bb'c}\,,\end{aligned} \quad (8.108)$$

where in the second line we used the Jacobi identity for the $f^{abc}$ (see App. C).

For the real-gluon contribution we have, using (8.53, 8.55),

$$\mathrm{Im}\, A^{(2)}_{\underline{8},\mathrm{real}}(s,t) = \frac{2\alpha_s^3}{\pi^2}\, 2\, \mathcal{C}^{(2)}_{\underline{8},\mathrm{real}}\, t^a_{ij} t^a_{kl}\, s \ln\left(\frac{s}{|t|}\right) \int \mathrm{d}^2\boldsymbol{k}_1^2 \int \mathrm{d}^2\boldsymbol{k}_2^2 \times \left[\frac{\boldsymbol{q}^2}{\boldsymbol{k}_1^2\boldsymbol{k}_2^2(\boldsymbol{k}_1-\boldsymbol{q})^2(\boldsymbol{k}_2-\boldsymbol{q})^2} - \frac{1}{\boldsymbol{k}_2^2(\boldsymbol{k}_1-\boldsymbol{q})^2(\boldsymbol{k}_1-\boldsymbol{k}_2)^2} - \frac{1}{\boldsymbol{k}_1^2(\boldsymbol{k}_2-\boldsymbol{q})^2(\boldsymbol{k}_1-\boldsymbol{k}_2)^2}\right]. \quad (8.109)$$

where the color factor is (we use (8.55) and omit the details of the derivation)

$$\mathcal{C}^{(2)}_{\underline{8},\mathrm{real}} = \frac{1}{N_c^2-1}\, \mathcal{P}^{ji}_{lk}(\underline{8})\, \mathcal{G}'_{\mathrm{real}} = \frac{N_c^2}{8}\,. \quad (8.110)$$

The virtual gluon contribution is

$$\mathrm{Im}\, A^{(2)}_{\underline{8},\mathrm{virt}}(s,t) = -\frac{N_c \alpha_s^3}{\pi^2}\, 2\, \mathcal{C}^{(2)}_{\underline{8},\mathrm{virt}}\, t^a_{ij} t^a_{kl}\, s \ln\left(\frac{s}{|t|}\right) \int \mathrm{d}^2\boldsymbol{k}_1 \int \mathrm{d}^2\boldsymbol{k}_2 \times \left[\frac{1}{\boldsymbol{k}_1^2(\boldsymbol{k}_2-\boldsymbol{q})^2(\boldsymbol{k}_1-\boldsymbol{k}_2)^2} + \frac{1}{\boldsymbol{k}_2^2(\boldsymbol{k}_1-\boldsymbol{q})^2(\boldsymbol{k}_1-\boldsymbol{k}_2)^2}\right], \quad (8.111)$$

with

$$\mathcal{C}^{(2)}_{\underline{8},\mathrm{virt}} = \frac{1}{N_c^2-1}\,\mathcal{P}^{ji}_{lk}(\underline{8})\,\mathcal{G}'_{\mathrm{virt}} = -\frac{N_c}{4}\,. \tag{8.112}$$

Due to the particular color-octet factors (8.110) and (8.112), summing (8.109) and (8.111) to get the full color-octet discontinuity we find that the virtual contribution exactly cancels the last two terms in the square bracket of (8.109), and we are left with

$$\begin{aligned}\mathrm{Im}\,A^{(2)}_{\underline{8}}(s,t) &= \mathrm{Im}\,A^{(2)}_{\underline{8},\mathrm{real}}(s,t) + \mathrm{Im}\,A^{(2)}_{\underline{8},\mathrm{virt}}(s,t)\\ &= \frac{N_c^2\alpha_s^3}{2\pi^3}\,t^a_{ij}t^a_{kl}\,s\,\ln\left(\frac{s}{|t|}\right)\\ &\quad\times\int \mathrm{d}^2\boldsymbol{k}_1\int \mathrm{d}^2\boldsymbol{k}_2\,\frac{1}{\boldsymbol{q}^2}\boldsymbol{k}_1^2\boldsymbol{k}_2^2(\boldsymbol{k}_1-\boldsymbol{q})^2(\boldsymbol{k}_2-\boldsymbol{q})^2\,,\end{aligned} \tag{8.113}$$

which can be rewritten as

$$\mathrm{Im}\,A^{(2)}_{\underline{8}}(s,t) = 8\pi^2\alpha_s\,t^a_{ij}t^a_{kl}\,\frac{s}{|t|}\,\ln\left(\frac{s}{|t|}\right)\,\epsilon^2(t)\,. \tag{8.114}$$

The corresponding real part is enhanced by a factor $\ln(s/|t|)$ and dominates over the imaginary part. Therefore, in leading $\ln s$ contribution, the color-octet amplitude is real and reads

$$A^{(2)}_{\underline{8}}(s,t) = 4\pi\alpha_s\,t^a_{ij}t^a_{kl}\,\frac{s}{t}\,\ln^2\left(\frac{s}{|t|}\right)\,\epsilon^2(t)\,. \tag{8.115}$$

Putting together the results of the first three orders of perturbation theory – (8.17), (8.61) and (8.115) –, we find that up to $\mathcal{O}(\alpha_s^3)$ the color-octet amplitude in LLA has the suggestive form

$$\begin{aligned}A_{\underline{8}}(s,t) &= 8\pi\alpha_s\,\frac{s}{t}\,t^a_{ij}t^a_{kl}\\ &\quad\times\left[1+\epsilon(t)\,\ln\left(\frac{s}{|t|}\right)+\frac{1}{2}\,\epsilon^2(t)\,\ln^2\left(\frac{s}{|t|}\right)+\ldots\right].\end{aligned} \tag{8.116}$$

It comes natural to conjecturate that these are the first three terms in the expansion of

$$\begin{aligned}A_{\underline{8}}(s,t) &= 8\pi\alpha_s\,t^a_{ij}t^a_{kl}\,\frac{s}{t}\left(\frac{s}{|t|}\right)^{\epsilon(t)}\\ &= 8\pi\alpha_s\,t^a_{ij}t^a_{kl}\left(\frac{s}{|t|}\right)^{\alpha_g(t)}\,,\end{aligned} \tag{8.117}$$

where

$$\alpha_g(t) = 1+\epsilon(t)\,. \tag{8.118}$$

In Sect. 8.5 we shall see that this conjecture is indeed true: the series (8.118), continued to all orders in $\alpha_s$ in the leading $\ln s$ approximation, builds a reggeized gluon with trajectory $\alpha_g(t)$.

**Color-Singlet Exchange.** Tha calculation of the color-singlet amplitude proceeds along the same lines as for color octet, but the result is drastically different. Due to its crossing property under $s \leftrightarrow u$ interchange, the amplitude for color-singlet exchange is purely imaginary in leading ln $s$ approximation. Thus we have

$$
\begin{aligned}
A^{(2)}_{\underline{1},\mathrm{real}}(s,t) &= \mathrm{i}\,\frac{2\alpha_s^3}{\pi^2}\,\mathcal{C}^{(2)}_{\underline{1},\mathrm{real}}\,\delta_{ij}\delta_{kl}\,s\,\ln\left(\frac{s}{|t|}\right)\int \mathrm{d}^2\boldsymbol{k}_1^2\int \mathrm{d}^2\boldsymbol{k}_2^2 \\
&\times\left[\frac{\boldsymbol{q}^2}{\boldsymbol{k}_1^2\boldsymbol{k}_2^2(\boldsymbol{k}_1-\boldsymbol{q})^2(\boldsymbol{k}_2-\boldsymbol{q})^2}-\frac{1}{\boldsymbol{k}_2^2(\boldsymbol{k}_1-\boldsymbol{q})^2(\boldsymbol{k}_1-\boldsymbol{k}_2)^2}\right. \\
&\left.-\frac{1}{\boldsymbol{k}_1^2(\boldsymbol{k}_2-\boldsymbol{q})^2(\boldsymbol{k}_1-\boldsymbol{k}_2)^2}\right]\,. \qquad (8.119)
\end{aligned}
$$

with

$$
\mathcal{C}^{(2)}_{\underline{1},\mathrm{real}} = \mathcal{P}^{ji}_{lk}(\underline{1})\,\mathcal{G}_{\mathrm{real}} = -\frac{N_c^2-1}{4}\,, \qquad (8.120)
$$

for the real-gluon contribution, and

$$
\begin{aligned}
A^{(2)}_{\underline{1},\mathrm{virt}}(s,t) &= -\mathrm{i}\,\frac{N_c\alpha_s^3}{\pi^2}\,\mathcal{C}^{(2)}_{\underline{1},\mathrm{virt}}\,\delta_{ij}\delta_{kl}\,s\,\ln\left(\frac{s}{|t|}\right)\int \mathrm{d}^2\boldsymbol{k}_1\int \mathrm{d}^2\boldsymbol{k}_2 \\
&\times\left[\frac{1}{\boldsymbol{k}_1^2(\boldsymbol{k}_2-\boldsymbol{q})^2(\boldsymbol{k}_1-\boldsymbol{k}_2)^2}+\frac{1}{\boldsymbol{k}_2^2(\boldsymbol{k}_1-\boldsymbol{q})^2(\boldsymbol{k}_1-\boldsymbol{k}_2)^2}\right]\,, \qquad (8.121)
\end{aligned}
$$

with

$$
\mathcal{C}^{(2)}_{\underline{1},\mathrm{virt}} = \mathcal{P}^{ji}_{lk}(\underline{1})\,\mathcal{G}_{\mathrm{virt}} = -\frac{N_c^2-1}{4N_c}\,, \qquad (8.122)
$$

for the virtual gluon contribution.

It is immediate to verify that there is no cancellation between the real and virtual radiative correction terms. Therefore, differently from the the color-octet case, the two-loop amplitude is not proportional to the one-loop amplitude. The situation is complicated, but, despite the appearance, not hopeless. It is still possible to resum the leading powers in ln $s$, although in a more difficult way. This will be done by deriving and solving an integral equation (the BFKL equation) which describes the evolution in ln $s$ of the LLA amplitudes. The color-octet solution will be found very easily, as one may figure out from the simplicity of the result obtained above. The color-singlet solution will require a slightly harder work, but can be put in a very elegant form as well.

### 8.3.5 Higher Orders: the BFKL Ladder

Let us now extend the results of the previous subsections to higher orders. Since the eikonal approximation is independent on the spin of the particle which emits the soft gluon, the quark lines in Fig. 8.13 can be replaced by

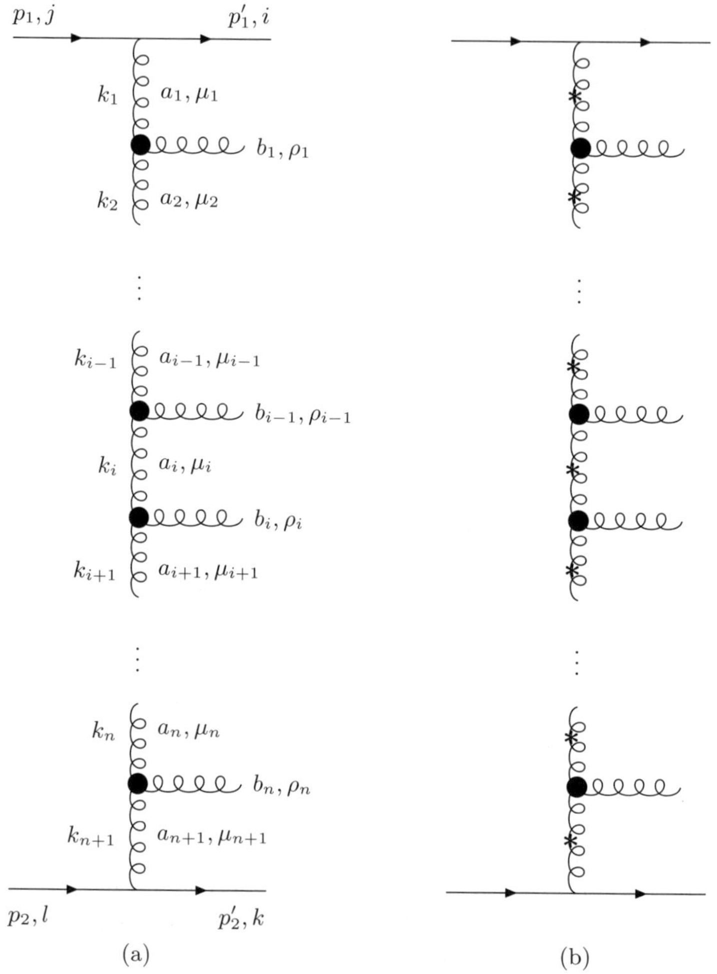


**Fig. 8.16.** Diagram for the process $qq \to qq{+}n$ gluons (**a**) at tree level and (**b**) with virtual radiative corrections. The blobs represent Lipatov vertices. The asterisks denote reggeized gluons.

gluons with strongly ordered longitudinal momenta. The generalization of the tree-level diagram of Fig. 8.13 is shown in Fig. 8.16a. We work in the so-called *multi-Regge kinematics*, which yields the leading $\ln s$ contributions. Adopting the usual Sudakov parametrization for the momenta of the $t$-channel gluons

$$k_1 = \alpha_1 p_1 + \beta_i p_2 + k_{i\perp} , \quad i = 1, 2, \ldots, n+1 , \tag{8.123}$$

the multi-Regge regime (that we already encountered in Sect. 5.9) corresponds to all transverse momenta being of the same order (and much smaller than $s$)

$$\boldsymbol{k}_1^2 \simeq \boldsymbol{k}_2^2 \simeq \ldots \simeq \boldsymbol{k}_n^2 \simeq \boldsymbol{k}_{n+1}^2 \simeq \boldsymbol{q}^2 , \tag{8.124}$$

and to a strong ordering of the longitudinal momenta

$$1 \gg \alpha_1 \gg \alpha_2 \gg \ldots \gg \alpha_{n+1} \gg \frac{\boldsymbol{q}^2}{s} , \tag{8.125}$$

$$1 \gg |\beta_{n+1}| \gg \ldots \gg |\beta_2| \gg |\beta_1| \gg \frac{\boldsymbol{q}^2}{s} . \tag{8.126}$$

Note that Eq. (8.125) implies also that the produced gluons are strongly ordered in rapidity. The kinematics is therefore the same as in the old multiperipheral model (see Sect. 5.9).

The tree-level amplitude for the production of $n$ gluons in multi-Regge kinematics generalizes Eq. (8.84) and reads (for the notations see Fig. 8.16a)

$$\begin{aligned} iA_{2\to n+2}^{\rho_1\ldots\rho_n} = {} & (-2i\,g_s)\,p_1^{\mu_1}\,t_{ij}^{a_1}\left(-\frac{i}{k_1^2}\right) \\ & \times g_s\,f_{a_1a_2b_1}\,\Gamma^{\rho_1}_{\mu_1\mu_2}(k_1,k_2)\left(-\frac{i}{k_2^2}\right) \\ & \times g_s\,f_{a_2a_3b_2}\,\Gamma^{\mu_2\rho_2}_{\mu_3}(k_2,k_3)\left(-\frac{i}{k_3^2}\right) \\ & \vdots \\ & \times g_s\,f_{a_na_{n+1}b_n}\,\Gamma^{\mu_n\rho_n}_{\mu_{n+1}}(k_n,k_{n+1})\left(-\frac{i}{k_{n+1}^2}\right) \\ & \times(-2i\,g_s)\,p_2^{\mu_{n+1}}t_{kl}^{a_{n+1}} . \end{aligned} \tag{8.127}$$

An intelligible derivation of (8.127) can be found in Forshaw and Ross (1997). A sample of graphs which do not contribute in leading logarithmic approximation is shown in Fig. 8.17.

Using

$$\begin{aligned} & p_1^{\mu_1}\,\Gamma^{\rho_1}_{\mu_1\mu_2}(k_1,k_2)\,\Gamma^{\mu_2\rho_2}_{\mu_3}(k_2,k_3)\ldots\Gamma^{\mu_n\rho_n}_{\mu_{n+1}}(k_n,k_{n+1})\,p_2^{\mu_{n+1}} \\ & \quad = \frac{s}{2}\,C^{\rho_1}(k_1,k_2)\,C^{\rho_2}(k_2,k_3)\ldots C^{\rho_n}(k_n,k_{n+1}) \end{aligned} \tag{8.128}$$

with

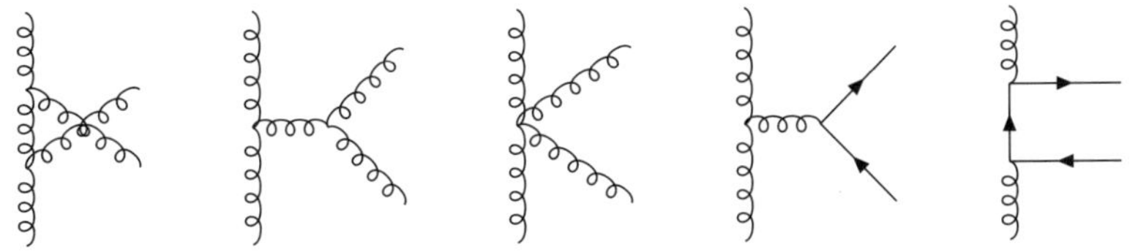

**Fig. 8.17.** Sections of ladder diagrams which do not contribute in leading ln $s$ approximation.

$$C^{\rho}(k_i, k_{i+1}) = \left(\alpha_i + \frac{2\,\boldsymbol{k}_i^2}{\beta_i s}\right) p_1^{\rho} + \left(\beta_{i+1} + \frac{2\,\boldsymbol{k}_{i+1}^2}{\alpha_i\, s}\right) p_2^{\rho} - (k_{i\perp}^{\rho} + k_{i+1\perp}^{\rho})\,, \tag{8.129}$$

the scattering amplitude (8.127) is rewritten as

$$\begin{aligned} A_{2\to n+2}^{\rho_1\ldots\rho_n} &= 2\mathrm{i}s\, g_s\, t_{ij}^{a_1} \left(\frac{i}{\boldsymbol{k}_1^2}\right) \\ &\times g_s f_{a_1 a_2 b_1}\, C^{\rho_1}(k_1, k_2) \left(\frac{i}{\boldsymbol{k}_2^2}\right) \\ &\times g_s f_{a_2 a_3 b_2}\, C^{\rho_2}(k_2, k_3) \left(\frac{i}{\boldsymbol{k}_3^2}\right) \\ &\vdots \\ &\times g_s f_{a_n a_{n+1} b_n}\, C^{\rho_n}(k_n, k_{n+1}) \left(\frac{i}{\boldsymbol{k}_{n+1}^2}\right) \\ &\times g_s\, t_{kl}^{a_{n+1}}\,. \end{aligned} \tag{8.130}$$

Equation (8.130) is only the *tree-level* amplitude. It does not take into account virtual radiative corrections, like those of Fig. 8.15. Following Lipatov (1976), Kuraev, Lipatov and Fadin (1976), we make the ansatz that in LLA the effect of these corrections, to all orders in $\alpha_s$, is to modify the $t$-channel gluon propagators as follows

$$\frac{-\mathrm{i}}{k_i^2} \to \frac{-\mathrm{i}}{k_i^2}\left(-\frac{s_i}{k_i^2}\right)^{\epsilon(k_i^2)} \simeq \frac{-\mathrm{i}}{k_i^2}\left(\frac{\alpha_{i-1}}{\alpha_i}\right)^{\epsilon(k_i^2)}\,, \tag{8.131}$$

where

$$s_i = (k_{i-1} - k_{i+1})^2 \simeq \frac{\alpha_{i-1}}{\alpha_i}\,(\boldsymbol{k}_i - \boldsymbol{k}_{i+1})^2 \tag{8.132}$$

is the c.m. energy in the $i$-th section of the ladder and $\epsilon(k_i^2)$ is the function defined in (8.45), that is

$$\epsilon(k_i^2) = \frac{N_c\,\alpha_s}{4\pi^2} \int \mathrm{d}^2\boldsymbol{\kappa}\, \frac{-\boldsymbol{k}_i^2}{\boldsymbol{\kappa}^2\,(\boldsymbol{\kappa} - \boldsymbol{k}_i)^2}\,. \tag{8.133}$$

As anticipated in Sect. 8.2, a gluon with a propagator (in Feynman gauge)

$$D_{\mu\nu}(s_i, k_i^2) = \frac{-\mathrm{i}g_{\mu\nu}}{k_i^2}\left(\frac{s}{\boldsymbol{k}^2}\right)^{\epsilon(t)}\,, \tag{8.134}$$

is called a *reggeized gluon*. The change (8.131) corresponds to replacing the gluons on the vertical lines of the diagram of Fig. 8.16a with reggeized gluons. This procedure automatically accounts for virtual radiative corrections. The modified diagram is shown in Fig. 8.16b (the reggeized gluons are denoted by asterisks). The function $\alpha_g(t) = 1 + \epsilon(t)$, as we shall see, is the *trajectory* of the reggeized gluon.

The ansatz (8.131) is equivalent to the conjecture (8.117). In fact, if we insert (8.131) into (8.17), we obtain the amplitude for the process $qq \to qq$ with the leading logarithmic contribution of the virtual corrections to all orders in $\alpha_s$

$$8\pi\alpha_s \, t^a_{ij} t^a_{kl} \, \frac{s}{t} \left(\frac{s}{|t|}\right)^{\epsilon(t)} , \tag{8.135}$$

which coincides with (8.117).

Using (8.131) in (8.130), we get the amplitude for the production of $n$ gluons, which incorporates the leading contribution of the virtual corrections to all orders in $\alpha_s$,

$$\begin{aligned}
A^{\rho_1 \dots \rho_n}_{2\to n+2} &= 2\mathrm{i}s \, g_s \, t^{a_1}_{ij} \left(\frac{\mathrm{i}}{\boldsymbol{k}_1^2}\right) \left(\frac{1}{\alpha_1}\right)^{\epsilon(k_1^2)} \\
&\quad \times g_s f_{a_1 a_2 b_1} \, C^{\rho_1}(k_1, k_2) \left(\frac{\mathrm{i}}{\boldsymbol{k}_2^2}\right) \left(\frac{\alpha_1}{\alpha_2}\right)^{\epsilon(k_2^2)} \\
&\quad \times g_s f_{a_2 a_3 b_2} \, C^{\rho_2}(k_2, k_3) \left(\frac{\mathrm{i}}{\boldsymbol{k}_3^2}\right) \left(\frac{\alpha_2}{\alpha_3}\right)^{\epsilon(k_3^2)} \\
&\quad \vdots \\
&\quad \times g_s f_{a_n a_{n+1} b_n} \, C^{\rho_n}(k_n, k_{n+1}) \left(\frac{\mathrm{i}}{\boldsymbol{k}_{n+1}^2}\right) \left(\frac{\alpha_1}{\alpha_2}\right)^{\epsilon(k_2^2)} \\
&\quad \times g_s \, t^{a_{n+1}}_{kl} \\
&= 2\mathrm{i}s \, g_s^2 \, t^{a_1}_{ij} t^{a_{n+1}}_{kl} \, \frac{\mathrm{i}}{\boldsymbol{k}_1^2} \left(\frac{1}{\alpha_1}\right)^{\epsilon(k_1^2)} \\
&\quad \times \prod_{i=1}^{n} \left\{ g_s \, f_{a_i a_{i+1} b_i} \, C^{\rho_i}(k_i, k_{i+1}) \, \frac{\mathrm{i}}{\boldsymbol{k}_{i+1}^2} \left(\frac{\alpha_i}{\alpha_{i+1}}\right)^{\epsilon(k_{i+1}^2)} \right\} .
\end{aligned} \tag{8.136}$$

A proof of Eq. (8.136) can be found in Appendix II of Lipatov (1989), where techniques developed by Bartels (1975) are used. Here we shall follow the original approach of Kuraev, Lipatov and Fadin (1976), namely we assume the validity of (8.136), as a consequence of the ansatz (8.131), and prove its self-consistency by using it to derive the elastic amplitude with the virtual radiative corrections in LLA, and showing that this amplitude is indeed given by (8.135).

The imaginary part of the amplitude for elastic $qq$ scattering with exchange of a gluon ladder, whose vertical lines are reggeized gluons, is obtained, as usual, using unitarity relations. Thus we have (Fig. 8.18)

$$\begin{aligned}
\operatorname{Im} A(s,t) &= \frac{1}{2} (-1)^n \, g_{\rho_1 \sigma_1} \cdots g_{\rho_n \sigma_n} \\
&\quad \times \int \mathrm{d}\Pi_{n+2} \, A^{\rho_1 \dots \rho_n}_{2\to n+2}(k_1, \dots, k_n) \, A^{\sigma_1 \dots \sigma_n \dagger}_{2\to n+2}(k_1 - q, \dots, k_n - q) ,
\end{aligned} \tag{8.137}$$

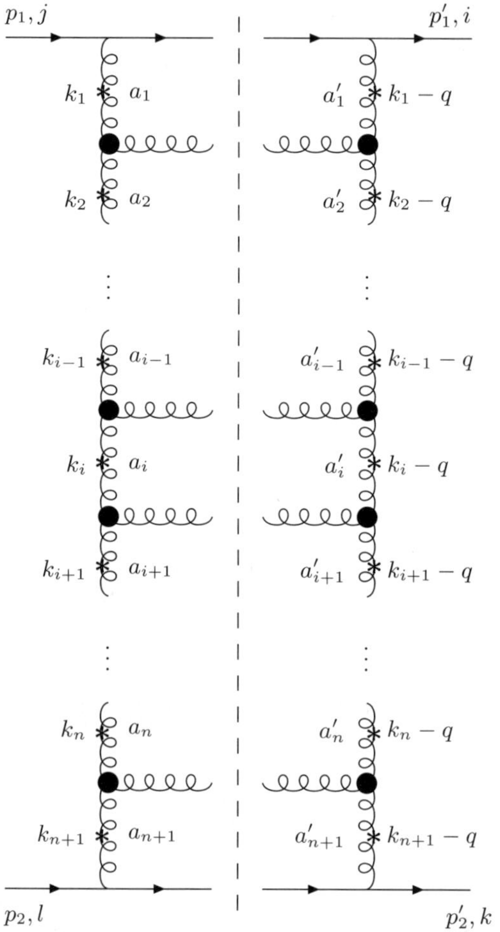


**Fig. 8.18.** The imaginary part of the quark–quark elastic scattering amplitude with exchange of a gluon ladder.

where $d\Pi_{n+2}$ is the volume element of the $(n+2)$-body phase space, which will be specified later. The sums over the helicitics of the intermediate gluons are performed by means of Eq. (8.90). The amplitude to the right of the cut of Fig. 8.18 is

$$\begin{aligned} A^{\sigma_1 \dots \sigma_n \dagger}_{2\to n+2} &= -2\mathrm{i}s\, g_s^2\, t^{a'_1}_{ij} t^{a'_{n+1}}_{kl} \frac{-\mathrm{i}}{\boldsymbol{k}_1^2} \left(\frac{1}{\alpha_1}\right)^{\epsilon((k_1-q)^2)} \\ &\times \prod_{i=1}^{n} \left\{ -\, g_s\, f_{a'_i a'_{i+1} b_i}\, C^{\sigma_i}(-(k_i - q), -(k_{i+1} - q)) \right. \\ &\times \left. \frac{-\mathrm{i}}{(\boldsymbol{k}_{i+1} - \boldsymbol{q})^2} \left(\frac{\alpha_i}{\alpha_{i+1}}\right)^{\epsilon((k_{i+1}-q)^2)} \right\} . \end{aligned} \qquad (8.138)$$

The contraction of the Lipatov vertices gives

$$C^{\rho_i}(k_i, k_{i+1})\, C_{\rho_i}(-k_i + q, -k_{i+1} + q)$$
$$= -2\left[\boldsymbol{q}^2 - \frac{\boldsymbol{k}_i^2\,(\boldsymbol{k}_{i+1}-\boldsymbol{q})^2}{(\boldsymbol{k}_i-\boldsymbol{k}_{i+1})^2} - \frac{\boldsymbol{k}_{i+1}^2\,(\boldsymbol{k}_i-\boldsymbol{q})^2}{(\boldsymbol{k}_i-\boldsymbol{k}_{i+1})^2}\right] \equiv -2\,K(\boldsymbol{k}_i, \boldsymbol{k}_{i+1})\,. \tag{8.139}$$

The color factors are evaluated by iterating the procedure used in Sect. 8.3.4 for the one-rung ladder. We finally get for the amplitudes $\mathcal{A}_R$ defined in (8.51), and corresponding to definite color configuration in the $t$-channel ($R = \underline{1}, \underline{8}$),

$$\begin{aligned}\operatorname{Im}\mathcal{A}_R(s,t) = \frac{1}{2}\sum_{n=0}^{\infty} 4s^2 g_s^4\,\mathcal{G}_R \int \mathrm{d}\Pi_{n+2}\, \frac{1}{\boldsymbol{k}_1^2(\boldsymbol{k}_1-\boldsymbol{q})^2}\left(\frac{1}{\alpha_1}\right)^{\epsilon(k_1^2)+\epsilon((k_1-q)^2)} \\ \times \prod_{i=1}^{n}\left\{\frac{g_s^2}{\boldsymbol{k}_{i+1}^2(\boldsymbol{k}_{i+1}-\boldsymbol{q})^2}\,(-2\eta_R)\,K(\boldsymbol{k}_i,\boldsymbol{k}_{i+1}) \right. \\ \left. \times \left(\frac{\alpha_i}{\alpha_{i+1}}\right)^{\epsilon(k_{i+1}^2)+\epsilon((k_{i+1}-q)^2)}\right\}\,,\end{aligned} \tag{8.140}$$

where

$$\mathcal{G}_{\underline{1}} = \frac{N_c^2-1}{4N_c}\,, \quad \mathcal{G}_{\underline{8}} = -\frac{N_c}{8}\,, \tag{8.141}$$

and

$$\eta_{\underline{1}} = \frac{N_c}{2}\,, \quad \eta_{\underline{8}} = N_c\,. \tag{8.142}$$

As we shall see, the factor $\frac{1}{2}$ difference between $\eta_{\underline{1}}$ and $\eta_{\underline{8}}$ plays a crucial rôle in determining the structure of the amplitude.

Once Eq. (8.140) has been computed, one can obtain the full amplitude $\mathcal{A}_R$ by using dispersion relations (Sect. 4.6), as we did at lowest orders. However, it will prove convenient to work in the complex angular momentum plane. Thus, instead of evaluating directly Eq. (8.140), we shall first calculate its Mellin tranform

$$f_R(\omega,t) = \int_1^{\infty} \mathrm{d}\left(\frac{s}{|t|}\right)\left(\frac{s}{|t|}\right)^{-\omega-1}\frac{\operatorname{Im}\mathcal{A}_R(s,t)}{s}\,. \tag{8.143}$$

Equation (8.143) is the Froissart-Gribov representation of the partial-wave amplitude $f_R(\omega,t)$ (see Sects. 4.7 and 5.6). Comparing (8.143) and (5.59), we can make the identification

$$\omega \equiv \ell - 1\,, \tag{8.144}$$

where $\ell$ is the complex angular momentum. The inverse Mellin transform of (8.143) is

$$\frac{\operatorname{Im} \mathcal{A}_R(s,t)}{s} = \frac{1}{2\pi \mathrm{i}} \int_{c-\mathrm{i}\infty}^{c+\mathrm{i}\infty} \mathrm{d}\omega \left(\frac{s}{|t|}\right)^{\omega} f_R(\omega,t) , \tag{8.145}$$

with the integration contour standing to the right of all $\omega$-plane singularities of $f_R(\omega,t)$.

In order to include the $u$-channel contribution we use the property of the discontinuity $\operatorname{Im} \mathcal{A}_R$ under $s \leftrightarrow u$ crossing, namely

$$\operatorname{Im} \mathcal{A}_R(s,t) = -\xi_R \operatorname{Im} \mathcal{A}_R(u,t) , \tag{8.146}$$

where the signature $\xi_R$ is

$$\xi_{\underline{1}} = +1 , \quad \xi_{\underline{8}} = -1 . \tag{8.147}$$

Since $u \simeq -s$, the $u$-channel term is taken into account by the replacement

$$f_R(\omega,t) \to (1 - \xi_R \, \mathrm{e}^{-\mathrm{i}\pi\omega}) \, f_R(\omega,t) , \tag{8.148}$$

as one can see using (8.146) in (8.143).

Finally, the partial-wave amplitude $f_R(\omega,t)$ is related to the amplitude $\mathcal{A}_R(s,t)$ by the Watson-Sommerfeld transform (see Sects. 5.4 and 5.6)

$$\mathcal{A}_R(s,t) = -\frac{1}{4\pi \mathrm{i}} \int_{c-\mathrm{i}\infty}^{c+\mathrm{i}\infty} \mathrm{d}\omega \left(\frac{s}{|t|}\right)^{\omega+1} \frac{\xi_R - \mathrm{e}^{-\mathrm{i}\pi\omega}}{\sin \pi\omega} f_R(\omega,t) . \tag{8.149}$$

In conclusion, our strategy is the following. We first determine $f_R(\omega,t)$ by solving the BFKL equation (see next section). Then we calculate the amplitude $\mathcal{A}_R(s,t)$ either by means of (8.145) and of a dispersion relation (keeping only leading $\ln s$ terms), or by using directly Eq. (8.149).

## 8.4 The BFKL Equation

The explicit form of the $(n+2)$-body phase space appearing in (8.140) is

$$\begin{aligned} \mathrm{d}\Pi_{n+2} = {} & \frac{s^{n+1}}{2^{n+1}\,(2\pi)^{3n+2}} \int \prod_{i=1}^{n+1} \mathrm{d}\alpha_i \, \mathrm{d}\beta_i \, \mathrm{d}^2\boldsymbol{k}_i \\ & \times \delta\left(-\beta_1(1-\alpha_1)s - \boldsymbol{k}_1^2\right) \, \delta\left(\alpha_{n+1}(1+\beta_{n+1})s - \boldsymbol{k}_{n+1}^2\right) \\ & \times \prod_{j=1}^{n} \delta\left((\alpha_j - \alpha_{j+1})(\beta_j - \beta_{j+1})s - (\boldsymbol{k}_j - \boldsymbol{k}_{j+1})^2\right) . \end{aligned} \tag{8.150}$$

In multi-Regge kinematics, integrating over $\beta_i$, one gets

$$\begin{aligned} \mathrm{d}\Pi_{n+2} = {} & \frac{1}{2^{n+1}\,(2\pi)^{3n+2}} \prod_{i=1}^{n} \int_{\alpha_{i+1}}^{1} \frac{\mathrm{d}\alpha_i}{\alpha_i} \int_0^1 \mathrm{d}\alpha_{n+1} \\ & \times \prod_{j=1}^{n+1} \int \mathrm{d}^2\boldsymbol{k}_j \, \delta\left(\alpha_{n+1}s - \boldsymbol{k}^2\right) . \end{aligned} \tag{8.151}$$

The nested integrations over $\alpha_i$ can be disentangled by taking the Mellin transform of the amplitude (8.140) and using the convolution property of Mellin transforms, as explained in Appendix B. That is why it is convenient to first compute the partial-wave amplitude defined in (8.143). Thus, inserting (8.140) into (8.143) and using (8.151), we obtain, with the help of the formulas of Appendix B,

$$
\begin{aligned}
f_R(\omega, \boldsymbol{q}^2) = (4\pi\alpha_s)^2\, \mathcal{G}_R \sum_{n=0}^{\infty} \prod_{i=1}^{n+1} \frac{\mathrm{d}^2 \boldsymbol{k}_i}{(2\pi)^2} \\
&\times \frac{1}{\boldsymbol{k}_1^2 (\boldsymbol{k}_1 - \boldsymbol{q})^2} \frac{1}{\omega - \epsilon(k_1^2) - \epsilon((k_1 - q)^2)} \\
&\times (-2\alpha_s \eta_R)\, K(\boldsymbol{k}_1, \boldsymbol{k}_2) \\
&\times \frac{1}{\boldsymbol{k}_2^2 (\boldsymbol{k}_2 - \boldsymbol{q})^2} \frac{1}{\omega - \epsilon(k_2^2) - \epsilon((k_2 - q)^2)} \\
&\vdots \\
&\times (-2\alpha_s \eta_R)\, K(\boldsymbol{k}_n, \boldsymbol{k}_{n+1}) \\
&\times \frac{1}{\boldsymbol{k}_{n+1}^2 (\boldsymbol{k}_{n+1} - \boldsymbol{q})^2} \frac{1}{\omega - \epsilon(k_{n+1}^2) - \epsilon((k_{n+1} - q)^2)} \,. \qquad (8.152)
\end{aligned}
$$

This can be rewritten as a recursive relation,

$$
f_R(\omega, \boldsymbol{q}^2) = (4\pi\alpha_s)^2\, \mathcal{G}_R \int \frac{\mathrm{d}^2 \boldsymbol{k}}{(2\pi)^2} \frac{\mathcal{F}_R(\omega, \boldsymbol{k}, \boldsymbol{q})}{\boldsymbol{k}^2 (\boldsymbol{k} - \boldsymbol{q})^2}\,, \qquad (8.153)
$$

where the function $\mathcal{F}_R(\omega, \boldsymbol{k}, \boldsymbol{q})$ satisfies the integral equation

$$
\begin{aligned}
\left[\omega - \epsilon(-\boldsymbol{k}^2) - \epsilon(-(\boldsymbol{k} - \boldsymbol{q})^2)\right] \mathcal{F}_R(\omega, \boldsymbol{k}, \boldsymbol{q}) \\
= 1 - \frac{2\alpha_s \eta_R}{4\pi^2} \int \mathrm{d}^2 \boldsymbol{\kappa}\, \frac{K(\boldsymbol{k}, \boldsymbol{\kappa})}{\boldsymbol{\kappa}^2 (\boldsymbol{\kappa} - \boldsymbol{q})^2} \mathcal{F}_R(\omega, \boldsymbol{\kappa}, \boldsymbol{q})\,. \qquad (8.154)
\end{aligned}
$$

This is the general form of the *BFKL equation* (Lipatov 1976; Kuraev, Lipatov and Fadin 1976, 1977; Balitsky and Lipatov 1978). It describes the leading logarithmic evolution of the gluon ladder in $\ln s$. On the right-hand side of (8.154), the kernel of the integral equation represents the contribution of the *real radiative corrections*. On the left-hand side, the $\epsilon$ terms represent the *virtual radiative corrections*.

We now proceed to solve the BFKL equation for color-octet and color-singlet exchange.

## 8.5 Color-Octet Exchange

In the specific case of color-octet exchange the BFKL equation (8.154) simplifies considerably. Replacing in (8.154) $K(\boldsymbol{k}, \boldsymbol{\kappa})$ as defined in (8.139), i.e.

$$K(\boldsymbol{k},\boldsymbol{\kappa}) = \boldsymbol{q}^2 - \frac{\boldsymbol{k}^2\,(\boldsymbol{\kappa}-\boldsymbol{q})^2}{(\boldsymbol{k}-\boldsymbol{\kappa})^2} - \frac{\boldsymbol{\kappa}^2\,(\boldsymbol{k}-\boldsymbol{q})}{(\boldsymbol{k}-\boldsymbol{\kappa})^2}\,, \tag{8.155}$$

and the explicit expressions of the reggeized gluon trajectories,

$$\epsilon(-\boldsymbol{k}^2) = -\frac{N_c\alpha_s}{4\pi^2}\int \mathrm{d}^2\boldsymbol{\kappa}\,\frac{\boldsymbol{k}^2}{\boldsymbol{\kappa}^2(\boldsymbol{k}-\boldsymbol{\kappa})^2}\,, \tag{8.156}$$

$$\epsilon(-(\boldsymbol{k}-\boldsymbol{q})^2) = -\frac{N_c\alpha_s}{4\pi^2}\int \mathrm{d}^2\boldsymbol{\kappa}\,\frac{(\boldsymbol{k}-\boldsymbol{q})^2}{(\boldsymbol{\kappa}-\boldsymbol{q})^2(\boldsymbol{k}-\boldsymbol{\kappa})^2}\,, \tag{8.157}$$

we find that, due to the peculiar color octet factor $\eta_{\underline{8}} = N_c/2$, the last two terms in (8.155) exactly cancel the virtual contributions (i.e., the $\epsilon$ terms), so that the equation for $\mathcal{F}_{\underline{8}}$ becomes

$$\omega\,\mathcal{F}_{\underline{8}}(\omega,\boldsymbol{k},\boldsymbol{q}) = 1 - \frac{N_c\alpha_s}{4\pi^2}\int \mathrm{d}^2\boldsymbol{k}\,\frac{\boldsymbol{q}^2}{\boldsymbol{\kappa}^2(\boldsymbol{\kappa}-\boldsymbol{q})^2}\,\mathcal{F}_{\underline{8}}(\omega,\boldsymbol{\kappa},\boldsymbol{q})\,. \tag{8.158}$$

This equation admits the $\boldsymbol{k}$-independent solution

$$\mathcal{F}_{\underline{8}}(\omega,\boldsymbol{q}) = \frac{1}{\omega-\epsilon(-\boldsymbol{q}^2)}\,, \tag{8.159}$$

and from (8.153) we get

$$f_{\underline{8}}(\omega,\boldsymbol{q}^2) = 2\pi^2\alpha_s\frac{\epsilon(-\boldsymbol{q}^2)}{\boldsymbol{q}^2}\,\frac{1}{\omega-\epsilon(-\boldsymbol{q}^2)}\,. \tag{8.160}$$

In terms of the complex angular momentum $\ell \equiv \omega+1$, the octet partial-wave amplitude behaves as

$$f_{\underline{8}}(\ell,t) \sim \frac{1}{\ell-\alpha_g(t)}\,, \tag{8.161}$$

where $\alpha_g(t) = 1+\epsilon(t)$. Therefore, $f_{\underline{8}}(\ell,t)$ has a pole singularity at $\ell = \alpha_g(t)$.

The inverse Mellin transform of (8.160) – see Eq. (8.143) – is

$$\mathrm{Im}\,\mathcal{A}_{\underline{8}}(s,t) = 2\pi^2\alpha_s\,\epsilon(t)\left(\frac{s}{|t|}\right)^{1+\epsilon(t)}\,. \tag{8.162}$$

By using dispersion relations, one finds that, up to corrections of order $\alpha_s$ (recall that $\epsilon(t) = \mathcal{O}(\alpha_s)$), (8.162) is the imaginary part of

$$\mathcal{A}_{\underline{8}}(s,t) = 2\pi\alpha_s\left(\frac{s}{t}\right)^{\alpha_g(t)}\,. \tag{8.163}$$

Adding the $u$-channel contribution, which is obtained by the replacement $s \to u \simeq -s$, with an overall minus sign coming from the color factors, we get

$$\begin{aligned}\mathcal{A}_{\underline{8}}(s,t) &= 2\pi\alpha_s \left[\left(\frac{s}{t}\right)^{\alpha_g(t)} - \left(-\frac{s}{t}\right)^{\alpha_g(t)}\right] \\ &= -2\pi\alpha_s \left[1 - \mathrm{e}^{-\mathrm{i}\pi\alpha_g(t)}\right] \left(\frac{s}{|t|}\right)^{\alpha_g(t)} . \qquad (8.164)\end{aligned}$$

Using (8.53) and (8.55), we finally get the amplitude of quark–quark elastic scattering via color octet exchange,

$$A_{\underline{8}}(s,t) = -4\pi\alpha_s\, t^a_{ij} t^a_{kl} \left[1 - \mathrm{e}^{-\mathrm{i}\pi\alpha_g(t)}\right] \left(\frac{s}{|t|}\right)^{\alpha_g(t)} . \qquad (8.165)$$

This is a Regge-type amplitude for the exchange of an odd-signature trajectory. Since $\epsilon(t=0)=0$, the intercept is at $\ell = 1$, which corresponds to a reggeized gluon. The same result could have been obtained using (8.160) in the Watson-Sommerfeld transform (8.149).

Some remarks are in order. First of all, note that, although we started from a discontinuity – Eqs. (8.140) and (8.162) – at $\mathcal{O}(\alpha_s^2)$, which takes into account the radiative corrections due to the insertion of a gluon ladder, the octet solution (8.165) is $\mathcal{O}(\alpha_s)$, because the octet exchange receives a contribution already at level of one-gluon exchange.

In the multi-Regge regime, where the collinear logarithms are not large, we have $\alpha_s \ln(\boldsymbol{q}^2/\mu^2) \ll 1$, and the slope of the reggeized gluon trajectory $\epsilon(t)$, given by (8.47), is a small quantity. Thus, setting $\alpha_g(t) \simeq 1$ in the signature factor of (8.165), we can approximate the amplitude by

$$A_{\underline{8}}(s,t) \simeq -8\pi\alpha_s\, t^a_{ij} t^a_{kl} \left(\frac{s}{|t|}\right)^{\alpha_g(t)} . \qquad (8.166)$$

This coincides with (8.135), or (8.117), which justifies *a posteriori* the ansatz (8.131). Once more, we observe that the color-octet amplitude is purely real.

It is interesting to recast (8.166) in the form

$$A_{\underline{8}}(s,t) \simeq 8\pi\alpha_s\, t^a_{ij} t^a_{kl}\, \frac{s}{t} \exp\left[-\frac{N_c\alpha_s}{2\pi} \ln\left(\frac{s}{|t|}\right) \ln\left(\frac{|t|}{\mu^2}\right)\right] , \qquad (8.167)$$

where we have used (8.47). The argument of the exponential in (8.167) is a negative product of a logarithm of type $\ln(s/|t|)$ and a collinear logarithm. This is an example of a Sudakov form factor (see, e.g., Collins 1989). Note that the octet amplitude (8.167) vanishes in the limit $\mu^2 \to 0$.

Summing (averaging) over the final (initial) colors we obtain from (8.166)

$$\left|\overline{A_{\underline{8}}}\right|^2 = \frac{N_c^2 - 1}{4N_c^2} (8\pi\alpha_s)^2 \left(\frac{s}{|t|}\right)^{2\alpha_g(t)} . \qquad (8.168)$$

The corresponding expression for gluon–gluon scattering via color-octet exchange has the same form as (8.168), except for a different color factor (a detailed presentation of gluon–gluon scattering in the leading $\ln s$ approximation can be found in Del Duca 1995). One finds

$$\left|\overline{A^{gg}_{\underline{8}}}\right|^2 = \frac{N_c^2}{N_c^2-1}(8\pi\alpha_s)^2 \left(\frac{s}{|t|}\right)^{2\alpha_g(t)} , \tag{8.169}$$

and therefore

$$\left|\overline{A^{gg}_{\underline{8}}}\right|^2 = \left(\frac{2N_c^2}{N_c^2-1}\right)^2 \left|\overline{A^{qq}_{\underline{8}}}\right|^2 = \left(\frac{C_A}{C_F}\right)^2 \left|\overline{A^{qq}_{\underline{8}}}\right|^2 . \tag{8.170}$$

For quark-gluon scattering one has

$$\left|\overline{A^{qg}_{\underline{8}}}\right|^2 = \frac{C_A}{C_F} \left|\overline{A^{qq}_{\underline{8}}}\right|^2 . \tag{8.171}$$

Equations (8.170, 8.171) are the same relations which hold at one-gluon-exchange level – cfr. (8.28, 8.29).

## 8.6 Color-Singlet Exchange

Let us now consider the exchange of a gluon ladder in a color-singlet configuration, which corresponds to the *perturbative QCD pomeron.* pomeron!hard Equation (8.154), with $\eta_{\underline{1}} = N_c$, reads

$$\left[\omega - \epsilon(-\boldsymbol{k}^2) - \epsilon(-(\boldsymbol{k}-\boldsymbol{q})^2)\right] \mathcal{F}_{\underline{1}}(\omega,\boldsymbol{k},\boldsymbol{q}) = 1 - \frac{N_c\alpha_s}{2\pi^2}\int \mathrm{d}^2\boldsymbol{\kappa}\, \frac{K(\boldsymbol{k},\boldsymbol{\kappa})}{\boldsymbol{\kappa}^2(\boldsymbol{\kappa}-\boldsymbol{q})^2}\, \mathcal{F}_{\underline{1}}(\omega,\boldsymbol{\kappa},\boldsymbol{q}) . \tag{8.172}$$

We introduce the function $F(\omega,\boldsymbol{k},\boldsymbol{k}',\boldsymbol{q})$ related to $F(\omega,\boldsymbol{k},\boldsymbol{k}',\boldsymbol{q})$ by

$$\mathcal{F}_{\underline{1}}(\omega,\boldsymbol{k},\boldsymbol{q}) = \int \frac{\mathrm{d}^2\boldsymbol{k}'}{\boldsymbol{k}'^2}\, \boldsymbol{k}^2\, F(\omega,\boldsymbol{k},\boldsymbol{k}',\boldsymbol{q}) . \tag{8.173}$$

From (8.153) and (8.173) one finds the following relation between the singlet partial-wave amplitude $f_{\underline{1}}(\omega,\boldsymbol{q}^2)$ and $\mathcal{F}_{\underline{1}}(\omega,\boldsymbol{k},\boldsymbol{q})$

$$f_{\underline{1}}(\omega,\boldsymbol{q}^2) = (8\pi^2\alpha_s)^2\, \frac{N_c^2-1}{4N_c} \int \frac{\mathrm{d}^2\boldsymbol{k}}{(2\pi)^2} \int \frac{\mathrm{d}^2\boldsymbol{k}'}{(2\pi)^2}\, \frac{F(\omega,\boldsymbol{k},\boldsymbol{k}',\boldsymbol{q})}{\boldsymbol{k}'^2(\boldsymbol{k}-\boldsymbol{q})^2} . \tag{8.174}$$

By means of (8.173) we can rewrite (8.172) as an integral equation for $F(\omega,\boldsymbol{k},\boldsymbol{k}',\boldsymbol{q})$, that is

$$\left[\omega - \epsilon(-\boldsymbol{k}^2) - \epsilon(-(\boldsymbol{k}-\boldsymbol{q})^2)\right] F(\omega,\boldsymbol{k},\boldsymbol{k}',\boldsymbol{q}) = \delta^2(\boldsymbol{k}-\boldsymbol{k}') - \frac{N_c\alpha_s}{2\pi^2}\int \mathrm{d}^2\boldsymbol{\kappa}\, \frac{K(\boldsymbol{k},\boldsymbol{\kappa})}{\boldsymbol{k}^2(\boldsymbol{\kappa}-\boldsymbol{q})^2}\, F(\omega,\boldsymbol{\kappa},\boldsymbol{k}',\boldsymbol{q}) . \tag{8.175}$$

In the singlet case no cancellation of terms occurs between the virtual and real radiative correction components, i.e. between the l.h.s. and r.h.s. of (8.175), and the BFKL equation turns out to be more complicated. Using (8.156, 8.157) and making the substitutions

$$\int \frac{\mathrm{d}^2\boldsymbol{\kappa}}{\boldsymbol{\kappa}^2(\boldsymbol{k}-\boldsymbol{\kappa})^2} = 2\int \frac{\mathrm{d}^2\boldsymbol{\kappa}}{(\boldsymbol{k}-\boldsymbol{\kappa})^2[\boldsymbol{\kappa}^2+(\boldsymbol{k}-\boldsymbol{\kappa})^2]}\,, \tag{8.176}$$

$$\int \frac{\mathrm{d}^2\boldsymbol{\kappa}}{(\boldsymbol{\kappa}-\boldsymbol{q})^2(\boldsymbol{k}-\boldsymbol{\kappa})^2} = 2\int \frac{\mathrm{d}^2\boldsymbol{\kappa}}{(\boldsymbol{k}-\boldsymbol{\kappa})^2[(\boldsymbol{\kappa}-\boldsymbol{q})^2+(\boldsymbol{k}-\boldsymbol{\kappa})^2]}\,, \tag{8.177}$$

Eq. (8.175) becomes

$$\begin{aligned} \omega\, F(\omega,\boldsymbol{k},\boldsymbol{k}',\boldsymbol{q}) &= \delta^2(\boldsymbol{k}-\boldsymbol{k}') \\ &+\frac{N_c\alpha_s}{2\pi^2}\int \mathrm{d}^2\boldsymbol{\kappa}\left\{\frac{-\boldsymbol{q}^2}{(\boldsymbol{\kappa}-\boldsymbol{q})^2\boldsymbol{k}^2}\,F(\omega,\boldsymbol{\kappa},\boldsymbol{k}',\boldsymbol{q})\right. \\ &+\frac{1}{(\boldsymbol{\kappa}-\boldsymbol{k})^2}\left[F(\omega,\boldsymbol{\kappa},\boldsymbol{k}',\boldsymbol{q})-\frac{\boldsymbol{k}^2\,F(\omega,\boldsymbol{k},\boldsymbol{k}',\boldsymbol{q})}{\boldsymbol{\kappa}^2+(\boldsymbol{k}-\boldsymbol{\kappa})^2}\right] \\ &+\frac{1}{(\boldsymbol{\kappa}-\boldsymbol{k})^2}\left[\frac{(\boldsymbol{k}-\boldsymbol{q})^2\,\boldsymbol{\kappa}^2\,F(\omega,\boldsymbol{\kappa},\boldsymbol{k}',\boldsymbol{q})}{(\boldsymbol{\kappa}-\boldsymbol{q})^2\,\boldsymbol{k}^2}\right. \\ &\left.\left.-\frac{(\boldsymbol{k}-\boldsymbol{q})^2\,F(\omega,\boldsymbol{k},\boldsymbol{k}',\boldsymbol{q})}{(\boldsymbol{\kappa}-\boldsymbol{q})^2+(\boldsymbol{k}-\boldsymbol{\kappa})^2}\right]\right\}\,. \end{aligned} \tag{8.178}$$

This is the standard form of the *color-singlet BFKL equation*[3]. The inverse Mellin transform of its solution, that we call, with a slight abuse of notation, $F(s,\boldsymbol{k},\boldsymbol{k}',\boldsymbol{q})$,

$$F(s,\boldsymbol{k},\boldsymbol{k}',\boldsymbol{q}) = \frac{1}{2\pi\mathrm{i}}\int_{c-\mathrm{i}\infty}^{c+\mathrm{i}\infty}\mathrm{d}\omega\left(\frac{s}{|t|}\right)^{\omega} F(\omega,\boldsymbol{k},\boldsymbol{k}',\boldsymbol{q})\,, \tag{8.179}$$

gives the imaginary part of the scattering amplitude, via (8.145) and (8.174),

$$\frac{\mathrm{Im}\,\mathcal{A}_{\underline{1}}(s,t)}{s} = (8\pi^2\alpha_s)^2\,\frac{N_c^2-1}{4N_c}\int\frac{\mathrm{d}^2\boldsymbol{k}}{(2\pi)^2}\int\frac{\mathrm{d}^2\boldsymbol{k}'}{(2\pi)^2}\,\frac{F(s,\boldsymbol{k},\boldsymbol{k}',\boldsymbol{q})}{\boldsymbol{k}'^2(\boldsymbol{k}-\boldsymbol{q})^2}\,. \tag{8.180}$$

Let us take a look at the properties of the BFKL equation (8.178). First of all, note that this equation is ultraviolet finite, as can be seen immediately by taking the limits $\boldsymbol{\kappa}^2\to\infty$ and $\boldsymbol{k}^2\to\infty$ in the integrand of (8.178). As far as the infrared behavior is concerned, Eq. (8.178) is regular for $\boldsymbol{\kappa}^2\to 0$, and at $\boldsymbol{k}=\boldsymbol{\kappa}$. In the latter case, the singularities of the $1/(\boldsymbol{\kappa}-\boldsymbol{k})^2$ factors are cancelled by the zeros of the terms in square brackets. For $\boldsymbol{k}^2\to 0$ there are infrared divergences arising from the virtual-gluon terms but, as shown by Lipatov (1986), in the physical situation of colorless particle scattering, these divergences are regulated by the confinement of quarks and gluons.

[3] From now on, when speaking of the 'BFKL equation', with no further specification, we shall refer to the color singlet equation (8.178).

For zero momentum transfer the BFKL equation (8.178) reduces to (we omit the argument $\boldsymbol{q} = 0$ in $F$)

$$\omega F(\omega, \boldsymbol{k}, \boldsymbol{k}') = \delta^2(\boldsymbol{k} - \boldsymbol{k}') + \frac{N_c \alpha_s}{\pi^2} \int \frac{\mathrm{d}^2 \boldsymbol{\kappa}}{(\boldsymbol{k} - \boldsymbol{\kappa})^2} \times \left[ F(\omega, \boldsymbol{\kappa}, \boldsymbol{k}') - \frac{\boldsymbol{k}^2}{\boldsymbol{\kappa}^2 + (\boldsymbol{k} - \boldsymbol{\kappa})^2} F(\omega, \boldsymbol{k}, \boldsymbol{k}') \right] . \quad (8.181)$$

This has the form

$$\omega F(\omega, \boldsymbol{k}, \boldsymbol{k}') = \delta^2(\boldsymbol{k} - \boldsymbol{k}') + \int \mathrm{d}^2 \boldsymbol{\kappa} \, \mathcal{K}(\boldsymbol{k}, \boldsymbol{\kappa}) \, F(\omega, \boldsymbol{\kappa}, \boldsymbol{k}') , \quad (8.182)$$

where the BFKL kernel $\mathcal{K}(\boldsymbol{k}, \boldsymbol{\kappa})$ is

$$\mathcal{K}(\boldsymbol{k}, \boldsymbol{\kappa}) = 2 \, \epsilon(-\boldsymbol{k}^2) \, \delta^2(\boldsymbol{k} - \boldsymbol{\kappa}) + \frac{N_c \alpha_s}{\pi^2} \frac{1}{(\boldsymbol{k} - \boldsymbol{\kappa})^2} . \quad (8.183)$$

In (8.183) we recognize two terms: the virtual-radiative-correction contribution

$$\mathcal{K}_{\mathrm{virt}}(\boldsymbol{k}, \boldsymbol{\kappa}) = 2 \, \epsilon(-\boldsymbol{k}^2) \, \delta^2(\boldsymbol{k} - \boldsymbol{\kappa}) , \quad (8.184)$$

and the real-radiative-correction contribution

$$\mathcal{K}_{\mathrm{real}}(\boldsymbol{k}, \boldsymbol{\kappa}) = \frac{N_c \alpha_s}{\pi^2} \frac{1}{(\boldsymbol{k} - \boldsymbol{\kappa})^2} . \quad (8.185)$$

The BFKL equation (8.181) can be transformed into an integro-differential equation describing the evolution of the BFKL amplitude $F(s, \boldsymbol{k}, \boldsymbol{k}') \equiv F(s, \boldsymbol{k}, \boldsymbol{k}', 0)$ in $\ln s$. Using

$$\frac{\partial F(s, \boldsymbol{k}, \boldsymbol{k}')}{\partial \ln(s/\boldsymbol{k}^2)} = \frac{1}{2\pi \mathrm{i}} \int_{c-\mathrm{i}\infty}^{c+\mathrm{i}\infty} \mathrm{d}\omega \left( \frac{s}{\boldsymbol{k}^2} \right)^{\omega} \omega F(\omega, \boldsymbol{k}, \boldsymbol{k}') , \quad (8.186)$$

we get from (8.181)

$$\frac{\partial F(s, \boldsymbol{k}, \boldsymbol{k}')}{\partial \ln(s/\boldsymbol{k}^2)} = \frac{N_c \alpha_s}{\pi^2} \int \frac{\mathrm{d}^2 \boldsymbol{\kappa}}{(\boldsymbol{k} - \boldsymbol{\kappa})^2} \times \left[ F(s, \boldsymbol{\kappa}, \boldsymbol{k}') - \frac{\boldsymbol{k}^2}{\boldsymbol{\kappa}^2 + (\boldsymbol{k} - \boldsymbol{\kappa})^2} F(s, \boldsymbol{k}, \boldsymbol{k}') \right] . \quad (8.187)$$

## 8.7 Solution of the BFKL Equation for $t = 0$

We now want to solve the BFKL equation for $t = 0$, that we rewrite symbolically as

$$\omega F = 1\!\!1 + \mathcal{K} \otimes F , \quad (8.188)$$

where $\mathcal{K}$ is the BFKL kernel (8.183). Solving (8.188) amounts to finding the eigenfunctions $\phi_\alpha$ of $K$

$$\mathcal{K} \otimes \phi_\alpha = \omega_\alpha \phi_\alpha \ , \tag{8.189}$$

These eigenfunctions are labelled by some discrete or continuous indices that we collectively denote by $\alpha$. The $\phi_\alpha$ satisfy a completeness relation

$$\sum_\alpha \phi_\alpha(\boldsymbol{k})\, \phi_\alpha^*(\boldsymbol{k}') = \delta^2(\boldsymbol{k} - \boldsymbol{k}') \ , \tag{8.190}$$

and hence the solution of (8.188) is

$$F(\omega, \boldsymbol{k}, \boldsymbol{k}') = \sum_\alpha \frac{\phi_\alpha(\boldsymbol{k})\, \phi_\alpha^*(\boldsymbol{k}')}{\omega - \omega_\alpha} \ . \tag{8.191}$$

Here $\sum_\alpha$ may incorporate an integral over a continuous variable.

Using polar variables $\boldsymbol{k} \equiv (|\boldsymbol{k}|, \vartheta)$ and $\boldsymbol{\kappa} \equiv (|\boldsymbol{\kappa}|, \varphi)$, the l.h.s. of (8.189) becomes

$$\begin{aligned} \mathcal{K} \otimes \phi_\alpha &\equiv \frac{N_c \alpha_s}{\pi^2} \int \frac{\mathrm{d}^2\boldsymbol{\kappa}}{(\boldsymbol{k} - \boldsymbol{\kappa})^2} \left[ \phi_\alpha(\boldsymbol{\kappa}) - \frac{\boldsymbol{k}^2\, \phi_\alpha(\boldsymbol{k})}{\boldsymbol{\kappa}^2 + (\boldsymbol{k} - \boldsymbol{\kappa})^2} \right] \\ &= \frac{N_c \alpha_s}{2\pi^2} \int_0^\infty \mathrm{d}\boldsymbol{\kappa}^2 \int_0^{2\pi} \mathrm{d}\varphi \, \frac{1}{\boldsymbol{k}^2 + \boldsymbol{\kappa}^2 - 2\,|\boldsymbol{k}|\,|\boldsymbol{\kappa}|\, \cos(\varphi - \vartheta)} \\ &\quad \times \left[ \phi_\alpha(|\boldsymbol{\kappa}|, \varphi) - \frac{\boldsymbol{k}^2\, \phi_\alpha(|\boldsymbol{k}|, \vartheta)}{\boldsymbol{k}^2 + 2\boldsymbol{\kappa}^2 - 2\,|\boldsymbol{k}|\,|\boldsymbol{\kappa}|\, \cos(\varphi - \vartheta)} \right] \ . \end{aligned} \tag{8.192}$$

It is not difficult to guess the form of the eigenfunctions of the BFKL kernel. If we write $\phi_\alpha(|\boldsymbol{k}|, \vartheta)$ as a Fourier series in $\vartheta$, its $|\boldsymbol{k}|$-dependent coefficients are easily shown, by dimensional arguments, to be powers of $k^2$. Imposing the completeness relation (8.190), such powers are constrained to have the form $(\boldsymbol{k}^2)^{-\frac{1}{2}+\mathrm{i}\nu}$, with $\nu$ real. Thus the eigenfunctions are

$$\phi_{n\nu}(|\boldsymbol{k}|, \vartheta) = \frac{1}{\pi\sqrt{2}} \, (\boldsymbol{k}^2)^{-\frac{1}{2}+\mathrm{i}\nu} \, \mathrm{e}^{\mathrm{i}n\vartheta} \ , \tag{8.193}$$

with the normalization

$$\int \mathrm{d}^2\boldsymbol{k}\, \phi_{n\nu}(\boldsymbol{k})\, \phi_{n'\nu'}(\boldsymbol{k}) = \delta_{nn'}\, \delta(\nu - \nu') \ . \tag{8.194}$$

Inserting (8.193) into (8.189), with $\alpha \to (n, \nu)$, leads to the following expression for the eigenvalues

$$\begin{aligned} \omega_n(\nu) &= \frac{N_c \alpha_s}{2\pi^2} \int_0^\infty \mathrm{d}\boldsymbol{\kappa}^2 \int_0^{2\pi} \mathrm{d}\chi \, \frac{1}{\boldsymbol{k}^2 + \boldsymbol{\kappa}^2 - 2\,|\boldsymbol{k}|\,|\boldsymbol{\kappa}|\, \cos\chi} \\ &\quad \times \left[ \left( \frac{\boldsymbol{\kappa}^2}{\boldsymbol{k}^2} \right)^{-\frac{1}{2}+\mathrm{i}\nu} \mathrm{e}^{\mathrm{i}n\chi} - \frac{\boldsymbol{k}^2}{\boldsymbol{k}^2 + 2\,\boldsymbol{\kappa}^2 - 2\,|\boldsymbol{k}|\,|\boldsymbol{\kappa}|\, \cos\chi} \right] \ . \end{aligned} \tag{8.195}$$

Performing the angular integrations one finds

$$\omega_n(\nu) = \frac{\alpha_s N_c}{\pi} \left\{ \int_0^{\boldsymbol{k}^2} d\boldsymbol{\kappa}^2 \left[ \frac{1}{\boldsymbol{k}^2 - \boldsymbol{\kappa}^2} \left( \frac{\boldsymbol{\kappa}^2}{\boldsymbol{k}^2} \right)^{\frac{|n|-1}{2} + i\nu} - \frac{\boldsymbol{k}^2}{\boldsymbol{\kappa}^2(\boldsymbol{k}^2 - \boldsymbol{\kappa}^2)} \right.\right.$$
$$\left. + \frac{\boldsymbol{k}^2}{\boldsymbol{\kappa}^2\sqrt{4\,\boldsymbol{\kappa}^4 + \boldsymbol{k}^4}} \right] + \int_{\boldsymbol{k}^2}^{\infty} d\boldsymbol{\kappa}^2 \left[ \frac{1}{\boldsymbol{\kappa}^2 - \boldsymbol{k}^2} \left( \frac{\boldsymbol{k}^2}{\boldsymbol{\kappa}^2} \right)^{\frac{|n|+1}{2} - i\nu} \right.$$
$$\left.\left. - \frac{\boldsymbol{k}^2}{\boldsymbol{\kappa}^2(\boldsymbol{k}^2 - \boldsymbol{\kappa}^2)} + \frac{\boldsymbol{k}^2}{\boldsymbol{\kappa}^2\sqrt{4\,\boldsymbol{\kappa}^4 + \boldsymbol{k}^4}} \right] \right\} . \tag{8.196}$$

We now set $x = \boldsymbol{\kappa}^2/\boldsymbol{k}^2$ in the first integral and $x = \boldsymbol{k}^2/\boldsymbol{\kappa}^2$ in the second integral, thus obtaining

$$\omega_n(\nu) = \frac{\alpha_s N_c}{\pi} \left\{ 2\,\mathrm{Re} \int_0^1 dx\, \frac{x^{\frac{|n|+1}{2} - i\nu} - 1}{1 - x} \right.$$
$$\left. - \int_0^1 dx \left[ \frac{1}{x} - \frac{1}{x\sqrt{1 + 4\,x^2}} \right] + \int_0^1 \frac{dx}{\sqrt{4 + x^2}} \right\} . \tag{8.197}$$

The last two integrals in (8.197) cancel each other and we are left with

$$\omega_n(\nu) = \frac{2\alpha_s N_c}{\pi} \mathrm{Re} \int_0^1 dx\, \frac{x^{\frac{|n|+1}{2} - i\nu} - 1}{1 - x} , \tag{8.198}$$

which can be expressed in terms of the digamma function $\psi(z)$

$$\psi(z) \equiv \frac{d \ln \Gamma(z)}{dz} = \int_0^1 dx\, \frac{x^{z-1}}{x - 1} , \tag{8.199}$$

as

$$\omega_n(\nu) = -\frac{2\alpha_s N_c}{\pi} \mathrm{Re} \left[ \psi\left( \frac{|n| + 1}{2} + i\nu \right) - \psi(1) \right] , \tag{8.200}$$

where $\psi(1) = -\gamma_E = -0.577215...$ . The solution of the BFKL equation for $t = 0$ therefore is

$$F(\omega, \boldsymbol{k}, \boldsymbol{k}') = \frac{1}{2\pi^2\,(\boldsymbol{k}^2\boldsymbol{k}'^2)^{\frac{1}{2}}} \sum_{n=0}^{\infty} e^{in(\vartheta - \vartheta')} \int_{\infty}^{+\infty} d\nu\, \frac{e^{i\nu \ln\left(\frac{\boldsymbol{k}^2}{\boldsymbol{k}'^2}\right)}}{\omega - \omega_n(\nu)} . \tag{8.201}$$

From this expression one can see that the scattering amplitude has a *cut*, rather than a pole, in the complex $\omega$ plane.

The leading $\ln s$ behavior of $F(s, \boldsymbol{k}, \boldsymbol{k}')$ is determined by the rightmost singularity of its Mellin transform $F(\omega, \boldsymbol{k}, \boldsymbol{k}')$ on the real $\omega$ axis. Since $\omega_n(\nu)$ decreases with $n$, we can retain only the $n = 0$ term. Thus, the solution (8.201) reduces to

$$F(\omega, \boldsymbol{k}, \boldsymbol{k}') = \frac{1}{2\pi^2 (\boldsymbol{k}^2 \boldsymbol{k}'^2)^{\frac{1}{2}}} \int_{-\infty}^{+\infty} \mathrm{d}\nu \, \frac{\mathrm{e}^{\mathrm{i}\,\nu\, \ln\left(\frac{\boldsymbol{k}^2}{\boldsymbol{k}'^2}\right)}}{\omega - \omega_0(\nu)} \,. \tag{8.202}$$

Let us now study the $\nu$-dependence of $\omega_0(\nu)$. This function has its maximum at $\nu = 0$. Expanding around this point we have

$$\omega_0(\nu) = \frac{N_c \alpha_s}{\pi} \left(4 \ln 2 - 14\,\zeta(3)\,\nu^2 + \ldots\right) , \tag{8.203}$$

with $\zeta(3) \simeq 1.202$. We rewrite Eq. (8.203) as

$$\omega_0(\nu) \simeq \lambda - \frac{1}{2}\lambda' \nu^2 , \tag{8.204}$$

where the constants

$$\lambda = \frac{N_c \alpha_s}{\pi}\, 4 \ln 2 \,, \quad \lambda' = \frac{N_c \alpha_s}{\pi}\, 28\,\zeta(3) \,, \tag{8.205}$$

will repeatedly appear in the remainder of the book.

Integrating (8.202) over $\nu$, with $\omega_0(\nu)$ given by (8.204), we find

$$F(\omega, \boldsymbol{k}, \boldsymbol{k}') \propto \frac{1}{\sqrt{\omega - \lambda}} , \tag{8.206}$$

which explicitly shows the existence of a branch point at $\omega = \lambda$.

We are mainly interested in the $s$-dependence of the amplitude, so we take the inverse Mellin transform of (8.202), which gives the leading $\ln s$ color-singlet amplitude in the form

$$F(s, \boldsymbol{k}, \boldsymbol{k}') = \frac{1}{2\pi^2 (\boldsymbol{k}^2 \boldsymbol{k}'^2)^{\frac{1}{2}}} \int_{-\infty}^{+\infty} \mathrm{d}\nu \, \left(\frac{s}{\boldsymbol{k}^2}\right)^{\omega_0(\nu)} \mathrm{e}^{\mathrm{i}\,\nu\, \ln\left(\frac{\boldsymbol{k}^2}{\boldsymbol{k}'^2}\right)} . \tag{8.207}$$

Now, using (8.204) in (8.207), and integrating over $\nu$ yields

$$F(s, \boldsymbol{k}, \boldsymbol{k}') = \frac{1}{\sqrt{2\pi^3 \lambda' \, \boldsymbol{k}^2 \boldsymbol{k}'^2}} \frac{1}{\sqrt{\ln(s/\boldsymbol{k}^2)}} \times \left(\frac{s}{\boldsymbol{k}^2}\right)^{\lambda} \exp\left[-\frac{\ln^2(\boldsymbol{k}^2/\boldsymbol{k}'^2)}{2\lambda' \ln(s/\boldsymbol{k}^2)}\right] . \tag{8.208}$$

This is the LLA pomeron solution of the BFKL equation.

The $1/\sqrt{\ln s}$ factor arises from the cut of the amplitude at $\omega = \omega_0$. The quantity

$$\alpha_{I\!P}(0) = 1 + \lambda = 1 + \frac{N_c \alpha_s}{\pi}\, 4 \ln 2 \,, \tag{8.209}$$

is the intercept of the perturbative QCD pomeron (the so-called *BFKL pomeron*) in leading logarithm approximation. Setting, quite arbitrarily, $\alpha_s \simeq 0.2$, one finds

$$\alpha_{I\!P}(0) - 1 \simeq 0.5 \,, \tag{8.210}$$

which is much larger than the value attributed to the soft pomeron intercept, $\alpha_{I\!P}^{\rm soft}(0) - 1 \simeq 0.1$.

Inserting the solution (8.208) into (8.180) gives the imaginary part of the color-singlet amplitude. Reconstructing the full amplitude (which must include also the crossed channel contribution) is straightforward. We already learned, in fact, that the color-singlet amplitude is purely imaginary, and is suppressed, order by order in $\alpha_s$, by a power of $\ln s$ compared to the color-octet amplitude. This is due to the even-signature factor $(1 + \mathrm{e}^{-\mathrm{i}\pi\alpha_{I\!P}(t)})$ coming from Eq. (8.149) – with $\xi_{\underline{1}} = 1$ and $\omega = \alpha_{I\!P}(t) - 1$. In conclusion, the amplitude for *quark–quark scattering* via *pomeron exchange* is given, in leading $\ln s$ approximation, by

$$\begin{aligned} A_{\underline{1}}(s,t) &= (8\pi^2\alpha_s)^2 \, \frac{N_c^2 - 1}{4N_c} \, \delta_{ij}\delta_{kl} \, \mathrm{i}s \\ &\times \int \frac{\mathrm{d}^2\boldsymbol{k}}{(2\pi)^2} \int \frac{\mathrm{d}^2\boldsymbol{k}'}{(2\pi)^2} \, \frac{F(s,\boldsymbol{k},\boldsymbol{k}',\boldsymbol{q})}{\boldsymbol{k}'^2(\boldsymbol{k}-\boldsymbol{q})^2} \,. \end{aligned} \tag{8.211}$$

(We used (8.52) and (8.54) to obtain $A_{\underline{1}}$ from $\mathcal{A}_{\underline{1}}$.) The diagrammatic representation of (8.211) is shown in Fig. 8.19.

In the case of *gluon–gluon scattering*, the color-singlet amplitude has the same structure as (8.211), but with a different color factor. One finds (see, e.g., Del Duca 1995)

$$\begin{aligned} A_{\underline{1}}^{gg}(s,t) &= (8\pi^2\alpha_s)^2 \, \frac{N_c^2}{N_c^2 - 1} \, \delta_{ab}\delta_{cd} \, \mathrm{i}s \\ &\times \int \frac{\mathrm{d}^2\boldsymbol{k}}{(2\pi)^2} \int \frac{\mathrm{d}^2\boldsymbol{k}'}{(2\pi)^2} \, \frac{F(s,\boldsymbol{k},\boldsymbol{k}',\boldsymbol{q})}{\boldsymbol{k}'^2(\boldsymbol{k}-\boldsymbol{q})^2} \,. \end{aligned} \tag{8.212}$$

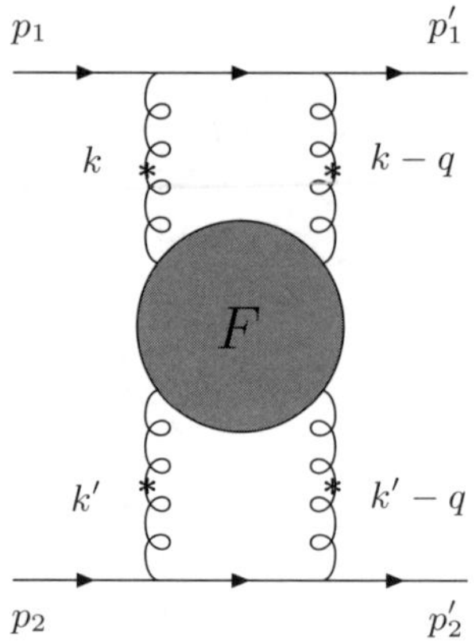


**Fig. 8.19.** BFKL diagram for quark–quark scattering.

Thus we have the following relation between quark–quark and gluon–gluon scattering amplitudes for color-singlet exchange

$$\left|\overline{A_{\underline{1}}^{gg}}\right|^2 = \left(\frac{2N_c^2}{N_c^2-1}\right)^4 \left|\overline{A_{\underline{1}}^{qq}}\right|^2 = \left(\frac{C_A}{C_F}\right)^4 \left|\overline{A_{\underline{1}}^{qq}}\right|^2 . \tag{8.213}$$

Comparing (8.213) to (8.170) we observe that there is an extra $(C_A/C_F)^2$ factor with respect to color-octet exchange. The reason for this difference is that in the color-singlet case *two* reggeized gluons are exchanged. Finally, let us give for completeness the quark-gluon elastic scattering amplitude, which is related to the quark–quark amplitude by

$$\left|\overline{A_{\underline{1}}^{qg}}\right|^2 = \left(\frac{C_A}{C_F}\right)^2 \left|\overline{A_{\underline{1}}^{qq}}\right|^2 . \tag{8.214}$$

## 8.8 Parton–Parton Scattering: Total Cross-sections

From (8.211) we can derive, via optical theorem, the total cross section for quark–quark scattering,

$$\begin{aligned}\sigma_{\text{tot}}^{qq} &= \frac{1}{s}\,\text{Im}\,A_{\underline{1}}(s,t=0) \\ &= 4\alpha_s^2\left(\frac{N_c^2-1}{4N_c^2}\right)\int \mathrm{d}^2\boldsymbol{k}\int \mathrm{d}^2\boldsymbol{k}'\,\frac{F(s,\boldsymbol{k},\boldsymbol{k}')}{\boldsymbol{k}^2\,\boldsymbol{k}'^2} ,\end{aligned} \tag{8.215}$$

where we used (8.52) and (8.54) to relate $A_{\underline{1}}$ to $\mathcal{A}_{\underline{1}}$.

Inserting the solution (8.207) into (8.215), we find that the integrals are infrared divergent. In physical situations, confinement of quarks, or some hard final state configuration, provide an infrared cutoff. For the moment, we just introduce by hand such a cutoff (that we call $\boldsymbol{k}_{\text{min}}^2$) and perform the integrations over $\boldsymbol{k}$ and $\boldsymbol{k}'$, thus obtaining

$$\sigma_{\text{tot}}^{qq} = \frac{N_c^2-1}{2N_c^2}\,\frac{\alpha_s^2}{\boldsymbol{k}_{\text{min}}^2}\int_{-\infty}^{+\infty}\mathrm{d}\nu\,\frac{\mathrm{e}^{\omega_0(\nu)y}}{\nu^2+\frac{1}{4}} , \tag{8.216}$$

where the rapidity variable is

$$y = \ln\frac{s}{\boldsymbol{k}_{\text{min}}^2} . \tag{8.217}$$

The integral over $\nu$, with $\omega_0(\nu)$ given by (8.204), can be calculated by elementary methods, and the final result is

$$\sigma_{\text{tot}}^{qq} = \frac{\pi\,(N_c^2-1)}{N_c^2}\,\frac{\alpha_s^2}{\boldsymbol{k}_{\text{min}}^2}\,\frac{\mathrm{e}^{\lambda y}}{\sqrt{\pi\lambda' y/8}} . \tag{8.218}$$

In terms of $s$, this cross section grows as

$$\sigma_{\mathrm{tot}}^{qq} \sim \frac{s^{\lambda}}{\sqrt{\ln s}} , \tag{8.219}$$

and violates asymptotically the Froissart-Martin bound (4.127), since $\lambda = N_c\alpha_s 4\ln 2/\pi > 1$.

The corresponding formula for gluon–gluon scattering is derived from (8.212) in a similar way, and reads

$$\sigma_{\mathrm{tot}}^{gg} = \frac{4\pi N_c^2}{N_c^2-1}\frac{\alpha_s^2}{\boldsymbol{k}_{\mathrm{min}}^2}\frac{\mathrm{e}^{\lambda y}}{\sqrt{\pi\lambda' y/8}} . \tag{8.220}$$

Note the usual ratio $(C_A/C_F)^2$ between (8.220) and (8.218),

$$\sigma_{\mathrm{tot}}^{gg} = \left(\frac{2N_c^2}{N_c^2-1}\right)^2 \sigma_{\mathrm{tot}}^{qq} = \left(\frac{C_A}{C_F}\right)^2 \sigma_{\mathrm{tot}}^{qq} . \tag{8.221}$$

We shall make use of Eqs. (8.218) and (8.220) when discussing high-$p_T$ jet production at large rapidities (Sect. 11.13.2). In that case, as we shall see, the infrared cutoff $\boldsymbol{k}_{\mathrm{min}}$ will be given by the minimum transverse momentum of jets.

## 8.9 Diffusion

An important feature of the BFKL solution (8.208) is its diffusion pattern. The BFKL amplitude $F(s,\boldsymbol{k},\boldsymbol{k}')$ has the form of a Gaussian distribution in $\ln(\boldsymbol{k}^2/\boldsymbol{k}'^2)$, with a width growing with $y \equiv \ln(s/\boldsymbol{k}^2)$, see (8.208). Therefore, as the energy increases, a wider and wider range of transverse momenta is probed, and the non-perturbative region is eventually entered.

This behavior is not accidental, since the BFKL equation can be written as a diffusion equation, with diffusion rate $\ln(\boldsymbol{k}^2/\boldsymbol{k}'^2) \sim y^{1/2}$, as we are now going to show (the presentation will follow quite closely Del Duca 1995).

The BFKL equation (8.181) can be put in the recursive form (recall (8.152–8.154))

$$\omega F^{(N)}(\omega,\boldsymbol{k}) = \frac{N_c\alpha_s}{\pi^2}\int \mathrm{d}^2\boldsymbol{k}_2 \left\{ \frac{F^{(N-1)}(\omega,\boldsymbol{k}_2) - (\boldsymbol{k}_1^2/\boldsymbol{k}_2^2)\, F^{(N-1)}(\omega,\boldsymbol{k}_1)}{(\boldsymbol{k}_1-\boldsymbol{k}_2)^2} + \frac{\boldsymbol{k}_1^2\, F^{(N-1)}(\omega,\boldsymbol{k}_1)}{\boldsymbol{k}_2^2\,[\boldsymbol{k}_2^2+(\boldsymbol{k}_1-\boldsymbol{k}_2)^2]} \right\} , \tag{8.222}$$

where the superscript $(N)$ denotes the iteration step. Since we are interested in high $N$ values, the inhomogeneous term of (8.181) is irrelevant, and we neglected it. In (8.222) we also omitted the third argument ($\boldsymbol{k}'$) of $F$.

Considering the LLA solution, we set

$$F^{(N)}(\omega, \boldsymbol{k}_i) \sim (\boldsymbol{k}_i^2)^{-\frac{1}{2}} \, \psi_N(\xi_i) \,, \tag{8.223}$$

with $\xi_i = \ln(\boldsymbol{k}_i^2/\boldsymbol{k}_0^2)$, and we transform (8.222) into an equation for $\psi_N$. In multi-Regge kinematics all produced gluons have comparable transverse momenta. Thus we can approximate the function $\psi_{N-1}(\xi_2)$, which appears on the right-hand side of (8.222), by

$$\psi_{N-1}(\xi_2) \simeq \psi_{N-1}(\xi_1) + \frac{1}{2}\,(\xi_2 - \xi_1)^2 \, \frac{\partial^2 \psi_{N-1}(\xi_1)}{\partial \xi_1^2} \,. \tag{8.224}$$

Replacing (8.223) in (8.222) and using (8.224), the recursive relation (8.222) becomes

$$\begin{aligned} \omega\, \psi_N(\xi) = \frac{N_c \alpha_s}{\pi} \int_0^\infty \mathrm{d}\xi' \Bigg\{ \left[ \frac{\mathrm{e}^{(\xi_2-\xi_1)/2} - 1}{|1 - \mathrm{e}^{\xi_2 - \xi_1}|} + \frac{1}{\sqrt{1 + 4\,\mathrm{e}^{2(\xi_2-\xi_1)}}} \right] \psi_{N-1}(\xi_1) \\ + \frac{(\xi_2 - \xi_1)^2}{2} \, \frac{\mathrm{e}^{(\xi_2-\xi_1)/2}}{|1 - \mathrm{e}^{\xi_2-\xi_1}|} \, \frac{\partial^2 \psi_{N-1}(\xi_1)}{\partial \xi_1^2} \Bigg\} \,. \end{aligned} \tag{8.225}$$

The integrations in (8.225) are similar to those performed in Sect. 8.7. Thus, without much effort we get

$$\omega\, \psi_N(\xi) = \lambda\, \psi_{N-1}(\xi) + \frac{\lambda'}{2} \, \frac{\partial^2 \psi_{N-1}(\xi)}{\partial \xi^2} \,, \tag{8.226}$$

where $\lambda, \lambda'$ are given in (8.205), and $\xi = \ln(\boldsymbol{k}^2/\boldsymbol{k}'^2)$, with $\boldsymbol{k}_0^2$ being an arbitrary scale of the order of all transverse momenta.

Taking the continuous limit for $N$, Eq. (8.226) becomes a partial differential equation in the two variables $\xi$ and $N$

$$\omega \, \frac{\partial \psi(N, \xi)}{\partial N} = (\lambda - \omega)\, \psi(N, \xi) + \frac{\lambda'}{2} \, \frac{\partial^2 \psi(N, \xi)}{\partial \xi^2} \,. \tag{8.227}$$

The solution of (8.227) is separable, $\psi(N, \xi) = f(N)\, g(\xi)$. At the singularity $\omega = \lambda$, (8.227) reduces to

$$\lambda \, \frac{\partial \psi(N, \xi)}{\partial N} = \frac{\lambda'}{2} \, \frac{\partial^2 \psi(N, \xi)}{\partial \xi^2} \,, \tag{8.228}$$

which is a typical *diffusion equation*, with $N$ playing the rôle of time.

If we assume that at "time" $N = 0$ the wave function $\psi(N, \xi)$ is a Gaussian,

$$\psi(0, \xi) = \frac{1}{(\pi \sigma^2)^{\frac{1}{4}}} \exp\left( -\frac{\xi^2}{2\sigma^2} \right) \,, \tag{8.229}$$

then at "time" $N$ the wave packet becomes

$$\psi(N, \xi) \sim \left( \frac{\lambda}{2\lambda' N} \right)^{\frac{1}{2}} \exp \left( -\frac{\lambda \xi^2}{2\lambda' N} \right) , \tag{8.230}$$

where we have neglected the initial width, $\lambda\sigma^2 \ll \lambda' N$. With the correspondence $N/\lambda \to y = \ln(s/\boldsymbol{k}^2)$ we find that the spreading of (8.230) is equivalent to the behavior (8.208).

In summary, the diffusive pattern of the BFKL solution tells us that, as the energy grows, the infrared region of transverse momenta becomes more and more relevant. This eventually entails the inapplicability of the perturbative treatment.

## 8.10 Running Coupling

The diffusion phenomenon seems to suggest that it would make sense to use a running coupling, instead of a fixed one, in the BFKL equation. We have seen, however, that neglecting in LLA the self-energy and vertex correction diagrams, which give rise to a running $\alpha_s$, implies for consistency that $\alpha_s$ should be treated as a constant.

One can of course adopt a very practical approach, and try to solve the BFKL equation in LLA with a *running coupling*. One of the motivations of this strategy is to see whether the singularity structure of the BFKL amplitude changes when $\alpha_s$ is let to depend on a scale of the order of the gluon transverse momenta. In his fundamental paper on the BFKL equation at $t \neq 0$, Lipatov (1986) showed by an approximate calculation that, under reasonable conditions, a running $\alpha_s$ leads to a discrete spectrum for the BFKL kernel. Therefore, the pomeron amplitude in the $\omega$-plane has a series of *isolated poles*, the leading one being identified with the pomeron.

Another important result is due to Collins and Kwieciński (1989), who derived and upper and lower limit for the intercept of the leading trajectory, using a running $\alpha_s$ frozen at some scale $\boldsymbol{k}_0^2$. They found that $\alpha_{I\!P}(0)$ lies in the interval

$$1 + 1.2\, \frac{N_c \alpha_s(\boldsymbol{k}_0^2)}{\pi} \leq \alpha_{I\!P}(0) \leq 1 + 4 \ln 2\, \frac{N_c \alpha_s(\boldsymbol{k}_0^2)}{\pi} \,. \tag{8.231}$$

The rigorous way to study the effect of the running coupling is obviously to go to the next-to-leading logarithm approximation (NLLA). This has been recently made possible by the work of Fadin and Lipatov (1998) and Camici and Ciafaloni (1998), see Sect. 8.17.

## 8.11 Soft vs. Hard Pomeron

The pomeron emerging from perturbative QCD is quite different from the pomeron of Regge theory (Sect. 5.8.3), and from the phenomenological pomeron found in the analyses *à la* Donnachie and Landshoff (Sect. 7.1). First of all, we saw that the leading singularity of the BFKL scattering amplitude in LLA is a cut, rather than a pole (accounting for the running of $\alpha_s$, however, may lead to a series of isolated poles, as mentioned in Sect. 8.10). Another crucial difference between the "hard", BFKL, pomeron and the "soft", phenomenological, pomeron is the intercept, which is very large for the former, and close to 1 for the latter:

$$\alpha_{I\!P}^{\mathrm{hard}}(0) \simeq 1.5 \ \ (\text{for } \alpha_s = 0.2) \, , \quad \alpha_{I\!P}^{\mathrm{soft}}(0) \simeq 1.1 \tag{8.232}$$

There arise two questions:

- Is the soft pomeron intrinsically distinct from the hard pomeron?
- How many pomerons exist?

From what we learned so far in this book, it should be easy to realize that these questions are actually not well posed. The answer to them, in fact, depends on the definition of pomeron one adopts. On the one hand, in perturbative QCD, "pomeron" is a synonymous of "ladder of interacting reggeized gluons". On the other hand, in phenomenological approaches, such as those from which the soft intercept comes out, the "pomeron" is not associated to a physical object, but is generically understood as "something" that *must lie* behind a successful, and amazingly simple, parametrization of a vast series of data. Quite obviously, comparing these two concepts of pomeron, and understanding their mutual relationships, is just impossible. The missing information is the physical picture underlying the soft pomeron. This is clearly determined by the non-perturbative structure of the pomeron, that we know very vaguely.

Essentially, there are two different viewpoints about the questions raised above. The first of them is based on the idea that there exists *only one pomeron*, with a smooth transition from "soft" to "hard" properties. Thus, focusing for instance on real and virtual photoproduction, the pomeron intercept would be a $Q^2$-dependent quantity $\alpha_{I\!P}^0(Q^2)$ which interpolates between a "soft" value $\sim 1.1$ at $Q^2 \approx 0$ and a "hard" value $\sim 1.4$ at large $Q^2$.

The second, opposite, viewpoint regards the soft and the hard pomeron as *two distinct objects*: while the former has a genuinely non-perturbative nature, the latter is a perturbative object which dominates at large $|t|$ (or large $Q^2$), but becomes irrelevant as $|t|$ (or $Q^2$) tends to zero.

It is clearly very important to study the interplay of soft and hard interactions. Experimentally, this has been made possible by the HERA measurements, which span a large variety of processes and various dynamical situations: from inclusive to diffractive DIS, from real to highly virtual photoproduction, etc. (see Chapts. 9 and 10).

## 8.12 Non-perturbative Effects

As discussed in the previous section, a full understanding of pomeron physics would require an insight – that we presently lack, at least at a fundamental level – into the non-perturbative aspects of QCD.

Various authors have tried to model the non-perturbative dynamics of quarks and gluons in order to explore its effect on the structure of pomeron. The first attempt in this direction was made by Landshoff and Nachtmann (1987), who proposed to derive some phenomenological features of the pomeron (in particular, the additive quark rule of its coupling to hadrons) from the properties of the QCD vacuum (see also Donnachie and Landshoff 1988/89). Their model of the pomeron – already mentioned in Sect. 8.1 – is the Low–Nussinov two–gluon exchange model with a non-perturbative propagator for the gluons, $D_{\mathrm{np}}(k^2)$, related to the vacuum correlation function for the field strengths by

$$-6\mathrm{i} \int \frac{\mathrm{d}^4 k}{(2\pi)^4} \, \mathrm{e}^{ik\cdot(x-y)} \, k^2 \, D_{\mathrm{np}}(k^2) = \langle 0| : F_{\mu\nu}(x) F^{\mu\nu}(y) : |0\rangle \,. \tag{8.233}$$

Thus $D_{\mathrm{np}}(k^2)$ depends on the gluon correlation length in the vacuum, and on the gluonic condensate $\langle 0|g_s^2 : F_{\mu\nu}(0)F^{\mu\nu}(0) : |0\rangle$. The non-perturbative scales make $D_{\mathrm{np}}(k^2)$ finite for $k^2 = 0$, whereas at large $k^2$ the perturbative $1/k^2$ behavior is regained.

A simpler way to model the non-perturbative gluon propagator is to insert directly an effective gluon mass $\mu_g$ into $D_{\mathrm{np}}(k^2)$, as done by Nikolaev, Zakharov and Zoller (1994a, 1994b)

$$D_{\mathrm{np}}(k^2) = \frac{1}{k^2 + \mu_g^2} \,. \tag{8.234}$$

Quite interestingly, the effect of introducing $\mu_g$ or, equivalently, a finite gluon correlation radius, is to *reduce* the hard pomeron intercept, which tends to decrease as $\mu_g$ gets larger. However, with realistic values of $\mu_g$, $\alpha_{I\!P}(0)$ is still much higher than the soft intercept 1.08-1.10. Thus, even though non-perturbative propagators go in the right direction, i.e. from the hard to the soft domain, something important is clearly missing in our discussion. A full non-perturbative treatment implies in fact much more than merely modifying the gluon propagators. Such a treatment, however, is beyond our present capabilities.

## 8.13 The Perturbative QCD Odderon

It is remarkable that QCD predicts the existence not only of the pomeron, but also of the *odderon*, the $C = P = -1$ trajectory that we encountered in Sects. 5.8.4 and 7.4.1 (for an updated review, see Vacca 2001).

In perturbative QCD the odderon is given by the exchange of three reggeized gluons. The integral equation describing the evolution of an $n$-gluon state and generalizing the BFKL equation valid for a two-gluon system is the so-called BKP equation (Bartels 1979, 1980; Kwieciński and Praszałowicz 1980). Recently, various properties of the BKP equation have been exploited in order to get the solution representing the odderon state. Using conformal invariance and integrability, Janik and Wosiek (1999) found a set of LLA eigenstates with a maximal intercept below 1. Another set of solutions was determined by Bartels, Lipatov and Vacca (2000) using the gluon reggeization property. The interesting outcome of this calculation is that the maximal intercept is found to be 1, and therefore the corresponding solution represents the dominant odderon at high energies.

## 8.14 Solution of the BFKL Equation for $t \neq 0$

Solving the BFKL equation for non zero momentum transfer, Eq. (8.178), is not an easy task. The solution was obtained by Lipatov (1986) by means of an ingenious trick, namely by working in the impact parameter space and using the conformal invariance properties of the BFKL equation. Here we shall outline Lipatov's derivation, omitting most of the details (for which we refer the reader to the original paper).

We introduce the transverse coordinates $\boldsymbol{r}_1, \boldsymbol{r}_1', \boldsymbol{r}_2, \boldsymbol{r}_2'$, conjugate to $\boldsymbol{k}, \boldsymbol{k}'$, $\boldsymbol{q}-\boldsymbol{k}, \boldsymbol{q}-\boldsymbol{k}'$, respectively. The BFKL Green function in the impact parameter space is defined as (with a slight abuse of notation, we use the same symbol $F$ to denote the amplitude in the $\boldsymbol{k}$-space and in the $\boldsymbol{r}$-space)

$$\begin{aligned} F(\omega, \boldsymbol{r}_1, \boldsymbol{r}_2, \boldsymbol{r}_1', \boldsymbol{r}_2') = & \int \mathrm{d}^2\boldsymbol{k}\, \mathrm{d}^2\boldsymbol{k}'\, \mathrm{d}^2\boldsymbol{q} \\ & \times \exp\{\mathrm{i}\, [\boldsymbol{k}\cdot\boldsymbol{r}_1 + (\boldsymbol{q}-\boldsymbol{k})\cdot\boldsymbol{r}_2 - \boldsymbol{k}'\cdot\boldsymbol{r}_1' - (\boldsymbol{q}-\boldsymbol{k}')\cdot\boldsymbol{r}_2']\} \\ & \times \frac{F(\omega, \boldsymbol{k}, \boldsymbol{k}', \boldsymbol{q})}{(\boldsymbol{k}-\boldsymbol{q})^2\, \boldsymbol{k}'^2} \,. \end{aligned} \tag{8.235}$$

The inverse relation is

$$\begin{aligned} \frac{F(\omega, \boldsymbol{k}, \boldsymbol{k}', \boldsymbol{q})}{(\boldsymbol{k}-\boldsymbol{q})^2\, \boldsymbol{k}'^2}\, \delta^2(\boldsymbol{q}-\boldsymbol{q}') = & \frac{1}{(2\pi)^8} \int \mathrm{d}^2\boldsymbol{r}_1\, \mathrm{d}^2\boldsymbol{r}_2\, \mathrm{d}^2\boldsymbol{r}_1'\, \mathrm{d}^2\boldsymbol{r}_2' \\ & \times \exp\{-\mathrm{i}\, [\boldsymbol{k}\cdot\boldsymbol{r}_1 + (\boldsymbol{q}-\boldsymbol{k})\cdot\boldsymbol{r}_2 - \boldsymbol{k}'\cdot\boldsymbol{r}_1' - (\boldsymbol{q}'-\boldsymbol{k}')\cdot\boldsymbol{r}_2']\} \\ & \times F(\omega, \boldsymbol{r}_1, \boldsymbol{r}_2, \boldsymbol{r}_1', \boldsymbol{r}_2') \,. \end{aligned} \tag{8.236}$$

The BFKL equation for $F(\omega, \boldsymbol{r}_1, \boldsymbol{r}_2, \boldsymbol{r}_1', \boldsymbol{r}_2')$ reads (we set $\boldsymbol{\nabla}_i \equiv \partial/\partial \boldsymbol{r}_i$, $\boldsymbol{r}_{ij} \equiv \boldsymbol{r}_i - \boldsymbol{r}_j$, and we omit the arguments $\boldsymbol{r}_1', \boldsymbol{r}_2'$ in $F$)

$$\begin{aligned}
&\omega\, \nabla_1^2\, \nabla_2^2\, F(\omega, \boldsymbol{r}_1, \boldsymbol{r}_2) = (2\pi)^4\, \delta^2(\boldsymbol{r}_1 - \boldsymbol{r}_1')\, \delta^2(\boldsymbol{r}_2 - \boldsymbol{r}_2') \\
&+ \frac{N_c \alpha_s}{2\pi^2} \left\{ (2\pi)^2\, \delta^2(\boldsymbol{r}_{12})\, (\nabla_1 + \nabla_2)^2 F(\omega, \boldsymbol{r}_1, \boldsymbol{r}_2) \right. \\
&+ \nabla_1^2 \int \frac{\mathrm{d}^2 \boldsymbol{r}_0}{\boldsymbol{r}_{10}^2} \left[ \nabla_2^2 F(\omega, \boldsymbol{r}_0, \boldsymbol{r}_2) - \frac{\boldsymbol{r}_{12}^2}{\boldsymbol{r}_{10}^2 + \boldsymbol{r}_{20}^2}\, \nabla_2^2 F(\omega, \boldsymbol{r}_1, \boldsymbol{r}_2) \right] \\
&\left. + \nabla_2^2 \int \frac{\mathrm{d}^2 \boldsymbol{r}_0}{\boldsymbol{r}_{20}^2} \left[ \nabla_1^2 F(\omega, \boldsymbol{r}_1, \boldsymbol{r}_0) - \frac{\boldsymbol{r}_{12}^2}{\boldsymbol{r}_{10}^2 + \boldsymbol{r}_{20}^2}\, \nabla_1^2 F(\omega, \boldsymbol{r}_1, \boldsymbol{r}_2) \right] \right\} .
\end{aligned} \tag{8.237}$$

A derivation of this equation from the momentum-space expression (8.178) can be found in Forshaw and Ross (1997).

Despite its very complicated form, Eq. (8.237) has some invariance properties which allow one to solve it analytically. Introducing the complex variables $\rho_i$ and $\rho_i'$, defined as

$$\rho_i = x_i + \mathrm{i}\, y_i\ , \quad \rho_i' = x_i' + \mathrm{i}\, y_i'\ , \quad i = 1, 2\ , \tag{8.238}$$

and their complex conjugates $\rho_i^*$ and $\rho_i'^*$, it is possible to show (Lipatov 1986) that Eq. (8.237) is invariant under the conformal transformation

$$\rho_i \to \frac{a\, \rho_i + b}{c \rho_i + d}\ , \tag{8.239}$$

with $ad - bc = 1$. Therefore, the eigenfunctions $\phi_\alpha(\boldsymbol{r}_1, \boldsymbol{r}_2)$ of the BFKL kernel $\mathcal{K}$, i.e. the solutions of the equation

$$\mathcal{K} \otimes \phi_\alpha = \omega_\alpha\, \phi_\alpha\ , \tag{8.240}$$

where ($\partial_i \equiv \partial/\partial\rho_i$, $\partial_i^* \equiv \partial/\partial\rho_i^*$, $\rho_{ij} \equiv \rho_i - \rho_j$)

$$\begin{aligned}
\mathcal{K} \otimes \phi_\alpha &= \frac{N_c \alpha_s}{2\pi^2} \left\{ \nabla_1^2 \int \frac{\mathrm{d}^2 \boldsymbol{r}_0}{\boldsymbol{r}_{10}^2} \left[ \nabla_2^2 \phi_\alpha(\boldsymbol{r}_0, \boldsymbol{r}_2) - \frac{\boldsymbol{r}_{12}^2}{\boldsymbol{r}_{10}^2 + \boldsymbol{r}_{20}^2}\, \nabla_2^2 \phi_\alpha(\boldsymbol{r}_1, \boldsymbol{r}_2) \right] \right. \\
&\quad \left. + \nabla_2^2 \int \frac{\mathrm{d}^2 \boldsymbol{r}_0}{\boldsymbol{r}_{20}^2} \left[ \nabla_1^2 \phi_\alpha(\boldsymbol{r}_1, \boldsymbol{r}_0) - \frac{\boldsymbol{r}_{12}^2}{\boldsymbol{r}_{10}^2 + \boldsymbol{r}_{20}^2}\, \nabla_1^2 \phi_\alpha(\boldsymbol{r}_1, \boldsymbol{r}_2) \right] \right\} \\
&= \frac{4 N_c \alpha_s}{\pi^2} \left\{ \partial_1 \partial_1^* \int \frac{\mathrm{d}\rho_0\, \mathrm{d}\rho_0^*}{|\rho_{10}|^2} \left[ \partial_2 \partial_2^*\, \phi_\alpha(\rho_0, \rho_2) \right. \right. \\
&\quad \left. - \frac{|\rho_{12}|^2}{|\rho_{10}|^2 + |\rho_{20}|^2}\, \partial_2 \partial_2^*\, \phi_\alpha(\rho_1, \rho_2) \right] \\
&\quad + \partial_2 \partial_2^* \int \frac{\mathrm{d}\rho_0\, \mathrm{d}\rho_0^*}{|\rho_{20}|^2} \left[ \partial_1 \partial_1^*\, \phi_\alpha(\rho_1, \rho_0) \right. \\
&\quad \left. \left. - \frac{|\rho_{12}|^2}{|\rho_{10}|^2 + |\rho_{20}|^2}\, \partial_1 \partial_1^*\, \phi_\alpha(\rho_1, \rho_2) \right] \right\}
\end{aligned} \tag{8.241}$$

generate an irreducible representation of the conformal group. The Casimir operators of this group are $\rho_{12}^2 \partial_1 \partial_2$ and ${\rho_{12}^*}^2 \partial_1^* \partial_2^*$, and have the following eigenvalues

$$\rho_{12}^2 \partial_1 \partial_2 \, \phi_{n,\nu} = h(h-1)\, \phi_{n,\nu} \,, \tag{8.242}$$

$$\rho_{12}^{*\,2} \partial_1^* \partial_2^* \, \phi_{n,\nu} = h'(h'-1)\, \phi_{n,\nu} \,, \tag{8.243}$$

where the conformal weights $h, h'$ are

$$h = \frac{1+n}{2} - \mathrm{i}\nu \,, \quad h' = \frac{1-n}{2} - \mathrm{i}\nu \,, \tag{8.244}$$

with integer $n$ and real $\nu$, so that

$$h(h-1) = \left(\frac{n}{2} - \mathrm{i}\nu\right)^2 - \frac{1}{4} \,, \tag{8.245}$$

$$h'(h'-1) = \left(\frac{n}{2} + \mathrm{i}\nu\right)^2 - \frac{1}{4} \,. \tag{8.246}$$

The explicit form of the eigenfunctions $\phi_{n,\nu}$ is

$$\begin{aligned} \phi_{n,\nu}(\rho_{10}, \rho_{20}) &= \left(\frac{\rho_{12}}{\rho_{10}\rho_{20}}\right)^{\frac{1-n}{2}+\mathrm{i}\nu} \left(\frac{\rho_{12}^*}{\rho_{10}^*\rho_{20}^*}\right)^{\frac{1+n}{2}+\mathrm{i}\nu} \\ &= \left|\frac{\rho_{12}}{\rho_{10}\rho_{20}}\right|^{1+2\mathrm{i}\nu-n} \left(\frac{\rho_{12}^*}{\rho_{10}^*\rho_{20}^*}\right)^{n} \,. \end{aligned} \tag{8.247}$$

Inserting $\phi_{n,\nu}$ into (8.241), one finds that the eigenvalues $\omega_n(\nu)$ are the same as in the $t = 0$ case, i.e. (8.200). As shown by Lipatov (1986), the completeness relation for the eigenfunctions $\phi_{n,\nu}$ is (we return to the vector notation)

$$\begin{aligned} &\sum_{n=-\infty}^{+\infty} \int_{-\infty}^{+\infty} \mathrm{d}\nu \int \mathrm{d}^2\boldsymbol{r}_0 \, 16 \left(\nu^2 + \frac{n^2}{4}\right) \frac{\phi_{n,\nu}(\boldsymbol{r}_{10}, \boldsymbol{r}_{20})\, \phi_{n,\nu}^*(\boldsymbol{r}_{1'0}, \boldsymbol{r}_{2'0})}{\boldsymbol{r}_{12}^2 \, \boldsymbol{r}_{1'2'}^2} \\ &\qquad = (2\pi)^4 \, \delta^2(\boldsymbol{r}_{11'})\, \delta^2(\boldsymbol{r}_{22'}) \,. \end{aligned} \tag{8.248}$$

If we now expand the solution of the non-forward BFKL equation in terms of the complete set of functions $\phi_{n,\nu}$ and make use of (8.248), we get the explicit form of $F(\omega, \boldsymbol{r}_1, \boldsymbol{r}_2, \boldsymbol{r}_1', \boldsymbol{r}_2')$

$$\begin{aligned} &F(\omega, \boldsymbol{r}_1, \boldsymbol{r}_2, \boldsymbol{r}_1', \boldsymbol{r}_2') \\ &\quad = \sum_{n=-\infty}^{+\infty} \int_{-\infty}^{+\infty} \mathrm{d}\nu \int \mathrm{d}^2\boldsymbol{r}_0 \frac{\left(\nu^2 + \frac{n^2}{4}\right)}{\left[\nu^2 + \left(\frac{n-1}{2}\right)^2\right] \left[\nu^2 + \left(\frac{n+1}{2}\right)^2\right]} \\ &\qquad \times \frac{\phi_{n,\nu}(\boldsymbol{r}_{10}, \boldsymbol{r}_{20})\, \phi_{n,\nu}^*(\boldsymbol{r}_{1'0}, \boldsymbol{r}_{2'0})}{\omega - \omega_n(\nu)} \,. \end{aligned} \tag{8.249}$$

(For $n \neq \pm 1$ the integral over $\nu$ is intended in the sense of its principal value).

The leading contribution is obtained by keeping only the $n = 0$ term in (8.249)

$$
\begin{aligned}
&F(\omega, \boldsymbol{r}_1, \boldsymbol{r}_2, \boldsymbol{r}_1', \boldsymbol{r}_2') \\
&\quad = \int_{-\infty}^{+\infty} \mathrm{d}\nu \int \mathrm{d}^2 \boldsymbol{r}_0 \frac{\nu^2}{\left(\nu^2 + \frac{1}{4}\right)^2} \frac{\phi_\nu(\boldsymbol{r}_{10}, \boldsymbol{r}_{20})\, \phi_\nu^*(\boldsymbol{r}_{1'0}, \boldsymbol{r}_{2'0})}{\omega - \omega_0(\nu)} , \qquad (8.250)
\end{aligned}
$$

where the functions $\phi_\nu \equiv \phi_{0,\nu}$ are

$$
\phi_\nu(\boldsymbol{r}_{10}, \boldsymbol{r}_{20}) = \left( \frac{\boldsymbol{r}_{12}^2}{\boldsymbol{r}_{10}^2\, \boldsymbol{r}_{20}^2} \right)^{\frac{1}{2} + \mathrm{i}\nu} . \qquad (8.251)
$$

The BFKL Green function in energy-momentum representation is obtained by computing the Mellin transform of (8.236) with respect to $\omega$, and using (8.250). After a change of variables to eliminate $\boldsymbol{r}_0$ one finally gets

$$
\begin{aligned}
\frac{F(y, \boldsymbol{k}, \boldsymbol{k}', \boldsymbol{q})}{(\boldsymbol{k} - \boldsymbol{q})^2\, \boldsymbol{k}'^2} &= \frac{1}{(2\pi)^6} \int \mathrm{d}^2 \boldsymbol{r}_1\, \mathrm{d}^2 \boldsymbol{r}_2\, \mathrm{d}^2 \boldsymbol{r}_1'\, \mathrm{d}^2 \boldsymbol{r}_2' \\
&\times \exp\left\{-\mathrm{i}\left[\boldsymbol{k} \cdot \boldsymbol{r}_1 + (\boldsymbol{q} - \boldsymbol{k}) \cdot \boldsymbol{r}_2 - \boldsymbol{k}' \cdot \boldsymbol{r}_1' - (\boldsymbol{q}' - \boldsymbol{k}') \cdot \boldsymbol{r}_2'\right]\right\} \\
&\times \int_{-\infty}^{+\infty} \mathrm{d}\nu \frac{\nu^2}{\left(\nu^2 + \frac{1}{4}\right)^2} \mathrm{e}^{\omega_0(\nu) y}\, \phi_\nu(\boldsymbol{r}_1, \boldsymbol{r}_2)\, \phi_\nu^*(\boldsymbol{r}_1', \boldsymbol{r}_2') , \qquad (8.252)
\end{aligned}
$$

where $y = \ln(s/|t|)$. It is convenient to rewrite (8.252) in the factorized form

$$
\begin{aligned}
\frac{F(y, \boldsymbol{k}, \boldsymbol{k}', \boldsymbol{q})}{(\boldsymbol{k} - \boldsymbol{q})^2\, \boldsymbol{k}'^2} &= \frac{1}{(2\pi)^6} \int_{-\infty}^{+\infty} \mathrm{d}\nu \frac{\nu^2}{\left(\nu^2 + \frac{1}{4}\right)^2} \mathrm{e}^{\omega_0(\nu) y} \\
&\times X_\nu(\boldsymbol{k}, \boldsymbol{q})\, X^*(\boldsymbol{k}', \boldsymbol{q}) , \qquad (8.253)
\end{aligned}
$$

where $X_\nu(\boldsymbol{k}, \boldsymbol{q})$ is the double Fourier transform of the leading-order BFKL eigenfunction

$$
X_\nu(\boldsymbol{k}, \boldsymbol{q}) = \int \mathrm{d}^2 \boldsymbol{r}_1 \int \mathrm{d}^2 \boldsymbol{r}_2\, \mathrm{e}^{-\mathrm{i}\boldsymbol{k} \cdot \boldsymbol{r}_1 - \mathrm{i}(\boldsymbol{q} - \boldsymbol{k}) \cdot \boldsymbol{r}_2}\, \phi_\nu(\boldsymbol{r}_1, \boldsymbol{r}_2) . \qquad (8.254)
$$

The solution of the non-forward BFKL equation has important applications in high-$|t|$ diffractive processes, as we shall see in Sect. 11.12.

## 8.15 Parton–Parton Elastic Scattering

Making use of the BFKL results derived above we can now write down explicitly the amplitude for elastic parton–parton scattering.

Let us start from quark–quark scattering via color singlet (i.e., pomeron) exchange. The corresponding BFKL diagram is shown in Fig. 8.19. The scattering amplitude is obtained from (8.211) and (8.51, 8.54), using the non-forward solution (8.253). We get (a color-conservation factor $\delta_{ij}\delta_{kl}$ is understood)

$$A_{\underline{1}}(s,t) = \frac{\mathrm{i}s}{(2\pi)^6}\,\frac{N_c^2-1}{4N_c^2}\int_{-\infty}^{+\infty}\mathrm{d}\nu\,\frac{\nu^2}{\left(\nu^2+\frac{1}{4}\right)^2}\,\mathrm{e}^{\omega_0(\nu)y}$$
$$\times\, I_\nu^q(\boldsymbol{q})\,I_\nu^{q*}(\boldsymbol{q}), \tag{8.255}$$

where $I_\nu^q(\boldsymbol{k},\boldsymbol{q})$ is (the superscript $q$ stands for 'quark')

$$I_\nu^q(\boldsymbol{q}) = 8\pi^2\alpha_s\int\frac{\mathrm{d}^2\boldsymbol{k}}{(2\pi)^2}\,X_\nu(\boldsymbol{k},\boldsymbol{q})$$
$$= 8\pi^2\alpha_s\int\frac{\mathrm{d}^2\boldsymbol{k}}{(2\pi)^2}\int\mathrm{d}^2\boldsymbol{r}_1\int\mathrm{d}^2\boldsymbol{r}_2$$
$$\times\,\mathrm{e}^{-\mathrm{i}\boldsymbol{k}\cdot\boldsymbol{r}_1-\mathrm{i}(\boldsymbol{q}-\boldsymbol{k})\cdot\boldsymbol{r}_2}\,\phi_\nu(\boldsymbol{r}_1,\boldsymbol{r}_2)\,. \tag{8.256}$$

From (8.255) one obtains the elastic cross section for color singlet quantum numbers in the $t$-channel

$$\frac{\mathrm{d}\hat{\sigma}_{\mathrm{sing}}}{\mathrm{d}t} = \frac{1}{16\pi s^2}\,|A_{\underline{1}}(s,t)|^2$$
$$= \frac{1}{16\pi\,(2\pi)^{12}}\left(\frac{N_c^2-1}{4N_c^2}\right)^2$$
$$\times\left|\int_{-\infty}^{+\infty}\mathrm{d}\nu\,\frac{\nu^2}{\left(\nu^2+\frac{1}{4}\right)^2}\,\mathrm{e}^{\omega_0(\nu)y}I_\nu^q(\boldsymbol{q})\,I_\nu^{q*}(\boldsymbol{q})\right|^2\,. \tag{8.257}$$

When the pomeron couples to individual quarks, which is the case under discussion, there is a subtlety concerning the BFKL eigenfunctions $\phi_\nu(\boldsymbol{r}_1,\boldsymbol{r}_2)$ to be used in (8.253). The point is that Lipatov's derivation of (8.253), outlined in Sect. 8.14, was not based on the direct evaluation of Feynman diagrams in momentum space but rather on the calculation of a conformally invariant scattering amplitude in impact-parameter representation. Thus, spurious contributions may arise when one goes from $\boldsymbol{r}$-space to $\boldsymbol{k}$-space. Indeed, as noticed by Mueller and Tang (1992), if we insert (8.251) into (8.254) we end up with $\delta^2(\boldsymbol{k})$ and $\delta^2(\boldsymbol{k}-\boldsymbol{q})$ terms which do not arise from Feynman graphs. Another way to see this is to observe that (8.237) is not an equation for $F$ but for $\boldsymbol{\nabla}_1^2\boldsymbol{\nabla}_2^2F$. Therefore, the solution we found in Sect. 8.14 is determined only up to the addition of any function independent of either $\boldsymbol{r}_1$ or $\boldsymbol{r}_2$. In momentum space this gives rise to terms proportional to $\delta^2(\boldsymbol{k})$ or $\delta^2(\boldsymbol{k}-\boldsymbol{q})$. Note that there is no problem in the case of hadron–hadron scattering, because the spurious terms do not contribute to the amplitude integrated over the impact factors of colorless particles (see Sect. 8.16).

Delta-function terms in parton–parton scattering can be removed by the replacement (Mueller and Tang 1992)

$$\left(\frac{\boldsymbol{r}_{12}^2}{\boldsymbol{r}_1^2\,\boldsymbol{r}_2^2}\right)^{\frac{1}{2}+\mathrm{i}\nu} \to \left(\frac{\boldsymbol{r}_{12}^2}{\boldsymbol{r}_1^2\,\boldsymbol{r}_2^2}\right)^{\frac{1}{2}+\mathrm{i}\nu} - \left(\frac{1}{\boldsymbol{r}_1^2}\right)^{\frac{1}{2}+\mathrm{i}\nu} - \left(\frac{1}{\boldsymbol{r}_2^2}\right)^{\frac{1}{2}+\mathrm{i}\nu}. \tag{8.258}$$

Use of this expression in (8.256) yields

$$\begin{aligned} I_\nu^q(\boldsymbol{q}) &= 8\pi^2\alpha_s \int \frac{\mathrm{d}^2\boldsymbol{k}}{(2\pi)^2} \int \mathrm{d}^2\boldsymbol{r}_1 \int \mathrm{d}^2\boldsymbol{r}_2\, \mathrm{e}^{-\mathrm{i}\boldsymbol{k}\cdot\boldsymbol{r}_1-\mathrm{i}(\boldsymbol{q}-\boldsymbol{k})\cdot\boldsymbol{r}_2} \\ &\quad\times \left[\left(\frac{\boldsymbol{r}_{12}^2}{\boldsymbol{r}_1^2\boldsymbol{r}_2^2}\right)^{\frac{1}{2}+\mathrm{i}\nu} - \left(\frac{1}{\boldsymbol{r}_1^2}\right)^{\frac{1}{2}+\mathrm{i}\nu} - \left(\frac{1}{\boldsymbol{r}_2^2}\right)^{\frac{1}{2}+\mathrm{i}\nu}\right] \\ &= -4\,(2\pi)^2 \int \mathrm{d}^2\boldsymbol{r}_1\, \mathrm{e}^{-\mathrm{i}\boldsymbol{q}\cdot\boldsymbol{r}_1} \left(\frac{1}{\boldsymbol{r}_1^2}\right)^{\frac{1}{2}+\mathrm{i}\nu} \\ &= -\frac{8\pi}{|\boldsymbol{q}|}\,(2\pi)^2\, 2^{-2\mathrm{i}\nu}\, \frac{\Gamma(1/2-\mathrm{i}\nu)}{\Gamma(1/2+\mathrm{i}\nu)}\,(\boldsymbol{q}^2)^{\mathrm{i}\nu}\,. \end{aligned} \tag{8.259}$$

It is now an easy task to compute the elastic cross section. From (8.257) and (8.259) we obtain

$$\frac{\mathrm{d}\hat{\sigma}_{\mathrm{sing}}}{\mathrm{d}t} = \frac{1}{\pi}\left(\frac{N_c^2-1}{N_c^2}\right)^2 \frac{\alpha_s^4}{t^2} \left| \int_{-\infty}^{+\infty} \mathrm{d}\nu\, \frac{\nu^2}{\left(\nu^2+\frac{1}{4}\right)^2}\, \mathrm{e}^{\omega_0(\nu)y} \right|^2 , \tag{8.260}$$

and the evaluation of the integral, with $\omega_0(\nu)$ given by (8.204), yields

$$\frac{\mathrm{d}\hat{\sigma}_{\mathrm{sing}}}{\mathrm{d}t} = \pi^3 \left(\frac{N_c^2-1}{N_c^2}\right)^2 \frac{\alpha_s^4}{t^2}\, \frac{\mathrm{e}^{2\lambda y}}{(\pi\lambda' y/8)^3}\,. \tag{8.261}$$

The elastic cross section for color-octet (i.e., reggeized gluon) exchange can be derived from (8.168), and reads

$$\frac{\mathrm{d}\hat{\sigma}_{\mathrm{oct}}}{\mathrm{d}t} = \pi \left(\frac{N_c^2-1}{N_c^2}\right) \frac{\alpha_s^2}{t^2}\, \mathrm{e}^{2\epsilon(t)y}\,. \tag{8.262}$$

Note that the singlet-to-octet cross section ratio is $\mathcal{O}(\alpha_s^2)$ and has a powerlike dependence on $s$ with exponent $2(\lambda-\epsilon(t))$,

$$\frac{\mathrm{d}\hat{\sigma}_{\mathrm{sing}}/\mathrm{d}t}{\mathrm{d}\hat{\sigma}_{\mathrm{oct}}/\mathrm{d}t} \sim \alpha_s^2\, \mathrm{e}^{2(\lambda-\epsilon(t))y} \sim \alpha_s^2 \left(\frac{s}{|t|}\right)^{2(\lambda-\epsilon(t))}\,. \tag{8.263}$$

In the high-energy limit, the cross sections for gluon–gluon scattering differ from (8.261) and (8.262) only by a color factor. They are – see (8.170) and (8.180),

$$\frac{\mathrm{d}\hat{\sigma}_{\mathrm{sing}}^{gg}}{\mathrm{d}t} = \left(\frac{C_A}{C_F}\right)^4 \frac{\mathrm{d}\hat{\sigma}_{\mathrm{sing}}^{qq}}{\mathrm{d}t} = \pi^3 \left(\frac{4N_c^2}{N_c^2-1}\right)^2 \frac{\alpha_s^4}{t^2}\, \frac{\mathrm{e}^{2\lambda y}}{(\pi\lambda' y/8)^3}\,, \tag{8.264}$$

$$\frac{\mathrm{d}\hat{\sigma}_{\mathrm{oct}}^{gg}}{\mathrm{d}t} = \left(\frac{C_A}{C_F}\right)^2 \frac{\mathrm{d}\hat{\sigma}_{\mathrm{oct}}^{qq}}{\mathrm{d}t} = 4\pi \left(\frac{N_c^2}{N_c^2-1}\right) \frac{\alpha_s^2}{t^2}\, \mathrm{e}^{2\epsilon(t)y}\,. \tag{8.265}$$

The BFKL analysis of parton–parton scattering will be applied to jet production in Sect. 11.13.

## 8.16 Hadron–Hadron Scattering

We now explore a more realistic situation: the scattering of two hadrons. Partons inside a hadron are slightly off-shell, and it is indeed this off-shellness which regulates the infrared divergencies arising from the integration over transverse momenta in the BFKL amplitude.

The internal structure of hadrons is taken into account in the BFKL amplitude by the insertion of a genuinely non-perturbative quantity, called *impact factor*. The amplitude for the elastic scattering of two hadrons $A$ and $B$ (see Fig. 8.20) via pomeron exchange is then written as

$$A(s,t) = \mathrm{i}s\,\mathcal{C} \int \frac{\mathrm{d}^2\boldsymbol{k}}{(2\pi)^2} \int \frac{\mathrm{d}^2\boldsymbol{k}'}{(2\pi)^2}\, \Phi_A(\boldsymbol{k},\boldsymbol{q})\, \Phi_B(\boldsymbol{k}',\boldsymbol{q})\, \frac{F(y,\boldsymbol{k},\boldsymbol{k}',\boldsymbol{q})}{\boldsymbol{k}'^2(\boldsymbol{k}-\boldsymbol{q})^2}\,, \tag{8.266}$$

where $\Phi_A$ and $\Phi_B$ are the impact factors of hadron $A$ and $B$, respectively, and $F$ is the usual BFKL Green function, Eq. (8.252). $\mathcal{C}$ is a color factor.

We can reexpress $A(s,t)$ in the form

$$A(s,t) = \frac{\mathrm{i}s}{(2\pi)^6}\,\mathcal{C} \int_{-\infty}^{+\infty} \mathrm{d}\nu\, \frac{\nu^2}{\left(\nu^2+\frac{1}{4}\right)^2}\, \mathrm{e}^{\omega_0(\nu)y}\, I_\nu^A(\boldsymbol{q})\, I_\nu^{B*}(\boldsymbol{q})\,, \tag{8.267}$$

where

$$\begin{aligned} I_\nu^{A,B}(\boldsymbol{q}) = &\int \frac{\mathrm{d}^2\boldsymbol{k}}{(2\pi)^2}\, \Phi_{A,B}(\boldsymbol{k},\boldsymbol{q}) \\ &\times \int \mathrm{d}^2\boldsymbol{r}_1 \int \mathrm{d}^2\boldsymbol{r}_2\, \mathrm{e}^{-\mathrm{i}\boldsymbol{k}\cdot\boldsymbol{r}_1 - \mathrm{i}(\boldsymbol{q}-\boldsymbol{k})\cdot\boldsymbol{r}_2}\, \phi_\nu(\boldsymbol{r}_1,\boldsymbol{r}_2)\,. \end{aligned} \tag{8.268}$$

The case of quark–quark scattering is recovered by setting $\mathcal{C} = (N_c^2-1)/4N_c^2$ and using for the quark impact factor

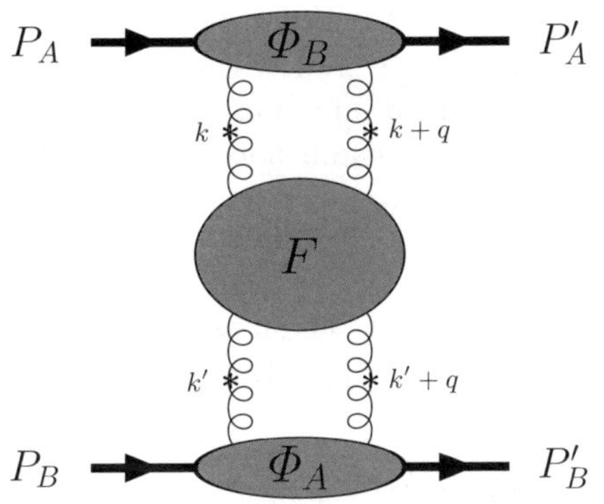


**Fig. 8.20.** BFKL picture of hadron–hadron scattering.

$$\Phi_q = 8\pi^2\alpha_s \,. \tag{8.269}$$

Equation (8.267), which is the result of a perturbative calculation, is applicable only at high $|t|$, or in presence of another hard scale. Needless to say, we lack a fundamental knowledge of the impact factors (except when hadrons are replaced by virtual photons, see Sect. 9.5.4), so they have to be modeled in some phenomenological way. We shall see how this program works in the case of diffractive vector meson production at high $|t|$ (Sect. 11.12).

## 8.17 The BFKL Pomeron at Next-to-Leading Order

The BFKL dynamics has been worked out so far in leading $\ln s$ approximation. Computing the next-to-leading contributions to color singlet exchange has been a formidable task to which many authors have participated in the past decade (for a list of references see Fadin 2000, 2001). This program was started some years ago by Lipatov and Fadin (1989) and completed by the same authors in 1998 (Fadin and Lipatov 1998; see also Camici and Ciafaloni 1998), with the calculation of the BFKL kernel at $t = 0$ and of the corresponding eigenvalues (a different approach to the problem of higher-order corrections has been pursued by Corianò and White 1996). Reporting in detail such a computational *tour de force* is beyond the scope of this book. We shall limit ourselves to presenting the main results.

In next-to-leading logarithmic approximation (NLLA) the BFKL kernel $\mathcal{K}(\boldsymbol{k}, \boldsymbol{\kappa})$ has a structure similar to that of LLA, namely

$$\mathcal{K}(\boldsymbol{k}, \boldsymbol{\kappa}) = 2\epsilon(-\boldsymbol{k}^2)\delta^2(\boldsymbol{k} - \boldsymbol{\kappa}) + \mathcal{K}_{\mathrm{real}}(\boldsymbol{k}, \boldsymbol{\kappa}) \,, \tag{8.270}$$

but the reggeized gluon trajectory $\epsilon(-\boldsymbol{k}^2)$ must be calculated in the two-loop approximation (Fadin, Fiore and Kotsky 1995, 1996a; Fadin, Fiore and Quartarolo 1996; Blümlein, Ravindran and van Neerven 1998; Del Duca and Glover 2001), and the real radiative term $\mathcal{K}_{\mathrm{real}}$ gets contributions from gluon production at one-loop level (Fadin and Lipatov 1993; Fadin, Fiore and Quartarolo 1994; Fadin, Fiore and Kotsky 1996b; Del Duca and Schmidt 1999) and from two-gluon and quark-antiquark production (Fadin and Lipatov 1989; Catani, Ciafaloni and Hautmann 1991; Fadin and Lipatov 1996; Del Duca 1996a, 1996b; Fadin, Kotsky and Lipatov 1997; Fadin et al. 1998). In NLLA, the amplitudes receive contributions also from diagrams like those of Fig. 8.17, which were neglected in LLA.

Referring to the original paper (Fadin and Lipatov 1998) for the explicit form of $\mathcal{K}$ in NLLA, we focus on $\omega(\gamma)$, defined as

$$\int \mathrm{d}^2\boldsymbol{\kappa}\, \mathcal{K}(\boldsymbol{k}, \boldsymbol{\kappa}) \left(\frac{\boldsymbol{\kappa}^2}{\boldsymbol{k}^2}\right)^{\gamma-1} = \omega(\gamma) \,. \tag{8.271}$$

(As we are going see, due to the running of $\alpha_s$, the eigenfunctions of the BFKL kernel are no more $(\boldsymbol{k}^2)^{\gamma-1}$.) The result for $\omega(\gamma)$ can be written as the sum of two terms

$$\omega(\gamma) = \frac{N_c \alpha_s(\boldsymbol{k}^2)}{\pi} \left[ \chi^{(0)}(\gamma) + \frac{N_c \alpha_s(\boldsymbol{k}^2)}{\pi} \chi^{(1)}(\gamma) \right] , \tag{8.272}$$

where $\chi^{(0)}(\gamma)$ is

$$\chi^{(0)}(\gamma) = 2\,\psi(1) - \psi(\gamma) - \psi(1-\gamma) \, . \tag{8.273}$$

and corresponds to the LLA contribution (8.200), with $\gamma = 1/2 + \mathrm{i}\nu$ and $n = 0$, whereas $\chi^{(1)}(\gamma)$ represents the NLO correction

$$\begin{aligned} \chi^{(1)}(\gamma) = -\frac{1}{4} \Bigg\{ & \frac{1}{2} \left( \frac{11}{3} - \frac{2n_f}{3N_c} \right) \left[ (\chi^{(0)}(\gamma))^2 - \psi'(\gamma) + \psi'(1-\gamma) \right] \\ & -6\,\zeta(3) + \frac{\pi^2 \cos \pi\gamma}{(\sin^2 \pi\gamma)(1-2\gamma)} \left[ 3 + \left( 1 + \frac{n_f}{N_c^3} \right) \frac{2 + 3\gamma\,(1-\gamma)}{(3-2\gamma)(1+2\gamma)} \right] \\ & - \left( \frac{67}{9} - \frac{\pi^2}{3} - \frac{10}{9} \frac{n_f}{N_c} \right) \chi^{(0)}(\gamma) - \psi''(\gamma) - \psi''(1-\gamma) \\ & - \frac{\pi^3}{\sin \pi\gamma} + 4\,\phi(\gamma) \Bigg\} \, . \end{aligned} \tag{8.274}$$

Here the apices denote differentiation and the function $\phi(\gamma)$ is

$$\begin{aligned} \phi(\gamma) &= - \int_0^1 \frac{\mathrm{d}x}{1+x} \left( x^{\gamma-1} + x^{-\gamma} \right) \int_x^1 \frac{\mathrm{d}t}{t} \ln(1-t) \\ &= \sum_{n=0}^{\infty} (-1)^n \left[ \frac{\psi(n+1+\gamma) - \psi(1)}{(n+\gamma)^2} + \frac{\psi(n+2-\gamma) - \psi(1)}{(n+1-\gamma)^2} \right] \end{aligned} \tag{8.275}$$

The strong coupling in (8.272) is

$$\alpha_s(\boldsymbol{k}^2) \simeq \alpha_s(\mu^2) \left[ 1 - \frac{\alpha_s(\mu^2)}{4\pi} \left( \frac{11}{3} N_c - \frac{2}{3} n_f \right) \ln \left( \frac{\boldsymbol{k}^2}{\mu^2} \right) \right] . \tag{8.276}$$

In $\omega(\gamma)$ the only term not symmetric with respect to the transformation $\gamma \leftrightarrow 1-\gamma$ is the one proportional to $\psi'(\gamma) - \psi'(1-\gamma)$. So we can write $\omega(\gamma)$ as

$$\omega(\gamma) = \omega_{\text{sym}}(\gamma) + \Delta(\gamma) \, , \tag{8.277}$$

where $\omega_{\text{sym}}(\gamma)$ is symmetric under $\gamma \leftrightarrow 1-\gamma$, and

$$\Delta(\gamma) = \frac{N_c^2\, \alpha_s^2(\mu^2)}{8\pi^2} \left( \frac{11}{3} - \frac{2n_f}{3N_c} \right) \left[ \psi'(\gamma) - \psi'(1-\gamma) \right] . \tag{8.278}$$

This term can be cancelled if we redefine the function $(\boldsymbol{k}^2)^{\gamma-1}$ in (8.271) by incorporating the factor $[\alpha_s(\boldsymbol{k}^2)/\alpha_s(\mu^2)]^{-\frac{1}{2}}$.

With respect to the LLA result we observe in (8.272) two types of corrections. One arises from the running of $\alpha_s$: the first term of (8.272) differs from the LLA result by

$$\frac{N_c \alpha_s(\mu^2)}{4\pi} \left( \frac{11}{3} - \frac{2n_f}{3N_c} \right) \ln \left( \frac{\boldsymbol{k}^2}{\mu^2} \right) . \tag{8.279}$$

The second, more important, correction is scale-independent and is due to the presence of $\chi^{(1)}(\gamma)$. It can be quantified by defining the ratio $r(\gamma) = -\chi^{(1)}(\gamma)/\chi^{(0)}(\gamma)$. In the point $\gamma = 1/2$, which corresponds to the largest eigenvalue in LLA, and hence to the pomeron intercept[4], we have

$$\begin{aligned} r(1/2) = & \left( \frac{11}{6} - \frac{n_f}{3N_c} \right) \ln 2 - \frac{67}{36} + \frac{\pi^2}{12} + \frac{5}{18} \frac{n_f}{N_c} \\ & + \frac{1}{\ln 2} \left[ \int_0^1 \frac{dt}{t} \tan^{-1} \sqrt(t) \ln \left( \frac{1}{1-t} \right) + \frac{11}{8} \zeta(3) \right. \\ & \left. + \frac{\pi^3}{32} \left( \frac{27}{16} + \frac{11}{16} \frac{n_f}{N_c^3} \right) \right] \\ \simeq & \ 6.46 + 0.05 \frac{n_f}{N_c} + 2.66 \frac{n_f}{N_c^3} . \end{aligned} \tag{8.280}$$

This correction is evidently very large, and this was a real surprise when the Fadin-Lipatov results were presented. Numerical estimates show that in the HERA regime the correction almost completely cancels the LLA term. Many papers have been – and are still – devoted to this problem, and various possible ways to cure it have been proposed. At the time of writing, however, the question remains unsettled (see a list of references in Fadin 2001).

The NLLA non-forward BFKL equation is also currently under study (see, e.g., Fadin and Fiore 1998; Fadin and Gorbachev 2000; Fadin et al. 2000a, 2000b). This is a highly technical work, still in progress, that we cannot report here. For an update and references we refer the reader to Fadin (2001).

[4] This is no longer true in NLLA, except when $\alpha_s$ is very small, see (8.278).

# 9. Deep Inelastic Scattering

Deep inelastic scattering (DIS) is the prototype of hard hadronic processes. As such, it provides an important - and very successful - test of perturbative QCD. It represents also the most direct way to explore the internal structure of hadrons.

In this book we are mainly concerned with those aspects of DIS that are more closely related to our main subject, diffraction. Thus, we shall focus on DIS at low $x$ and devote a relevant fraction of the next chapter to diffractive DIS and related phenomena. The presentation of the general features of DIS (parton model, QCD evolution, experimental results, etc.) will necessarily be rather sketchy. Some topics, such as sum rules, polarized DIS, electroweak DIS, higher-twist effects, etc., will be completely ignored. For a more detailed discussion of DIS we refer the reader to the existing treatises on the subject, e.g., Field (1989), Roberts (1990), Leader and Predazzi (1996), Ellis, Stirling and Webber (1996).

## 9.1 Kinematics

Deep inelastic scattering (DIS) is the scattering of a charged or neutral lepton off a hadron (typically a nucleon) at high momentum transfer, with measurement of the energy and scattering angle of the outgoing lepton, that is

$$l(\ell) + N(P) \to l'(\ell') + X(P_X) \,. \tag{9.1}$$

Here $X$ is an undetected hadronic system and we have written in brackets the four-momenta. The reaction is *fully inclusive* with respect to the hadronic state. In the neutral current case ($l = l' = e, \mu$) DIS is dominated – when the momentum transfer is not ultra-high – by one-photon-exchange, and is represented by the diagram in Fig. 9.1.

The inclusive reaction (9.1) is described by three kinematic variables. One of them (the incoming lepton energy $E$, or alternatively the c.m. energy squared $s = (\ell + P)^2$) is fixed by the experimental conditions. The other two independent variables are usually chosen among the following invariants ($m_N$ is the nucleon mass)

$$q^2 \equiv -Q^2 = (\ell - \ell')^2 \quad \text{(momentum transfer squared)} \,, \tag{9.2}$$

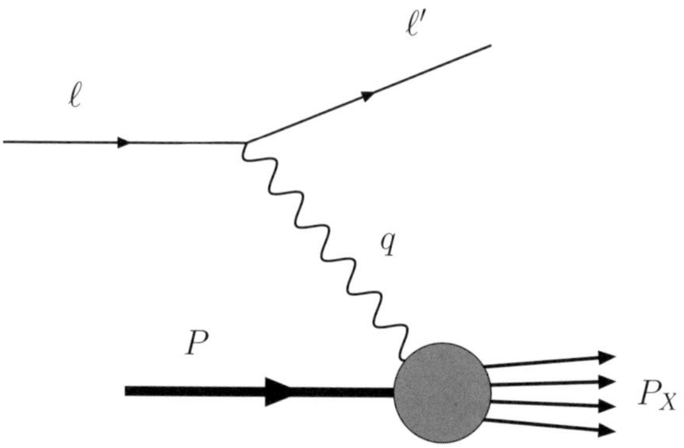


**Fig. 9.1.** Deep inelastic scattering

$$W^2 = (P+q)^2 \,, \tag{9.3}$$

$$\nu = \frac{P \cdot q}{m_N} = \frac{W^2 + Q^2 - m_N^2}{2m_N} \,, \tag{9.4}$$

$$x = \frac{Q^2}{2P \cdot q} = \frac{Q^2}{2m_N \nu} = \frac{Q^2}{Q^2 + W^2 - m_N^2} \quad \text{(Bjorken's } x) \,, \tag{9.5}$$

$$y = \frac{P \cdot q}{P \cdot \ell} = \frac{W^2 + Q^2 - m_N^2}{s - m_N^2} \,. \tag{9.6}$$

In the target rest frame $\nu$ is the transferred energy, that is $\nu = E - E'$ (we denote the initial and final lepton energies by $E$ and $E'$, respectively), and $y$ (sometimes called "inelasticity") is the fraction of the incoming lepton energy carried by the exchanged photon, $y = \nu/E$. A useful relation connecting $x$, $y$ and $Q^2$ is

$$xy = \frac{Q^2}{s - m_N^2} \simeq \frac{Q^2}{s} \,. \tag{9.7}$$

Since $W^2 \geq m_N^2$ ($W^2$ is the the c.m. energy squared of the $\gamma^* N$ system, that is the invariant mass squared of the hadronic system $X$) the Bjorken variable $x$ takes values between 0 and 1 (and so does $y$). We call *deep inelastic* the kinematic regime where both $m_N \nu$ and $Q^2$ are much larger than $m_N^2$, with $x$ fixed and finite. In this regime one can safely neglect the nucleon mass with respect to the other (large) scales of the problem.

The cross section for the process (9.1) reads

$$\mathrm{d}\sigma = \frac{1}{4\,(\ell \cdot P)} \frac{1}{2} \sum_{s_l, s_l'} \frac{1}{2} \sum_S \sum_X \int \frac{\mathrm{d}P_X}{(2\pi)^3\, 2P_X^0}$$
$$\times (2\pi)^4\, \delta^4(P + \ell - P_X - \ell')\, |\mathcal{M}|^2\, \frac{\mathrm{d}^3\ell'}{(2\pi)^3\, 2E'} \,. \tag{9.8}$$

As we are considering unpolarized DIS, in (9.8) we averaged over the initial lepton spin and the nucleon spin, and summed over the final lepton spin. The squared amplitude in (9.8) is

$$|\mathcal{M}|^2 = \frac{e^4}{q^4} \left[\overline{u}_{l'}(\ell', s_l')\gamma_\mu u_l(\ell, s_l)\right]^* \left[\overline{u}_{l'}(\ell', s_l')\gamma_\nu u_l(\ell, s_l)\right] \\ \times \langle X|J^\mu(0)|P,S\rangle^* \langle X|J^\nu(0)|P,S\rangle \,. \tag{9.9}$$

We introduce now the hadronic tensor $W^{\mu\nu}$ as

$$W^{\mu\nu} \equiv \frac{1}{(2\pi)} \frac{1}{2} \sum_S \sum_X \int \frac{\mathrm{d}^3 P_X}{(2\pi)^3 \, 2P_X^0} (2\pi)^4 \, \delta^4(P + q - P_X) \\ \times \langle P,S|J^\mu(0)|X\rangle \langle X|J^\nu(0)|P,S\rangle \\ = \frac{1}{2\pi} \int \mathrm{d}^4 z \, \mathrm{e}^{iq\cdot z} \, \langle N|J^\mu(z) J^\nu(0)|N\rangle \,, \tag{9.10}$$

where the spin average has been absorbed in the nucleon state $|N\rangle$. The leptonic tensor $L_{\mu\nu}$ is defined as (lepton masses are neglected)

$$L_{\mu\nu} \equiv \frac{1}{2} \sum_{s_l, s_l'} \left[\overline{u}_{l'}(\ell', s_l')\gamma_\mu u_l(\ell, s_l)\right]^* \left[\overline{u}_{l'}(\ell', s_l')\gamma_\nu u_l(\ell, s_l)\right] \\ = \frac{1}{2} \, \mathrm{Tr} \left[\not\ell \, \gamma_\mu \, \not\ell' \, \gamma_\nu\right] \\ = 2 \left(\ell_\mu \ell_\nu' + \ell_\nu \ell_\mu' - g_{\mu\nu} \, \ell \cdot \ell'\right) . \tag{9.11}$$

If we insert (9.9) into (9.8) and use the definitions (9.10) and (9.11), the differential cross section, written in the target rest frame, where $(\ell \cdot P) = m_N E$, takes the form ($\Omega \equiv (\vartheta, \varphi)$ is the solid angle identifying the direction of the outgoing lepton)

$$\frac{\mathrm{d}\sigma}{\mathrm{d}E' \, \mathrm{d}\Omega} = \frac{\alpha_{\mathrm{em}}^2}{2 \, m_N \, Q^4} \frac{E'}{E} L_{\mu\nu} W^{\mu\nu} \,. \tag{9.12}$$

The hadronic tensor $W_{\mu\nu}$ can be parametrized as

$$\frac{1}{2m_N} W_{\mu\nu} = \left(-g_{\mu\nu} + \frac{q_\mu q_\nu}{q^2}\right) W_1(P\cdot q, q^2) \\ + \frac{1}{m_N^2} \left[\left(P_\mu - \frac{P\cdot q}{q^2} q_\mu\right)\left(P_\nu - \frac{P\cdot q}{q^2} q_\nu\right)\right] W_2(P\cdot q, q^2), \tag{9.13}$$

and the (unpolarized) DIS cross section is then expressed in terms of the two *structure functions* $W_1$ and $W_2$ ($\mathrm{d}\Omega = \mathrm{d}\cos\vartheta \, \mathrm{d}\varphi$)

$$\frac{\mathrm{d}\sigma}{\mathrm{d}E' \, \mathrm{d}\Omega} = \frac{4\alpha_{\mathrm{em}}^2 E'^2}{Q^4} \left[2 \, W_1 \, \sin^2 \frac{\vartheta}{2} + W_2 \, \cos^2 \frac{\vartheta}{2}\right] \,. \tag{9.14}$$

As one can see, in the unpolarized case the DIS cross section depends on the scattering angle $\vartheta$, but not on the azimuthal angle $\varphi$ which can be integrated over.

It is customary to introduce the dimensionless structure functions

$$F_1(x,Q^2) \equiv m_N\, W_1(\nu,Q^2)\,,$$
$$F_2(x,Q^2) \equiv \nu\, W_2(\nu,Q^2)\,.$$

Bjorken (1969) argued that in the limit

$$\nu, Q^2 \to \infty\,, \qquad x = \frac{Q^2}{2m_N\nu} \text{ fixed}\,, \tag{9.15}$$

now called the *Bjorken limit*, $F_1$ and $F_2$ should (approximately) scale, i.e. depend on $x$ only. While this is true in the parton model (see Sect. 9.2), in QCD there is a mild, logarithmical, dependence on $Q^2$ (Sect. 9.3).

In terms of $F_1$ and $F_2$, the hadronic tensor reads

$$W_{\mu\nu} = 2\left(-g_{\mu\nu} + \frac{q_\mu q_\nu}{q^2}\right) F_1(x,Q^2) + \frac{2}{(P\cdot q)}\left[\left(P_\mu - \frac{P\cdot q}{q^2}\,q_\mu\right)\left(P_\nu - \frac{P\cdot q}{q^2}\,q_\nu\right)\right] F_2(x,Q^2)\,, \tag{9.16}$$

and the DIS cross section, written as a function of $x$ and $y$, is

$$\frac{d\sigma}{dx\,dy} = \frac{4\pi\alpha_{\rm em}^2\, s}{Q^4}\left\{xy^2\, F_1(x,Q^2) + \left(1 - y - \frac{xym_N^2}{s}\right) F_2(x,Q^2)\right\}\,. \tag{9.17}$$

It is useful to relate, via optical theorem, the DIS structure functions to the total cross sections of virtual photoabsorption. These are given by

$$\sigma_\lambda^{\gamma^* N}(x,Q^2) = \frac{2\pi^2\alpha_{\rm em}}{m_N\sqrt{\nu^2+Q^2}}\,\varepsilon_\mu^{(\lambda)}\,\varepsilon_\nu^{(\lambda)*}\,W^{\mu\nu}\,, \tag{9.18}$$

where $\varepsilon_\mu^{(\lambda)}$ is the polarization four-vector of a virtual photon of helicity $\lambda$, and we adopted Gilman's choice for the flux factor, which is not uniquely defined for a virtual particle (for more details, see e.g., Leader and Predazzi 1996). Let us introduce the helicity projectors of the photon

$$d_{\mu\nu}^{(\Sigma)} = \sum_{\lambda=0,\pm 1} \varepsilon_\mu^{(\lambda)}\,\varepsilon_\nu^{(\lambda)*} = -\left(g_{\mu\nu} + \frac{q_\mu q_\nu}{Q^2}\right)\,, \tag{9.19}$$

$$d_{\mu\nu}^{(L)} = \varepsilon_\mu^{(0)}\,\varepsilon_\nu^{(0)*} = \frac{Q^2}{m_N^2\,(\nu^2+Q^2)}\left(P_\mu + \frac{P\cdot q}{Q^2}\,q_\mu\right)\left(P_\nu + \frac{P\cdot q}{Q^2}\,q_\nu\right), \tag{9.20}$$

$$d_{\mu\nu}^{(T)} = \frac{1}{2}\left[\varepsilon_\mu^{(+1)}\,\varepsilon_\nu^{(+1)*} + \varepsilon_\mu^{(-1)}\,\varepsilon_\nu^{(-1)*}\right] = \frac{1}{2}\left[d_{\mu\nu}^{(\Sigma)} - d_{\mu\nu}^{(L)}\right]\,. \tag{9.21}$$

Using these expressions we can calculate the longitudinal and transverse virtual photoabsorption cross sections, which turn out to be

$$\begin{aligned}\sigma_L^{\gamma^* N} &= \frac{2\pi^2\alpha_{\rm em}}{m_N\sqrt{\nu^2+Q^2}}\, d_{\mu\nu}^{(L)}\, W^{\mu\nu} \\ &= \frac{4\pi^2\alpha_{\rm em}}{\sqrt{\nu^2+Q^2}}\left[-W_1+\left(1+\frac{\nu^2}{Q^2}\right)W_2\right], \qquad (9.22)\end{aligned}$$

$$\begin{aligned}\sigma_T^{\gamma^* N} &= \frac{2\pi^2\alpha_{\rm em}}{m_N\sqrt{\nu^2+Q^2}}\, d_{\mu\nu}^{(T)}\, W^{\mu\nu} \\ &= \frac{4\pi^2\alpha_{\rm em}}{\sqrt{\nu^2+Q^2}}\, W_1 . \qquad (9.23)\end{aligned}$$

Neglecting $Q^2$ with respect to $\nu^2$, that is ignoring target mass correction terms $\sim x^2 m_N^2/Q^2$, and introducing the structure functions $F_1$ and $F_2$, Eqs. (9.22, 9.23) take the form

$$\sigma_L^{\gamma^* N} = \frac{4\pi^2\alpha_{\rm em}}{Q^2}\,(F_2 - 2x\,F_1)\,, \qquad (9.24)$$

$$\sigma_T^{\gamma^* N} = \frac{4\pi^2\alpha_{\rm em}}{Q^2}\,2xF_1\,. \qquad (9.25)$$

Finally, if we define the longitudinal and transverse structure functions as

$$F_T = 2x\,F_1\,, \qquad (9.26)$$

$$F_L = F_2 - 2x\,F_1\,, \qquad (9.27)$$

the $\gamma^* N$ cross sections can be reexpressed as

$$\sigma_{L,T}^{\gamma^* N}(x,Q^2) = \frac{4\pi^2\alpha_{\rm em}}{Q^2}\,F_{L,T}(x,Q^2)\,. \qquad (9.28)$$

Note that $F_2 = F_L + F_T$ and the (longitudinal + transverse) virtual photoabsorption cross section $\sigma^{\gamma^* N}$ is proportional to $F_2$

$$\sigma^{\gamma^* N}(x,Q^2) = \frac{4\pi^2\alpha_{\rm em}}{Q^2}\,F_2(x,Q^2)\,. \qquad (9.29)$$

## 9.2 Parton Model

The parton model is based on the assumption that the virtual photon scatters incoherently off the internal constituents of the nucleon, which are treated as free particles. The hadronic tensor $W^{\mu\nu}$ is then represented by the handbag diagram shown in Fig. 9.2 and reads (to simplify the discussion, for the moment being we shall consider quarks only)

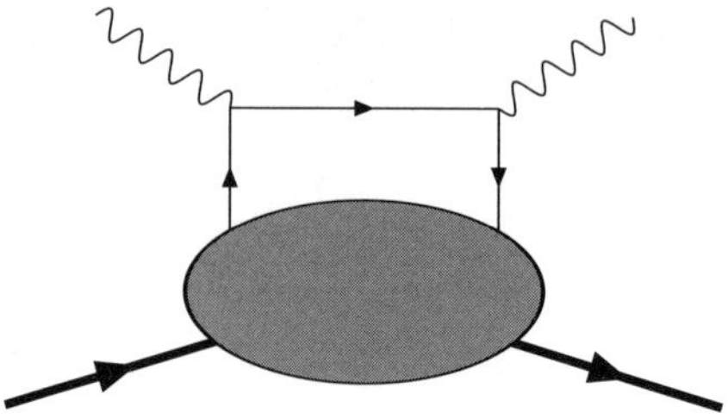

**Fig. 9.2.** The handbag diagram.

$$
\begin{aligned}
W^{\mu\nu} = \frac{1}{2\pi} \sum_a e_a^2 \sum_X \int \frac{\mathrm{d}^3 \boldsymbol{P}_X}{(2\pi)^3\, 2P_X^0} \int \frac{\mathrm{d}^4 k}{(2\pi)^4} \int \frac{\mathrm{d}^4 \kappa}{(2\pi)^4}\, \delta(\kappa^2) \\
\times \left[\overline{u}(\kappa)\gamma^\mu \phi(k,P)\right]^* \left[\overline{u}(\kappa)\gamma^\nu \phi(k,P)\right] \\
\times (2\pi)^4\, \delta^4(P-k-P_X)\,(2\pi)^4\, \delta^4(k+q-\kappa)\,, \qquad (9.30)
\end{aligned}
$$

where $\sum_a$ is a sum over the quark flavors, $e_a$ is the quark charge, and we have introduced the quark-nucleon vertex functions

$$
\phi(k,P) = \langle X|\psi(0)|N\rangle\,. \qquad (9.31)
$$

An average over the nucleon spin is understood in (9.30).

We define the quark–quark correlation matrix $\Phi(k,P)$ as

$$
\Phi(k,P) = \sum_X \int \frac{\mathrm{d}^3 \boldsymbol{P}_X}{(2\pi)^3\, 2P_X^0}\,(2\pi)^4\,\delta^4(P-k-P_X)\,\phi(k,P)\,\overline{\phi}(k,P)\,, \quad (9.32)
$$

so that the hadronic tensor can be rewritten in the form (we neglect quark masses)

$$
\begin{aligned}
W^{\mu\nu} &= \sum_a e_a^2 \int \frac{\mathrm{d}^4 k}{(2\pi)^4} \int \frac{\mathrm{d}^4 \kappa}{(2\pi)^4}\, \delta(\kappa^2)\,(2\pi)^4\,\delta^4(k+q-\kappa)\,\mathrm{Tr}\,[\Phi\,\gamma^\mu\, \not\kappa\, \gamma^\nu] \\
&= \sum_a e_a^2 \int \frac{\mathrm{d}^4 k}{(2\pi)^4}\,\delta((k+q)^2)\,\mathrm{Tr}\,[\Phi\,\gamma^\mu\,(\not k + \not q)\,\gamma^\nu]\,. \qquad (9.33)
\end{aligned}
$$

Using the definition (9.31) and the completeness of the $|X\rangle$ states allows us to reexpress the correlation matrix $\Phi$ in the more transparent form

$$
\Phi_{ij}(k,P) = \int \mathrm{d}^4\xi\; \mathrm{e}^{ik\cdot\xi}\,\langle N|\overline{\psi}_j(0)\psi_i(\xi)|N\rangle\,. \qquad (9.34)
$$

It is important to remind that the matrix elements in (9.34) are connected.

Let us now introduce two Sudakov vectors $p^\mu$ and $n^\mu$, satisfying $p^2 = n^2 = 0$, $p\cdot n = 1$, $p^- = n^+ = 0$, so that a generic four-vector $A^\mu$ can be written as (see App. A.2)

$$A^\mu = \alpha\, p^\mu + \beta\, n^\mu + A_\perp^\mu \\ \equiv (A \cdot n)\, p^\mu + (A \cdot p)\, n^\mu + A_\perp^\mu \,, \tag{9.35}$$

where $A_\perp^\mu = (0, \boldsymbol{A}_\perp, 0)$ lies in the plane perpendicular to the $\gamma^* N$ direction, which we take to be the $z$ axis.

In terms of $p$ and $n$ the nucleon momentum $P$ can be decomposed as

$$P^\mu = p^\mu + \frac{m_N^2}{2}\, n^\mu \,. \tag{9.36}$$

Neglecting the nucleon mass, as we may do in the deep inelastic limit, it coincides with $p$

$$P^\mu = p^\mu \,. \tag{9.37}$$

The momentum of the virtual photon can be written as

$$q^\mu \simeq (P \cdot q)\, n^\mu - x\, p^\mu \,, \tag{9.38}$$

where we are implicitly ignoring terms $\mathcal{O}(m_N^2/Q^2)$. Finally, the Sudakov decomposition of the quark momentum is

$$k^\mu = \alpha\, p^\mu + \frac{k^2 + \boldsymbol{k}_\perp^2}{2\alpha}\, n^\mu + k_\perp^\mu \,. \tag{9.39}$$

In the parton model one assumes that the handbag diagram contribution to the hadronic tensor is dominated by small values of $k^2$ and $\boldsymbol{k}_\perp^2$. Thus we can approximately write $k^\mu$ as

$$k^\mu \simeq \alpha\, p^\mu \,. \tag{9.40}$$

Notice that a natural frame for DIS is the infinite momentum frame ($P^+ \to \infty$), where $n^\mu = \mathcal{O}(1/P^+)$ is suppressed by a factor $(P^+)^2$ with respect to $p^\mu = \mathcal{O}(P^+)$, and $k^\mu$ reduces indeed to (9.40).

The on-shellness of the outgoing quark, in the light of (9.40), implies

$$\delta((k+q)^2) \simeq \delta(-Q^2 + 2\,\alpha\, P \cdot q) = \frac{1}{2\, P \cdot q}\, \delta(\alpha - x) \,, \tag{9.41}$$

that is

$$k^\mu \simeq x\, P^\mu \,. \tag{9.42}$$

Thus the Bjorken variable $x$ emerges as the fraction of the longitudinal proton momentum carried by the struck quark:

$$x = \frac{k^+}{P^+} \,. \tag{9.43}$$

Coming back to the hadronic tensor (9.33), the identity

$$\gamma^\mu \gamma^\rho \gamma^\nu = (g^{\mu\rho} g^{\nu\sigma} + g^{\mu\sigma} g^{\nu\rho} - g^{\mu\nu} g^{\rho\sigma}) \gamma_\sigma - \mathrm{i}\varepsilon^{\mu\rho\nu\sigma}\, \gamma_\sigma \gamma^5 \,, \tag{9.44}$$

allows splitting $W^{\mu\nu}$ into a symmetric (S) and an antisymmetric (A) part under $\mu \leftrightarrow \nu$ interchange. Only $W^{(S)}_{\mu\nu}$ contributes to unpolarized DIS, since in this case the leptonic tensor is symmetric, see (9.11). $W^{(S)}_{\mu\nu}$ is given by

$$\begin{aligned} W^{(S)}_{\mu\nu} = & \frac{1}{2(P\cdot q)} \sum_a e_a^2 \int \frac{\mathrm{d}^4 k}{(2\pi)^4}\, \delta\left(x - \frac{k^+}{P^+}\right) \\ & \times \left[(k_\mu + q_\mu)\, \mathrm{Tr}\,(\Phi\,\gamma_\nu) + (k_\nu + q_\nu)\, \mathrm{Tr}\,(\Phi\,\gamma_\mu) \right. \\ & \left. - g_{\mu\nu}(k^\rho + q^\rho)\, \mathrm{Tr}\,(\Phi\,\gamma_\rho)\right] . \end{aligned} \tag{9.45}$$

From (9.38) and (9.42) we get $k_\mu + q_\mu \simeq (P\cdot q)\, n_\mu$ and (9.45) becomes

$$\begin{aligned} W^{(S)}_{\mu\nu} = & \frac{1}{2} \sum_a e_a^2 \int \frac{\mathrm{d}^4 k}{(2\pi)^4}\, \delta\left(x - \frac{k^+}{P^+}\right) \\ & \times \left[n_\mu\, \mathrm{Tr}\,(\Phi\gamma_\nu) + n_\nu\, \mathrm{Tr}\,(\Phi\gamma_\mu) - g_{\mu\nu} n^\rho\, \mathrm{Tr}\,(\Phi\gamma_\rho)\right] . \end{aligned} \tag{9.46}$$

It is convenient to introduce the following notation. We call $\langle\Gamma\rangle$, where $\Gamma$ is a Dirac matrix, the quantity

$$\begin{aligned} \langle\Gamma\rangle \equiv & \int \frac{\mathrm{d}^4 k}{(2\pi)^4}\, \delta\left(x - \frac{k^+}{P^+}\right) \mathrm{Tr}\,(\Gamma\,\Phi) \\ = & P^+ \int \frac{\mathrm{d}\xi^-}{2\pi}\, \mathrm{e}^{\mathrm{i}xP^+\xi^-}\, \langle N|\overline{\psi}(0)\, \Gamma\, \psi(0,\xi^-,\mathbf{0}_\perp)|N\rangle \\ = & \int \frac{\mathrm{d}\tau}{2\pi}\, \mathrm{e}^{\mathrm{i}\tau x}\, \langle N|\overline{\psi}(0)\, \Gamma\, \psi(\tau n)|N\rangle . \end{aligned} \tag{9.47}$$

Note that $\langle\Gamma\rangle$ is a function of the Bjorken variable $x$. Hence $W^{(S)}_{\mu\nu}$ reads

$$W^{(S)}_{\mu\nu} = \sum_a \frac{e_a^2}{2} \left[n_\mu\, \langle\gamma_\nu\rangle + n_\nu\, \langle\gamma_\mu\rangle - g_{\mu\nu}\, n^\rho\, \langle\gamma_\rho\rangle\right] . \tag{9.48}$$

We now have to parametrize $\langle\gamma^\mu\rangle$, which is a vector quantity. At leading twist, that is considering contributions $\mathcal{O}(P^+)$ in the infinite momentum frame, the only vector at our disposal is $p^\mu = P^\mu$ (recall that $n^\mu = \mathcal{O}(1/P^+)$ and $k^\mu \simeq xP^\mu$). Thus we can write

$$\begin{aligned} \langle\gamma^\mu\rangle \equiv & \int \frac{\mathrm{d}^4 k}{(2\pi)^4}\, \delta\left(x - \frac{k^+}{P^+}\right) \mathrm{Tr}\,(\gamma^\mu\,\Phi) \\ = & \int \frac{\mathrm{d}\tau}{2\pi}\, \mathrm{e}^{\mathrm{i}\tau x}\, \langle N|\overline{\psi}(0)\, \gamma^\mu\, \psi(\tau n)|N\rangle \\ = & \, 2\, f_q(x)\, P^\mu , \end{aligned} \tag{9.49}$$

where the coefficient of $P^\mu$, that we called $f_q(x)$, is the number density of quarks, as it will become clear later on. Multiplication of (9.49) by $n_\mu$ gives $f_q(x)$ in the explicit form

$$f_q(x) = \int \frac{\mathrm{d}\xi^-}{4\pi}\, \mathrm{e}^{\mathrm{i}xP^+\xi^-} \langle N|\overline{\psi}(0)\gamma^+\psi(0,\xi^-,0_\perp)|N\rangle\,, \tag{9.50}$$

where $\gamma^+ = (\gamma^0 + \gamma^3)/\sqrt{2}$. In QCD, there are two important things to add to (9.50). First of all, in order to make the expression (9.49) gauge invariant, a path-dependent link operator

$$\mathcal{L}(0,\xi) = \mathcal{P}\exp\left(-\mathrm{i}g \int_0^\xi \mathrm{d}s_\mu\, A^\mu(s)\right), \tag{9.51}$$

where $\mathcal{P}$ denotes path-ordering, must be inserted between the quark fields. Since distribution functions involve separations $\xi$ of the form $(0,\xi^-,0_\perp)$, if one works in the axial gauge $A^+ = 0$ and chooses an appropriate path, $\mathcal{L}$ can be reduced to unity. For details, we refer the reader to Collins and Soper (1982). Hereafter, when writing parton distribution functions, we shall always assume that the link operator is unity and just omit it.

The other extra ingredient coming from QCD is the scale dependence of distribution functions. Since the operator products appearing in (9.50) are ultraviolet divergent, they have to be renormalized at some scale $\mu$, and the parton distributions become $\mu$-dependent: $f_q(x) \to f_q(x,\mu)$. Note that, in so doing, a dependence on the renormalization scheme is also introduced.

Inserting (9.49) in (9.48) yields (remember that $P\cdot n = 1$)

$$W^{(\mathrm{S})}_{\mu\nu} = \sum_a e_a^2\,(n_\mu p_\nu + n_\nu p_\mu - g_{\mu\nu})\, f_a(x)\,. \tag{9.52}$$

The structure functions $F_1$ and $F_2$ can be extracted from $W^{\mu\nu}$ by means of two projectors (terms of order $1/Q^2$ are neglected)

$$F_1 = \mathcal{P}_1^{\mu\nu} W_{\mu\nu} = \frac{1}{4}\left(\frac{4x^2}{Q^2}\, P^\mu P^\nu - g^{\mu\nu}\right) W_{\mu\nu}\,, \tag{9.53}$$

$$F_2 = \mathcal{P}_2^{\mu\nu} W_{\mu\nu} = \frac{x}{2}\left(\frac{12x^2}{Q^2}\, P^\mu P^\nu - g^{\mu\nu}\right) W_{\mu\nu}\,. \tag{9.54}$$

Since $(P^\mu P^\nu/Q^2)\, W_{\mu\nu} = \mathcal{O}(M^2/Q^2)$ we find that $F_1$ and $F_2$ are proportional to each other (the so-called Callan-Gross relation) and are given by

$$\begin{aligned} F_2(x) = 2xF_1(x) &= -\frac{x}{2}\, g^{\mu\nu} W_{\mu\nu} \\ &= \sum_q e_q^2\, x\, f_q(x)\,, \end{aligned} \tag{9.55}$$

which is the well known parton model expression for the unpolarized structure functions. This justifies the identification of (9.50) with the unpolarized quark distribution function (i.e., the number density of quarks). To get the full expression of $F_1$ and $F_2$, one should simply add to (9.54) the antiquark distributions $f_{\bar{q}}(x)$, which were left aside in the above discussion. They read (the role of $\psi$ and $\overline{\psi}$ is interchanged with respect to the quark distributions)

$$f_{\bar{q}}(x) = \int \frac{\mathrm{d}\xi^-}{4\pi}\, \mathrm{e}^{\mathrm{i}xP^+\xi^-} \langle N|\mathrm{Tr}\left[\gamma^+\psi(0)\overline{\psi}(0,\xi^-,0_\perp)\right]|N\rangle \,, \tag{9.56}$$

and the structure functions $F_1, F_2$ are

$$F_2(x) = 2xF_1(x) = \sum_q e_q^2\, x\, \left[f_q(x) + f_{\bar{q}}(x)\right] \,. \tag{9.57}$$

Before ending this section, let us derive the parton model expressions for the structure functions by a different method, which opens the way to the QCD treatment of DIS that will occupy the rest of the chapter.

As we have seen, in the parton model the DIS cross section is the incoherent sum of cross sections for scattering off the individual components of the target. Thus, if we introduce from the beginning the quark distribution functions $f_q(\xi)$ as the probability densities to find a quark $q$ carrying a fraction $\xi$ of the longitudinal momentum of the target, we can write the DIS differential cross section $\mathrm{d}\sigma$ in the *factorized* form

$$\mathrm{d}\sigma = \sum_{q,\bar{q}} \int \mathrm{d}\xi\, f_q(\xi)\, \mathrm{d}\hat{\sigma}\left(\frac{x}{\xi}\right) \,. \tag{9.58}$$

where $\mathrm{d}\hat{\sigma}$ is the cross section for the elementary subprocess $l\,q\,(\bar{q}) \to l\,q\,(\bar{q})$, namely ($\hat{s} = \xi s$)

$$\frac{\mathrm{d}\hat{\sigma}}{\mathrm{d}y} = \frac{1}{16\pi\,\hat{s}^2}\,\frac{1}{4} \sum_{\lambda\lambda'\eta\eta'} \mathcal{M}_{\lambda\eta\lambda'\eta'}\, \mathcal{M}^*_{\lambda\eta\lambda'\eta'} \,, \tag{9.59}$$

where $\mathcal{M}$ is the amplitude of $\ell\, q(\bar{q})$ scattering (see Fig. 9.3), and an average (sum) is performed over the initial (final) helicities. Working out the energy-momentum conservation, Eq. (9.58) can be rewritten as

$$\begin{aligned} \frac{\mathrm{d}\sigma}{\mathrm{d}x\,\mathrm{d}y} &= \sum_{q,\bar{q}} \int \mathrm{d}\xi\, f_a(\xi)\, \frac{\mathrm{d}\hat{\sigma}}{\mathrm{d}y}\, \delta(\xi - x) \\ &= \sum_{q,\bar{q}} f_a(x)\, \frac{\mathrm{d}\hat{\sigma}}{\mathrm{d}y} \,. \end{aligned} \tag{9.60}$$

The only non vanishing amplitudes are

$$\mathcal{M}_{++++} = \mathcal{M}_{----} = 2\mathrm{i}\, e^2 e_q\, \frac{1}{y} \,, \tag{9.61}$$

$$\mathcal{M}_{+-+-} = \mathcal{M}_{-+-+} = 2\mathrm{i}\, e^2 e_q\, \frac{1-y}{y} \,. \tag{9.62}$$

Substituting these expressions in (9.59) yields ($\hat{s} = xs$)

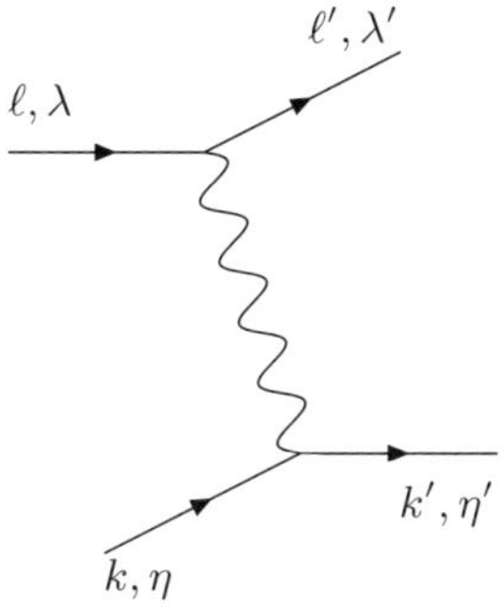


**Fig. 9.3.** Lepton-quark scattering ($\lambda, \lambda', \eta, \eta'$ are helicities).

$$\frac{\mathrm{d}\hat{\sigma}}{\mathrm{d}y} = \frac{2\pi\alpha_{\mathrm{em}}^2\, e_q^2}{\hat{s}\, y^2}\,[1 + (1-y)^2]\,, \tag{9.63}$$

and for the DIS cross section we get

$$\frac{\mathrm{d}\sigma}{\mathrm{d}x\,\mathrm{d}y} = \frac{4\pi\alpha_{\mathrm{em}}^2 s}{Q^4}\,\frac{1}{2}\,\left[1 + (1-y)^2\right] \sum_q e_q^2\, x\, [f_q(x) + f_{\bar{q}}(x)]\,, \tag{9.64}$$

which is the result already obtained in (9.17), with $m_N$ set to zero and the structure functions $F_1, F_2$ given by (9.55).

It turns out that QCD preserves the general structure of Eq. (9.58), except that the parton densities acquire a scale dependence and the elementary cross sections $\mathrm{d}\hat{\sigma}$ are written as a power series in the strong coupling constant $\alpha_s$. We shall now proceed to see how all this comes about.

## 9.3 Structure Functions in QCD

The parton model is only the zero-th order approximation to the real world. After all, the partonic constituents of hadrons are not free objects. They are described by QCD, the interacting theory of quarks and gluons. From an empirical point of view, one observes that the scaling predicted by the parton model is violated. Structure functions appear to depend on $Q^2$, although in a relatively mild way – logarithmically. We shall now see how this behavior arises from perturbative QCD. We inform the reader that the discussion below will be limited to a summary of few essential results. For the derivation of these results and more details we refer to the textbooks on deep inelastic scattering and QCD (for instance, Field 1989; Roberts 1990; Leader and Predazzi, 1996; Ellis, Stirling and Webber, 1996).

We take as a starting point the parton model factorization formula (9.58), that we rewrite for the structure function $F_2$ as

$$F_2(x) = \sum_{q,\bar{q}} \int_x^1 \mathrm{d}\xi \, f_q(\xi) \, \hat{F}_2^q \left( \frac{x}{\xi} \right) . \tag{9.65}$$

Here $\hat{F}_2^q$ is the elementary structure function of quarks, i.e., apart from a constant factor, the virtual photoabsorption cross section for $\gamma^* q$ scattering. In the parton model this process is simply $\gamma^* q\,(\bar{q}) \to q\,(\bar{q})$ (see Fig. 9.4a) and $\hat{F}_2^q$ is given by

$$\hat{F}_2^q(z) = e_q^2 \, \delta(1-z) . \tag{9.66}$$

Inserting this into (9.65) leads to the well known parton model expression $F_2 = x \sum_q e_q^2 \, [f_q(x) + f_{\bar{q}}(x)]$.

This would be the end of the story if quarks (and antiquarks) were really free particles. But they are not. They interact by emitting and absorbing gluons. Thus there other processes contributing to the $\gamma^* q\,(\bar{q})$ cross section. At order $\alpha_s^1$ they are shown in Fig. 9.4:

- real-gluon emission: $\hat{t}$-channel (Fig. 9.4b) and $\hat{s}$-channel (Fig. 9.4c);
- virtual-gluon radiation: vertex corrections (Fig. 9.4d) and self-energy insertions (Fig. 9.4e,f).

Consider first the real-gluon emission diagrams. They develop two types of singularities: *i)* a collinear singularity, arising from the $\hat{t}$-channel diagram in the limit $\hat{t} \to 0$, that is when the gluon is emitted parallel to the quark; (remember that $\hat{t} \propto (1 - \cos\theta)$, where $\theta$ is the scattering angle in the CM frame); *ii)* a singularity due to soft gluon emission. It turns out that soft divergences cancel out in summing the contributions of real gluon and virtual gluon diagrams. Hence, the only surviving divergence is the collinear one. We

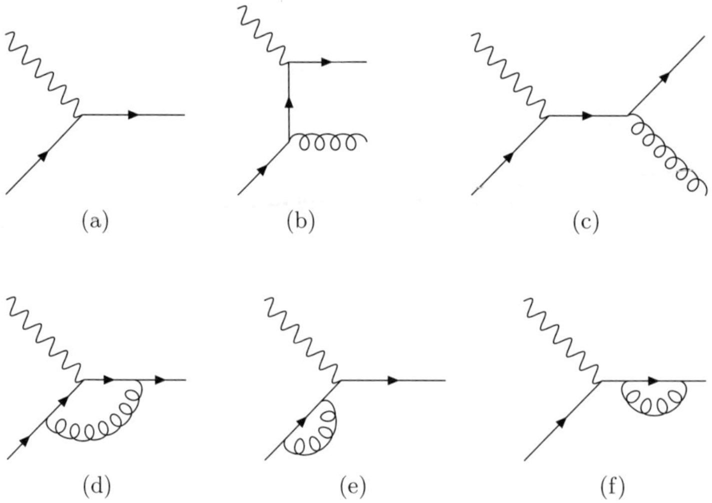


**Fig. 9.4a–f.** Processes contributing to the $\gamma^* q$ scattering.

can regularize it by setting a lower cutoff $\kappa_0^2$ on the transverse momentum $\kappa_\perp^2$ of the struck quark. It finally turns out that the $\mathcal{O}(\alpha_s)$ diagrams of Fig. 9.4 contribute to $\hat{F}_2^q$ a term of the form

$$\hat{F}_2^q(z, Q^2) = \frac{\alpha_s}{2\pi}\, e_q^2\, z \left[ P(z) \ln \frac{Q^2}{\kappa_0^2} + h(z) \right] , \tag{9.67}$$

where $P(z)$ and $h(z)$ are finite functions. Note that a scale dependence appears in $\hat{F}_2^q$ at order $\alpha_s^1$. Let us then rewrite the convolution formula (9.65) introducing a $Q^2$ dependence in $\hat{F}_2^q$ and renaming $f_q^0(x)$ the quark number densities, with the superscript 0 signaling that they are the *bare distributions*

$$F_2(x, Q^2) = \sum_{q,\bar{q}} \int \mathrm{d}\xi\, f_q^0(\xi)\, \hat{F}_2^q \left( \frac{x}{\xi}, Q^2 \right) . \tag{9.68}$$

Adding (9.67) to the $\mathcal{O}(\alpha_s^0)$ contribution (9.66) gives for the nucleon structure function

$$\begin{aligned} F_2(x, Q^2) = \sum_{q,\bar{q}} e_q^2\, x \Bigg\{ f_q^0(x) + \frac{\alpha_s}{2\pi} \int_x^1 \frac{\mathrm{d}\xi}{\xi}\, f_q^0(\xi) \\ \times \left[ P\left(\frac{x}{\xi}\right) \ln \frac{Q^2}{\kappa_0^2} + h\left(\frac{x}{\xi}\right) \right] + \ldots \Bigg\} , \end{aligned} \tag{9.69}$$

where the dots denote higher-order terms. We introduce now a factorization scale $\mu^2$, so that the divergent logarithm in (9.69) can be split as

$$\ln \frac{Q^2}{\kappa_0^2} = \ln \frac{Q^2}{\mu^2} + \ln \frac{\mu^2}{\kappa_0^2} . \tag{9.70}$$

We also separate arbitrarily the finite function $h(z)$ in two parts

$$h(z) = \widetilde{h}(z) + h'(z) \tag{9.71}$$

and absorb the singularity $\ln(\mu^2/\kappa_0^2)$ and the term $h'(z)$ in a redefinition of the quark distribution. We thus define the renormalized distribution functions as

$$q(x, \mu^2) = f_q^0(x) + \frac{\alpha_s}{2\pi} \int_x^1 \frac{\mathrm{d}\xi}{\xi}\, f_q^0(\xi) \left[ P\left(\frac{x}{\xi}\right) \ln \frac{\mu^2}{\kappa_0^2} + h'\left(\frac{x}{\xi}\right) \right] + \ldots \tag{9.72}$$

The separation (9.71) defines the factorization scheme. In terms of $q(x, \mu^2)$ the structure function $F_2$ reads

$$F_2(x, Q^2) = \sum_{q,\bar{q}} e_q^2\, x \int_x^1 \frac{\mathrm{d}\xi}{\xi}\, q(\xi, \mu^2)\, C\left( \frac{x}{\xi}, Q^2, \mu^2 \right) , \tag{9.73}$$

where $C(x/\xi, Q^2, \mu^2)$, called *coefficient function*, is given by

$$C(z, Q^2, \mu^2) = \delta(1-z) + \frac{\alpha_s}{2\pi}\left[P(z)\ln\frac{Q^2}{\mu^2} + \widetilde{h}(z)\right] + \ldots \tag{9.74}$$

The coefficient functions are essentially the regularized, and properly subtracted, partonic structure functions.

There is a more clever way to handle infrared and ultraviolet divergences, that is *dimensional regularization*. Working in a $(4+\epsilon)$-dimensional space collinear singularities emerge as $1/\epsilon$ poles. These poles are reabsorbed in the definition of the parton distributions. In order to have a dimensionless coupling constant in $4+\epsilon$ dimensions, one has to rescale the strong coupling $g_s$ as

$$g_s \to g_s\, \mu_{\mathrm{r}}^{-\epsilon/2} . \tag{9.75}$$

This introduces the renormalization scale $\mu_{\mathrm{r}}$, that we can identify with the factorization scale. The result for $F_2(x, Q^2)$ is

$$\begin{aligned} F_2(x, Q^2) = \sum_{q,\bar{q}} e_q^2\, x \left\{ f_q^0(x) + \frac{\alpha_s}{2\pi}\int_x^1 \frac{\mathrm{d}\xi}{\xi}\, f_q^0(\xi) \right. \\ \left. \times \left[P\left(\frac{x}{\xi}\right)\left(\ln\frac{Q^2}{\mu^2} + \frac{2}{\epsilon}\right) + h\left(\frac{x}{\xi}\right)\right] + \ldots \right\} , \end{aligned} \tag{9.76}$$

where the finite function $h$ has the structure

$$h(z) = \widetilde{h}(z) + (\gamma_E - \ln 4\pi)\, P(z) , \tag{9.77}$$

and $\gamma_E = 0.5772\ldots$ is the Euler constant.

Absorbing the term $(2/\epsilon + \gamma_E - \ln 4\pi)$ in the renormalized distribution functions $q(x, \mu^2)$ defines the so-called $\overline{\mathrm{MS}}$ (modified minimal subtraction) scheme,

$$F_2(x, Q^2) = \sum_{q,\bar{q}} e_q^2\, x \int_x^1 \frac{\mathrm{d}\xi}{\xi}\, q(\xi, \mu^2)\, C_{\overline{\mathrm{MS}}}\left(\frac{x}{\xi}, Q^2, \mu^2\right) , \tag{9.78}$$

with

$$\begin{aligned} C_{\overline{\mathrm{MS}}}(z, Q^2, \mu^2) = \delta(1-z) + \frac{\alpha_s}{2\pi}\left\{ P(z)\ln\frac{Q^2}{\mu^2} \right. \\ \left. + \left[\widetilde{h}(z) - (\gamma_E - \ln 4\pi)\, P(z)\right] \right\} + \ldots \end{aligned} \tag{9.79}$$

Since $F_2(x, Q^2)$ is a physical observable which cannot depend on the unphysical quantity $\mu^2$, differentiating (9.73) or (9.76) with respect to $\ln\mu^2$ leads to an equation governing the scale dependence of the quark distributions

$$\frac{\partial\, q(x, \mu^2)}{\partial \ln \mu^2} = \frac{\alpha_s}{2\pi}\int_x^1 \frac{\mathrm{d}y}{y}\, P\left(\frac{x}{y}\right) q(y, \mu^2) . \tag{9.80}$$

This integro-differential equation is known as Altarelli-Parisi, or DGLAP equation (after Dokshitzer 1977, Gribov and Lipatov 1972, Altarelli and Parisi 1977). The splitting function $P(x)$ represents the probability for a quark to emit another quark with momentum fraction $x$. It can be expanded as a power series in $\alpha_s$

$$P(x) = \sum_n \alpha_s^n \, P^{(n)}(x) \, . \tag{9.81}$$

At leading order, that is $\mathcal{O}(\alpha_s^0)$ in the splitting functions and $\mathcal{O}(\alpha_s^1)$ in the coefficient functions, the Altarelli-Parisi equation effectively resums contributions of the type $(\alpha_s \ln Q^2)^n$.

So far, we restricted our treatment to quarks and antiquarks only. Taking also gluons into account, (9.78) becomes

$$\begin{aligned} F_2(x,Q^2) = \sum_{q,\bar{q}} e_q^2 \, x \int_x^1 \frac{\mathrm{d}\xi}{\xi} \Big[ & q(\xi,\mu^2) \, C^q_{\overline{\mathrm{MS}}}\left(\frac{x}{\xi}, Q^2, \mu^2\right) \\ & + g(\xi,\mu^2) \, C^g_{\overline{\mathrm{MS}}}\left(\frac{x}{\xi}, Q^2, \mu^2\right) \Big] \, . \end{aligned} \tag{9.82}$$

The gluonic coefficient function $C^g_{\overline{\mathrm{MS}}}$ arises from photon–gluon fusion diagrams (Fig. 9.5) and hence starts at order $\alpha_s^1$. The field theoretical definition of the gluon distribution is (Collins and Soper 1982)

$$g(x,\mu^2) = \frac{1}{x\,P^+} \int \frac{\mathrm{d}\xi^-}{2\pi} \, \mathrm{e}^{ixP^+\xi^-} \, \langle PS | F_a^{+\nu}(0) \, F_{a\nu}^{+}(0,\xi^-,0_\perp) | PS \rangle, \tag{9.83}$$

where $F_a^{\mu\nu}$ is the gluonic field.

The definition of the scale-dependent quark distributions is now

$$\begin{aligned} q(x,\mu^2) = f_q^0(x) &+ \frac{\alpha_s}{2\pi} \int_x^1 \frac{\mathrm{d}\xi}{\xi} \, F_q^0(\xi) \left[ P_{qq}\left(\frac{x}{\xi}\right) \ln \frac{\mu^2}{\kappa_0^2} + h'_q\left(\frac{x}{\xi}\right) \right] \\ &+ \frac{\alpha_s}{2\pi} \int_x^1 \frac{\mathrm{d}\xi}{\xi} \, f_g^0(\xi) \left[ P_{qg}\left(\frac{x}{\xi}\right) \ln \frac{\mu^2}{\kappa_0^2} + h'_g\left(\frac{x}{\xi}\right) \right] + \ldots \end{aligned} \tag{9.84}$$

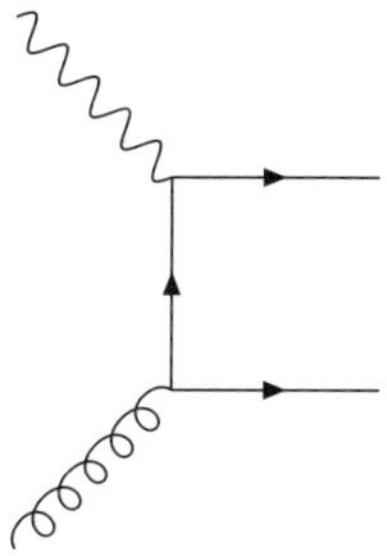

**Fig. 9.5.** The photon–gluon fusion diagram.

The explicit forms of the coefficient functions can be found, for instance, in Field (1989), or Ellis, Stirling and Webber (1996).

If we define the non-singlet distribution

$$q_{NS}(x,Q^2) = q(x,Q^2) - \bar{q}(x,Q^2) \, , \tag{9.85}$$

and the singlet combination

$$\Sigma(x,Q^2) = \sum_i [q_i(x,Q^2) + \bar{q}_i(x,Q^2)] \, , \tag{9.86}$$

where $\sum_i$ is a sum over flavors, the DGLAP equations take the form (with $t = \ln \frac{Q^2}{\mu^2}$)

$$\frac{\partial q_{NS}(x,t)}{\partial t} = \frac{\alpha_s(t)}{2\pi} \int_x^1 \frac{\mathrm{d}y}{y} P_{qq}\left(\frac{x}{y}\right) q_{NS}(y,t) \tag{9.87}$$

and

$$\frac{\partial}{\partial t} \begin{pmatrix} \Sigma(x,t) \\ g(x,t) \end{pmatrix} = \frac{\alpha_s(t)}{2\pi} \int_x^1 \frac{\mathrm{d}y}{y} \begin{pmatrix} P_{qq}\left(\frac{x}{y}\right) & 2n_f P_{qg}\left(\frac{x}{y}\right) \\ P_{gq}\left(\frac{x}{y}\right) & P_{gg}\left(\frac{x}{y}\right) \end{pmatrix} \begin{pmatrix} \Sigma(y,t) \\ g(y,t) \end{pmatrix} , \tag{9.88}$$

where $n_f$ is the number of flavors.

At leading order the splitting functions are given by

$$P_{qq}^{(0)}(x) = C_F \left[ \frac{1+x^2}{(1-x)_+} + \frac{3}{2}\delta(1-x) \right] \tag{9.89}$$

$$P_{qg}^{(0)}(x) = \frac{1}{2} \left[ x^2 + (1-x)^2 \right] \tag{9.90}$$

$$P_{gq}^{(0)}(x) = C_F \left[ \frac{1+(1-x)^2}{x} \right] \tag{9.91}$$

$$P_{gg}^{(0)}(x) = 2\,C_A \left[ \frac{x}{(1-x)_+} + \frac{1-x}{x} + x(1-x) \right] + \frac{11\,C_A - 2\,n_f}{6}\,\delta(1-x) \, , \tag{9.92}$$

where $C_F$ and $C_A$ are related to the number of colors $N_c$ by $C_F = (N_c^2 - 1)/2N_c$, $C_A = N_c$, and the "plus" distributions are defined so that, given a smooth function $f(x)$, one has

$$\int_0^1 \mathrm{d}x \, \frac{f(x)}{(1-x)_+} = \int_0^1 \mathrm{d}x \, \frac{f(x) - f(1)}{1-x} \, . \tag{9.93}$$

It is useful to introduce the *moments* $f(N,t)$ of the parton distribution functions $f(x,t)$, defined as the Mellin transforms of $f(x,t)$

$$f(N,t) = \int_0^1 \mathrm{d}x \, x^{N-1} f(x,t) \, . \tag{9.94}$$

From $f(N,t)$ the distributions in $x$-space can be obtained by inverse Mellin transform

$$f(x,t) = \frac{1}{2\pi i} \int_{c-\mathrm{i}\infty}^{c+\mathrm{i}\infty} \mathrm{d}N \, x^{-N} \, f(N,t) \; , \tag{9.95}$$

where $c$ is such that the integration contour lies to the right of all singularities of the integrand. The moments of the splitting functions are the so-called *anomalous dimensions*

$$\gamma(N, \alpha_S) = \int_0^1 \mathrm{d}x \, x^{N-1} \, P(x, \alpha_S) \; . \tag{9.96}$$

Since the Mellin transform of a convolution of two functions is the product of Mellin trasforms of those functions, the evolution equations become algebraic equations when reexpressed in terms of the moments of distributions. Equation (9.88) for instance becomes in the $N$-space

$$\frac{\partial}{\partial t} \begin{pmatrix} \Sigma(N,t) \\ g(N,t) \end{pmatrix} = \frac{\alpha_S(t)}{2\pi} \begin{pmatrix} \gamma_{qq} & 2n_f \gamma_{qg} \\ \gamma_{gq} & \gamma_{gg} \end{pmatrix} \begin{pmatrix} \Sigma(N,t) \\ g(N,t) \end{pmatrix} \; . \tag{9.97}$$

The leading order anomalous dimensions are

$$\gamma_{qq}^{(0)}(N) = C_F \left[ -\frac{1}{2} + \frac{1}{N(N+1)} - 2 \sum_{k=2}^{N} \frac{1}{k} \right] \tag{9.98}$$

$$\gamma_{qg}^{(0)}(N) = \frac{1}{2} \left[ \frac{2+N+N^2}{N(N+1)(N+2)} \right] \tag{9.99}$$

$$\gamma_{gq}^{(0)}(N) = C_F \left[ \frac{2+N+N^2}{N(N^2-1)} \right] \tag{9.100}$$

$$\gamma_{gg}^{(0)}(N) = 2\, C_A \left[ -\frac{1}{12} + \frac{1}{N(N-1)} + \frac{1}{(N+1)(N+2)} - \sum_{k=2}^{N} \frac{1}{k} \right] - \frac{n_f}{3} \; . \tag{9.101}$$

Note that due to

$$\int_0^1 \mathrm{d}x \, x^{N-1} \frac{1}{x} = \frac{1}{N-1} \tag{9.102}$$

poles of the splitting functions in $x = 0$ appear as poles of the anomalous dimensions at $N = 1$.

In the following we will be interested in the structure functions at low-$x$. In this limit, structure functions are governed by the behavior of the splitting functions as $x \to 0$, which is related in turn to the $N \to 1$ behavior of the corresponding anomalous dimensions.

## 9.4 Phenomenology of DIS

The experimental information accumulated so far on deep inelastic scattering is just impressive. About three decades of measurements have led to an extremely rich and precise knowledge of the internal structure of the proton. Perturbative QCD has been put, more or less directly, to test, with a great success in its realm of application (see, e.g., Forte 2000).

The kinematic ranges covered by various experiments are shown in Fig. 9.6. Note the difference between the fixed target kinematics (NMC, SLAC, BCDMS, etc.) and the *ep* collider kinematics (HERA). The HERA experiments (H1, ZEUS, HERMES) have considerably extended the explored region of DIS towards very low $x$ and very high $Q^2$. The range covered by HERA is: $0.2 \lesssim Q^2 \lesssim 10^4$ GeV$^2$ and $10^{-5} \lesssim x \lesssim 10^{-1}$. In particular, at HERA it has been possible for the first time to investigate DIS at very low $x$ in the perturbative regime, that is $x \lesssim 10^{-4}$ with $Q^2 \gtrsim$ few GeV$^2$.

In terms of the resolving power $d = \hbar c/Q$, the HERA kinematic domain corresponds to a variation of two orders of magnitude, $d \simeq (10^{-16} - 10^{-14})$ cm. So, at HERA the proton is resolved down to one thousandth of its size.

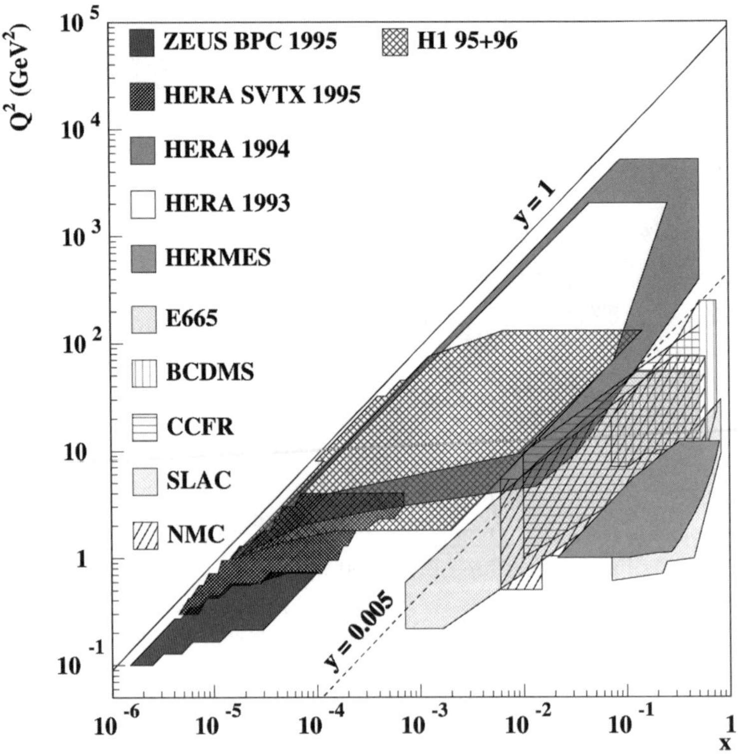


**Fig. 9.6.** Kinematic ranges of various DIS experiments.

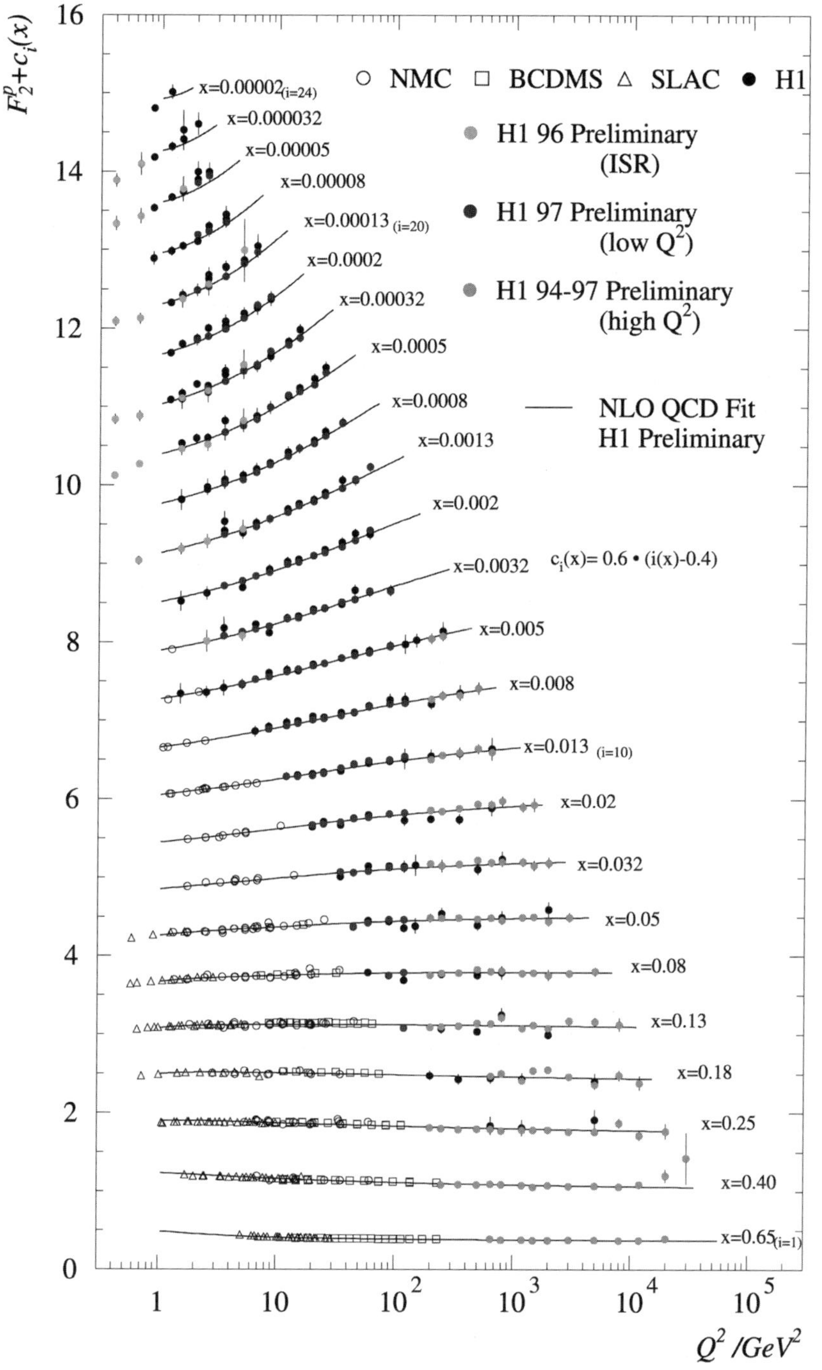


**Fig. 9.7.** The $F_2$ structure function as measured by H1 and earlier DIS experiments. The lines are the result of a NLO DGLAP fit. From Marage (1999).

The QCD analyses of DIS data aim at extracting from cross sections, or structure functions, the parton distribution functions (pdf's). This is usually done by parametrizing the pdf's at some input scale $Q_0^2 \sim (1-2)$ GeV$^2$ and then evolving them by means of next-to-leading order DGLAP equations. The functional form of the pdf's at the starting scale, that is their $x$ dependence, is simply guessed, since QCD is not able to predict it. On the other hand, the $Q^2$ dependence is completely governed by the evolution equations, and represents a genuine test of perturbative QCD.

The result of a QCD analysis of DIS is shown in Fig. 9.7, together with the data on $F_2$. As one can see, the precision of the data is such that their errors are hardly visible, and the agreement with the QCD fit is excellent. QCD appears to predict correctly the $Q^2$ dependence of structure functions over four orders of magnitude. The data on the longitudinal structure function $F_L$ are much less precise (see Fig. 9.8), but also in this case QCD seems to work well.

As for the parton distributions, Fig. 9.9 shows the results of the ZEUS extraction of the gluon and singlet densities (Breitweg et al. 1999b): while the latter is rather well determined, the former is more uncertain. DIS alone is not able to constrain completely $g(x)$ in the whole $x$ range. Other hard processes, such as high-$E_T$ jet and prompt photon production in $pp$ interactions, are needed in order to get a more exhaustive information on the gluon density. Unfortunately, these reactions suffer of many experimental and theoretical problems.

The various global analyses of DIS (and related data) performed so far are all based on NLO DGLAP equations, but differ in some respects: the choice

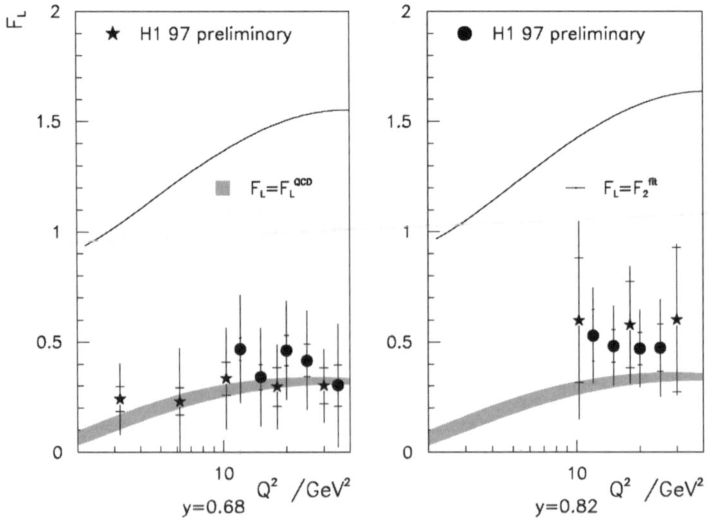


**Fig. 9.8.** The longitudinal structure function. From Marage (1999).

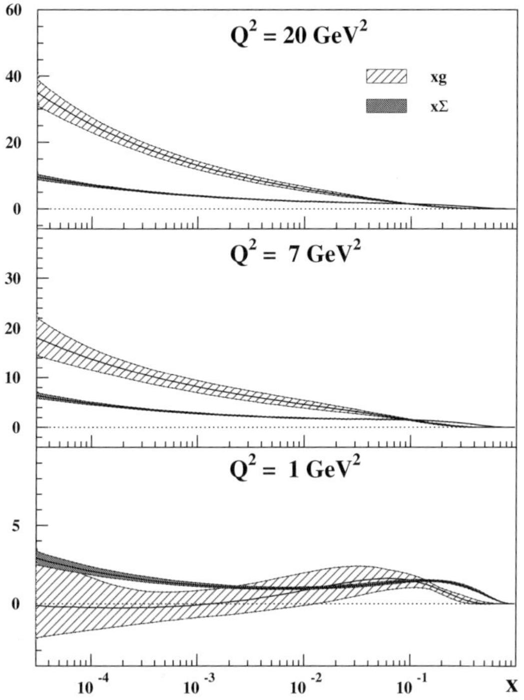


**Fig. 9.9.** The gluon and singlet densities extracted by ZEUS (Breitweg et al. 1999b).

of the functional form of pdf's at the input scale, the selection of data, the value of $\alpha_s$, the treatment of experimental errors, the massive quark scheme adopted for charm and heavier flavors. A comparison between the results of different QCD global fits is presented in Fig. 9.10. The discrepancy of the curves can give a rough idea of the present phenomenological uncertainty in the extraction of the pdf's. As one can see, the $u$ and $d$ distributions are known with a considerably high degree of accuracy (except perhaps in the very large-$x$ region), whereas the strange sea density and the gluon density are still affected by relatively large uncertainties.

We shall come back to the HERA measurements of structure functions in the next section, where our attention will be directed to the low-$x$ domain of DIS.

## 9.5 DIS at Low-$x$

The DGLAP equations presented in Sect. 9.3 resum logarithms of the type $\ln(Q^2/\mu^2)$. When $x$ is very small, another class of large logarithms, of the

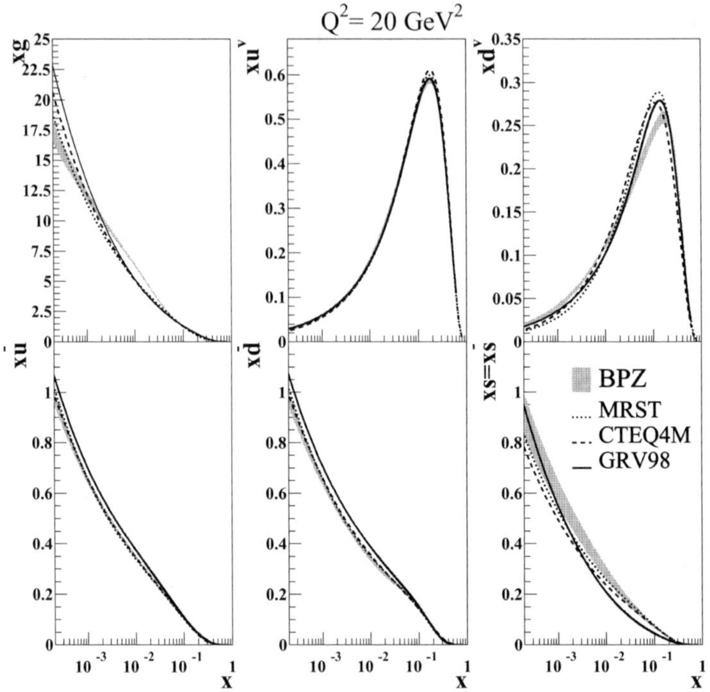


**Fig. 9.10.** Parton distribution functions from various fits. From Barone, Pascaud and Zomer (2000).

form $\ln(1/x)$, become important, and a proper resummation of these logarithms is needed. Recalling that $x = Q^2/(Q^2 + W^2)$, the limit of low $x$ corresponds to $W^2 \gg Q^2$, that is to the Regge limit. Since $Q^2 \gg \Lambda_{\mathrm{QCD}}$, DIS at low $x$ probes the Regge limit of virtual photoabsorption cross sections in perturbative QCD. This makes the low-$x$ domain of DIS highly interesting from a theoretical viewpoint.

As we have already mentioned, in the last decade HERA experiments allowed to access extremely small values of $x$, at moderately large values $Q^2$, which makes perturbative QCD applicable (see Fig. 9.6). The behavior of DIS structure functions is now known with a high degree of precision, down to $x \simeq 10^{-4} - 10^{-5}$. In the impossibility of quoting all relevant experimental papers, we refer the reader to some recent reviews, where a list of references can be found: Marage (1999), De Roeck (2000).

One of the most important results at HERA is the observation of a (relatively) steep rise of the $F_2$ structure function at low $x$ (see Fig. 9.11), attributed to an increase of the gluon density. Parametrizing $F_2$ for $x < 0.1$ in the form $F_2(x, Q^2) = A(Q^2)\, x^{-\lambda}$, one finds $\lambda \approx 0.1$ at small $Q^2$ ($\lesssim 1$ GeV$^2$), and $\lambda \approx 0.25 - 0.35$ at large $Q^2$ ($\sim (10 - 100)$ GeV$^2$). One must recall however that at different $Q^2$ different $x$ regions are probed (the larger $Q^2$, the larger

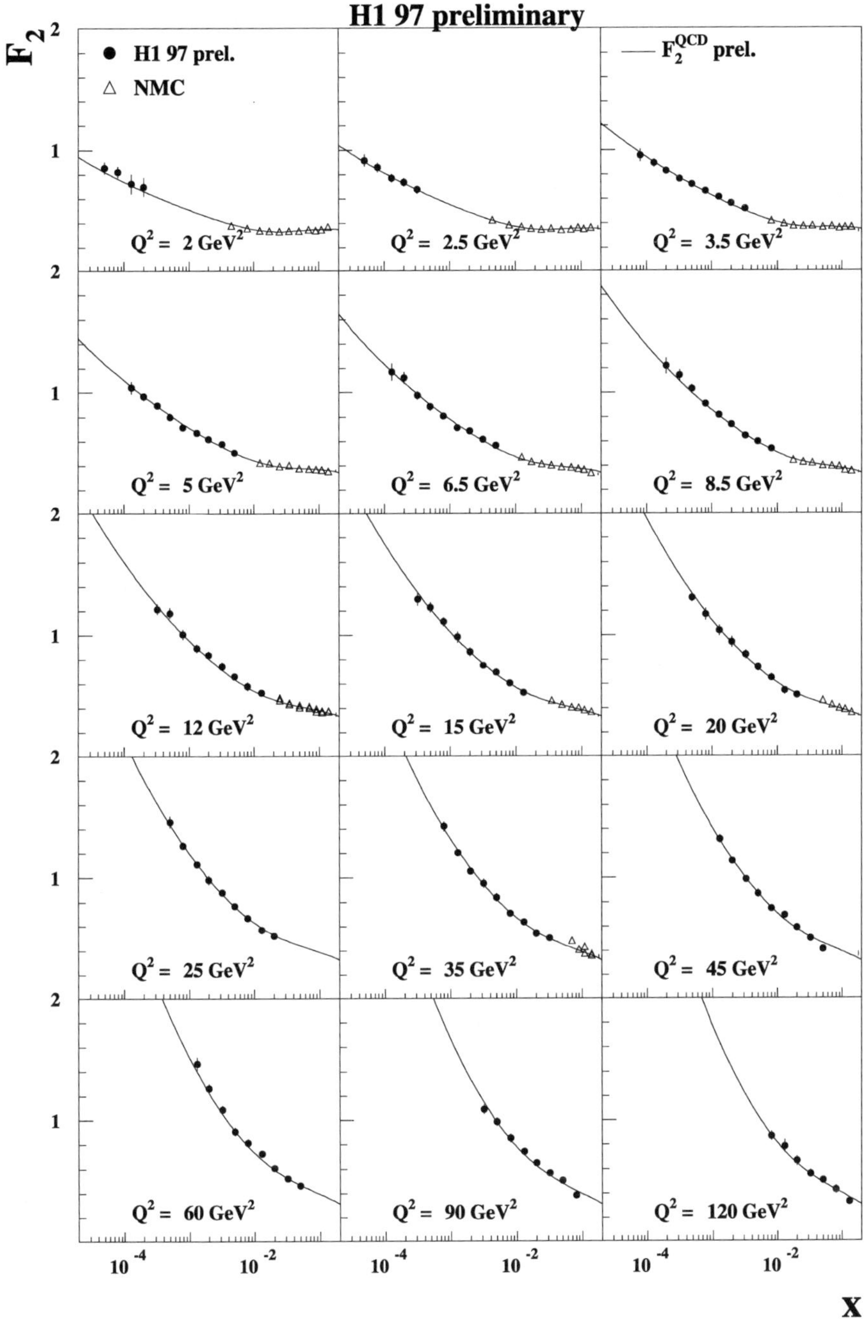


**Fig. 9.11.** Measurement of the $F_2$ structure function by the H1 and NMC collaborations as a function of $x$ in bins of $Q^2$. The lines show the result of a NLO DGLAP fit. From Marage (1999).

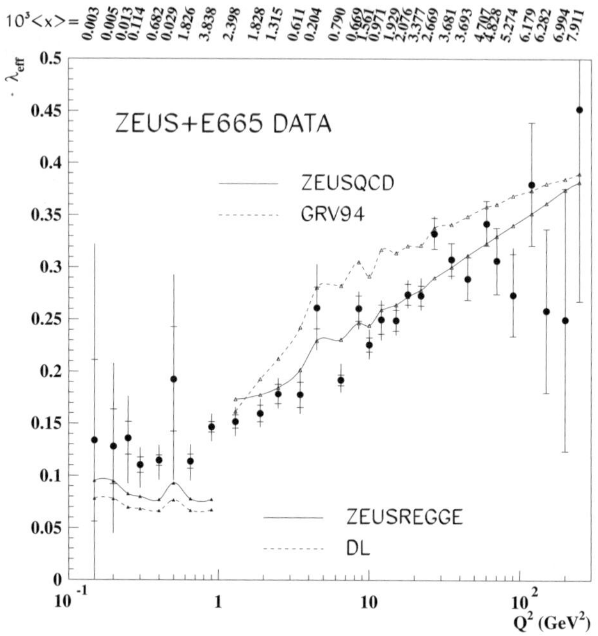


**Fig. 9.12.** ZEUS measurement of the parameter $\lambda = \mathrm{d}\ln F_2/\mathrm{d}\ln(1/x)$ as a function of $Q^2$. From Breitweg et al. (1999b).

the lowest $x$ value), so the comparison between the $\lambda$ exponents at various $Q^2$ must be made with some caution (the situation is summarized in Fig. 9.12). What seems to emerge quite clearly from the HERA measurements is that the DGLAP equations describe remarkably well the data in the *whole* kinematic range covered by them, down to very low $x$, as one can see from the quality of the fits in Figs. 9.7 and 9.11.

Let us come now to the theoretical frameworks. Before presenting the perturbative QCD approaches to low-$x$ DIS, we shall see how Regge theory describes this process.

### 9.5.1 Regge Theory Predictions

Low $x$ means large $W^2$. In this limit Regge theory predicts that the total cross section for virtual photoabsorption behaves as (see Eq. (5.87))

$$\sigma^{\gamma^* N} \sim \mathcal{A}_{I\!P}\,(W^2)^{\alpha_{I\!P}(0)-1} + \mathcal{A}_{I\!R}\,(W^2)^{\alpha_{I\!R}(0)-1}, \tag{9.103}$$

where $I\!R$ denotes (secondary) Regge trajectories like $\rho, \omega, a_2, f_2$, etc. Rewritten for the structure function $F_2 = \frac{Q^2}{4\,\pi^2\,\alpha_{\mathrm{em}}}\sigma^{\gamma^* N}$, (9.103) becomes ($x = Q^2/W^2$)

$$F_2(x) \sim \mathcal{A}_{I\!P}\,x^{1-\alpha_{I\!P}(0)} + \mathcal{A}_{I\!R}\,x^{1-\alpha_{I\!R}(0)}. \tag{9.104}$$

With the typical intercepts

$$\alpha_{I\!P}(0) = 1, \quad \alpha_{I\!R}(0) = \frac{1}{2}, \tag{9.105}$$

the low-$x$ behavior of $F_2$ is

$$F_2(x) \sim \mathcal{A}_{I\!P}\, x^0 + \mathcal{A}_{I\!R}\, x^{\frac{1}{2}}. \tag{9.106}$$

Using $F_2(x) \sim x\, q(x)$, the quark distributions $q(x)$ behave as

$$q(x) \sim \mathcal{A}_{I\!P}\, x^{-1} + \mathcal{A}_{I\!R}\, x^{-\frac{1}{2}}. \tag{9.107}$$

Since the pomeron has isospin $I = 0$, the nonsinglet part of $q(x)$, that is the valence $V(x) = \sum_q [q(x) - \bar{q}(x)]$, contains only the $I\!R$ contribution, namely

$$V(x) \sim x^{-\frac{1}{2}}, \tag{9.108}$$

and the proton-neutron structure function difference is expected for $x \to 0$ to vanish as

$$F_2^p(x) - F_2^n(x) \sim x^{\frac{1}{2}}. \tag{9.109}$$

On the other hand, the singlet quark distributions, like the sea $S(x)$, and the gluon density, which drives the structure function at low $x$, are predicted to diverge as

$$S(x), g(x) \sim x^{-1}. \tag{9.110}$$

Two questions now arise: *i)* at what $Q^2$ are Regge theory expectations supposed to be valid? *ii)* how to reconcile these expectations with QCD, which predicts that, if $F_2$ is parametrized as $F_2 \propto x^{-\lambda}$, the exponent of $x$ should change with $Q^2$?

The answer is that Regge theory is believed to apply to the low-$Q^2$ region (see Fig. 9.13). Thus one expects (9.106–9.110) to hold at some $Q^2 \lesssim 1$

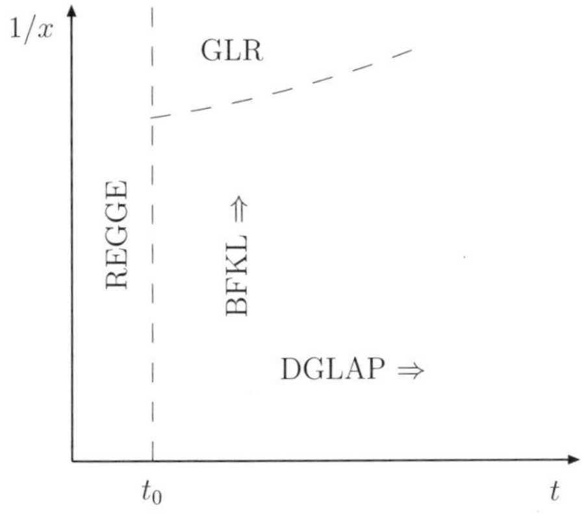


**Fig. 9.13.** Regions of validity of Regge theory and DGLAP equations.

GeV$^2$. At higher $Q^2$ the ordinary QCD evolution takes over and modifies the $x$-dependence of the structure functions.

We have seen above that the HERA data give, at $Q^2 \sim 1$ GeV$^2$, $F_2^p(x) \sim x^{-0.1}$, to be compared with (9.106). Note that the observed behavior of $F_2^p(x)$ at low $Q^2$ is close to the one dictated by the soft pomeron of Donnachie and Landshoff, that is $F_2(x) \sim x^{-0.08}$ (see Sect. 7.1). On the other hand, as soon as one enters the perturbative regime (i.e., for $Q^2 \gtrsim 1$ GeV$^2$), the low-$x$ structure functions are well described by the DGLAP equations. Unfortunately, from the theoretical point of view the transition from the Regge picture to the DGLAP picture is not well understood as yet (for an attempt to describe phenomenologically this transition see Capella et al. 1994).

### 9.5.2 Resummations in QCD

Before coming to the details of the QCD description of low-$x$ DIS, let us briefly outline the resummation procedures in perturbative QCD.

Every physical observable can be written, in perturbative QCD, as a power series in $\alpha_s$. In these series the coupling constant is accompanied by logarithms which are, or may become, large, and hence need to be resummed. The use of renormalization group equations in DIS allows resumming large logarithms of the type $\ln Q^2$ and gives rise to the DGLAP equations. At low $x$ there is another class of important logarithms, of the type $\ln(1/x)$, which must be carefully dealt with. According to the type and to the powers of logarithms that are effectively resummed one gets different evolution equations.

In the *leading* $\ln Q^2$ *approximation* (LLA), terms of the type $\alpha_s^n \ln^n Q^2$ are resummed. At each perturbative order only the highest power in $\ln Q^2$ is retained. Symbolically we can write (we omit the coefficients of all terms)

$$\text{LLA}: \sum_n \alpha_s^n \ln^n Q^2 \left( \ln^n \frac{1}{x} + \ln^{n-1} \frac{1}{x} + \cdots \right). \tag{9.111}$$

Diagrammatically, in LLA one considers, in a physical gauge, ladder graphs (see Fig. 9.14a) with *strong ordering* in the transverse momenta of the cascade

$$Q^2 \gg \boldsymbol{\kappa}_\perp^2 \gg \boldsymbol{k}_{1\perp}^2 \gg \boldsymbol{k}_{2\perp}^2 \gg \cdots \gg \boldsymbol{k}_{n\perp}^2 \tag{9.112}$$

The longitudinal momenta are *normally ordered* ($x_i \equiv k_i^+/P^+$)

$$x < x_1 < x_2 < \cdots < x_n < 1 . \tag{9.113}$$

If we keep subdominant powers in $\ln Q^2$, that is terms of the type $\alpha_s^n \ln^{n-1} Q^2$, we get the *next-to-leading* $\ln Q^2$ *approximation* (NLLA). The LLA and NLLA yield the leading-order and next-to-leading order DGLAP equations, respectively.

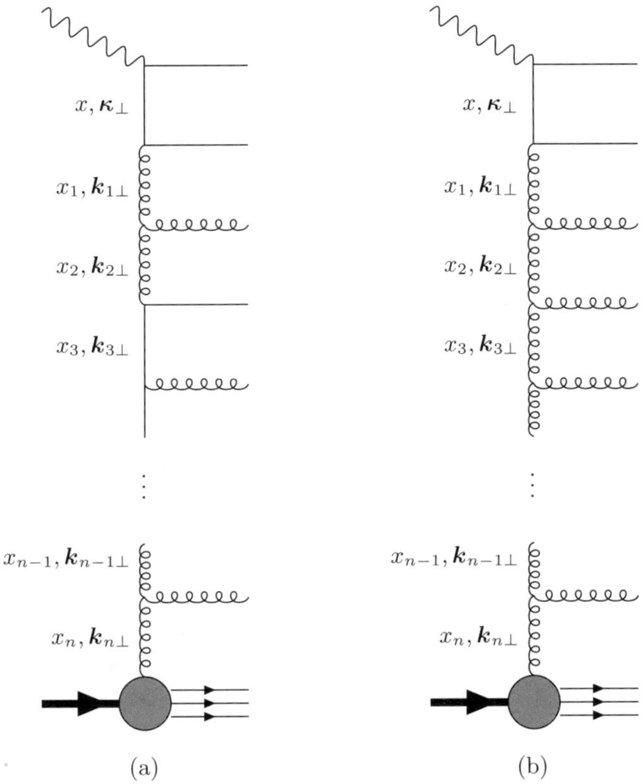


**Fig. 9.14a,b.** Ladder diagrams.

In the limit of small $x$ one can retain in the LLA only the dominant terms in $\ln(1/x)$, thus getting the *double leading log approximation* (DLLA):

$$\text{DLLA}: \quad \sum_n \alpha_s^n \ln^n Q^2 \ln^n \frac{1}{x}. \tag{9.114}$$

Diagrammatically, this amounts to considering, in a physical gauge, gluon ladders (gluons dominate at low $x$) – see Fig. 9.14b – with strong ordering both in $\boldsymbol{k}_{i\perp}^2$ and $x_i$, that is (9.112) and

$$x \ll x_1 \ll x_2 \ll \cdots \ll x_n \ll 1\,. \tag{9.115}$$

In the LLA and DLLA only the dominant powers of $\ln(1/x)$ accompanied by dominant powers of $\ln Q^2$ are considered. In practice, at very small $x$ the momentum transfer $Q^2$ is not so large. Thus one can think of resumming leading terms in $\ln(1/x)$ which are not coupled to leading powers of $\ln Q^2$. This yields the *leading* $\ln(1/x)$ *approximation* (LL$_x$A), which is represented symbolically by

$$\mathrm{LL}_x\mathrm{A}: \sum_n \alpha_s^n \ln^n \frac{1}{x} \left( \ln^n Q^2 + \ln^{n-1} Q^2 + \cdots \right). \tag{9.116}$$

Note that there is a symmetry $Q^2 \leftrightarrow 1/x$ between the LLA and the $\mathrm{LL}_x\mathrm{A}$. The latter leads to the BFKL equation (Lipatov 1976, Kuraev, Lipatov and Fadin 1977, Balitsky and Lipatov 1978) which resums gluon ladders (Fig. 9.14b) with strong ordering in the longitudinal momenta (9.115), and no ordering in the transverse momenta

$$\boldsymbol{\kappa}_\perp^2 \simeq \boldsymbol{k}_{1\perp}^2 \simeq \boldsymbol{k}_{2\perp}^2 \simeq \cdots \simeq \boldsymbol{k}_{n\perp}^2. \tag{9.117}$$

In the DIS context, the BFKL equation is an integro-differential evolution equation for the unintegrated gluon distribution

$$f(x, \boldsymbol{k}_\perp^2) \equiv \frac{\partial \left[ x g(x, \boldsymbol{k}_\perp^2 \right]}{\partial \ln \boldsymbol{k}_\perp^2}, \tag{9.118}$$

which is the probability to find a gluon with longitudinal momentum fraction $x$ and transverse momentum $\boldsymbol{k}_\perp^2$. The BFKL equation can be used to obtain the unintegrated gluon density at low $x$, once a starting distribution at some $x_0$ value is assumed. The applicability of the leading order BFKL equation does not extend to very large $Q^2$ (see Fig. 9.13) since the subdominant contributions in $\ln(1/x)$, including those accompanied by large $\ln Q^2$ terms, are neglected. One can improve the leading $\ln(1/x)$ approximation by considering subdominant powers in $\ln(1/x)$. This gives the next-to-leading $\ln(1/x)$ approximation, which has been recently worked out by Fadin and Lipatov (1998) (see also Camici and Ciafaloni 1998).

The approximations described above are summarized in Fig. 9.15. In the limit $1/x \to \infty$ the relevant resummations are the DLLA and the $\mathrm{LL}_x\mathrm{A}$. We now turn to describe them in some detail.

### 9.5.3 DLLA and Double Scaling

The low-$x$ evolution of structure functions is governed by the rightmost singularity of their anomalous dimensions, which is located at $N = 1$. From (9.98–9.101) one sees that $\gamma_{gq}^{(0)}$ and $\gamma_{gg}^{(0)}$ have poles in $1/(N-1)$ and for behave for $N \to 1$ as

$$\gamma_{gq}^{(0)} \underset{N\to 1}{\to} \frac{2\,C_F}{N-1}, \quad \gamma_{gg}^{(0)} \underset{N\to 1}{\to} \frac{2\,C_A}{N-1}, \tag{9.119}$$

whereas $\gamma_{qg}^{(0)} = \mathcal{O}(1)$ and $\gamma_{qq}^{(0)} = \mathcal{O}(N-1)$. Correspondingly, in the limit $x \to 0$ the splitting functions $P_{gq}^{(0)}$ and $P_{gg}^{(0)}$ behave as (see (9.89–9.92))

$$P_{gq}^{(0)} \underset{x\to 0}{\to} \frac{2\,C_F}{x}, \quad P_{gg}^{(0)} \underset{x\to 0}{\to} \frac{2\,C_A}{x}, \tag{9.120}$$

whereas $P_{qq}^{(0)}$ and $P_{qg}^{(0)}$ are not singular (at NLO, also $P_{qq}$ and $P_{qg}$ develop $1/x$ singularities).

Hence, at low $x$ QCD evolution is dominated by gluons and the structure functions tend to grow with increasing $Q^2$ due to the abundant production of sea from the $g \to q\bar{q}$ process. If we retain in the $x \to 0$ limit only the most singular terms in the leading-order splitting functions and coefficient functions, we get the so-called double leading-log approximation (DLLA), which effectively resums terms of the type $\alpha_s^n \ln^n Q^2 \ln^n(1/x)$. The DLLA is clearly appropriate in the double asymptotic limit $Q^2 \to \infty$ and $1/x \to \infty$. In this domain it allows to derive the behavior of structure functions directly (and analytically) from the Altarelli-Parisi equations. We shall now do so in a simplified case.

Consider only gluons, with no mixing to quarks. The LO Altarelli-Parisi equation in moment space then reads

$$\frac{\partial}{\partial t} g(N,t) = \frac{\alpha_s(t)}{2\pi} \gamma_{gg}^{(0)}(N)\, g(N,t) \,, \tag{9.121}$$

where the LO coupling constant is

$$\alpha_s(t) = \frac{4\pi}{b_0 t} \,, \quad b_0 = \frac{33 - 2\,n_f}{3} \,, \tag{9.122}$$

In the low-$x$ limit, i.e. $N \to 1$, we take the dominant term in $\gamma_{gg}^{(0)}(N)$

$$\gamma_{gg}^{(0)}(N) \sim \frac{2\,N_c}{N-1} \,. \tag{9.123}$$

The solution of (9.121) is therefore

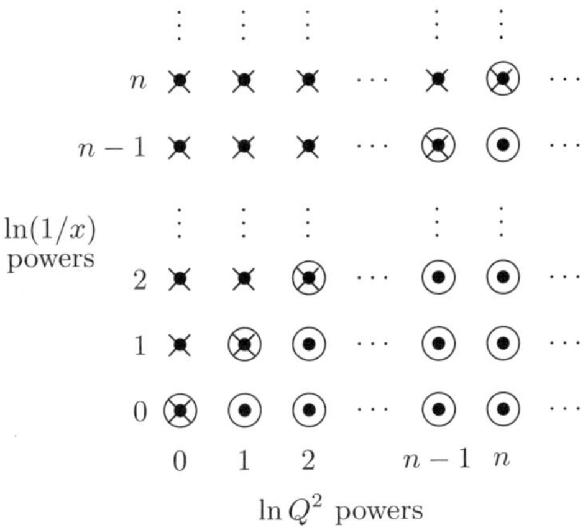


**Fig. 9.15.** The powers of $\ln Q^2$ and $\ln(1/x)$ resummed, up to the $n$-th perturbative order, by the LLA (circles), the LL$_x$A (crosses) and the DLLA (circles and crosses).

$$g(N,t) = g(N,t_0)\,\exp\left(\frac{\gamma^2}{N-1}\,\ln\frac{t}{t_0}\right)\,, \tag{9.124}$$

with $\gamma = 2\sqrt{N_c/b_0}$. In order to get the gluon distribution in the $x$ space, we have to compute the inverse Mellin transform of (9.124), that is

$$x\,g(x,t) = \frac{1}{2\pi i}\int_{c-i\infty}^{c+i\infty} \mathrm{d}N\,x^{-(N-1)}\,g(N,t)\,. \tag{9.125}$$

The integration contour in (9.125) stands to the right of all singularities of the integrand. Inserting (9.124) in (9.125) gives

$$x\,g(x,t) = \frac{1}{2\pi i}\int_{c-i\infty}^{c+i\infty} \mathrm{d}N\,\exp\left[\frac{\gamma^2\,\zeta}{N-1} + (N-1)\,\xi\right]\,g(N,t_0)\,, \tag{9.126}$$

where we have introduced the variables

$$\xi \equiv \ln\frac{x_0}{x}\,, \qquad \zeta \equiv \ln\frac{t}{t_0}\,, \tag{9.127}$$

and the parameter $x_0$ represents the onset of the low-$x$ region, $x \ll x_0$.

In the double asymptotic limit $\zeta, \xi \to \infty$ we can evaluate approximately the integral (9.126) by a saddle point method. The saddle point is

$$N_0 = 1 + \gamma\sqrt{\frac{\zeta}{\xi}}\,, \tag{9.128}$$

and, under the assumption that the possible singularities of $g(N,t_0)$ are located to the left of $N_0$, the asymptotic behavior of $g(x,t)$ resulting from (9.126) is

$$x\,g(x,t) \simeq g(N_0,t_0)\,\exp(2\,\gamma\,\sqrt{\zeta\xi})\,, \tag{9.129}$$

that is, explicitly

$$x\,g(x,t) \sim \exp\left\{2\left[\frac{4N_c}{b_0}\,\ln\left(\frac{t}{t_0}\right)\,\ln\left(\frac{x_0}{x}\right)\right]^{\frac{1}{2}}\right\}\,. \tag{9.130}$$

It should be noticed that the initial gluon distribution $g(x,t_0)$ determines only the normalization of $g(x,t)$, not its $x$-dependence. Since at low $x$ structure functions are driven by the gluon distribution, Eq. (9.130) gives also the asymptotic behavior of $F_2$.

The above procedure for obtaining (9.130) relies on the condition that the integration contour of (9.126), and therefore the saddle point (9.128), are located to the right of the singularities of $g(N,t_0)$. This may not be the case if $g(x,t_0)$ is too steep as $x \to 0$. If it behaves as $x^{-\eta}$, with $\eta > N_0 - 1$, its $N$-th moment has a pole to the right of the saddle point and the derivation of (9.130) is no more valid. In this case, the low-$x$ behavior of $g(x,t)$ is more complicated.

Studying the subdominant corrections to the behavior (9.130), Ball and Forte (1994a, 1994b) discovered a remarkable phenomenon: *double asymptotic scaling* (see also De Rujula et al. 1974). Let us retain the constant terms in the $N \to 1$ limit of the anomalous dimensions

$$\gamma_{gg}^{(0)} \underset{N\to 1}{\to} \frac{2\,C_A}{N-1} - \left(\frac{11}{6}\,C_A + \frac{1}{3}\,n_f\right) , \tag{9.131}$$

$$\gamma_{gq}^{(0)} \underset{N\to 1}{\to} \frac{2\,C_F}{N-1} - 3\,C_F , \tag{9.132}$$

$$\gamma_{qg}^{(0)} \underset{N\to 1}{\to} \frac{1}{3} , \tag{9.133}$$

$$\gamma_{qq}^{(0)} \underset{N\to 1}{\to} 0 . \tag{9.134}$$

Taking the inverse Mellin transforms of (9.131–9.134), one finds for the splitting functions

$$P_{gg}^{(0)}(x) \underset{x\to 0}{\to} \frac{2\,C_A}{x} - \left(\frac{11}{6}\,C_A + \frac{1}{3}\,n_f\right) , \tag{9.135}$$

$$P_{gq}^{(0)}(x) \underset{x\to 0}{\to} \frac{2\,C_F}{x} - 3\,C_F\,\delta(1-x) , \tag{9.136}$$

$$P_{qg}^{(0)}(x) \underset{x\to 0}{\to} \frac{1}{3}\,\delta(1-x) , \tag{9.137}$$

$$P_{qq}^{(0)}(x) \underset{x\to 0}{\to} 0 , \tag{9.138}$$

where the terms proportional to $\delta(1-x)$ correspond to the constant terms in (9.131–9.134) and account for ladders which are strongly ordered only in $t$ (the singular terms – we recall – account for ladders strongly ordered both in $t$ and $x$, and give rise to the DLLA). Using the splitting functions (9.135–9.138) in the DGLAP equations, and differentiating with respect to $\ln(1/x)$ leads to a two-dimensional wave equation for the gluon distribution $G(\xi,\zeta) \equiv x\,g(x,t)$

$$\left[\frac{\partial^2}{\partial\xi\,\partial\zeta} + \delta\,\frac{\partial}{\partial\xi} - \gamma^2\right] G(\xi,\zeta) = 0 , \tag{9.139}$$

where $\delta = (11 + 2\,n_f/27)/b_0$. Defining the new variables

$$\sigma = \left[\ln\left(\frac{x_0}{x}\right)\,\ln\left(\frac{t}{t_0}\right)\right]^{1/2} , \quad \rho = \left[\frac{\ln(x_0/x)}{\ln(t/t_0)}\right]^{1/2} , \tag{9.140}$$

one finds, in the limit $\sigma \to \infty$ at fixed (but large) $\rho$, and with reasonably soft boundary conditions, that is $xg(x,t_0) \sim x^{-\lambda}$ and $g(x_0,t) \sim t^{-\eta}$, with sufficiently small $\lambda$ and $\eta$ (Ball and Forte 1994a, 1994b)

$$G(\sigma,\rho) \underset{\sigma\to\infty}{\sim} A\left(\frac{\gamma}{\rho}\right)\frac{1}{\sqrt{4\pi\gamma\sigma}}\exp\left(2\gamma\rho-\frac{\delta\sigma}{\rho}\right)\left[1+\mathcal{O}\left(\frac{1}{\rho}\right)\right] . \quad (9.141)$$

Thus, perturbative QCD predicts in a parameter-free way (i.e., quite independently of the precise form of the boundary conditions) a universal growth of the glue distribution, and hence of the $F_2$ structure function, at large $t$ and small $x$: a growth which is faster than any power of $\ln(1/x)$ but slower than any inverse power of $x$. Note that dropping the sub-asymptotic term proportional to $\delta$ in (9.141) yields the leading behavior previously derived, Eq. (9.129).

The structure function $F_2$ is driven at small $x$, by the gluon distribution, so its behavior is similar to (9.141). To be precise, one finds

$$F_2(\sigma,\rho) \underset{\sigma\to\infty}{\sim} A'\left(\frac{\gamma}{\rho}\right)\sqrt{\frac{\gamma}{\sigma\,\rho^2}}\exp\left(2\gamma\rho-\frac{\delta\sigma}{\rho}\right)\left[1+\mathcal{O}\left(\frac{1}{\rho}\right)\right] . \quad (9.142)$$

Equations (9.141) and (9.142) exhibits a *double asymptotic scaling* in $\sigma$ and $\rho$. In fact, $\ln F_2$, at fixed $\rho$, is asymptotically a linear function of $\sigma$, with a slope independent of $\rho$ (up to a $1/\rho$ term), while at fixed $\sigma$ it is asymptotically a flat function of $\rho$. Therefore, in the double asymptotic limit of large $\sigma$ and $\rho$, the product $\sigma^{-1}\ln F_2$ is independent of both $\sigma$ and $\rho$.

This behavior has been checked at HERA and is in excellent agreement with the experimental findings. In Fig. 9.16 the H1 data for the quantity $R'_F\,F_2$ is plotted on a log scale, where $R'_F = N\,\rho\,\sqrt{\sigma}\,\exp(\delta\sigma/\rho)$. All data, which are taken at different values of $\rho$, lie on the same line. If the full asymptotic prediction is scaled away, i.e. the data on $F_2$ are rescaled by the

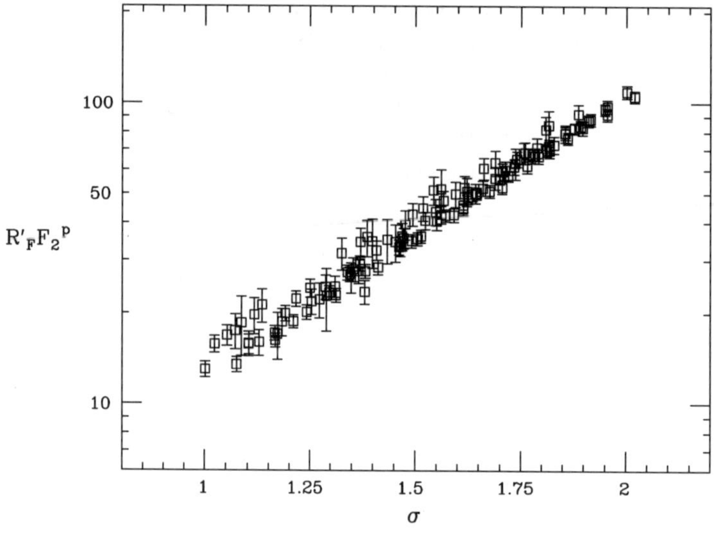


**Fig. 9.16.** The H1 data for $R'_F F_2$ as a function of $\sigma$. From Forte (1997).

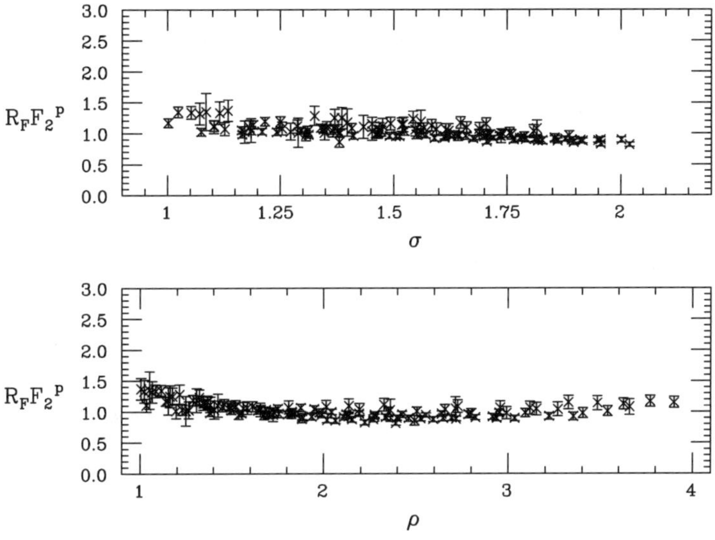


**Fig. 9.17.** The H1 data for $R_F F_2$ as a function of $\sigma$ and $\rho$. From Forte (1997).

factor $R_F = N\,\rho\,\sqrt{\sigma}\,\exp(-2\gamma\rho+\delta\sigma/\rho)$, an horizontal line is expected. Figure 9.17 shows the experimental result. The agreement with the QCD predictions is very good. The double scaling analyses of HERA data exclude that the observed $F_2$ behaves as a scale-independent power of $x$ (i.e., $F_2 \sim x^{-\lambda}$ at any $Q^2$), or as a very singular power of $x$ at some initial scale (i.e., $F_2 \sim x^{-\lambda}$ at $Q_0^2$, with $\lambda \gtrsim 0.3$).

### 9.5.4 $\boldsymbol{k}_\perp$-Factorization

In Sect. 9.3 we encountered the collinear factorization theorem (9.73), which states that the structure functions of hadrons are given, at all orders, by the convolution of some intrinsically non-perturbative, but universal, quantities – the parton densities –, with perturbatively calculable partonic structure functions.

In the large-$W^2$ (i.e., low-$x$) limit, a new form of factorization holds (Catani, Ciafaloni and Hautmann 1990, 1991). This factorization is $\boldsymbol{k}_\perp$-dependent ($\boldsymbol{k}_\perp$ is the transverse momentum of the gluons, which are the dominant partons at small $x$), and provides the low-$x$ structure functions as convolutions over $x$ and $\boldsymbol{k}_\perp$ of the unintegrated gluon distribution $f(x, \boldsymbol{k}_\perp^2)$ – Eq. (9.118) – with perturbatively calculable gluonic structure functions describing the $\gamma^* g$ scattering. Differently from the collinear case, in the $\boldsymbol{k}_\perp$-factorization scheme no ordering in $\boldsymbol{k}_\perp$ is assumed in the gluon ladders.

The $\boldsymbol{k}_\perp$-factorization formula for the $\gamma^* p$ cross section reads[1]

[1] For simplicity, we drop hereafter the subscripts $\perp$ from all transverse vectors: it is intended that boldface vectors are transverse two-vectors.

$$\sigma_\lambda^{\gamma^* p}(x, Q^2) = \int \frac{\mathrm{d}\boldsymbol{k}^2}{\boldsymbol{k}^2} \int_x^1 \frac{\mathrm{d}x'}{x'} f\left(\frac{x}{x'}, \boldsymbol{k}^2\right) \hat{\sigma}_\lambda^{\gamma^* g}(x', \boldsymbol{k}^2, Q^2) , \tag{9.143}$$

where $\lambda$ denotes the polarization of the virtual photon and $\hat{\sigma}_\lambda^{\gamma^* g}$ is the gluonic cross section. For convenience, we recall the definition of the unintegrated gluon distribution

$$f(x, \boldsymbol{k}^2) = \frac{\partial\, [xg(x, \boldsymbol{k}^2)]}{\partial \ln \boldsymbol{k}^2} , \tag{9.144}$$

from which the ordinary, i.e. $\boldsymbol{k}$-integrated, gluon distribution is obtained as

$$g(x, Q^2) = \int^{Q^2} \frac{\mathrm{d}\boldsymbol{k}^2}{\boldsymbol{k}^2} f(x, \boldsymbol{k}^2) . \tag{9.145}$$

At leading $\ln(1/x)$, which is the approximation we are adopting, we can write

$$f\left(\frac{x}{x'}, \boldsymbol{k}^2\right) \simeq f(x, \boldsymbol{k}^2) , \tag{9.146}$$

since

$$\ln^n \left(\frac{x}{x'}\right) = \ln^n x \left[1 + \mathcal{O}\left(\frac{1}{\ln x}\right)\right] . \tag{9.147}$$

Thus (9.143) becomes (setting to zero the lower limit of the integration on $x'$)

$$\sigma_\lambda^{\gamma^* p}(x, Q^2) = \int \frac{\mathrm{d}\boldsymbol{k}^2}{\boldsymbol{k}^2} f(x, \boldsymbol{k}^2) \int_x^1 \frac{\mathrm{d}x'}{x'} \sigma_\lambda^{\gamma^* g}(x', \boldsymbol{k}^2, Q^2) , \tag{9.148}$$

To keep contact with the formalism of Sect. 8.16, we write the $\gamma^* p$ cross section in the large $W^2$ limit as

$$\sigma_\lambda^{\gamma^* p}(x, Q^2) = \frac{1}{(2\pi)^4} \int \frac{\mathrm{d}^2\boldsymbol{k}}{\boldsymbol{k}^2} \int \frac{\mathrm{d}^2\boldsymbol{k}'}{\boldsymbol{k}'^2} \Phi_\lambda(\boldsymbol{k}^2, Q^2)\, \Phi_p(\boldsymbol{k}'^2)\, F(x, \boldsymbol{k}, \boldsymbol{k}') , \tag{9.149}$$

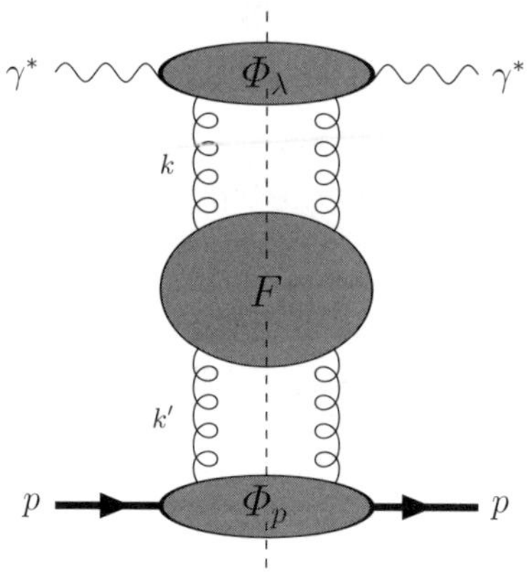


**Fig. 9.18.** DIS at low $x$ in the impact factor formalism.

where $\Phi_\lambda$ is the photon impact factor, $\Phi_p$ is the proton impact factor, and $F$ is the BFKL amplitude, containing the reggeized gluon ladder (Fig. 9.18). The color factors are incorporated into $\Phi_\lambda$ and $\Phi_p$. The unintegrated gluon distribution is related to $\Phi_p(\boldsymbol{k}')$ and $F(x, \boldsymbol{k}, \boldsymbol{k}')$ by

$$f(x, \boldsymbol{k}^2) = \frac{1}{(2\pi)^3} \int \frac{\mathrm{d}^2\boldsymbol{k}'}{\boldsymbol{k}'^2} \Phi_p(\boldsymbol{k}'^2)\, \boldsymbol{k}^2\, F(x, \boldsymbol{k}, \boldsymbol{k}') \,. \tag{9.150}$$

Using (9.150), and comparing (9.149) and (9.148), leads to the following relation between the photon impact factor and the $\gamma^* g$ cross sections

$$\Phi_\lambda(\boldsymbol{k}^2, Q^2) = 2\boldsymbol{k}^2 \int_0^1 \frac{\mathrm{d}x'}{x'}\, \hat{\sigma}_\lambda^{\gamma^* g}(x', \boldsymbol{k}^2, Q^2) \,, \tag{9.151}$$

and to a new form of the $\boldsymbol{k}_\perp$-factorization formula

$$\sigma_\lambda^{\gamma^* p}(x, Q^2) = \frac{1}{2\pi} \int \frac{\mathrm{d}^2\boldsymbol{k}}{\boldsymbol{k}^4}\, f(x, \boldsymbol{k}^2)\, \Phi_\lambda(\boldsymbol{k}^2, Q^2) \,. \tag{9.152}$$

Summarizing, the non-perturbative information on the structure of the proton is contained in $\Phi_p$, i.e. in $f(x, \boldsymbol{k}^2)$. The perturbative evaluation of $F$ allows knowing the $x$-evolution of $f(x, \boldsymbol{k}^2)$, once an input distribution $f(x_0, \boldsymbol{k}^2)$, i.e. a specific form of $\Phi_p$, is assumed. Finally, the impact factor $\Phi_\lambda$, or equivalently the gluonic cross section $\hat{\sigma}^{\gamma^* p}$, can be obtained by a one-loop calculation of photon–gluon diagrams.

Let us now see how (9.143) is derived. At low $x$ the dominant contribution to DIS comes from the gluon ladder diagrams shown in Fig. 9.14b. The $\gamma^* N$ cross section is obtained by squaring these diagrams and summing over the final states. Let us separate the fusion of the virtual photon with the off-shell upper gluon from the rest of the gluonic ladder, which is included in the unintegrated gluon distribution. We thus get the diagram of Fig. 9.19.

The virtual photoproduction cross section reads (we omit for the moment the photon polarization indices)

$$\sigma^{\gamma^* p}(x, Q^2) = \frac{1}{2W^2} \int \mathrm{d}\Pi\, |\overline{M}|^2 \,. \tag{9.153}$$

The squared matrix element in (9.153) is

$$|\overline{M}|^2 = \overline{\sum} \mathcal{A}^\mu\, \mathcal{G}_\mu\, \mathcal{A}^{\nu *}\, \mathcal{G}_\nu^* \,, \tag{9.154}$$

where $\overline{\sum}$ is an average over the initial states and a sum over the final states. The scattering amplitudes $\mathcal{A}^\mu$ and $\mathcal{G}^\mu$ are represented in Fig. 9.19. The gluon propagator is contained in $\mathcal{G}^\mu$. Using the explicit phase space expression (for the labeling of momenta see Fig. 9.19)

$$\begin{aligned}\mathrm{d}\Pi &= \frac{\mathrm{d}^4\kappa_1}{(2\pi)^3}\,\delta(\kappa_1^2)\,\frac{\mathrm{d}^4\kappa_2}{(2\pi)^3}\,\delta(\kappa_2^2)\,\mathrm{d}\Pi_X \\ &= \frac{\mathrm{d}^4 k}{(2\pi)^4}\left[\frac{\mathrm{d}^4\kappa_1}{(2\pi)^2}\,\delta(\kappa_1^2)\,\delta(\kappa_2^2)\right]\mathrm{d}\Pi_X\ , \end{aligned} \tag{9.155}$$

where

$$\mathrm{d}\Pi_X = \frac{\mathrm{d}^3\boldsymbol{P}_X}{(2\pi)^3\,2P_X^0}\,(2\pi)^4\,\delta^4(P-P_X-k)\ , \tag{9.156}$$

we can rearrange (9.153) as

$$\sigma^{\gamma^* p}(x,Q^2) = \frac{1}{2\,W^2}\int\frac{\mathrm{d}^4 k}{(2\pi)^4}\,A^{\mu\nu}(q,k)\,G_{\mu\nu}(P,k)\ , \tag{9.157}$$

with

$$A^{\mu\nu}(q,k) = \overline{\sum}\int\frac{\mathrm{d}^4\kappa_1}{(2\pi)^2}\,\delta(\kappa_1^2)\,\delta(\kappa_2^2)\,\mathcal{A}^\mu\,\mathcal{A}^{\nu *} \tag{9.158}$$

$$G^{\mu\nu}(P,k) = \overline{\sum}\int\mathrm{d}\Pi_X\,\mathcal{G}^\mu\mathcal{G}^{\nu *}\ . \tag{9.159}$$

$A^{\mu\nu}$ and $G^{\mu\nu}$ are the absorptive parts of $\gamma^*$-gluon and gluon-$N$ scattering, respectively.

Let us now introduce a Sudakov representation for the momenta. The null vectors are $P$ (the nucleon four-momentum) and $q' = q + xP$. Neglecting masses, one has in fact $P^2 = q'^2 = 0$, and, in the limit of large $W^2$, $q'\cdot P = q\cdot P = W^2/2$. We shall systematically neglect terms $\mathcal{O}(x)$ or $\mathcal{O}(1/W^2)$.

The exchanged gluon has four-momentum

$$k = \beta\,P - \alpha\,q' + k_\perp = \beta\,P + \frac{k^2+\boldsymbol{k}^2}{\beta\,W^2}\,q' + k_\perp\ , \tag{9.160}$$

where $k_\perp = (0,\boldsymbol{k},0)$ and $k_\perp^2 = -\boldsymbol{k}^2$.

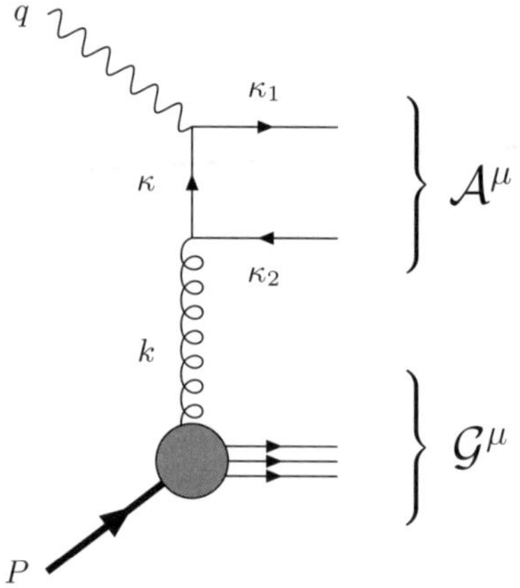


**Fig. 9.19.** Diagram contributing to DIS at low $x$.

It turns out that the dominant contribution to the integral over $k$ in (9.157) comes from the region of small $\beta$, fixed $\boldsymbol{k}^2 \ll W^2$ and $k^2 \simeq -\boldsymbol{k}^2$. Thus $\boldsymbol{k}^2/\beta W^2 = \mathcal{O}(1)$ and $(k^2+\boldsymbol{k}^2)/\beta W^2$ is negligible, so that we can write

$$k \simeq \beta P + k_\perp \,. \tag{9.161}$$

The tensor $A^{\mu\nu}$ appearing in (9.157) is a gauge-invariant conserved current, hence it satisfies

$$k_\mu A^{\mu\nu} = A^{\mu\nu} k_\nu = 0 \,, \tag{9.162}$$

and can be decomposed in general as

$$\begin{aligned} A^{\mu\nu}(q,k) &= A_1 \left(-g^{\mu\nu} + \frac{k^\mu k^\nu}{k^2}\right) \\ &\quad - \frac{1}{k^2} A_2 \left(k^\mu - \frac{k^2}{q\cdot k} q^\mu\right)\left(k^\nu - \frac{k^2}{q\cdot k} q^\nu\right) . \end{aligned} \tag{9.163}$$

For $k^2$, i.e. for on-shell gluons, we must have $A_1 = A_2$, so that the spurious $k^2$ pole in (9.163) cancels out. Note that the cross section for the scattering of the virtual photon off an on-shell gluon would be

$$\begin{aligned} \hat{\sigma}^{\gamma^* g}_{\text{on shell}} &= \frac{1}{\text{flux}}(-g^{\mu\nu})\, A^{\mu\nu} \\ &= \frac{1}{2\,\beta\, W^2}(-g^{\mu\nu})\, A^{\mu\nu} = \frac{1}{\beta\, W^2} A_2 \,, \quad (k^2 = 0) \,. \end{aligned} \tag{9.164}$$

In the limit of small $\beta$ and large $W^2$, which dominates the cross section at high energies, $A^{\mu\nu}$ tends to

$$A^{\mu\nu} \to \frac{4\,\boldsymbol{k}^2}{\beta^2 W^4} q^\mu\, q^\nu\, A_2 \,. \tag{9.165}$$

Inserting (9.165) in (9.157) and using $\int \mathrm{d}^4k = (W^2/2) \int \mathrm{d}\alpha\, \mathrm{d}\beta\, \mathrm{d}^2\boldsymbol{k}$, we get

$$\sigma^{\gamma^* p}_\lambda(x,Q^2) = \int \frac{\mathrm{d}\alpha\, \mathrm{d}\beta\, \mathrm{d}^2\boldsymbol{k}}{(2\pi)^4} \frac{\boldsymbol{k}^2}{\beta^2\, W^4} A_2(q,k)\, q^\mu\, q^\nu\, G_{\mu\nu}(P,k) \,. \tag{9.166}$$

In analogy with (9.164), we introduce the cross section for off-shell gluons

$$\hat{\sigma}^{\gamma^* g}_\lambda(\beta, \boldsymbol{k}^2, Q^2) = \frac{1}{\beta\, W^2} A_2 = \frac{\beta}{W^2\, \boldsymbol{k}^2} P_\mu P_\nu\, A^{\mu\nu} \,, \tag{9.167}$$

where, to obtain the second equality, we inverted (9.165) by means of the projector $P_\mu P_\nu$. The use of (9.167) in (9.166) yields

$$\sigma^{\gamma^* p}_\lambda(x,Q^2) = \int \frac{\mathrm{d}\boldsymbol{k}^2}{\boldsymbol{k}^2} \int \frac{\mathrm{d}\beta}{\beta} f(\beta, \boldsymbol{k}^2)\, \hat{\sigma}^{\gamma^* g}_\lambda(\beta, \boldsymbol{k}^2, Q^2) \,, \tag{9.168}$$

with the unintegrated gluon distribution related to $G^{\mu\nu}$ by

$$f(\beta, \boldsymbol{k}^2) = \int \mathrm{d}\alpha \, \frac{\pi \, \boldsymbol{k}^4}{W^2} \, q^\mu q^\nu G_{\mu\nu}(P, k) \, . \tag{9.169}$$

Equation (9.168) is the $\boldsymbol{k}_\perp$-factorization formula anticipated in (9.143). Recall that $\beta \equiv x/x'$ is the fraction of the longitudinal momentum of the proton carried by the gluon.

We compute now the $\gamma^* g$ cross section. First of all, we introduce the photon polarization indices. Thus, $A^{\mu\nu}$ becomes $A^{\mu\nu\alpha\beta}$ and (9.167) is rewritten as

$$\hat{\sigma}_\lambda^{\gamma^* g}(\beta, \boldsymbol{k}^2, Q^2) = \frac{\beta}{W^2 \, \boldsymbol{k}^2} \, \varepsilon_\alpha^{(\lambda)} \varepsilon_\beta^{(\lambda)*} \, P_\mu P_\nu \, A^{\mu\nu\alpha\beta} \, , \tag{9.170}$$

where $\varepsilon_\alpha^{(\lambda)}$ is the polarization vector of the virtual photon, and

$$\begin{aligned} A^{\mu\nu\alpha\beta}(q, k) &= \overline{\sum} \int \frac{\mathrm{d}^4\kappa_1}{(2\pi)^2} \, \delta(\kappa_1^2) \, \delta(\kappa_2^2) \, \mathcal{A}^{\alpha\mu} \, \mathcal{A}^{\beta\nu *} \\ &= \overline{\sum} \int \frac{\mathrm{d}^4\kappa}{(2\pi)^2} \, \delta((q-\kappa)^2) \, \delta((k+\kappa)^2) \, \mathcal{A}^{\alpha\mu} \, \mathcal{A}^{\beta\nu *} \, . \end{aligned} \tag{9.171}$$

In the last equality we introduced the four-momentum

$$\kappa = q - \kappa_1 = \kappa_2 - k \, . \tag{9.172}$$

The amplitude $\mathcal{A}^{\alpha\mu}$ describes the $\gamma^*(q) + g(k) \to q(\kappa_1) + \bar{q}(\kappa_2)$ process and is the sum of two terms, the $t$-channel and the $u$-channel contributions (Fig. 9.20)

$$\mathcal{A}^{\alpha\mu} = \mathcal{A}_{(t)}^{\alpha\mu} + \mathcal{A}_{(u)}^{\alpha\mu} \, , \tag{9.173}$$

where

$$\mathcal{A}_{(t)}^{\alpha\mu} = -\mathrm{i} \, g_s \, e_q \, e \, t_{ij}^a \, \bar{u}(\kappa_1) \, \gamma^\alpha \, \frac{\not{k} - \not{\kappa}_2}{(k-\kappa_2)^2} \, \gamma^\mu \, v(\kappa_2) \, , \tag{9.174}$$

$$\mathcal{A}_{(u)}^{\alpha\mu} = -\mathrm{i} \, g_s \, e_q \, e \, t_{ij}^a \, \bar{u}(\kappa_1) \, \gamma^\mu \, \frac{\not{\kappa}_1 - \not{k}}{(\kappa_1 - k)^2} \, \gamma^\alpha \, v(\kappa_2) \, . \tag{9.175}$$

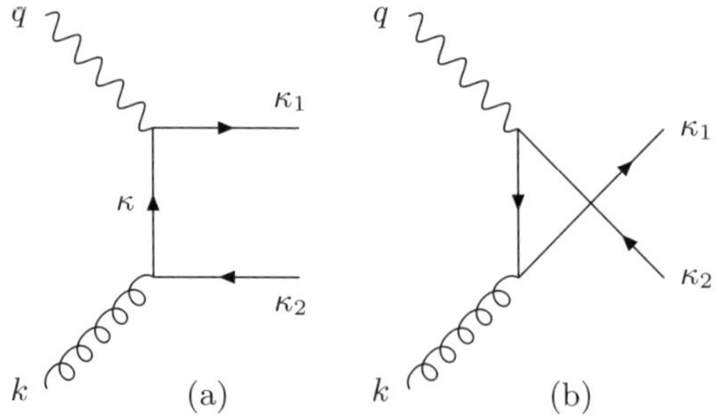


**Fig. 9.20.** Diagrams for $\gamma^* g \to q\bar{q}$: (**a**) $t$-channel; (**b**) $u$-channel.

Writing the Sudakov decomposition of $\kappa$

$$\kappa = -w\, P + z\, q' + \kappa_\perp \,, \tag{9.176}$$

and using $\mathrm{d}^4\kappa = (W^2/2)\, \mathrm{d}w\, \mathrm{d}z\, \mathrm{d}^2\boldsymbol{\kappa}$, the integral in (9.171) can be rewritten as

$$\begin{aligned} A^{\mu\nu\alpha\beta} &= \frac{1}{8\pi^2 W^2} \overline{\sum} \int \frac{\mathrm{d}z}{z\,(1-z)}\, \mathrm{d}^2\boldsymbol{\kappa} \\ &\quad \times \delta\left(\beta - \frac{z\boldsymbol{\kappa}^2 + (1-z)\,(\boldsymbol{\kappa}+\boldsymbol{k})^2 + \varepsilon^2}{z(1-z)\, W^2}\right) \mathcal{A}^{\alpha\mu}\, \mathcal{A}^{\beta\nu *} \,, \end{aligned} \tag{9.177}$$

with

$$\varepsilon^2 \equiv Q^2\, z\,(1-z) \,. \tag{9.178}$$

Squaring $\mathcal{A}^{\alpha\mu}$ we get

$$\mathcal{A}^{\mu\nu\alpha\beta} \equiv \overline{\sum} \mathcal{A}^{\alpha\mu}\, \mathcal{A}^{\beta\nu *} = 2\left(\mathcal{A}^{\mu\nu\alpha\beta}_{(a)} + \mathcal{A}^{\mu\nu\alpha\beta}_{(b)}\right) , \tag{9.179}$$

where $\mathcal{A}^{\mu\nu\alpha\beta}_{(a)}$ and $\mathcal{A}^{\mu\nu\alpha\beta}_{(b)}$ refer to the two diagrams in Fig. 9.21, and the factor 2 comes from the other two diagrams with reversed fermion lines. We have

$$\begin{aligned} \mathcal{A}^{\mu\nu\alpha\beta}_{(a)} &= -T_R\,(4\pi)^2\,\alpha_s\,\alpha_{\mathrm{em}}\,e_q^2 \\ &\quad \times \frac{\mathrm{Tr}\,[\not\kappa\gamma^\mu(\not\kappa + \not k)\gamma^\nu\, \not\kappa\gamma^\beta(\not\kappa - \not q)\gamma^\alpha]}{\kappa^4} \,, \end{aligned} \tag{9.180}$$

$$\begin{aligned} \mathcal{A}^{\mu\nu\alpha\beta}_{(b)} &= -T_R\,(4\pi)^2\,\alpha_s\,\alpha_{\mathrm{em}}\,e_q^2 \\ &\quad \times \frac{\mathrm{Tr}\,[\not\kappa\gamma^\mu(\not\kappa + \not k)\gamma^\beta(\not\kappa + \not k - \not q)\gamma^\nu(\not\kappa - \not q)\gamma^\alpha]}{\kappa^2\,(k+\kappa-q)^2} \,, \end{aligned} \tag{9.181}$$

where $T_R = 1/2$.

The denominators in (9.180–9.181) are

$$\kappa^2 = -\frac{\boldsymbol{\kappa}^2 + \varepsilon^2}{1-z} \,, \tag{9.182}$$

$$(\kappa + k - q)^2 = -\frac{(\boldsymbol{\kappa}+\boldsymbol{k})^2 + \varepsilon^2}{z} \,. \tag{9.183}$$

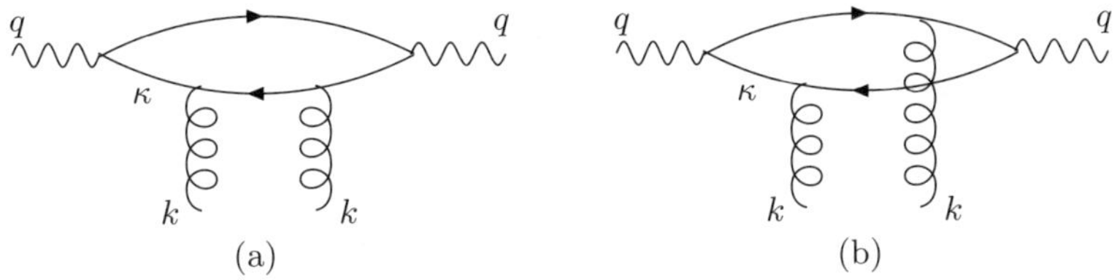


**Fig. 9.21a,b.** Two of the four diagrams contributing to $\mathcal{A}^{\mu\nu\alpha\beta}$. The other two diagrams are obtained by reversing the fermion lines.

Combining (9.167) and (9.177) leads to

$$\sigma_\lambda^{\gamma^* g} = \frac{\beta}{8\pi^2 W^4 \boldsymbol{k}^2} \int \frac{\mathrm{d}z}{z(1-z)} \int \mathrm{d}^2\boldsymbol{\kappa} \times \delta\left(\beta - \frac{z\boldsymbol{\kappa}^2 + (1-z)(\boldsymbol{\kappa}+\boldsymbol{k})^2 + \varepsilon^2}{z(1-z)W^2}\right) d_{\alpha\beta}^{(\lambda)} P_\mu P_\nu \mathcal{A}^{\alpha\beta\mu\nu} , \tag{9.184}$$

where the projectors onto the photon polarizations are – see Eqs. (9.19–9.21)

$$d_{\alpha\beta}^{(L)} \equiv \varepsilon_\alpha^L \varepsilon_\beta^{L*} \simeq \frac{4Q^2}{W^2}\left[P_\alpha P_\beta + \frac{W^2}{2Q^2}(P_\alpha q_\beta + P_\beta q_\alpha) + \frac{W^4}{4Q^4} q_\alpha q_\beta\right] , \tag{9.185}$$

$$d_{\alpha\beta}^{(\Sigma)} \equiv \sum_{\lambda=0,\pm} \varepsilon_\alpha^{(\lambda)} \varepsilon_\beta^{(\lambda)*} = -\left[g_{\alpha\beta} + \frac{q_\alpha q_\beta}{Q^2}\right] , \tag{9.186}$$

$$d_{\alpha\beta}^{(T)} \equiv \frac{1}{2}\left[\varepsilon_\alpha^- \varepsilon_\beta^{-*} + \varepsilon_\alpha^+ \varepsilon_\beta^{+*}\right] = \frac{1}{2}\left[d_{\alpha\beta}^{(\Sigma)} + d_{\alpha\beta}^{(L)}\right] . \tag{9.187}$$

Note that the terms proportional to $q_\alpha$ and $q_\beta$, when contracted with $A^{\mu\nu\alpha\beta}$, give a vanishing contribution because of current conservation.

The impact factors are obtained by integrating (9.184) over $\beta$ and read, see (9.151)

$$\Phi_\lambda = \frac{1}{4\pi^2 W^4} \int \frac{\mathrm{d}z}{z(1-z)} \int \mathrm{d}^2\boldsymbol{\kappa}\, d_{\alpha\beta}^{(\lambda)} P_\mu P_\nu \mathcal{A}^{\alpha\beta\mu\nu} . \tag{9.188}$$

In computing

$$H_\lambda \equiv d_{\alpha\beta}^{(\lambda)} P_\mu P_\nu \mathcal{A}^{\alpha\beta\mu\nu} \tag{9.189}$$

from the traces (9.180–9.181), one may take advantage of the fact that, due to the eikonal projector $P_\mu P_\nu$, only the terms proportional to $q'^\mu q'^\nu$ in $\mathcal{A}^{\mu\nu\alpha\beta}$ contribute. This considerably simplifies the calculation. The result is

$$H_\lambda = 8\pi^2 \alpha_s \alpha_{\mathrm{em}} e_q^2 \frac{z(1-z)W^4}{Q^2} \times \left\{ \frac{N_\lambda(\boldsymbol{\kappa},\boldsymbol{\kappa})}{(\boldsymbol{\kappa}^2+\varepsilon^2)^2} + \frac{N_\lambda(\boldsymbol{\kappa}+\boldsymbol{k},\boldsymbol{\kappa}+\boldsymbol{k})}{[(\boldsymbol{\kappa}+\boldsymbol{k})^2+\varepsilon^2]^2} - \frac{2N_\lambda(\boldsymbol{\kappa},\boldsymbol{\kappa}+\boldsymbol{k})}{(\boldsymbol{\kappa}^2+\varepsilon^2)[(\boldsymbol{\kappa}+\boldsymbol{k})^2+\varepsilon^2]} \right\} \tag{9.190}$$

with

$$N_L(\boldsymbol{k}_1,\boldsymbol{k}_2) = 4z^2(1-z)^2 Q^4 , \tag{9.191}$$

$$N_T(\boldsymbol{k}_1,\boldsymbol{k}_2) = Q^2\left[z^2 + (1-z)^2\right] \boldsymbol{k}_1 \cdot \boldsymbol{k}_2 . \tag{9.192}$$

Inserting $H_\lambda$ into (9.184), and then $\hat{\sigma}_\lambda^{\gamma^* g}$ into the factorization formula (9.168), yields finally

$$\sigma_{L,T}^{\gamma^* p}(x, Q^2) = \frac{\alpha_{\rm em}}{Q^2} \sum_q e_q^2 \int \frac{{\rm d}\boldsymbol{k}^2}{\boldsymbol{k}^4} \int_0^1 {\rm d}z \int {\rm d}^2\boldsymbol{\kappa}\, \alpha_s(\mu^2)\, f(\beta, \boldsymbol{k}^2) \times \left\{ \frac{N_\lambda(\boldsymbol{\kappa}, \boldsymbol{\kappa})}{(\boldsymbol{\kappa}^2 + \varepsilon^2)^2} + \frac{N_\lambda(\boldsymbol{\kappa}+\boldsymbol{k}, \boldsymbol{\kappa}+\boldsymbol{k})}{[(\boldsymbol{\kappa}+\boldsymbol{k})^2 + \varepsilon^2]^2} - \frac{2\, N_\lambda(\boldsymbol{\kappa}, \boldsymbol{\kappa}+\boldsymbol{k})}{(\boldsymbol{\kappa}^2 + \varepsilon^2)[(\boldsymbol{\kappa}+\boldsymbol{k})^2 + \varepsilon^2]} \right\} , \tag{9.193}$$

where a sum over the flavors has been introduced, and $\beta$ is constrained by the delta function in (9.184). The corresponding structure functions $F_{L,T}(x, Q^2)$ can be obtained by multiplying the cross sections by $Q^2/4\pi^2\, \alpha_{\rm em}$.

In (9.193) the strong coupling has been written as $\alpha_s(\mu^2)$. Although there are some phenomenological arguments in favor of a running coupling (see Sect. 9.6), one should recall that in the leading $\ln(1/x)$ approximation $\alpha_s$ is a fixed parameter. In this limit we can put $f(\beta, \boldsymbol{k}^2) \simeq f(x, \boldsymbol{k}^2)$, introduce the impact factors, and (9.193) reduces to (9.152) with $\Phi_{L,T}$ explicitly given by

$$\Phi_L = 8\, \alpha_s\, \alpha_{\rm em} \sum_q e_q^2 \int {\rm d}z\, Q^2\, z^2\, (1-z)^2 \times \int {\rm d}^2\boldsymbol{\kappa} \left\{ \frac{1}{\boldsymbol{\kappa}^2 + \varepsilon^2} - \frac{1}{(\boldsymbol{\kappa}+\boldsymbol{k})^2 + \varepsilon^2} \right\} , \tag{9.194}$$

$$\Phi_T = 2\, \alpha_s\, \alpha_{\rm em} \sum_q e_q^2 \int {\rm d}z\, [z^2 + (1-z)^2] \times \int {\rm d}^2\boldsymbol{\kappa} \left\{ \frac{\boldsymbol{k}^2 + 2\, Q^2\, z(1-z)}{(\boldsymbol{\kappa}^2 + \varepsilon^2)[(\boldsymbol{\kappa}+\boldsymbol{k})^2 + \varepsilon^2]} - \frac{Q^2\, z(1-z)}{(\boldsymbol{\kappa}^2 + \varepsilon^2)^2} - \frac{Q^2\, z(1-z)}{[(\boldsymbol{\kappa}+\boldsymbol{k})^2 + \varepsilon^2]^2} \right\} . \tag{9.195}$$

The $\boldsymbol{\kappa}$-integration in (9.194–9.195) can be, at least partially, performed, using

$$\int {\rm d}^2\boldsymbol{\kappa} \frac{1}{(\boldsymbol{\kappa}^2 + \varepsilon^2)^2} = \frac{\pi}{Q^2\, z(1-z)} , \tag{9.196}$$

and introducing a Feynman parameter $\zeta$ in the other type of integral

$$\int {\rm d}^2\boldsymbol{\kappa}\, \frac{1}{(\boldsymbol{\kappa}^2 + \varepsilon^2)[(\boldsymbol{\kappa}+\boldsymbol{k})^2 + \varepsilon^2]} = \int_0^1 {\rm d}\zeta\, \frac{\pi}{Q^2\, z(1-z) + \boldsymbol{k}^2\, \zeta(1-\zeta)} . \tag{9.197}$$

We then get for the impact factors

$$\Phi_L = 16\pi\, \alpha_{\rm em}\, \alpha_s \sum_q e_q^2 \times \int_0^1 {\rm d}z \int_0^1 {\rm d}\zeta\, \frac{z(1-z)\, \zeta(1-\zeta)\, \boldsymbol{k}^2}{Q^2\, z(1-z) + \boldsymbol{k}^2\, \zeta(1-\zeta)} . \tag{9.198}$$

$$\Phi_T = 2\pi\, \alpha_{\rm em}\, \alpha_s \sum_q e_q^2 \times \int_0^1 {\rm d}z \int_0^1 {\rm d}\zeta\, \frac{[z^2 + (1-z)^2]\,[\zeta^2 + (1-\zeta)^2]\, \boldsymbol{k}^2}{Q^2\, z(1-z) + \boldsymbol{k}^2\, \zeta(1-\zeta)}\,. \tag{9.199}$$

Use of (9.197–9.199) in (9.152) leads to the following expressions for the structure functions $F_2 = F_L + F_T$

$$F_2(x, Q^2) = \frac{Q^2}{4\pi^2}\, \alpha_s \sum_q e_q^2 \int \frac{{\rm d}^2\boldsymbol{k}}{\boldsymbol{k}^2}\, f(x, \boldsymbol{k}^2) \int_0^1 {\rm d}z \int_0^1 {\rm d}\zeta \times \frac{1 - 2z(1-z) - 2\zeta(1-\zeta) + 12\, z(1-z)\, \zeta(1-\zeta)}{Q^2\, z(1-z) + \boldsymbol{k}^2\, \zeta(1-\zeta)}\,, \tag{9.200}$$

and for $F_L$

$$F_L(x, Q^2) = \frac{2\,Q^2}{\pi^2}\, \alpha_s \sum_q e_q^2 \int \frac{{\rm d}^2\boldsymbol{k}}{\boldsymbol{k}^2}\, f(x, \boldsymbol{k}^2) \times \int_0^1 {\rm d}z \int_0^1 {\rm d}\zeta \frac{z(1-z)\, \zeta(1-\zeta)}{Q^2\, z(1-z) + \boldsymbol{k}^2\, \zeta(1-\zeta)}\,. \tag{9.201}$$

In the limit $\boldsymbol{k}^2 \ll Q^2$, if we differentiate (9.200) with respect to $\ln Q^2$ and perform the $z$-integration, we get, as expected, the DLLA result for the logarithmic derivative of $F_2$

$$\begin{aligned} \frac{\partial F_2(x, Q^2)}{\partial \ln Q^2} &= 2 \sum_q e_q^2 \left(\frac{\alpha_s}{2\pi}\right) \int_0^1 {\rm d}\zeta\, P_{qg}(\zeta)\, x\, g(x, Q^2) \\ &= \sum_q e_q^2 \frac{\alpha_s}{3\pi}\, x\, g(x, Q^2)\,, \end{aligned} \tag{9.202}$$

where $P_{qg}(\zeta) = \frac{1}{2}\,[\zeta^2 + (1-\zeta)^2]$ is the $g \to q\bar{q}$ splitting function. Equation (9.202) is what we would get from the leading-order expression $F_2(x, Q^2) = \sum_q e_q^2 x[q(x, Q^2) + \bar{q}(x, Q^2)]$, by differentiating with respect to $\ln Q^2$ and using the Altarelli-Parisi equations in the limit $x \to 0$, which is dominated by gluon splitting.

Incidentally, note that recognizing in (9.202) the well-known DLLA result supports *a posteriori* the identification of (9.145), i.e. of the $\boldsymbol{k}$-integral of $f(x, \boldsymbol{k}^2)$, with the ordinary gluon distribution.

## 9.6 The BFKL Equation in DIS

The unintegrated gluon distribution (9.144) satisfies the BFKL equation, that we are now going to study in the context of DIS.

First of all, note that, since the leading $\ln(1/x)$ BFKL amplitude $F(x,\boldsymbol{k},\boldsymbol{k}')$ – Eq. (8.208) with $x \simeq Q^2/s$ – does not depend on the azimuthal angles of $\boldsymbol{k}$ and $\boldsymbol{k}'$, we can perform the angular integrations in the BFKL equation (8.187), thus obtaining

$$\frac{\partial F(x,\boldsymbol{k}^2,\boldsymbol{k}'^2)}{\partial \ln(1/x)} = \frac{N_c\alpha_s}{\pi}\int \frac{\mathrm{d}\boldsymbol{\kappa}^2}{\boldsymbol{\kappa}^2} \times \left[\frac{\boldsymbol{\kappa}^2\, F(x,\boldsymbol{\kappa}^2,\boldsymbol{k}'^2) - \boldsymbol{k}^2\, F(x,\boldsymbol{k}^2,\boldsymbol{k}'^2)}{|\boldsymbol{k}^2-\boldsymbol{\kappa}^2|} + \frac{\boldsymbol{k}^2\, F(x,\boldsymbol{k}^2,\boldsymbol{k}'^2)}{(4\,\boldsymbol{\kappa}^4+\boldsymbol{k}^4)^{\frac{1}{2}}}\right]. \quad (9.203)$$

Using the relation (9.150) between $f(x,\boldsymbol{k}^2)$ and $F(x,\boldsymbol{k},\boldsymbol{k}')$, Eq. (9.203) translates into the following integro-differential equation for $f(x,\boldsymbol{k}^2)$ (we set $N_c = 3$)

$$\frac{\partial f(x,\boldsymbol{k}^2)}{\partial \ln(1/x)} = \frac{3\alpha_s}{\pi}\,\boldsymbol{k}^2\int_0^\infty \frac{\mathrm{d}\boldsymbol{\kappa}^2}{\boldsymbol{\kappa}^2} \times \left[\frac{f(x,\boldsymbol{\kappa}^2)-f(x,\boldsymbol{k}^2)}{|\boldsymbol{\kappa}^2-\boldsymbol{k}^2|} + \frac{f(x,\boldsymbol{k}^2)}{(4\,\boldsymbol{\kappa}^4+\boldsymbol{k}^4)^{1/2}}\right]. \quad (9.204)$$

This is the BFKL equation of deep inelastic scattering, which resums the leading powers in $\alpha_s\ln(1/x)$. In the leading $\ln(1/x)$ approximation, the coupling constant $\alpha_s$ appearing in (9.204) is a fixed parameter.

Solving the BFKL equation with a starting distribution $f(x_0,\boldsymbol{k}^2)$ gives the unintegrated gluon distribution $f(x,\boldsymbol{k}^2)$ "evolved" to lower $x$ values. Recall that, although the integration in (9.204) is over an infinite range between 0 and $\infty$, the BFKL equation for a physical quantity such as $f(x,\boldsymbol{k}^2)$ is free from both infrared and ultraviolet divergences.

From (9.204) one can derive analytically, and quite independently from the input distribution, the low-$x$ behavior of $f(x,\boldsymbol{k}^2)$. To solve (9.204) we adopt a procedure slightly different from (but equivalent to) that of Sect. 8.7. The first step is to take the Mellin transform of $f(x,\boldsymbol{k}^2)$ with respect to $\boldsymbol{k}^2$

$$f(x,\gamma) = \int_1^\infty \mathrm{d}\left(\frac{\boldsymbol{k}^2}{\boldsymbol{k}_0^2}\right)\left(\frac{\boldsymbol{k}^2}{\boldsymbol{k}_0^2}\right)^{-\gamma-1} f(x,\boldsymbol{k}^2)\,, \quad (9.205)$$

where we have introduced a fixed arbitrary scale $\boldsymbol{k}_0^2$ for dimensional reasons. The corresponding inverse transform is

$$f(x,\boldsymbol{k}^2) = \frac{1}{2\pi\mathrm{i}}\int_{c-\mathrm{i}\infty}^{c+\mathrm{i}\infty} \mathrm{d}\gamma\left(\frac{\boldsymbol{k}^2}{\boldsymbol{k}_0^2}\right)^{\gamma} f(x,\gamma)\,. \quad (9.206)$$

In terms of $f(x,\gamma)$ the BFKL equation reads

$$\frac{\partial f(x,\gamma)}{\partial \ln(1/x)} = K(\gamma)\, f(x,\gamma)\,, \quad (9.207)$$

where the kernel $K(\gamma)$ is given by

$$K(\gamma) = \frac{3\,\alpha_s}{\pi} \int_0^\infty \frac{\mathrm{d}u}{u} \left[\frac{u^\gamma - 1}{|u-1|} + \frac{1}{(4\,u^2+1)^{1/2}}\right] = \frac{3\,\alpha_s}{\pi} \left[2\,\psi(1) - \psi(\gamma) - \psi(1-\gamma)\right] . \tag{9.208}$$

The solution of (9.207) is

$$f(x,\gamma) = f(x_0,\gamma) \left(\frac{x}{x_0}\right)^{-K(\gamma)} , \tag{9.209}$$

from which, using (9.206), we get

$$f(x,\boldsymbol{k}^2) = \frac{1}{2\pi \mathrm{i}} \int_{c-\mathrm{i}\infty}^{c+\mathrm{i}\infty} \mathrm{d}\gamma \left(\frac{\boldsymbol{k}^2}{\boldsymbol{k}_0^2}\right)^\gamma f(x_0,\gamma) \left(\frac{x}{x_0}\right)^{-K(\gamma)} . \tag{9.210}$$

To compute this integral, we first observe that $K(\gamma)$ is symmetric around $\gamma = 1/2$ and, along the contour $\gamma = 1/2 + \mathrm{i}\nu$, with $-\infty < \nu < \infty$, has a maximum at $\gamma = 1/2$. We then set $\gamma = 1/2 + \mathrm{i}\nu$ in (9.210) and write $(x/x_0)^{-K(\gamma)}$ as $\exp\left[-K(\gamma)\ln(x/x_0)\right]$, so that the integral becomes

$$f(x,\boldsymbol{k}^2) = \frac{1}{2\pi} \left(\frac{\boldsymbol{k}^2}{\boldsymbol{k}_0^2}\right)^{1/2} \int_{-\infty}^{+\infty} \mathrm{d}\nu\, f(x_0, 1/2+\mathrm{i}\nu) \times \exp\left[\mathrm{i}\nu \ln\left(\frac{\boldsymbol{k}^2}{\boldsymbol{k}_0^2}\right) + K(1/2+\mathrm{i}\nu)\ln\left(\frac{x_0}{x}\right)\right] . \tag{9.211}$$

We now expand $K(\gamma)$ around $\nu = 0$

$$K(1/2+\mathrm{i}\nu) = \lambda - \frac{1}{2}\lambda'\,\nu^2 + \mathcal{O}(\nu^4) , \tag{9.212}$$

where $\lambda = 3\alpha_s\, 4\ln 2/\pi$ and $\lambda' = 3\alpha_s 28\zeta(3)/\pi$.

The expansion of $f(x_0, 1/2+\mathrm{i}\nu)$ gives

$$\begin{aligned} f(x_0, 1/2+\mathrm{i}\nu) &= f(x_0,1/2) + \nu \left.\frac{\mathrm{d}f}{\mathrm{d}\nu}\right|_{\nu=0} + \ldots \\ &= f(x_0,1/2)\left(1 + \nu \left.\frac{\mathrm{d}\ln f}{\mathrm{d}\nu}\right|_{\nu=0}\right) + \ldots \\ &\simeq f(x_0,1/2)\exp\left(\nu \left.\frac{\mathrm{d}\ln f}{\mathrm{d}\nu}\right|_{\nu=0}\right) . \end{aligned} \tag{9.213}$$

Using (9.212) and (9.213) in (9.211) yields

$$f(x,\boldsymbol{k}^2) \simeq \frac{1}{2\pi}\left(\frac{\boldsymbol{k}^2}{\boldsymbol{k}_0^2}\right)^{1/2} f(x_0,1/2)\left(\frac{x}{x_0}\right)^{-\lambda} \times \int_{-\infty}^{+\infty} \mathrm{d}\nu \exp\left[\mathrm{i}\nu\ln\left(\frac{\boldsymbol{k}^2}{\tilde{\boldsymbol{k}}^2}\right) - \frac{1}{2}\lambda'\,\nu^2\ln\left(\frac{x_0}{x}\right)\right] , \tag{9.214}$$

where

$$\ln \widetilde{\boldsymbol{k}}^2 = \ln \boldsymbol{k}_0^2 + \mathrm{i} \left. \frac{\mathrm{d} \ln f}{\mathrm{d}\nu} \right|_{\nu=0} . \tag{9.215}$$

Finally, performing the $\nu$ integration we get

$$f(x, \boldsymbol{k}^2) \sim \left( \frac{x}{x_0} \right)^{-\lambda} \left[ \frac{(\boldsymbol{k}^2/\boldsymbol{k}_0^2)}{\ln(x_0/x)} \right]^{1/2} \exp\left[ -\frac{\ln^2(\boldsymbol{k}^2/\widetilde{\boldsymbol{k}}^2)}{2\,\lambda' \ln(x_0/x)} \right] . \tag{9.216}$$

Thus the dominant behavior of the unintegrated gluon distribution at low $x$ is

$$f(x, \boldsymbol{k}^2) \sim x^{-\lambda} , \tag{9.217}$$

with $\lambda \approx 0.5$ for $\alpha_s = 0.2$. This is the celebrated BFKL behavior of low-$x$ structure functions, that we already encountered in Sect. 8.7 as the leading $\ln s$ behavior of total cross sections. The resulting pomeron intercept, as we know, is $\alpha_{I\!P}(0) = 1+\lambda \approx 1.5$, corresponding to the "hard" or BFKL pomeron.

Let us come back to (9.216) and look at the $\boldsymbol{k}^2$-dependence of the solution of the BFKL equation. One sees that $f(x, \boldsymbol{k}^2)$ is a Gaussian distribution in $\ln(\boldsymbol{k}^2/\boldsymbol{k}_0^2)$, with a width growing as $\sqrt{\ln(x_0/x)}$ when $x \to 0$. Hence the evolution in $x$ broadens the $\ln \boldsymbol{k}^2$ distribution of $f(x, \boldsymbol{k}^2)$. The position of the maximum and the normalization of the Gaussian distribution depends on the input $f(x_0, \boldsymbol{k}^2)$, but the rate of diffusion is independent of the boundary conditions, depending only on $\lambda' \ln(x_0/x)$.

The evolution of $f(x, \boldsymbol{k}^2)/(\boldsymbol{k}^2)^{1/2}$, obtained by introducing an infrared cutoff in the BFKL equation, is shown in Fig. 9.22. Both the diffusion in $\boldsymbol{k}^2$ and the growth of the type $x^{-\lambda}$ are clearly visible.

If we take a double Mellin transform of $f(x, \boldsymbol{k}^2)$ with respect to both $\boldsymbol{k}^2$ and $x$, i.e.

$$f(N, \gamma) = \int_1^\infty \mathrm{d}\left( \frac{\boldsymbol{k}^2}{\boldsymbol{k}_0^2} \right) \left( \frac{\boldsymbol{k}^2}{\boldsymbol{k}_0^2} \right)^{-\gamma-1} \int_0^1 \mathrm{d}\left( \frac{x}{x_0} \right) \left( \frac{x}{x_0} \right)^{N-1} f(x, \boldsymbol{k}^2) , \tag{9.218}$$

then, using (9.209), the solution of the BFKL equation takes the suggestive form

$$f(N, \gamma) = \frac{f(x_0, \gamma)}{N - K(\gamma)} , \tag{9.219}$$

which shows that the unintegrated gluon distribution has a pole in the $N$-plane at $N = K(\gamma)$.

In principle, one can solve the BFKL equation with some input distribution, taken from DIS global fits, and then use the $\boldsymbol{k}_\perp$-factorization formula (9.143) to compute the structure functions at low $x$. In practice, the predictions are affected by many uncertainties. The point is that the diffusion in $\ln \boldsymbol{k}^2$ of the BFKL solution leads to an increasingly large contribution from the non-perturbative region of $\boldsymbol{k}^2$, where the gluon distribution is unknown.

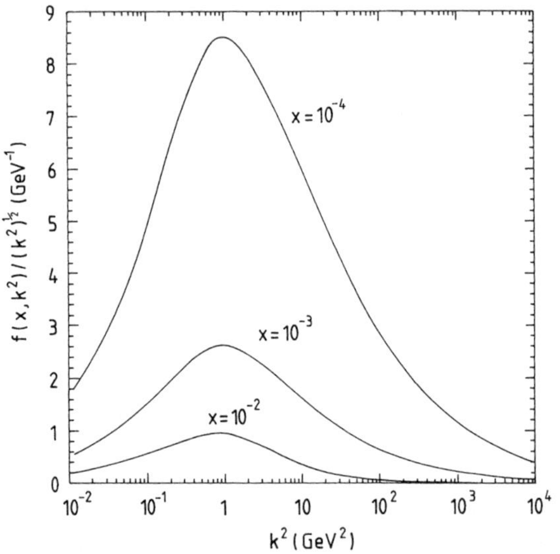


**Fig. 9.22.** The BFKL evolution of $f(x, \boldsymbol{k}^2)/(\boldsymbol{k}^2)^{1/2}$. An infrared cutoff is introduced by hand. From Askew et al. (1994).

So, it is not enough to assume a starting distribution $f(x_0, \boldsymbol{k}^2)$ at large $\boldsymbol{k}^2$. One is also forced to introduce an infrared cutoff $\boldsymbol{k}_0^2$ in the BFKL equation, and in the $\boldsymbol{k}_\perp$-factorization formula, in order to exclude the small-$\boldsymbol{k}^2$ region.

An alternative, and more sophisticated, procedure, is to extend $f(x_0, \boldsymbol{k}^2)$ to the non-perturbative domain, assuming some particular behavior as $\boldsymbol{k}^2 \to 0$, such as, for instance (Askew et al. 1994; Bojak and Ernst 1996)

$$f(x, \boldsymbol{k}^2) = \frac{\boldsymbol{k}_c^2 + \boldsymbol{k}_a^2}{\boldsymbol{k}_c^2} \frac{\boldsymbol{k}^2}{\boldsymbol{k}^2 + \boldsymbol{k}_a^2} f(x, \boldsymbol{k}_c^2) , \quad (\boldsymbol{k}^2 < \boldsymbol{k}_c^2) , \tag{9.220}$$

where $\boldsymbol{k}_c^2$ gives the onset of the infrared region and $\boldsymbol{k}_a^2$ is another parameter which controls the infrared behavior of $f(x, \boldsymbol{k}^2)$. Note that the assumption (9.220) ensures $f(x, \boldsymbol{k}^2) \propto \boldsymbol{k}^2$ as $\boldsymbol{k}^2 \to 0$, as required by gauge invariance. In practical calculations, an ultraviolet parameter $\boldsymbol{k}_{\rm max}^2$ is also required.

A further subtle point concerns the strong coupling. This should be taken as a fixed parameter in the BFKL equation. However, if we want to get from BFKL the double-leading log approximation in the limit of strongly ordered transverse momenta, we have to take a running coupling $\alpha_s(\boldsymbol{k}^2)$. The introduction of a running $\alpha_s$ has the effect of suppressing the importance of the ultraviolet cutoff, but introduces one more scale in the problem, the "freezing" scale $\boldsymbol{k}_b^2$

$$\alpha_s \to \alpha_s(\boldsymbol{k}^2 + \boldsymbol{k}_b^2) . \tag{9.221}$$

It turns out that all the parameters one has to introduce in order to keep the non-perturbative and the ultraviolet regions of $\boldsymbol{k}^2$ under control do not

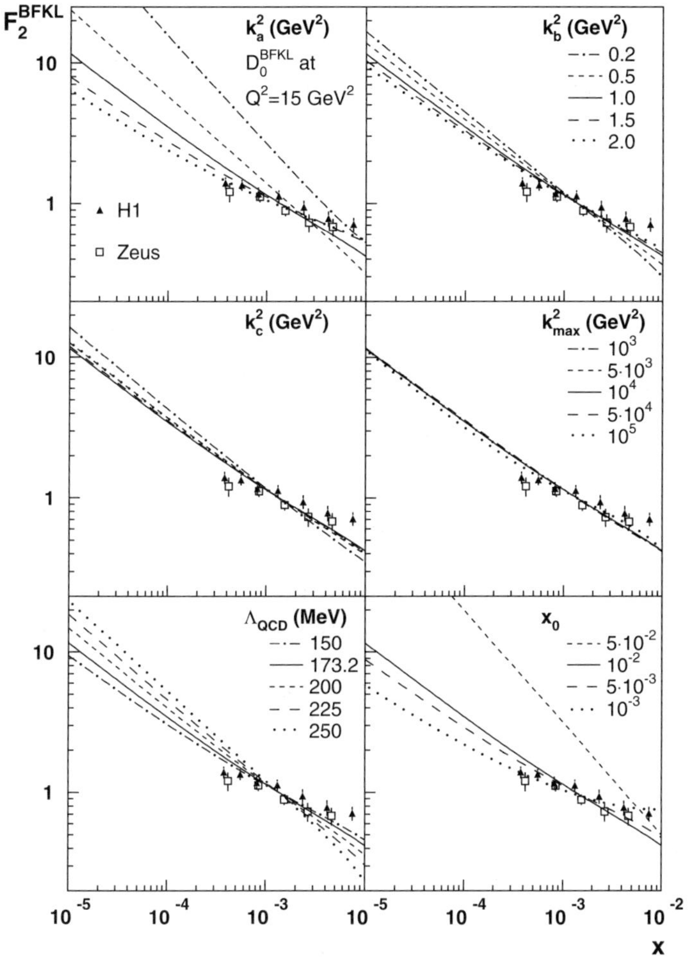


**Fig. 9.23.** BFKL predictions for the $F_2$ structure function compared to HERA data. From Bojak and Ernst (1996).

change the leading low-$x$ behavior $(x/x_0)^{-\lambda}$, but sensibly affect the normalization of the structure functions, which is therefore extremely uncertain. In Fig. 9.23 we show the results of a calculation of $F_2$, with various choices of the parameters defined above.

## 9.7 The CCFM Equation

The BFKL equation, which resums the leading $\ln(1/x)$ contributions, has a limited range of validity. This is restricted to the region $\alpha_s \ln(1/x) \sim \mathcal{O}(1)$

and $\alpha_s \ln(Q^2/\mu^2) \ll 1$, where $\mu^2 \sim 1$ GeV$^2$ sets the bondary of the non-perturbative domain (see Fig. 9.13). On the other hand, the DGLAP equation, which resums the leading $\alpha_s \ln(Q^2/\mu^2)$ contributions is not expected to be valid at very low $x$ where the $\ln(1/x)$ powers become relevant.

Clearly, it would be important to have a unified treatment of the QCD evolution of structure functions throughout the $(x, Q^2)$ plane. A theoretical framework which provides such a treatment has been developed by Ciafaloni (1988), Catani, Fiorani and Marchesini (1990a, 1990b). It leads to an evolution equation, usually called the *CCFM equation*, which reduces to the BFKL equation in the leading $\ln(1/x)$ approximation, and is equivalent to the DGLAP equation at moderate $x$ (solutions of the CCFM equation are presented by Kwieciński, Martin and Sutton 1995).

The CCFM equation is based on the coherent branching of gluons along a ladder like that shown in Fig. 9.14b. The emissions are coherent in the sense that there is an angular ordering $\theta_i > \theta_{i-1}$ along the tower, where $\theta_i$ is the angle that the $i$-th gluon forms with the original direction. Outside this kinematic region there is a destructive interference such that the multigluon contributions vanish to leading order. Referring the reader to the original papers for details, we simply recall that the CCFM equation is an integral equation for the $Q^2$-dependent unintegrated gluon density $f(x, \boldsymbol{k}^2, Q^2)$. The additional scale $Q^2$ (besides the gluon transverse momentum $\boldsymbol{k}^2$) arises from the angular ordering and specifies the maximum angle of gluon emission. At small $x$, the angular ordering does not provide any constraint on the transverse momenta along the gluon ladder, so that $f(x, \boldsymbol{k}^2, Q^2)$ becomes independent of $Q^2$, and one recovers the BFKL equation for $f(x, \boldsymbol{k}^2)$. At larger $x$, the angular ordering becomes an ordering in the gluon transverse momenta and, by integrating over $\boldsymbol{k}^2$, one reobtains the DGLAP evolution for $g(x, Q^2)$.

## 9.8 Gluon Recombination Effects

The steep rise of structure functions at $x \to 0$ predicted by BFKL must eventually stop, in order not to violate unitarity. One expects that, at extremely small values of $x$, the probability of interactions between partons (gluons, in particular) becomes so large that they begin to recombine with each other. This is a shadowing correction, proportional to the square of the gluon distribution per unit transverse area, i.e. to

$$\text{gluon fusion} \propto \frac{[x\, g(x, Q^2)]^2}{R^2} , \tag{9.222}$$

where $R$ is the size of the region inside the proton where the gluons are concentrated.

Resumming a special class of diagrams, which incorporate gluon recombination (the so-called "fan" diagrams), the BFKL equation turns out to be

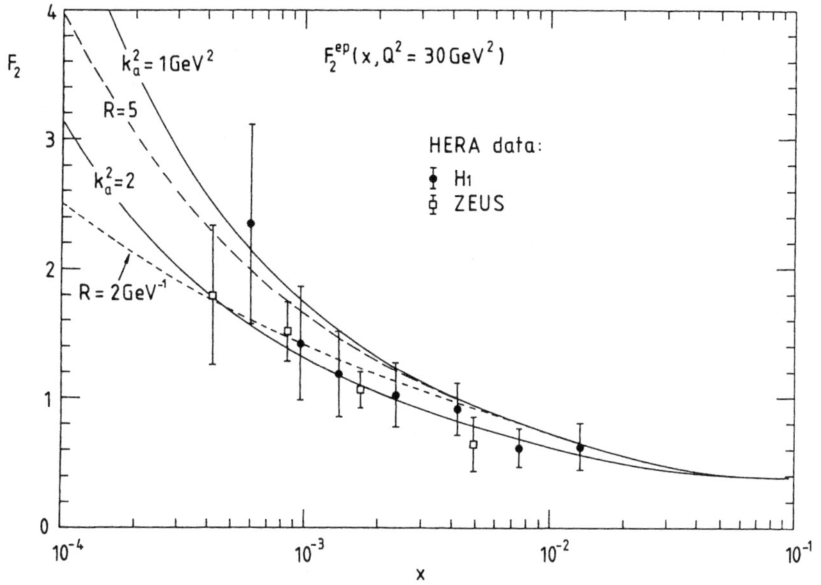


**Fig. 9.24.** BFKL predictions for $F_2(x, Q^2)$. The dashed curves show the suppression caused by shadowing corrections for different values of $R$. From Askew et al. 1994.

modified by a negative non-linear terms, and becomes (Gribov, Levin and Ryskin 1983)

$$\frac{\partial f(x, \boldsymbol{k}^2)}{\partial \ln(1/x)} = K \otimes f - \frac{C}{\boldsymbol{k}^2 R^2} \alpha_s^2(\boldsymbol{k}^2) \left[x\, g(x, \boldsymbol{k}^2)\right]^2 , \qquad (9.223)$$

where $K \otimes f$ is the r.h.s. of (9.204) and $C$ is a constant ($C = 81/16$). The crucial (and unknown) parameter in the GLR equation (9.223) is the radius $R$. If $R$ is small, i.e., if the gluons are concentrated in "hot spots" inside the proton, the effects of shadowing might be visible even at the presently accessible values of $x$. If, on the contrary, $R$ is large, of the order of the proton size, non-linear effects are pushed down to the $x < 10^{-4}$ region. An evaluation of gluon fusion effects on structure functions is shown in Fig. 9.24. At the moment, there is no evidence for non-linear effects in the evolution of structure functions at HERA.

## 9.9 The Color Dipole Picture of DIS

We present now a different approach to DIS, which exhibits some interesting features: *i)* it provides an intuitive quantum mechanical picture of low-$x$ DIS; *ii)* it establishes a bridge between low-$x$ DIS and diffraction; *iii)* it will allow us to apply to DIS the unitarization procedures of $s$-channel models. We are talking about the so-called *color dipole picture* of DIS[2] (Nikolaev and Zakharov 1991b, 1992, 1994b; Mueller 1994). The equivalence between this picture and the QCD factorization schemes discussed earlier in this chapter has been verified to leading order.

The color dipole approach describes low-$x$ DIS in the proton rest frame. In this frame, when $x \to 0$, the virtual photon transforms into a quark-antiquark pair at very large distances upstream the target. Then, after quite a long time, the $q\bar{q}$ pair scatters off the proton. Since the interaction time is much shorter than the formation time of the pair, the transverse size of the $q\bar{q}$ dipole is approximately frozen during the scattering process. Let us see how this happens.

In the target rest frame the photon momentum is

$$q = (\nu, 0, 0, \sqrt{\nu^2 + Q^2}) \, . \tag{9.224}$$

Using light-cone variables this reads

$$q = \left( q^+, -\frac{Q^2}{2q^+}, \mathbf{0} \right) \, , \tag{9.225}$$

with $q^+ \simeq \sqrt{2}\,\nu$ in the Bjorken limit. If we call $\kappa$ ($\kappa'$) the momentum of the quark (antiquark) of the pair, we have

$$\kappa = \left( zq^+, \frac{\boldsymbol{\kappa}^2}{2zq^+}, \boldsymbol{\kappa} \right) \, , \tag{9.226}$$

$$\kappa' = \left( (1-z)q^+, \frac{\boldsymbol{\kappa}^2}{2\,(1-z)q^+}, -\boldsymbol{\kappa} \right) \, , \tag{9.227}$$

where $z$ is the fraction of the light-cone momentum of the photon carried by the quark, and we put the quark and the antiquark on mass shell. The invariant mass squared of the pair is

$$M^2 = (\kappa + \kappa')^2 = \frac{\boldsymbol{\kappa}^2}{z(1-z)} \, . \tag{9.228}$$

To evaluate the formation time of the pair $\tau_f$ we compute the energy difference $\Delta E$ between the $q\bar{q}$ pair and the virtual photon. By the uncertainty principle it will be $\tau_f \sim 1/\Delta E$. From (9.226–9.227) the energy of the pair turns out to be

$$E_{\text{pair}} = \frac{1}{\sqrt{2}} \left( q^+ + \frac{\boldsymbol{\kappa}^2}{2z\,(1-z)q^+} \right) \, , \tag{9.229}$$

whereas the energy of the photon is

$$E_{\gamma^*} = \frac{1}{\sqrt{2}} \left( q^+ - \frac{Q^2}{2q^+} \right) \, . \tag{9.230}$$

---

[2] A similar approach was used by Bjorken, Kogut and Soper (1971) to calculate dimuon electroproduction off an external electromagnetic field.

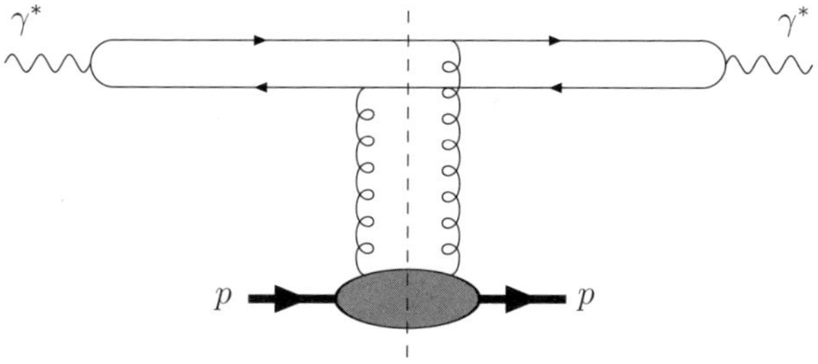


**Fig. 9.25.** DIS at low $x$: the virtual photon transforms into a $q\bar{q}$ pair at large distances from the target.

Thus we get

$$\Delta E = E_{\text{pair}} - E_{\gamma^*} = \frac{1}{2\sqrt{2}\,q^+}\left(Q^2 + \frac{\boldsymbol{\kappa}^2}{z\,(1-z)}\right) , \tag{9.231}$$

that is

$$\Delta E \simeq \frac{Q^2}{\sqrt{2}\,q^+} = m_N x , \tag{9.232}$$

to the extent that $\boldsymbol{\kappa}^2 \lesssim z\,(1-z)\,Q^2$, that is $M^2 \lesssim Q^2$ (see below).

The formation time of the $q\bar{q}$ pair is therefore $\tau_f \sim 1/\Delta E \sim 1/m_N x$, and, at very low $x$, is much larger than the typical interaction time $\tau_{\text{int}} \sim R_p$, where $R_p$ is the radius of the proton. (Equivalently, we can say that the pair travels a long distance $\ell \sim 1/m_N x$ before scattering off the target). Consequently, the transverse size of the pair is frozen during the interaction with the proton, and we can interpret low-$x$ DIS as the scattering of a fixed-size $q\bar{q}$ color dipole off a nucleon.

We can make this picture more quantitative as follows. We start from the cross sections (9.193), with fixed $\alpha_s$ and $f(\beta, \boldsymbol{k}^2) \simeq f(x, \boldsymbol{k}^2)$, and we rewrite them in the impact parameter representation, by resorting to the following identities

$$\begin{aligned} &\int \frac{\mathrm{d}^2\boldsymbol{\kappa}}{(\boldsymbol{\kappa}^2+\varepsilon^2)\,[(\boldsymbol{\kappa}+\boldsymbol{k})^2+\varepsilon^2]} \\ &\quad = \frac{1}{(2\pi)^2}\int \mathrm{d}^2\boldsymbol{\kappa}_1 \int \mathrm{d}^2\boldsymbol{\kappa}_2 \int \mathrm{d}^2\boldsymbol{\rho}\, \frac{\mathrm{e}^{\mathrm{i}\,\boldsymbol{\rho}\cdot(\boldsymbol{\kappa}_1+\boldsymbol{k}) - \mathrm{i}\,\boldsymbol{\rho}\cdot\boldsymbol{\kappa}_2}}{(\boldsymbol{\kappa}_1^2+\varepsilon^2)\,(\boldsymbol{\kappa}_2^2+\varepsilon^2)} \\ &\quad = \int \mathrm{d}^2\boldsymbol{\rho}\,\mathrm{e}^{\mathrm{i}\boldsymbol{\rho}\cdot\boldsymbol{k}} \left| \frac{1}{2\pi}\int \mathrm{d}^2\boldsymbol{\kappa}\,\frac{\mathrm{e}^{\mathrm{i}\,\boldsymbol{\kappa}\cdot\boldsymbol{\rho}}}{\boldsymbol{\kappa}^2+\varepsilon^2}\right|^2 , \end{aligned} \tag{9.233}$$

and ($\boldsymbol{\nabla} \equiv \partial/\partial\boldsymbol{\rho}$)

$$\begin{aligned} &\int \mathrm{d}^2\boldsymbol{\kappa}\,\frac{\boldsymbol{\kappa}^2+\boldsymbol{\kappa}\cdot\boldsymbol{k}}{(\boldsymbol{\kappa}^2+\varepsilon^2)\,[(\boldsymbol{\kappa}+\boldsymbol{k})^2+\varepsilon^2]} \\ &\quad = \frac{1}{(2\pi)^2}\int \mathrm{d}^2\boldsymbol{\kappa}_1 \int \mathrm{d}^2\boldsymbol{\kappa}_2 \int \mathrm{d}^2\boldsymbol{\rho}\, \frac{(\boldsymbol{\nabla}\mathrm{e}^{\mathrm{i}\,\boldsymbol{\rho}\cdot(\boldsymbol{\kappa}_1+\boldsymbol{k})})\,(\boldsymbol{\nabla}\mathrm{e}^{-\mathrm{i}\,\boldsymbol{\rho}\cdot\boldsymbol{\kappa}_2})}{(\boldsymbol{\kappa}_1^2+\varepsilon^2)\,(\boldsymbol{\kappa}_2^2+\varepsilon^2)} . \end{aligned} \tag{9.234}$$

Substituting (9.233) and (9.234) in (9.193), performing the angular integrations, and making use of the integral (Abramowitz and Stegun 1965, p. 488)

$$\int_0^\infty \mathrm{d}t \, \frac{t^{\nu+1} \, J_\nu(at)}{t^2+y^2} = b^\nu \, K_\nu(ay) \, , \tag{9.235}$$

the virtual photoabsorption cross sections take the form

$$\sigma_{L,T}^{\gamma^* p}(x, Q^2) = \int_0^1 \mathrm{d}z \int \mathrm{d}^2\boldsymbol{\rho} \, |\Psi_{L,T}(z,\rho)|^2 \, \sigma(x,\rho) \, , \tag{9.236}$$

where $\sigma(x,\rho)$ is the total cross section of interaction of a $q\bar{q}$ pair of transverse size $\rho$ with the proton, given by (Nikolaev and Zakharov 1991; Barone et al. 1994)

$$\sigma(x,\rho) = \frac{4\pi}{3} \int \frac{\mathrm{d}^2\boldsymbol{k}}{\boldsymbol{k}^4} \, \alpha_s \, f(x, \boldsymbol{k}^2) \left(1 - \mathrm{e}^{i\,\boldsymbol{k}\cdot\boldsymbol{\rho}}\right) \, . \tag{9.237}$$

$\Psi_{L,T}$ are the wave functions of the $q\bar{q}$ fluctuations of virtual longitudinal and transverse photons. They read

$$|\Psi_L(z,\rho)|^2 = \frac{6\alpha_{\mathrm{em}}}{(2\pi)^2} \sum_q 4\, e_q^2 \, Q^2 \, z^2(1-z)^2 \, K_0^2(\varepsilon\rho) \, , \tag{9.238}$$

$$|\Psi_T(z,\rho)|^2 = \frac{6\alpha_{\mathrm{em}}}{(2\pi)^2} \sum_q e_q^2 \, [z^2 + (1-z)^2] \, \varepsilon^2 \, K_1^2(\varepsilon\rho) \, , \tag{9.239}$$

where $\varepsilon^2 = Q^2 z(1-z)$. Comparing (9.237) with (9.152) we find that the photon impact factors $\Phi_{L,T}$ are related to the wave functions $\Psi_{L,T}$ by

$$\Phi_\lambda = \frac{8\pi^2\alpha_s}{3} \int \mathrm{d}z \int \mathrm{d}^2\boldsymbol{\rho} \, |\Psi_\lambda|^2 \, (1 - \mathrm{e}^{\mathrm{i}\boldsymbol{k}\cdot\boldsymbol{\rho}}) \, . \tag{9.240}$$

The physical interpretation of (9.236) is the one we anticipated above: in the low-$x$ limit, DIS is due to the interaction of a $q\bar{q}$ dipole of fixed size with the proton.

We now discuss some important properties of $\sigma(x,\rho)$ and $\Psi_{L,T}(z,\rho)$. By expanding the exponential in (9.237), one finds

$$\sigma(x,\rho) \sim \rho^2 \quad \text{at small } \rho \, . \tag{9.241}$$

Thus, small-size pairs interact very little with the proton (this property is called *color transparency*, since it implies that nuclear matter is nearly transparent for small-size pairs, see Sect. 11.9.3). A more accurate evaluation of $\sigma(x,\rho)$ gives, at small $\rho$ (Nikolaev and Zakharov 1991b)

$$\sigma(x,\rho) \sim \rho^2 \, \alpha_s(\rho) \, \ln\left(\ln \frac{\rho^2}{\rho_0^2}\right) \, . \tag{9.242}$$

Once inserted into (9.236), this behavior leads to a transverse cross section

$$\sigma_T^{\gamma^* p} \sim \frac{1}{Q^2} \ln^2 \left( \ln \frac{Q^2}{\Lambda^2} \right) , \tag{9.243}$$

which signals the equivalence, in the $\rho \to 0$ limit, of the color dipole picture to the leading $\ln Q^2$ approximation of QCD. At large $\rho$, on the other hand, due to confinement, the cross section $\sigma(x, \rho)$ saturates at some typical hadron-nucleon cross section $\sigma_0$

$$\sigma(x, \rho) \sim \sigma_0 \quad \text{for large } \rho . \tag{9.244}$$

An important relation between the dipole cross section and the gluon density is obtained by rewriting (9.237) as

$$\sigma(x, \rho) = \frac{\pi}{3} \rho^2 \int \frac{\mathrm{d}^2 \boldsymbol{k}}{\boldsymbol{k}^2} \alpha_s \frac{4\,[1 - J_0(k\rho)]}{(k\rho)^2} \frac{\partial\, xg(x, \boldsymbol{k}^2)}{\partial \ln \boldsymbol{k}^2} . \tag{9.245}$$

Since $4\,[1 - J_0(\xi^2)]/\xi^2$ is approximated by the step function $\theta(A - \ln \xi^2)$ with $A \approx 10$, Eq. (9.245) gives (Barone et al. 1993, 1994; Nikolaev and Zakharov 1994b)

$$\sigma(x, \rho) \sim \frac{\pi^2}{3} \rho^2 \alpha_s(\rho)\, xg(x, A/\rho^2) . \tag{9.246}$$

The photon wave functions are not normalized to unity. Actually, for transverse photons, the normalization integral $\int \mathrm{d}z \int \mathrm{d}^2\boldsymbol{\rho}\, |\Psi_T(z, \rho)|^2$ diverges logarithmically in the ultraviolet limit, i.e. in the limit of small distances (recall that $K_1(y) \sim 1/y$ as $y \to 0$). This causes no problem since $\sigma(x, \rho) \sim \rho^2$ for small $\rho$. Therefore, the physical cross sections are finite.

In order to understand the scaling properties of the cross sections, let us look in greater detail at the structure of (9.236).

Since $K_0(y)$ and $K_1(y)$ fall exponentially at large $y$, the dominant contribution to $\sigma_{L,T}^{\gamma^* p}$ comes from $q\bar{q}$ pairs of size

$$\rho^2 \sim \frac{1}{\varepsilon^2} = \frac{1}{Q^2\, z(1 - z)} . \tag{9.247}$$

(Note that, since $\rho^2 \sim 1/\boldsymbol{\kappa}^2$, Eqs. (9.247) and (9.228) imply $M^2 \sim Q^2$).

Thus, *asymmetric* pairs, i.e. pairs with $z \approx 0$ or $z \approx 1$, in which one of the parton carries most of the momentum (the so-called "aligned-jet" configuration), have large sizes $\rho \gtrsim R \gg 1/Q$, where $R \sim 1$ fm is a typical confinement radius (Fig. 9.26a).

On the contrary, *symmetric* pairs, i.e. pairs with $z \approx 1/2$, in which the quark and the antiquark carry an equal fraction of the longitudinal momentum, have small sizes $\rho \lesssim 1/Q$ (Fig. 9.26b).

Consider the contribution of small–size, symmetric, pairs ($z \sim 1/2$, $\rho \sim 1/Q$) to the $\gamma^* p$ cross sections. Due to their exponential behavior, the modified Bessel functions $K_0$ and $K_1$ introduce an effective cutoff in the integral over $\rho$. To get a rough estimate of (9.236), we can approximate $K_0$

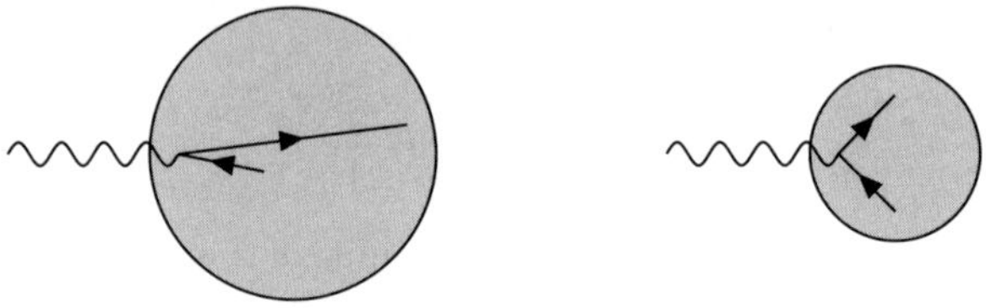

(a) (b)

**Fig. 9.26.** Large-size asymmetric pairs (**a**) and small-size symmetric pairs (**b**).

and $K_1$ by theta functions. In particular, we set $K_0(\varepsilon\rho) \sim \theta(1-\varepsilon\rho)$ and $K_1(\varepsilon\rho) \sim \theta(1-\varepsilon\rho)/(\varepsilon\rho)$. Thus, we get for the transverse cross section, using (9.236) and (9.239)

$$\sigma_T^{\gamma^* p} \sim \int \mathrm{d}z\,[z^2+(1-z)^2] \int_0^{1/Q^2} \mathrm{d}\rho^2\, \frac{\sigma(\rho)}{\rho^2} \sim \int \mathrm{d}z\,[z^2+(1-z)^2]\, \frac{1}{Q^2} \sim \frac{1}{Q^2}\,. \tag{9.248}$$

For the longitudinal cross section, from (9.236) and (9.238), we obtain

$$\sigma_L^{\gamma^* p} \sim Q^2 \int \mathrm{d}z\, z^2(1-z)^2 \int_0^{1/Q^2} \mathrm{d}\rho^2\, \sigma(\rho) \sim Q^2 \int \mathrm{d}z\, z^2(1-z)^2\, \frac{1}{Q^4} \sim \frac{1}{Q^2}\,. \tag{9.249}$$

We conclude that small–size pairs give a $1/Q^2$ contribution to both transverse and longitudinal cross sections, and therefore lead to a scaling behavior for the transverse and longitudinal structure functions.

Consider now the large–size, asymmetric, pairs ($z \sim \mu^2/Q^2$ and $\rho \sim 1/\mu$, where $\mu \sim 1/R$). In this case we find for the transverse cross section

$$\sigma_T^{\gamma^* p} \sim \int \mathrm{d}z\,[z^2+(1-z)^2] \int_0^{1/\mu^2} \mathrm{d}\rho^2\, \frac{\sigma(\rho)}{\rho^2} \sim \int \mathrm{d}z\,[z^2+(1-z)^2]\, \frac{1}{\mu^2} \sim \frac{\mu^2}{Q^2}\,\frac{1}{\mu^2} = \frac{1}{Q^2}\,, \tag{9.250}$$

and for the longitudinal cross section

$$\sigma_L^{\gamma^* p} \sim Q^2 \int \mathrm{d}z\, z^2(1-z)^2 \int_0^{1/\mu^2} \mathrm{d}\rho^2\, \sigma(\rho) \sim Q^2 \int \mathrm{d}z\, z^2(1-z)^2\, \frac{1}{\mu^4} \sim Q^2\,\frac{\mu^6}{Q^6}\,\frac{1}{\mu^4} = \frac{\mu^2}{Q^4}\,. \tag{9.251}$$

We see that large–size pairs behave quite differently from small–size ones. The aligned-jet configurations give a scaling contribution to $\sigma_T^{\gamma^* p}$, but their contribution to $\sigma_L^{\gamma^* p}$ is suppressed by a factor $1/Q^2$: the longitudinal cross section is dominated by small dipoles.

By modeling the dipole cross section $\sigma(x,\rho)$ with the constraints (9.241, 9.244), it is possible to describe the DIS data at low $x$. Golec-Biernat and Wüsthoff (1999a) obtained a good fit with the form

$$\sigma(x,\rho) = \sigma_0 \left\{ 1 - \exp\left[ -\frac{\rho^2}{4R_0^2(x)} \right] \right\} , \tag{9.252}$$

where the $x$-dependent radius $R_0(x)$ is

$$R_0(x) = \frac{1}{Q_0} \left( \frac{x}{x_0} \right)^{\lambda/2} , \tag{9.253}$$

and $\sigma_0, x_0, \lambda$ are free parameters ($Q_0 = 1$ GeV is introduced for dimensional reasons). The cross section (9.252) saturates for $\rho \gtrsim 2R_0$. At low $Q^2$, DIS is dominated by large dipoles – recall (9.247) – and we are in the saturation regime, where (9.252) tends to $\sigma_0$. In the opposite limit of large $Q^2$, the dominant contribution comes from small dipole configurations with $\rho \sim (1/Q) \ll R_0$, and $\sigma(x,\rho) \propto \rho^2$. It is important to note that, due to the $x$-dependence of $R_0$, at very low $x$, saturation effects become relevant for $Q^2$ values not so small ($\sim 1-2$ GeV$^2$ at HERA energies). The critical line defined by

$$R_0^2(x) = \frac{1}{Q^2} \tag{9.254}$$

divides the $(x,Q^2)$ plane into two regions: the scaling region, where $\sigma(x,\rho) \sim \rho^2$ and $\sigma_T^{\gamma^* p} \sim 1/Q^2$, and the saturation region, where $\sigma(x,\rho) \sim \sigma_0$ and $\sigma_T^{\gamma^* p}$ is approximately $Q^2$-independent. The quality of the fit based on (9.252) is shown in Fig. 9.27 (Golec-Biernat and Wüsthoff 1999a).

Using (9.237) one can easily check that the dipole cross section (9.252) is reproduced by the following expression of the unintegrated gluon density

$$f(x,\boldsymbol{k}^2) = \frac{3\sigma_0}{4\pi^2\alpha_s} R_0^2(x)\, \boldsymbol{k}^4 \exp\left[ -R_0^2(x)\boldsymbol{k}^2 \right] . \tag{9.255}$$

At large $Q^2$ the gluon distribution $xg(x,Q^2)$ is obtained by integrating (9.255). We find

$$\begin{aligned} xg(x,Q^2) &= \int_0^{Q^2} \frac{\mathrm{d}\boldsymbol{k}^2}{\boldsymbol{k}^2} f(x,\boldsymbol{k}^2) \\ &= \frac{3}{4\pi^2\alpha_s} \frac{\sigma_0}{R_0^2(x)} \left[ 1 - (1 + Q^2R_0^2)\, \mathrm{e}^{-R_0^2 Q^2} \right] \\ &\simeq \frac{3}{4\pi^2\alpha_s} \frac{\sigma_0}{R_0^2(x)} = \frac{3\sigma_0 Q_0^2}{4\pi^2\alpha_s} \left( \frac{x_0}{x} \right)^{\lambda} , \quad \text{for } Q^2R_0^2 \gg 1 , \end{aligned} \tag{9.256}$$

that is a power-law behavior with the exponent $\lambda$. In the fit of Golec-Biernat and Wüsthoff (1999a) the $\lambda$ parameter is found to be 0.288.

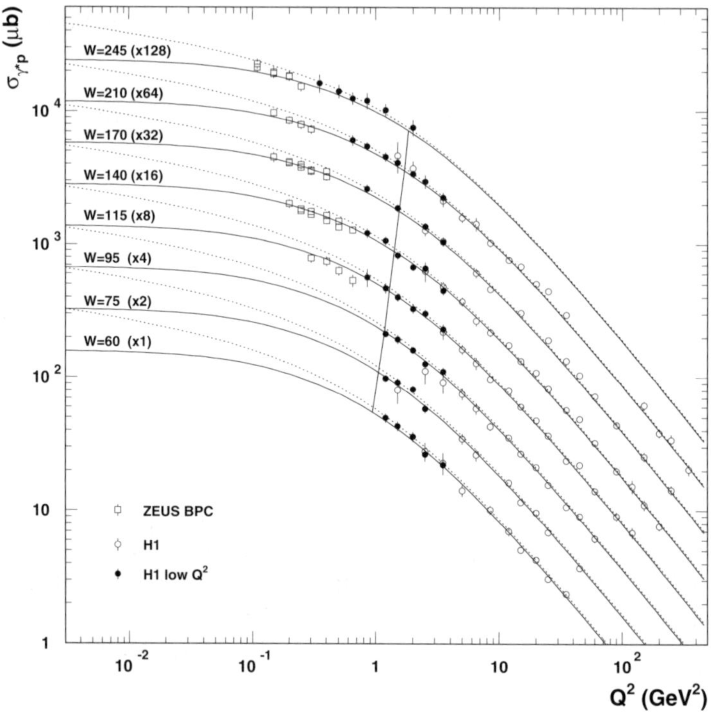


**Fig. 9.27.** The $\gamma^* p$ cross section fitted in the saturation model of Golec-Biernat and Wüsthoff (1999a). Solid lines: light-quark mass of 140 MeV; dotted lines: zero quark mass.

## 9.10 The BFKL Equation in the Color Dipole Formalism

The BFKL equation can be derived in a very enlightening way in the color dipole formalism, as shown by Nikolaev and Zakharov (1992, 1994b) and Mueller (1994) (see also Nikolaev, Zakharov and Zoller 1994a, 1994b; Mueller and Patel 1994). The idea is that the emission of soft gluons by an initial quark–antiquark pair[3] gives rise to a cascade of secondary dipoles which build up the BFKL gluonic tower. This is incorporated into the wave function of a state containing the primary $q\bar{q}$ pair and an arbitrary number of soft gluons. Symbolically, the equivalence of this approach to the traditional BFKL formalism can be represented, for the specific case of $\gamma^* p$ scattering, as

$$\Phi_{\gamma^*}^{q\bar{q}} \otimes F \otimes \Phi_p \leftrightarrow \Phi_{\gamma^*}^{q\bar{q}gg\cdots} \otimes \Phi_p \,, \tag{9.257}$$

where the $\Phi$'s are the impact factors, and $\otimes$ denotes a convolution over transverse momenta.

[3] In literature this quark–antiquark state is often called *onium* (Mueller 1994).

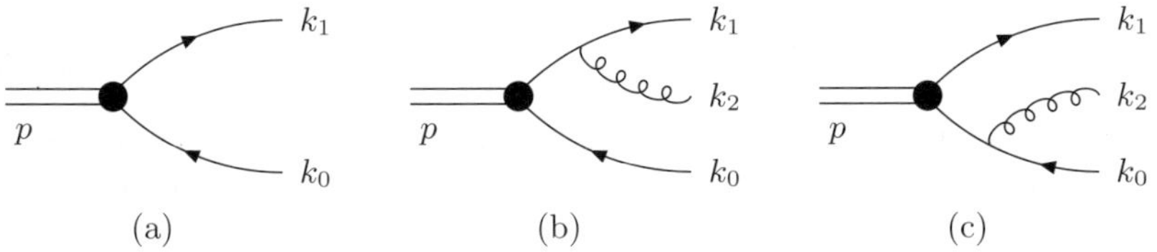


**Fig. 9.28.** A $q\bar{q}$ bound state (**a**). Emission of a soft gluon from the quark (**b**) and from the antiquark (**c**) of the pair.

We shall now derive the BFKL equation for the unintegrated gluon distribution following the approach of Mueller (1994). The Nikolaev–Zakharov version of the BFKL equation will be presented in detail in Chap. 11.10.2.

Let us start from the wave function $\Psi^{(0)}(k_1)$ of a $q\bar{q}$ bound state (Fig. 9.28a). We call $k_1$ and $k_0$ the momenta of the quark and of the antiquark, respectively. The momentum of the quarkonium is $p = k_0 + k_1$. For simplicity we omit spinor, polarization and color indices. Introducing the Sudakov vectors $p$ and $n$ (the first of which can be identified with the momentum of the dipole in the limit $p^+ \to \infty$), and imposing the mass shell condition, $k_1$ can be parametrized as

$$k_1 = z_1 p + \frac{\boldsymbol{k}_1^2}{2z_1} n + k_{1\perp} \,, \tag{9.258}$$

It will be convenient to work in the impact parameter space, where the dipole wave function is defined according to

$$\Psi^{(0)}(z_1, \boldsymbol{b}_1) = \int \frac{\mathrm{d}^2\boldsymbol{k}_1}{(2\pi)^2} \, \mathrm{e}^{\mathrm{i}\,\boldsymbol{k}_1\cdot\boldsymbol{b}_1} \, \Psi^{(0)}(z_1, \boldsymbol{k}_1) \tag{9.259}$$

The normalization is

$$\int \frac{\mathrm{d}^2\boldsymbol{k}_1}{(2\pi)^2} \int_0^1 \mathrm{d}z_1 \, \Xi(z_1, \boldsymbol{k}_1) = \int \mathrm{d}^2\boldsymbol{b}_1 \int_0^1 \mathrm{d}z_1 \, \Xi^{(0)}(z_1, \boldsymbol{b}_1) = 1 \,, \tag{9.260}$$

where (a sum over spinor indices is understood)

$$\Xi^{(0)}(z_1, \boldsymbol{k}_1) = |\Psi^{(0)}(z_1, \boldsymbol{k}_1)|^2 \,, \quad \Xi^{(0)}(z_1, \boldsymbol{b}_1) = |\Psi^{(0)}(z_1, \boldsymbol{b}_1)|^2 \,. \tag{9.261}$$

What makes the impact parameter picture useful, and much simpler than the description in terms of momenta, is the fact, already discussed in Sect. 9.9, that the transverse size of a dipole is frozen during the time of emission of a soft gluon.

Suppose now that the quark, or the antiquark of the pair, emits a soft gluon 2 (Fig. 9.28b,c). The momentum $k_2$ of this gluon is parametrized as

$$k_2 = z_2 p + \frac{\boldsymbol{k}_2^2}{2z_2} n + k_{2\perp} \,, \tag{9.262}$$

and "soft" means

$$z_2 \ll z_1, 1 - z_1 \ . \tag{9.263}$$

The $q\bar{q}g$ wave function $\Psi^{(1)}(k_1, k_2)$ (the superscript is the number of soft gluons) can be computed from the diagrams of Fig. 9.28b,c. Using eikonal couplings, we get

$$\Psi^{(1)}(k_1, k_2) = -\mathrm{i}\, g_s\, t^a \left[ \Psi^{(0)}(k_1 + k_2) \frac{k_1 \cdot \varepsilon_2}{k_1 \cdot k_2} - \Psi^{(0)}(k_1) \frac{k_0 \cdot \varepsilon_2}{k_0 \cdot k_2} \right] , \tag{9.264}$$

where $\varepsilon_2$ is the polarization vector of the gluon, that we can write as

$$\varepsilon_2 = \frac{\boldsymbol{\varepsilon}_2 \cdot \boldsymbol{k}_2}{x_2} n + \boldsymbol{\varepsilon}_2 \ . \tag{9.265}$$

In the limit (9.263) we have

$$k_1 \cdot \varepsilon_2 \simeq \frac{x_1}{x_2} (\boldsymbol{\varepsilon}_2 \cdot \boldsymbol{k}_2) \, , \quad k_1 \cdot k_2 \simeq \frac{x_1\, \boldsymbol{k}_2^2}{2\, x_2} \, , \tag{9.266}$$

and the $q\bar{q}g$ wave function (9.264) becomes

$$\begin{aligned} &\Psi^{(1)}(x_1, \boldsymbol{k}_1, x_2, \boldsymbol{k}_2) \\ &\quad = -2\,\mathrm{i}\, g_s\, t^a \left[ \Psi^{(0)}(x_1, \boldsymbol{k}_1 + \boldsymbol{k}_2) - \Psi^{(0)}(x_1, \boldsymbol{k}_1) \right] \frac{\boldsymbol{k}_2 \cdot \boldsymbol{\varepsilon}_2}{\boldsymbol{k}_2^2} \ . \end{aligned} \tag{9.267}$$

In the impact parameter space we define

$$\begin{aligned} &\Psi^{(1)}(x_1, \boldsymbol{b}_1, x_2, \boldsymbol{b}_2) \\ &\quad = \int \frac{\mathrm{d}^2 \boldsymbol{k}_1}{(2\pi)^2} \int \frac{\mathrm{d}^2 \boldsymbol{k}_2}{(2\pi)^2}\, \mathrm{e}^{\mathrm{i}\,(\boldsymbol{k}_1 \cdot \boldsymbol{b}_1 + \boldsymbol{k}_2 \cdot \boldsymbol{b}_2)}\, \Psi^{(1)}(x_1, \boldsymbol{k}_1, x_2, \boldsymbol{k}_2) \ . \end{aligned} \tag{9.268}$$

If we insert (9.267) into (9.258), and use the identity

$$\int \frac{\mathrm{d}^2 \boldsymbol{k}}{(2\pi)^2}\, \mathrm{e}^{\mathrm{i}\, \boldsymbol{k} \cdot \boldsymbol{b}} \frac{k^i}{\boldsymbol{k}^2} = -\frac{1}{2\pi \mathrm{i}} \frac{b^i}{\boldsymbol{b}^2} \, , \tag{9.269}$$

we find

$$\Psi^{(1)}(x_1, \boldsymbol{b}_1, x_2, \boldsymbol{b}_2) = \frac{g_s}{\pi}\, t^a \left( \frac{\boldsymbol{b}_{21} \cdot \boldsymbol{\varepsilon}_2}{\boldsymbol{b}_{21}^2} - \frac{\boldsymbol{b}_{20} \cdot \boldsymbol{\varepsilon}_2}{\boldsymbol{b}_{20}^2} \right) \Psi^{(0)}(x_1, \boldsymbol{b}_1) \, , \tag{9.270}$$

where $\boldsymbol{b}_{ij} \equiv \boldsymbol{b}_i - \boldsymbol{b}_j$, with $\boldsymbol{b}_0 = \boldsymbol{0}$. Squaring $\Psi^{(1)}$ corresponds to computing the four diagrams in Fig. 9.29. We get (a sum over colors and polarizations is understood)

$$|\Psi^{(1)}|^2 = 4\, C_F \frac{\alpha_s}{\pi} \frac{\boldsymbol{b}_{10}^2}{\boldsymbol{b}_{20}^2\, \boldsymbol{b}_{21}^2} |\Psi^{(0)}|^2 \, , \tag{9.271}$$

and

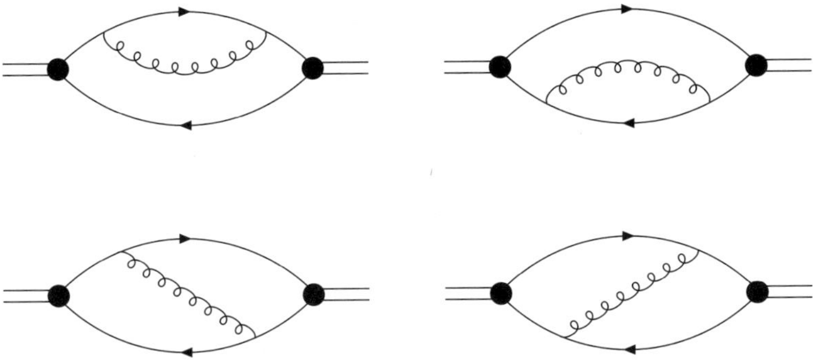

**Fig. 9.29.** Diagrams contributing to $|\Psi^{(1)}|^2$.

$$\begin{aligned}\Xi^{(1)}(z_1, \boldsymbol{b}_1) &= \int \mathrm{d}^2\boldsymbol{b}_2 \int^{z_1} \frac{\mathrm{d}z_2}{z_2} \frac{C_F \alpha_s}{\pi^2} \frac{\boldsymbol{b}_{10}^2}{\boldsymbol{b}_{20}^2\, \boldsymbol{b}_{21}^2}\, \Xi^{(0)}(z_1, \boldsymbol{b}_1) \\ &= \frac{C_F \alpha_s}{\pi^2} \ln\left(\frac{z_1}{z_0}\right) \int \mathrm{d}^2\boldsymbol{b}_2 \frac{\boldsymbol{b}_{10}^2}{\boldsymbol{b}_{20}^2\, \boldsymbol{b}_{21}^2}\, \Xi^{(0)}(z_1, \boldsymbol{b}_1)\ . \end{aligned} \tag{9.272}$$

In the $z_2$-integration we have introduced a cutoff $z_0$ which does not apper in the physical quantities.

The $\boldsymbol{b}$-dependent factor in (9.271) is obtained from

$$\frac{\boldsymbol{b}_{10}^2}{\boldsymbol{b}_{20}^2\, \boldsymbol{b}_{21}^2} = \frac{1}{\boldsymbol{b}_{21}^2} + \frac{1}{\boldsymbol{b}_{20}^2} - 2\,\frac{\boldsymbol{b}_{21}\cdot \boldsymbol{b}_{20}}{\boldsymbol{b}_{21}^2\, \boldsymbol{b}_{20}^2}\ , \tag{9.273}$$

where the first two terms come from the upper diagrams in Fig. 9.29, whereas the third, crossed, term comes from the two interference diagrams in the bottom of Fig. 9.29.

Let us go further, and consider a state with two soft gluons 2 and 3. The second gluon 3 has longitudinal momentum fraction $z_3$ and transverse coordinate $\boldsymbol{b}_3$. We assume strong ordering

$$z_3 \ll z_2 \ll z_1, 1 - z_1\ . \tag{9.274}$$

Now, 3 can be emitted in many different ways: from the quark, from the antiquark, from the first gluon, before or after the emission of the first gluon, etc. At higher orders, the number of possible processes is so high that an exact treatment of the problem becomes impossible. Things simplify considerably in the large-$N_c$ limit. In this case, gluons can be represented by $q\bar{q}$ double lines (Fig. 9.30), and the non-planar diagrams are suppressed with respect to the planar ones by powers of $1/N_c$. As a consequence, the number of graphs to be considered is strongly limited. When 3 is emitted from the original quark (1) or from the antiquark line of 2, it must be readsorbed only by the original quark or by the antiquark part of 2. The contribution of this process (pictorially shown in Fig. 9.31a) to the wave function squared $\left|\Psi^{(2)}\right|^2$ of the $q\bar{q}gg$ state is then (using $C_F \to N_c/2$, as $N_c \to \infty$)

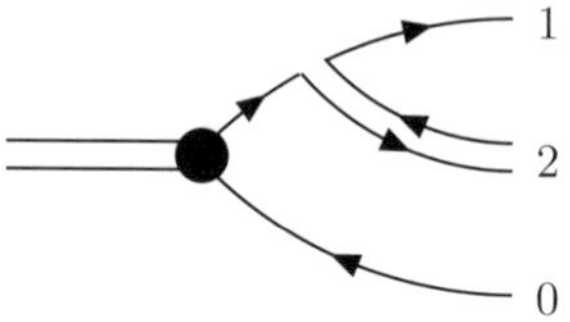


**Fig. 9.30.** Emission of a gluon as a $q\bar{q}$ pair in the large $N_c$ limit

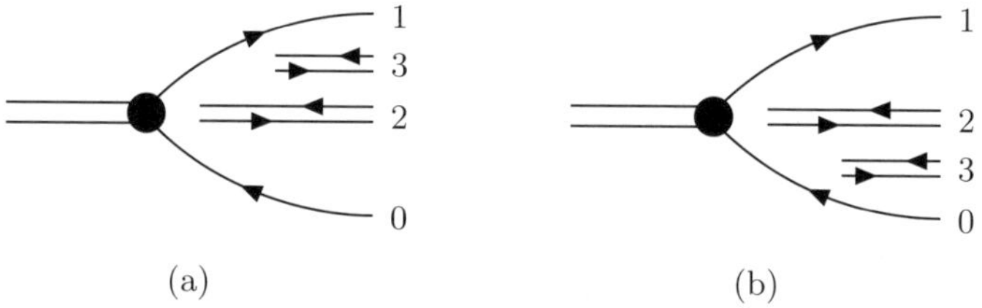


**Fig. 9.31a,b.** Emission of two gluons

$$2\,\frac{N_c\alpha_s}{\pi}\,\frac{\boldsymbol{b}_{21}^2}{\boldsymbol{b}_{31}^2\,\boldsymbol{b}_{32}^2}\,. \tag{9.275}$$

Analogously, when 3 is emitted from the original antiquark (0) or from the quark line of 2, it must be readsorbed only by the original antiquark or by the quark part of 2. The contribution of this process is

$$2\,\frac{N_c\alpha_s}{\pi}\,\frac{\boldsymbol{b}_{20}^2}{\boldsymbol{b}_{30}^2\,\boldsymbol{b}_{32}^2}\,. \tag{9.276}$$

Summing the two terms, the wave function of the $q\bar{q}gg$ state turns out to be

$$|\Psi^{(2)}|^2 = 2\,\frac{N_c\alpha_s}{\pi}\left(\frac{\boldsymbol{b}_{21}^2}{\boldsymbol{b}_{31}^2\,\boldsymbol{b}_{32}^2} + \frac{\boldsymbol{b}_{20}^2}{\boldsymbol{b}_{30}^2\,\boldsymbol{b}_{32}^2}\right)|\Psi^{(1)}|^2\,. \tag{9.277}$$

Higher-order contributions are treated in a similar way: any additional gluon is emitted by one of the dipoles corresponding to pairs of partons which are adjacent in the planar diagram.

From (9.272) we can single out the differential probability $\mathrm{d}\Pi(z_2, \boldsymbol{b}_{10}, \boldsymbol{b}_{20})$ for the emission of a gluon with longitudinal momentum fraction $z_2$ and transverse separation $\boldsymbol{b}_{20}$ from the antiquark of the parent dipole (10)

$$\mathrm{d}\Pi(z_2, \boldsymbol{b}_{10}, \boldsymbol{b}_{20}) = \frac{N_c\alpha_s}{2\pi^2}\,\frac{\boldsymbol{b}_{10}^2}{\boldsymbol{b}_{20}^2\,\boldsymbol{b}_{21}^2}\,\mathrm{d}^2\boldsymbol{b}_{20}\,\frac{\mathrm{d}z_2}{z_2}\,. \tag{9.278}$$

Introducing the density of gluons $f(x, \boldsymbol{b}_{10}^2, z_1)$ in the dipole state (10), the variation of this density with $z_1$ is determined by the emission of a gluon 2 which creates two new dipoles (20) and (21) in place of the original one. Thus we have, using (9.278)

$$\mathrm{d}f(x, z_1, \boldsymbol{b}_{10}^2) = \frac{N_c\alpha_s}{2\pi^2}\frac{\mathrm{d}z_1}{z_1}\int \mathrm{d}^2\boldsymbol{b}_{20}\,\frac{\boldsymbol{b}_{10}^2}{\boldsymbol{b}_{20}^2\,\boldsymbol{b}_{21}^2} \times \left[f(x, \boldsymbol{b}_{21}^2, z_1) + f(x, \boldsymbol{b}_{20}^2, z_1) - f(x, \boldsymbol{b}_{10}^2, z_1)\right] . \quad (9.279)$$

Invariance under longitudinal boosts implies that $f$ should be a function of $x/z_1$, so that (9.279) finally gives

$$\frac{\partial f(x, \boldsymbol{b}_{10}^2)}{\partial \ln(1/x)} = \frac{N_c\alpha_s}{2\pi^2}\int \mathrm{d}^2\boldsymbol{b}_{20}\,\frac{\boldsymbol{b}_{10}^2}{\boldsymbol{b}_{20}^2\,\boldsymbol{b}_{21}^2} \times \left[f(x, \boldsymbol{b}_{21}^2) + f(x, \boldsymbol{b}_{20}^2) - f(x, \boldsymbol{b}_{10}^2)\right] . \quad (9.280)$$

This is the BFKL equation for the unintegrated gluon distribution, written in the impact-parameter space. We can put it in a simpler form by some manipulations. The integration over the azimuthal angle of $\boldsymbol{b}_{20}$ yields

$$\int \frac{\mathrm{d}^2\boldsymbol{b}_{20}}{\boldsymbol{b}_{21}^2} = \pi\int_0^\infty \frac{\mathrm{d}\boldsymbol{b}_{20}^2}{|\boldsymbol{b}_{20}^2 - \boldsymbol{b}_{10}^2|} , \quad (9.281)$$

and with the redefinitions $\boldsymbol{b}_{10}^2 = \boldsymbol{b}^2$, $\boldsymbol{b}_{20}^2 = u\boldsymbol{b}^2$ (when $\boldsymbol{b}_{20}^2 < \boldsymbol{b}^2$), $\boldsymbol{b}_{20}^2 = \boldsymbol{b}^2/u$ (when $\boldsymbol{b}_{20}^2 > \boldsymbol{b}^2$), Eq. (9.280) becomes

$$\frac{\partial f(x, \boldsymbol{b}^2)}{\partial \ln(1/x)} = \frac{N_c\alpha_s}{2\pi^2}\int_0^1 \frac{\mathrm{d}u}{1-u} \times \left[\frac{1}{u}\,f(x, u\boldsymbol{b}^2) + f(x, \boldsymbol{b}^2/u) - 2\,f(x, \boldsymbol{b}^2)\right] . \quad (9.282)$$

The solution of this equation can be found proceeding as in Sect. 9.6 (the Mellin transform must now be taken with respect to the variable $\boldsymbol{b}^2$).

The DLLA approximation is obtained from (9.280) by imposing a strong ordering of the parton separations

$$\boldsymbol{b}_{20}^2 \ll \boldsymbol{b}_{10}^2 \quad \text{or} \quad \boldsymbol{b}_{21}^2 \ll \boldsymbol{b}_{10}^2 , \quad (9.283)$$

that is requiring that the emitted gluon be very close to the parent quark or antiquark. The reader will easily verify that in the limit (9.283), corresponding in momentum space to the strong ordering of transverse momenta, the solution of (9.280) behaves asymptotically as

$$f(x, \boldsymbol{b}^2) \sim \exp\sqrt{\frac{4N_c\alpha_s}{\pi}\ln\frac{x_0}{x}\ln\frac{\boldsymbol{b}_0^2}{\boldsymbol{b}^2}} . \quad (9.284)$$

## 9.11 Unitarization of Structure Functions in the Color Dipole Approach

We have mentioned, among the advantages of the color dipole approach, the possibility that it offers to apply the familiar $s$-channel unitarization procedures to DIS structure functions. As seen in Sect. 9.9, the virtual photoabsorption cross sections are given by quantum mechanical averages of the color dipole cross section $\sigma(x,\rho)$ in the $q\bar{q}$ Fock state of the photon. Due to the low-$x$ (i.e., large-$W^2$) powerlike growth of the gluon distribution, one expects $\sigma(x,\rho)$ to eventually violate the Froissart–Martin bound. In order to restore $s$-channel unitarity, it comes quite natural to apply the eikonal unitarization scheme to $\sigma(x,\rho)$. Let us see how this works (Barone et al. 1994).

Assuming a purely imaginary scattering amplitude, and an exponential $t$-dependence with a slope $B$, the profile function corresponding to $\sigma(x,\rho)$ is

$$\Gamma(\boldsymbol{b}) = \frac{\sigma(x,\rho)}{4\pi B} \exp\left(-\frac{\boldsymbol{b}^2}{2B}\right) . \tag{9.285}$$

The eikonal unitarization consists in replacing $\Gamma(\boldsymbol{b})$ by $\widetilde{\Gamma}(\boldsymbol{b})$ defined as

$$\widetilde{\Gamma}(\boldsymbol{b}) = 1 - \exp[-\Gamma(\boldsymbol{b})] , \tag{9.286}$$

so that the unitarized dipole cross section $\widetilde{\sigma}(x,\rho)$ is obtained as

$$\begin{aligned}\widetilde{\sigma}(x,\rho) &= 2 \int \mathrm{d}^2\boldsymbol{b}\, \widetilde{\Gamma}(\boldsymbol{b}) \\ &= 4\pi B \left[\ln(\eta(x,\rho)) + E_1(\eta(x,\rho)) + \gamma_E\right] ,\end{aligned} \tag{9.287}$$

where $E_1$ is the integral exponential function, $\gamma_E$ is the Euler–Mascheroni constant and $\eta(x,\rho) \equiv \sigma(x,\rho)/4\pi B$. The quantity $\eta(x,\rho)$ controls the effects of unitarization: for $\eta(x,\rho) \ll 1$ one has $\widetilde{\sigma}(x,\rho) \simeq \sigma(x,\rho)$, whereas for $\eta(x,\rho) \gg 1$ the unitarization suppresses the cross section, $\widetilde{\sigma}(x,\rho) \ll \sigma(x,\rho)$. It turns out that unitarization effects are important only at large $\rho$, where $\eta(x,\rho)$ exceeds unity.

Keeping only the first-order correction in the eikonal series, we get for the unitarized virtual photoabsorption cross section $\sigma^{\gamma^* N}$ ($\langle\cdot\rangle$ denotes average over the $q\bar{q}$ wave function of the photon)

$$\begin{aligned}\sigma^{\gamma^* N}(x,Q^2) &= 2\left\langle \int \mathrm{d}^2\boldsymbol{b}\, \widetilde{\Gamma}(\boldsymbol{b}) \right\rangle \\ &= 2\left\langle \int \mathrm{d}^2\boldsymbol{b}\, \Gamma(\boldsymbol{b}) \right\rangle - \left\langle \int \mathrm{d}^2\boldsymbol{b}\, \Gamma^2(\boldsymbol{b}) \right\rangle \\ &= \langle \sigma(x,\rho) \rangle - \frac{1}{16\pi B} \langle \sigma^2(x,\rho) \rangle .\end{aligned} \tag{9.288}$$

The first term in (9.288) is the non unitarized DIS cross section, while the second term represents the leading unitarity correction. As we shall see in Sect. 11.2, the latter is proportional to the cross section of diffractive dissociation of virtual photons into $q\bar{q}$ pairs. Therefore unitarity corrections to DIS structure functions are intimately related to diffractive dissociation. The color dipole approach allows one to derive such a connection in a very simple way.

# 10. Phenomenology of Hard Diffraction

Hard diffraction is a relatively new research field. With some historical approximation, we can say that it was born in 1985, when Ingelman and Schlein (1985) suggested to investigate diffractive production of high-$p_T$ jets as a way to probe the partonic structure of the pomeron (soon after, Fritzsch and Streng (1985) discussed a related process, heavy-quark diffractive production; see also Berger et al. 1987). The next important step was taken by Bjorken (1993), who proposed to use large rapidity gaps as a signature of hard hadronic processes dominated by color–singlet exchange, and gave an estimate for the fraction of events with such a final–state topology (the idea of using the rapidity-gap signature had been also suggested, some years earlier, by Dokshitzer, Troyan and Khoze 1987).

The first observation of jet production events with large rapidity gaps was reported by the UA8 Collaboration in 1988 (Bonino et al. 1988), and marks the birth of the experimental study of hard diffraction. In recent years, with the HERA and Tevatron experiments, this study has entered its mature phase and achieved its main results.

In particular, the discovery, due to the ZEUS and H1 Collaborations at HERA (Derrick et al. 1993; Ahmed et al. 1994), of a large diffractive contribution to deep inelastic scattering, which came somewhat as a surprise, considerably triggered the interest of the high-energy physics community in the "new" diffraction. The Tevatron measurements of various hard diffractive cross sections have brought further information and raised new questions.

This chapter is devoted to the phenomenology of hard diffraction. We shall cover two classes of processes: *i)* diffractive deep inelastic scattering and *ii)* hard diffractive scattering of hadrons.

## 10.1 Diffractive Deep Inelastic Scattering

In a certain fraction of deep inelastic scattering events (about 5–10 %), the target proton remains (nearly) intact. We speak, in this case, of *diffractive deep inelastic scattering* (DDIS). The process, depicted in Fig. 10.1, is a semi-inclusive reaction (we put in brackets the four-momenta)

$$l\,(\ell) + p\,(P) \to l'\,(\ell') + p'\,(P') + X\,(P_X)\,, \qquad (10.1)$$

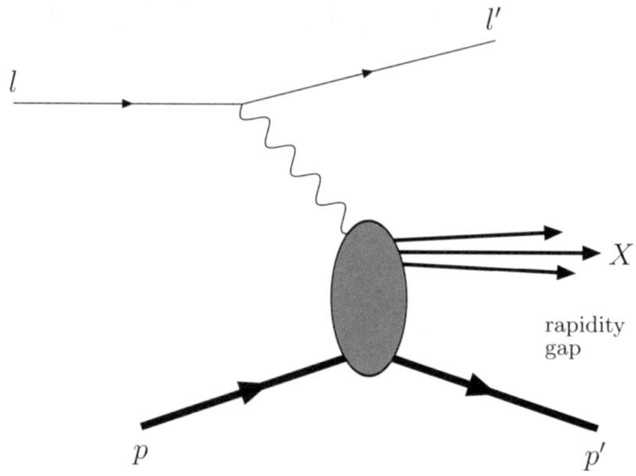


**Fig. 10.1.** Diffractive deep inelastic scattering.

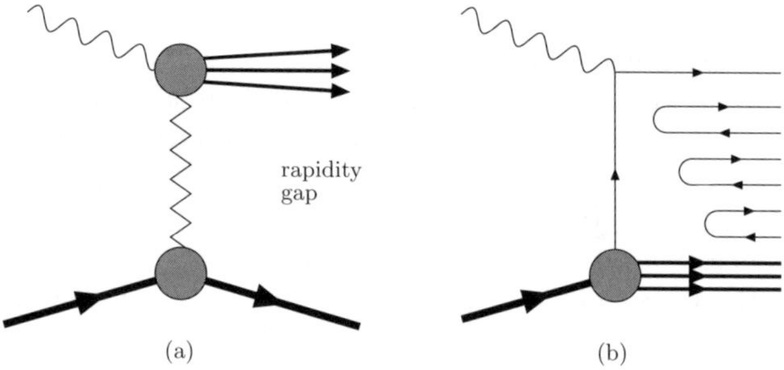


**Fig. 10.2.** Diffractive (**a**) vs. inclusive (**b**) $\gamma^* p$ scattering. Diffractive scattering is mediated by a pomeron, which carries vacuum quantum numbers and produces a rapidity gap. In the inclusive case the rapidity space is completely filled.

characterized by a particular final state configuration, wherein the presence of a rapidity gap between the scattered proton and the hadronic final state $X$ signals that no quantum numbers are exchanged between the virtual photon and the incoming proton. This justifies the term *diffractive* used for this reaction, according to the definitions of Sect. 1.1.

One can describe the situation by saying that an object, carrying the quantum numbers of vacuum, is exchanged between $\gamma^*$ and $p$. This object, represented in Fig. 10.2a by a zigzag line, is the pomeron. We should keep in mind, however, that this is only a theoretical intepretation of what one observes experimentally, namely a rapidity gap between $X$ and the outgoing proton. The hadronic system $X$ has the same quantum numbers, $J^{PC} = 1^{--}$, as the photon. In particular, when $X$ is a vector meson, one has another important diffractive process, vector meson photo- or lepto-production, that will be studied in Sect. 10.10. In Fig. 10.2 diffractive DIS is compared to inclusive DIS. In the latter case, the color string connecting the struck parton to the proton remnant produces a partonic cascade which fills the whole rapidity space.

## 10.2 Kinematics of DDIS

Diffractive DIS (and, more in general, semi-inclusive DIS) is described by 5 kinematic variables, besides the center-of-mass energy squared $s$ of the $lp$ system (Fig. 10.3). Two of them are the same variables appearing in DIS, that is the Bjorken $x$

$$x = \frac{Q^2}{2P \cdot q} = \frac{Q^2}{W^2 + Q^2 - m_N^2} \simeq \frac{Q^2}{W^2 + Q^2} \tag{10.2}$$

and the squared momentum transfer at the lepton vertex

$$Q^2 = -q^2 = -(\ell - \ell')^2 \,, \tag{10.3}$$

or, alternatively,

$$y = \frac{P \cdot q}{P \cdot \ell} = \frac{W^2 + Q^2 - m_N^2}{s - m_N^2} \simeq \frac{Q^2}{x\, s} \,. \tag{10.4}$$

where $W^2 = (P + q)^2$ is the center-of-mass energy squared of the process $\gamma^* p \to X p'$. In writing the approximate equalities in (10.2, 10.4) we neglected, as usual, the proton mass $m_N$.

The new kinematic variables, typical of diffractive DIS, are the components of the three-momentum $\boldsymbol{P}'$ of the outgoing proton. As far as unpolarized scattering is considered there is no dependence on the azimuthal angle

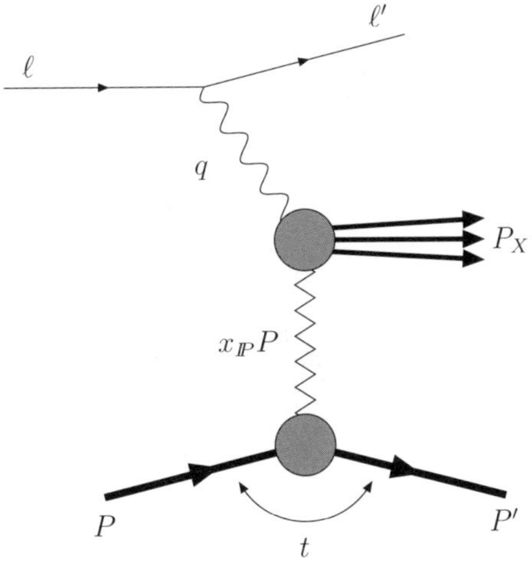


**Fig. 10.3.** Kinematics of diffractive DIS.

of $\boldsymbol{P}'$. Hence we can integrate over this angle and we are left with only two variables: $P'_z$ and $\boldsymbol{P}'^2_\perp$. It is convenient to define and use invariant quantities, such as

$$t = -(P' - P)^2 \simeq -\frac{\boldsymbol{P}'^2_\perp}{x_F} , \tag{10.5}$$

$$x_{I\!P} = \frac{(P - P') \cdot q}{P \cdot q} = \frac{M^2 + Q^2 - t}{W^2 + Q^2 - m_N^2} \simeq \frac{M^2 + Q^2}{W^2 + Q^2} = 1 - x_F , \tag{10.6}$$

where $M^2$ is the invariant mass of the $X$ system and $x_F$ is the Feynman variable (see Sect. 3.3.1). Note that in a single-diffractive process $|t|$ is typically rather small ($\sim 1$ GeV$^2$) and can safely be neglected with respect to the large energy scales of the problem ($M^2, Q^2, W^2$). The variable $x_{I\!P} = 1 - x_F$, which is sometimes called $\xi$ in literature, is the fraction of the proton longitudinal momentum carried by the pomeron.

It is customary to introduce another invariant quantity, denoted by $\beta$ and defined as

$$\beta = \frac{Q^2}{2q \cdot (P - P')} = \frac{Q^2}{M^2 + Q^2 - t} \simeq \frac{Q^2}{M^2 + Q^2} . \tag{10.7}$$

We shall see that $\beta$ can be interpreted as the momentum fraction of the struck parton inside the pomeron. Comparing (10.2), (10.6) and (10.7), one finds the following relation between $x$, $x_{I\!P}$ and $\beta$

$$x = \beta x_{I\!P} . \tag{10.8}$$

By virtue of their definitions, the variables $x$, $x_{I\!P}$ and $\beta$ have all range $[0, 1]$.

It may happen that the proton does not remain exactly intact, but produces an hadronic system $Y$ with the same quantum numbers as the proton and an invariant mass $M_Y^2$ (see Fig. 10.4). This is still a diffractive event. When two hadronic states $X$ and $Y$, separated by a rapidity gap, are produced, the process is called *double diffraction.* In particular, events with $M_Y^2 \approx 1$ GeV$^2$ are hardly distinguishable from single-diffractive events with a leading proton.

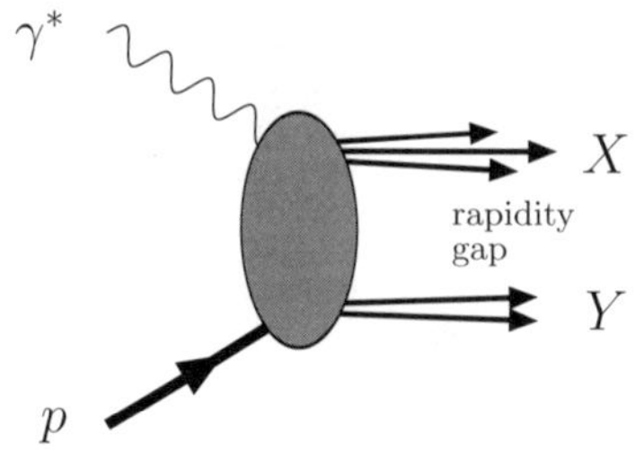


**Fig. 10.4.** Diffractive DIS with a proton remnant $Y$.

## 10.3 Diffractive Structure Functions

In perfect analogy with inclusive DIS, the diffractive DIS differential cross section can be written in terms of two structure functions, $F_1^{D(4)}$ and $F_2^{D(4)}$, which depend on four variables, $x$, $Q^2$, $x_{I\!P}$ and $t$ (this is the reason of the superscript 4). Introducing the longitudinal and transverse diffractive structure functions $F_L^{D(4)} = F_2^{D(4)} - 2x\,F_1^{D(4)}$ and $F_T^{D(4)} = 2x\,F_1^{D(4)}$, and the longitudinal–to–transverse ratio $R^{D(4)} = F_L^{D(4)}/F_T^{D(4)}$, the DDIS cross section reads (note the analogy with Eq. (9.17))

$$\frac{\mathrm{d}\sigma^D_{\gamma^* p}}{\mathrm{d}x\,\mathrm{d}Q^2\,\mathrm{d}x_{I\!P}\,\mathrm{d}t} = \frac{4\pi\alpha_{\mathrm{em}}^2}{x\,Q^4}\left\{1 - y + \frac{y^2}{2\left[1 + R^{D(4)}(x, Q^2, x_{I\!P}, t)\right]}\right\} \times F_2^{D(4)}(x, Q^2, x_{I\!P}, t)\,. \tag{10.9}$$

Since the data are taken predominantly at small $y$, there is very little sensitivity to $R^{D(4)}$. Moreover, we expect $F_L^{D(4)}$ to be much smaller than $F_T^{D(4)}$ for $\beta \lesssim 0.8 - 0.9$, so in this range we can safely neglect $R^{D(4)}$ in (10.9) and write

$$\frac{\mathrm{d}\sigma^D_{\gamma^* p}}{\mathrm{d}x\,\mathrm{d}Q^2\,\mathrm{d}x_{I\!P}\,\mathrm{d}t} = \frac{4\pi\alpha_{\mathrm{em}}^2}{x\,Q^4}\left(1 - y + \frac{y^2}{2}\right) F_2^{D(4)}(x, Q^2, x_{I\!P}, t)\,. \tag{10.10}$$

$F_2^{D(4)}$ is proportional to the cross section for diffractive $\gamma^* p$ scattering

$$F_2^{D(4)}(x, Q^2, x_{I\!P}, t) = \frac{Q^2}{4\pi^2\alpha_{\mathrm{em}}}\,\frac{\mathrm{d}\sigma^D_{\gamma^* p}}{\mathrm{d}x_{I\!P}\,\mathrm{d}t}\,. \tag{10.11}$$

Note that $F_2^{D(4)}$ is a dimensional quantity. Thus, it would be more appropriate to denote it as a differential quantity

$$F_2^{D(4)} \equiv \frac{\mathrm{d}F_2^D(x, Q^2, x_{I\!P}, t)}{\mathrm{d}x_{I\!P}\,\mathrm{d}t}\,, \tag{10.12}$$

where the diffractive structure function $F_2^D$ introduced here is dimensionless.

When the outgoing proton is not detected, there is no measurement of $t$ and only the cross section integrated over $t$ is obtained, that is

$$\frac{\mathrm{d}\sigma^D_{\gamma^* p}}{\mathrm{d}x\,\mathrm{d}Q^2\,\mathrm{d}x_{I\!P}} = \frac{4\pi\alpha_{\mathrm{em}}^2}{x\,Q^4}\left(1 - y + \frac{y^2}{2}\right) F_2^{D(3)}(x, Q^2, x_{I\!P})\,. \tag{10.13}$$

where the structure function $F_2^{D(3)}$ is defined as

$$F_2^{D(3)}(x, Q^2, x_{I\!P}) = \int_0^\infty \mathrm{d}|t|\; F_2^{D(4)}(x, Q^2, x_{I\!P}, t)\,. \tag{10.14}$$

## 10.4 Diffractive Parton Distributions

It has been proven in perturbative QCD (Collins 1998; see also Trentadue and Veneziano 1994; Berera and Soper 1996) that, similarly to what happens in inclusive DIS, a factorization theorem holds for diffractive structure functions. These can be written in terms of some quantities, the *diffractive parton distributions*, which represent the probability to find a parton in a hadron $h$, under the condition the $h$ undergoes a diffractive scattering.

The QCD factorization formula for $F_2^D$ is

$$\frac{\mathrm{d}F_2^D(x, Q^2, x_{I\!P}, t)}{\mathrm{d}x_{I\!P}\,\mathrm{d}t} = \sum_i \int_x^{x_{I\!P}} \mathrm{d}\xi\, \frac{\mathrm{d}f_i(\xi, \mu^2, x_{I\!P}, t)}{\mathrm{d}x_{I\!P}\,\mathrm{d}t}\, \hat{F}_2^i\left(\frac{x}{\xi}, Q^2, \mu^2\right) . \quad (10.15)$$

Here $\mathrm{d}f_i(\xi, \mu^2, x_{I\!P}, t)/\mathrm{d}x_{I\!P}\,\mathrm{d}t$ is the diffractive distribution of parton $i$, i.e. the probability to find, in a proton, a parton of type $i$ carrying momentum fraction $\xi$, under the requirement that the proton remains intact, except for a momentum transfer quantified by $x_{I\!P}$ and $t$. The perturbatively calculable coefficients $\hat{F}_2^i(x/\xi, Q^2, \mu^2)$ are the same partonic structure functions appearing in the DIS factorization formula (see Sect. 9.3). Thus, the only difference between DIS and DDIS resides in the parton distributions. The parameter $\mu^2$ appearing in $\hat{F}_2^i$ and $\mathrm{d}f_i/\mathrm{d}x_{I\!P}\mathrm{d}t$ is a factorization scale. Since (10.15) holds at all orders, and the l.h.s. is a physical observable, the $\mu$-dependences of the parton distributions and of the partonic structure functions must cancel each other.

At lowest order in $\alpha_s$, the coefficient functions are $\hat{F}_2^q(z, Q^2, \mu^2) = e_q^2\,\delta(1-z)$ and $\hat{F}_2^g = 0$, and therefore the diffractive structure function is a superposition of quark and antiquark distributions

$$\frac{\mathrm{d}F_2^D(x, Q^2, x_{I\!P}, t)}{\mathrm{d}x_{I\!P}\,\mathrm{d}t} = \sum_{q,\bar{q}} e_q^2\, x\, \frac{\mathrm{d}f_q(x, \mu^2, x_{I\!P}, t)}{\mathrm{d}x_{I\!P}\,\mathrm{d}t} . \quad (10.16)$$

Diffractive parton distributions satisfy the same renormalization group equations that govern the evolution of the DIS parton densities, i.e. the DGLAP equations. Thus we have

$$\frac{\partial}{\partial \ln \mu^2} \frac{\mathrm{d}f_i(\xi, \mu^2, x_{I\!P}, t)}{\mathrm{d}x_{I\!P}\,\mathrm{d}t} = \sum_j \int_\xi^1 \frac{\mathrm{d}\zeta}{\zeta}\, P_{ij}\left(\frac{\xi}{\zeta}, \alpha_s(\mu)\right) \frac{\mathrm{d}f_j(\xi, \mu^2, x_{I\!P}, t)}{\mathrm{d}x_{I\!P}\,\mathrm{d}t} . \quad (10.17)$$

Veneziano and Trentadue (1994) introduced a quantity related to $\mathrm{d}f_i/\mathrm{d}x_{I\!P}\mathrm{d}t$, and called "fracture function". Specifically, a fracture functions is

$$\frac{\mathrm{d}f_i(\xi, \mu^2, x_{I\!P})}{\mathrm{d}x_{I\!P}} = \int_{\frac{x_{I\!P}^2 m_N^2}{1-x_{I\!P}}}^{\infty} \mathrm{d}|t|\, \frac{\mathrm{d}f_i(\xi, \mu^2, x_{I\!P}, t)}{\mathrm{d}x_{I\!P}\,\mathrm{d}t} , \quad (10.18)$$

that is a diffractive parton distribution integrated over $t$. This integration introduces other ultraviolet divergences and hence the renormalization group equations get modified. For more details about fracture functions we refer the reader to the original papers (Veneziano and Trentadue 1994; Grazzini, Trentadue and Veneziano 1998).

An important point that should be borne in mind is that in the context of collinear factorization higher-twist terms are subleading. However, a direct calculation of the diffractive structure functions (Nikolaev and Zakharov 1994a, Golec-Biernat and Wüsthoff 1999b), based on the color dipole picture (see Sect. 11.4), shows that there is a significant higher-twist longitudinal contribution $F_L^D$ at large $\beta = x/x_{I\!P}$, which is disregarded by the collinear factorization approach. Thus, the use of Eqs. (10.15) and (10.17) in the analysis of the data must be accompanied by some caution. Moreover, it has been also demonstrated (Collins 1998) that factorization is violated in diffractive hadron–hadron scattering. Therefore we cannot use the diffractive parton distributions extracted from DDIS to make predictions for diffractive processes at hadron colliders.

For completeness we give, in conclusion, the field theoretical definitions of the diffractive quark distributions (see the analogue for DIS quark distributions, Eq. (9.50), and Fig. 10.5 for a diagrammatic representation),

$$\frac{\mathrm{d}f_q(\xi,\mu^2,x_{I\!P},t)}{\mathrm{d}x_{I\!P}\,\mathrm{d}t} = \frac{1}{(4\pi)^3}\sum_X \int \mathrm{d}z^- \,\mathrm{e}^{\mathrm{i}\xi P^+ z^-}\,\langle PS|\bar{\psi}(0)|P'S';X\rangle \times \gamma^+\,\langle P'S';X|\psi(0,z^-,0_\perp)|PS\rangle\,, \tag{10.19}$$

and of the diffractive gluon distribution (the DIS analogue is (9.83))

$$\frac{\mathrm{d}f_g(\xi,\mu^2,x_{I\!P},t)}{\mathrm{d}x_{I\!P}\,\mathrm{d}t} = \frac{1}{4\,(2\pi)^3\,\xi P+}\sum_X \int \mathrm{d}z^- \,\mathrm{e}^{\mathrm{i}\xi P^+ z^-}\,\langle PS|F_a^{+\nu}(0)|P'S';X\rangle \times \gamma^+\,\langle P'S';X|F_{a\nu}^{+}(0,z^-,0_\perp)|PS\rangle\,. \tag{10.20}$$

Path-ordered exponentials should be inserted into (10.19) and (10.20) in order to ensure the gauge invariance of the expressions.

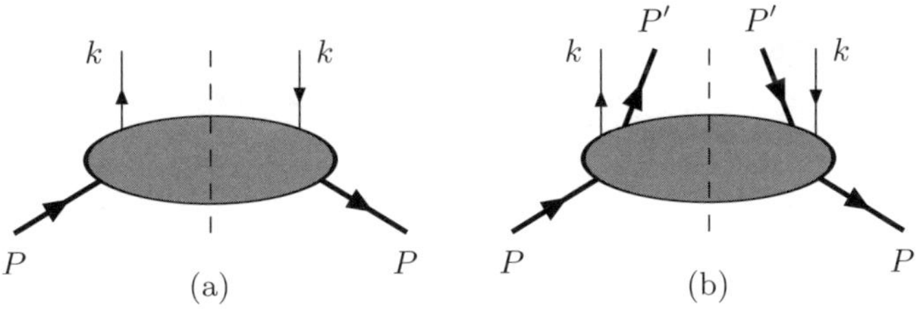


**Fig. 10.5.** Inclusive (**a**) and diffractive (**b**) parton distributions.

## 10.5 Regge Theory of DDIS

A straightforward generalization of the results on single-inclusive processes of Sec. 5.8.6 leads to the Regge theory predictions for diffractive DIS. The correspondence between the general reaction $1 + 2 \to 3 + X$ and DDIS is shown in Fig. 10.6. Particles 1 and 3 are now the incoming and outgoing proton, respectively; particle 2 is the virtual photon.

The c.m. energy squared of $\gamma^* p$ scattering is $W^2$ and the replacement $M^2 \to M^2 + Q^2$ must be made in the formulas of Sec. 5.8.6, to take the photon virtuality into account.

For $W^2 \gg M^2, Q^2 \gg t$, diffractive deep inelastic scattering is described by the triple Regge diagrams (Fig. 10.7). There are two contributions: the triple-pomeron $I\!PI\!PI\!P$ diagram and the pomeron–pomeron–reggeon $I\!PI\!PI\!R$ diagram (Fig. 10.8). The DDIS cross section then reads (as usual, powers of a reference energy scale $s_0$ should be inserted in order to have the correct dimensions)

$$W^2 \frac{\mathrm{d}\sigma^D_{\gamma^* p}}{\mathrm{d}M^2\,\mathrm{d}t} = A_{I\!P}(t)\,(W^2)^{2\alpha_{I\!P}(t)-1}\,(M^2+Q^2)^{\alpha_{I\!P}(0)-2\alpha_{I\!P}(t)}$$
$$+A_{I\!R}(t)\,(W^2)^{2\alpha_{I\!P}(t)-1}\,(M^2+Q^2)^{\alpha_{I\!R}(0)-2\alpha_{I\!P}(t)}, \quad (10.21)$$

where the functions $A_{I\!P,I\!R}$ incorporate all couplings.

In particular, for large $M^2$, triple pomeron dominates and the cross section becomes

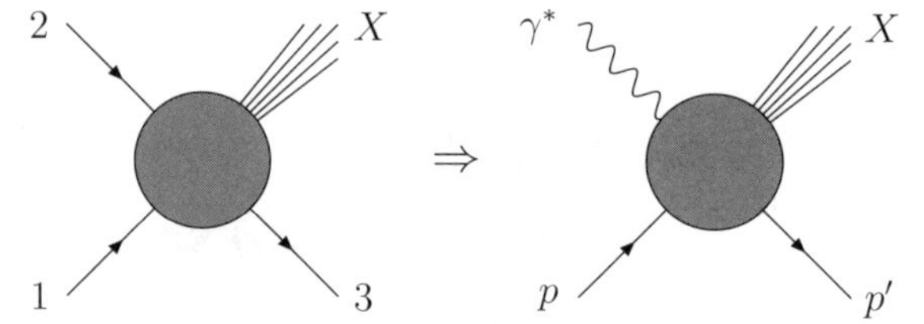


**Fig. 10.6.** From hadronic diffractive dissociation to diffractive DIS.

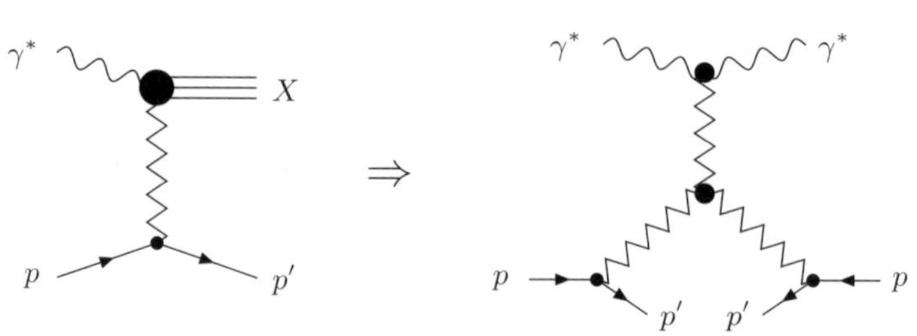


**Fig. 10.7.** Diffractive $\gamma^* p$ scattering and the triple Regge diagram that describes it in the limit $W^2 \gg M^2, Q^2 \gg t$.

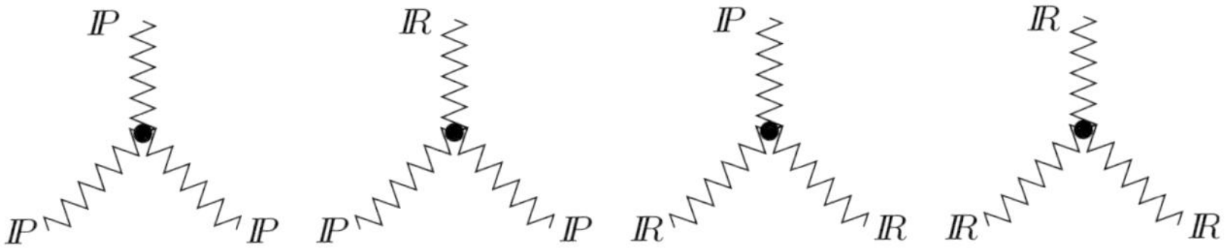


**Fig. 10.8.** Triple-Regge diagrams contributing to diffraction dissociation.

$$W^2 \frac{\mathrm{d}\sigma^D_{\gamma^* p}}{\mathrm{d}M^2\,\mathrm{d}t} = \frac{1}{16\pi^2}\,|g_{I\!P}(t)|^2 \left(\frac{W^2}{M^2+Q^2}\right)^{2\alpha_{I\!P}(t)-1}$$
$$\times g_{3I\!P}(t)\, g_{I\!P}(0)\,(M^2+Q^2)^{\alpha_{I\!P}(0)-1}\,, \qquad (10.22)$$

where we have assumed for simplicity that the pomeron couples in the same way to the proton and to the virtual photon, and $g_{3I\!P}$ is the triple-pomeron coupling. The triple-pomeron mass spectrum is

$$\left.\frac{\mathrm{d}\sigma^D_{\gamma^* p}}{\mathrm{d}M^2\,\mathrm{d}t}\right|_{t=0} \sim \frac{1}{(Q^2+M^2)^{\alpha_{I\!P}(0)}} \sim \frac{1}{(M^2)^{\alpha_{I\!P}(0)}}\,. \qquad (10.23)$$

Since $t$ is limited and the triple-pomeron coupling does not depend much on it we can make the approximation $g_{3I\!P}(t) \simeq g_{3I\!P}(0)$ and, introducing the variable $x_{I\!P} = (M^2+Q^2)/(W^2+Q^2) \simeq M^2/W^2$ (remember that we are considering the large $M$ region), Eq. (10.22) can be rewritten as

$$\frac{\mathrm{d}\sigma^D_{\gamma^* p}}{\mathrm{d}x_{I\!P}\,\mathrm{d}t} = f_{I\!P}(x_{I\!P},t)\,\sigma_{\gamma^* I\!P}(M^2)\,, \qquad (10.24)$$

where

$$f_{I\!P}(x_{I\!P},t) = \frac{1}{16\pi^2}\,|g_{I\!P}(t)|^2\,x_{I\!P}^{1-2\alpha_{I\!P}(t)} \qquad (10.25)$$

is the *pomeron flux factor* and

$$\sigma_{\gamma^* I\!P}(M^2) = g_{3I\!P}(0)\,g_{I\!P}(0)\,(M^2)^{\alpha_{I\!P}(0)-1} \qquad (10.26)$$

is the total cross section of $\gamma^* I\!P$ scattering. It must be stressed that since the pomeron is *not* a particle the separation of the flux from the pomeron cross section is quite arbitrary, and the normalization of the pomeron flux is ambiguous.

Assuming a linear pomeron trajectory $\alpha_{I\!P}(t) = \alpha_{I\!P}(0) + \alpha'_{I\!P}\,t$ and the typical exponential behavior for the $t$-dependence of the pomeron coupling

$$g_{I\!P}(t) = g_{I\!P}(0)\,\mathrm{e}^{b_0 t/2}, \qquad (10.27)$$

we can write the pomeron flux as

$$f_{I\!P}(x_{I\!P},t) = \frac{1}{16\pi^2}\,|g_{I\!P}(0)|^2\,x_{I\!P}^{1-2\alpha_{I\!P}(0)}\,\exp\left[\left(b_0 + 2\,\alpha'_{I\!P}\,\ln\frac{1}{x_{I\!P}}\right)t\right]. \qquad (10.28)$$

Note that the slope of the $t$-distribution of (10.28) and hence of (10.24) increases with $\ln(1/x_{I\!P})$, that is with $\ln W^2$. This shrinkage is a well known prediction of Regge theory, that we already encountered in Sect. 5.8.2.

If we use the variable $\beta = Q^2/(M^2 + Q^2) \simeq Q^2/M^2$ (at large $M^2$) Eq. (10.24) becomes

$$\frac{d\sigma^D_{\gamma^* p}}{dx_{I\!P}\, dt} = f_{I\!P}(x_{I\!P}, t)\, \sigma_{\gamma^* I\!P}(\beta, Q^2) \,, \tag{10.29}$$

where the $\gamma^* I\!P$ cross section takes the form

$$\sigma_{\gamma^* I\!P}(\beta, Q^2) = A_{I\!P}(Q^2)\, \beta^{1-\alpha_{I\!P}(0)} \,, \tag{10.30}$$

valid at small $\beta$. Note that the $Q^2$ dependence, that we embodied in $A_{I\!P}(Q^2)$, is actually irrelevant in the context of Regge theory. In this theory the virtuality of the photon is a fixed parameter (a mass) and one cannot predict how the cross section depends on $Q^2$. What Regge theory does predict is the $\beta$ dependence of $\sigma_{\gamma^* I\!P}$ at fixed $Q^2$.

The interesting feature of Eq. (10.29) is the factorization of the $x_{I\!P}$ dependence from the $\beta$ dependence, the so-called *Regge factorization.* This is an important and highly non-trivial prediction of Regge theory, as we saw in Sect. 5.8.1. The behavior in $x_{I\!P}$, i.e. in $W^2$, is completely determined by the flux factor. With $\alpha_{I\!P}(0) = 1 + \epsilon$ the diffractive cross section behaves as

$$\left.\frac{d\sigma^D_{\gamma^* p}}{dx_{I\!P}\, dt}\right|_{t=0} \sim \frac{1}{x_{I\!P}^{1+2\epsilon}} \tag{10.31}$$

In terms of the diffractive structure function $F_2^{D(4)}$ we can rewrite (10.29) as

$$F_2^{D(4)}(x, Q^2, x_{I\!P}, t) = f_{I\!P}(x_{I\!P}, t)\, F_2^{I\!P}(\beta, Q^2) \,, \tag{10.32}$$

where $F_2^{I\!P}$, defined as

$$F_2^{I\!P}(\beta, Q^2) = \frac{Q^2}{4\pi^2 \alpha_{\rm em}}\, \sigma_{\gamma^* I\!P}(\beta, Q^2) \tag{10.33}$$

is the so-called *pomeron structure function.* The Regge theory expectation for the $\beta$ dependence of $F_2^{I\!P}$ is that at small $\beta$, where triple pomeron is dominant, it should behave as

$$F_2^{I\!P}(\beta, Q^2) \sim \beta^{1-\alpha_{I\!P}(0)} \,, \tag{10.34}$$

with $\alpha_{I\!P}(0) = 1 + \lambda$.

Suppose now that the invariant mass of the $X$ system is not exceedingly large, i.e., that $\beta$ is not too small. In this case the $I\!P I\!P I\!R$ diagram of Fig. 10.8 may become important. According to Eq. (10.21) it yields a mass spectrum (with $\alpha_{I\!R}(0) = 1/2$)

$$\left.\frac{\mathrm{d}\sigma^D_{\gamma^* p}}{\mathrm{d}M^2\,\mathrm{d}t}\right|_{t=0} \sim \left(\frac{1}{M^2+Q^2}\right)^{\frac{3}{2}} . \tag{10.35}$$

The factorization formula (10.29) still holds but the $\gamma^* I\!P$ cross section now has two contributions

$$\sigma_{\gamma^* I\!P}(\beta, Q^2) = A_{I\!P}(Q^2)\,\beta^{1-\alpha_{I\!P}(0)} + A_{I\!R}(Q^2)\,\beta^{1-\alpha_{I\!R}(0)} . \tag{10.36}$$

As $\beta$ gets larger, the $I\!P I\!P I\!R$ term becomes important and the pomeron structure function acquires a $\sim \beta^{\frac{1}{2}}$ contribution. An analysis of the HERA data on DDIS based on Regge theory has been performed by Capella et al. (1995, 1996).

When the $\gamma^* p$ energy $W^2$ does not reach very high values, that is $x_{I\!P}$ is not very close to zero, there may be important non-diffractive corrections to the $\gamma^* p \to Xp$ cross sections, arising from $I\!R I\!R I\!P$ and $I\!R I\!R I\!R$ diagrams (Fig. 10.8). The non-diffractive (ND) contribution is

$$\begin{aligned} W^2 \frac{\mathrm{d}\sigma^{ND}_{\gamma^* p}}{\mathrm{d}M^2\,\mathrm{d}t} &= A_{I\!R I\!R I\!P}(t)\,(W^2)^{2\alpha_{I\!R}(t)-1}\,(M^2+Q^2)^{\alpha_{I\!P}(0)-2\alpha_{I\!R}(t)} \\ &+ A_{I\!R I\!R I\!R}(t)\,(W^2)^{2\alpha_{I\!R}(t)-1}\,(M_X^2+Q^2)^{\alpha_{I\!R}(0)-2\alpha_{I\!R}(t)}, \end{aligned} \tag{10.37}$$

and in terms of the variable $x_{I\!P}$ (whose name is now quite inappropriate, since what is exchanged between the photon and the proton is a subleading reggeon, not a pomeron)

$$\begin{aligned} \frac{\mathrm{d}\sigma^{ND}_{\gamma^* p}}{\mathrm{d}x_{I\!P}\,\mathrm{d}t} &= \frac{1}{16\pi^2}\,|g_{I\!R}(t)|^2\,x_{I\!P}^{1-2\alpha_{I\!R}(t)} \\ &\times \left[\mathcal{A}_{I\!R I\!R I\!P}(Q^2)\,\beta^{1-\alpha_{I\!P}(0)} + \mathcal{A}_{I\!R I\!R I\!R}(Q^2)\,\beta^{1-\alpha_{I\!R}(0)}\right] . \end{aligned} \tag{10.38}$$

Using for definiteness $\alpha_{I\!R}(0) = 1/2$ we see that the energy dependence of the non-diffractive terms is

$$\left.\frac{\mathrm{d}\sigma^{ND}_{\gamma^* p}}{\mathrm{d}x_{I\!P}\,\mathrm{d}t}\right|_{t=0} \sim x^0_{I\!P}\,, \quad i.e.\quad W^2 \left.\frac{\mathrm{d}\sigma^{ND}}{\mathrm{d}M^2\,\mathrm{d}t}\right|_{t=0} \sim (W^2)^0\,, \tag{10.39}$$

and therefore these terms become (relatively) important when $x_{I\!P}$ departs sensibly from zero.

Let us come now to the diffractive structure function $F_2^{D(3)}$. Integrating (10.32) over $t$, $F_2^{D(3)}$ can be put in the factorized form

$$F_2^{D(3)}(x_{I\!P}, \beta, Q^2) = \overline{f}_{I\!P}(x_{I\!P})\,F_2^{I\!P}(\beta, Q^2)\,, \tag{10.40}$$

where $\overline{f}_{I\!P}$ is the $t$-integrated pomeron flux

$$\overline{f}_{I\!P}(x_{I\!P}) = \int_0^\infty \mathrm{d}|t|\,f_{I\!P}(x_{I\!P}, t)\,. \tag{10.41}$$

Using (10.28) we get for $\overline{f}_{I\!P}$

$$\overline{f}_{I\!P}(x_{I\!P}) \propto \frac{1}{\left(b_0 + 2\,\alpha'_{I\!P}\,\ln\frac{1}{x_{I\!P}}\right)}\,x_{I\!P}^{1-2\alpha_{I\!P}(0)} . \tag{10.42}$$

In practical measurements the $t$-integration in (10.41) has a limited range due to experimental conditions (see the discussion in Sect. 10.8).

## 10.6 The Partonic Structure of the Pomeron

It is quite usual, but nevertheless not free from possible conceptual dangers, to introduce a partonic structure for $F_2^{I\!P}$ (Ingelman and Schlein 1985). At leading order the pomeron structure function (10.33) is written as a superposition of quark and antiquark distributions in the pomeron

$$F_2^{I\!P}(\beta, Q^2) = \sum_{q,\bar{q}} e_q^2\,\beta\,q^{I\!P}(\beta, Q^2) . \tag{10.43}$$

The variable $\beta = x/x_{I\!P}$ is interpreted as the fraction of the pomeron momentum carried by its partonic constituents and $q^{I\!P}(\beta, Q^2)$ is the probability of finding, inside the pomeron, a quark $q$ with momentum fraction $\beta$ (see Fig. 10.9). Clearly, this interpretation makes sense only insofar as we can specify unambigously the probability of finding a pomeron in the proton (i.e., the pomeron flux), and assume the pomeron to be a real particle. Since it is not so, the whole picture must be taken as a purely phenomenological one.

Comparing (10.43) and (10.16), that is combining Regge factorization and QCD factorization, we find the following relationships between diffractive quark distributions and quark distributions of the pomeron

$$\frac{\mathrm{d}f_q(\beta, Q^2, x_{I\!P}, t)}{\mathrm{d}x_{I\!P}\mathrm{d}t} = \frac{1}{16\pi^2}\,|g_{I\!P}(t)|^2\,x_{I\!P}^{-2\alpha_{I\!P}(t)}\,q^{I\!P}(\beta, Q^2) . \tag{10.44}$$

Similarly, one introduces the gluon distribution of the pomeron $g^{I\!P}(\beta, Q^2)$ related to $\mathrm{d}f_g/\mathrm{d}X_{I\!P}\mathrm{d}t$ by

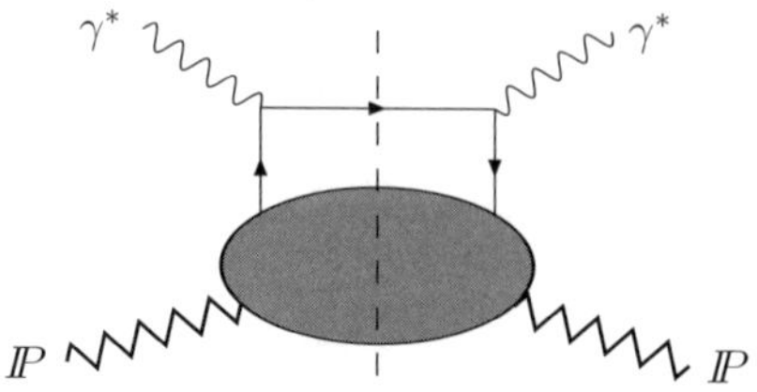


**Fig. 10.9.** The handbag diagram for $\gamma^* I\!P$ scattering.

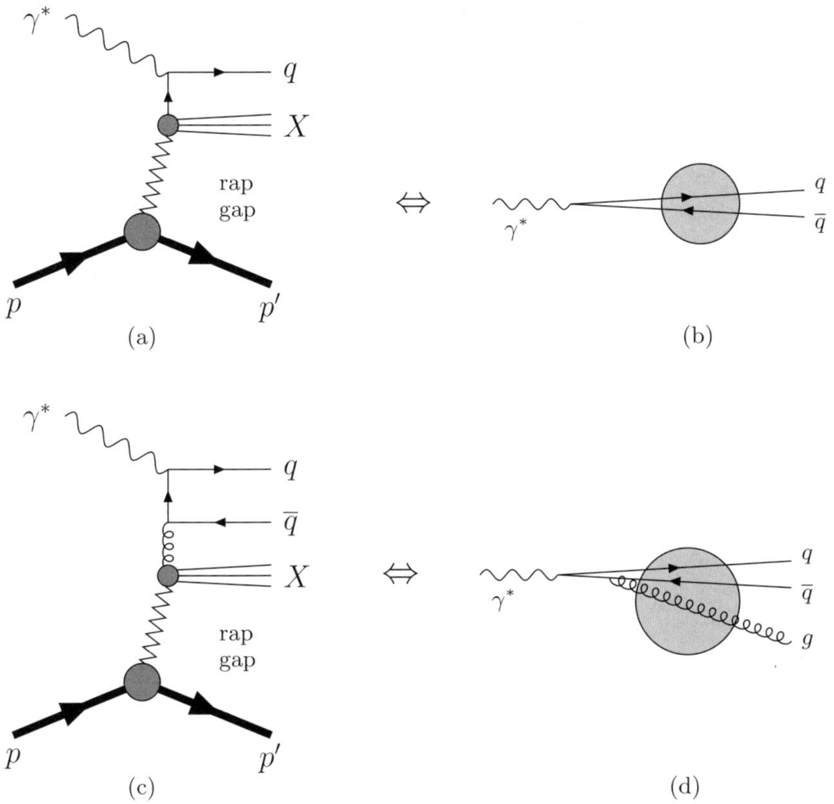


**Fig. 10.10.** Partonic structure of the pomeron (infinite-momentum frame picture, (**a**, **c**)) vs. color dipole fluctuations of the photon (proton rest-frame picture (**b**, **d**)).

$$\frac{\mathrm{d}f_g(\beta, Q^2, x_{I\!P}, t)}{\mathrm{d}x_{I\!P}\mathrm{d}t} = \frac{1}{16\pi^2}\,|g_{I\!P}(t)|^2\, x_{I\!P}^{-2\alpha_{I\!P}(t)}\, g^{I\!P}(\beta, Q^2)\,. \tag{10.45}$$

At next-to-leading order the pomeron structure function acquires a term containing $g^{I\!P}(\beta, Q^2)$. The situation is pictorially shown in Fig. 10.10a,c, where the infinite-momentum frame description of DDIS, with the pomeron resolved in its partonic components, is represented.

By virtue of (10.44, 10.45), and of what we saw in Sect. 10.4, the $Q^2$-dependence of $q^{I\!P}(\beta, Q^2)$ and $g^{I\!P}(\beta, Q^2)$ is governed by the DGLAP equations. One should recall, however, that, since the pomeron is not a particle, its parton distributions do not satisfy energy-momentum conservation. Once more, we stress that the pomeron flux appearing in (10.44) and (10.45) is ambiguous and so is the definition of the parton distributions of the pomeron.

We can adopt a different point of view (Fig. 10.10b,d) and describe diffractive DIS in the proton rest-frame using the color dipole approach already employed in Sect. 9.9 to model inclusive DIS. From this perspective, probing the quark and antiquark distributions of the pomeron corresponds to considering the $q\bar{q}$ excitations of the virtual photon and their interaction with the proton via two-gluon (or BFKL ladder) exchange. The gluonic contribution to the pomeron structure function is in turn reinterpreted in terms of $q\bar{q}g$ fluctuations of the photon. The correspondence between the infinite-momentum frame picture of diffractive DIS (wherein the internal structure of the pomeron is resolved) and the proton rest-frame description (wherein the hadronic fluctuations of the photon are involved) is sketched in Fig. 10.10. Note however that, as already remarked in Sect. 10.4, the collinear factorization formulas defining the diffractive parton distributions, do not take into account higher-twist contributions, which instead emerge in the color dipole approach and turn out to be non-negligible in some kinematic regions (see Sect. 11.4).

## 10.7 Experimental Signatures of DDIS

A typical diffractive DIS event is shown in Fig. 10.11. Note the absence of energy deposit in the forward direction (the proton remnant goes into the beam pipe), and the corresponding (pseudo-)rapidity gap of several units.

The situation is schematically reproduced in Fig. 10.12. In general, in the fully inclusive DIS case, there are two main groups of particles, besides the recoil electron: particles produced at high rapidity in the fragmentation region of the proton, and, in the opposite emisphere, particles arising from

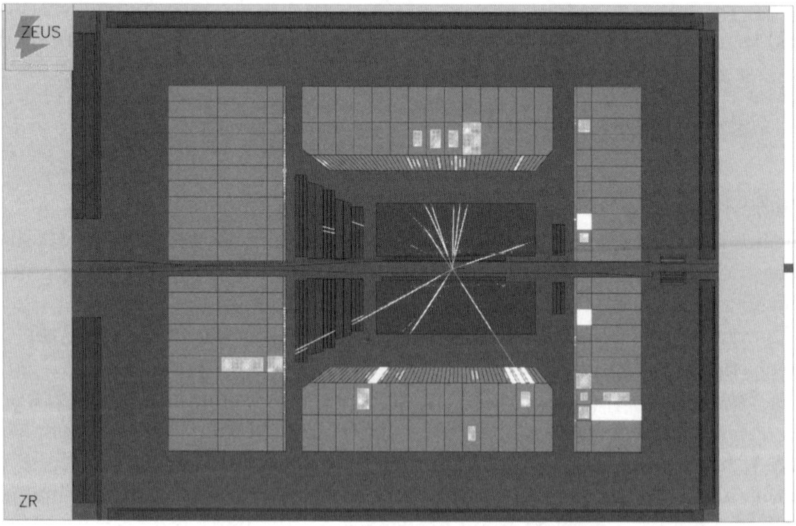


**Fig. 10.11.** A typical diffractive DIS event in the ZEUS detector. Note the absence of signals in the forward direction (which points to the left).

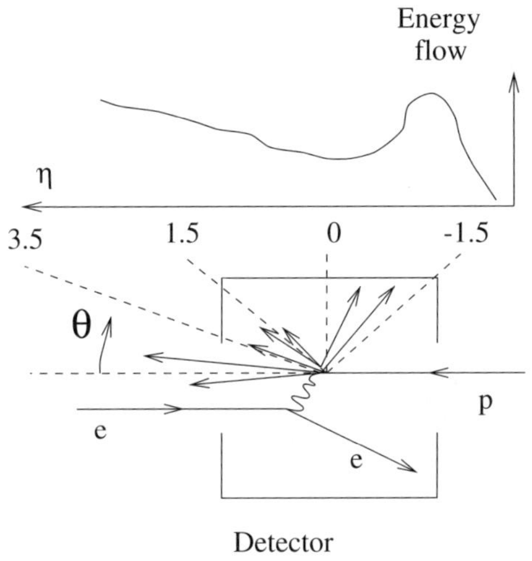


**Fig. 10.12.** Schematic view of *non-diffractive* DIS at HERA. From Cartiglia (1997).

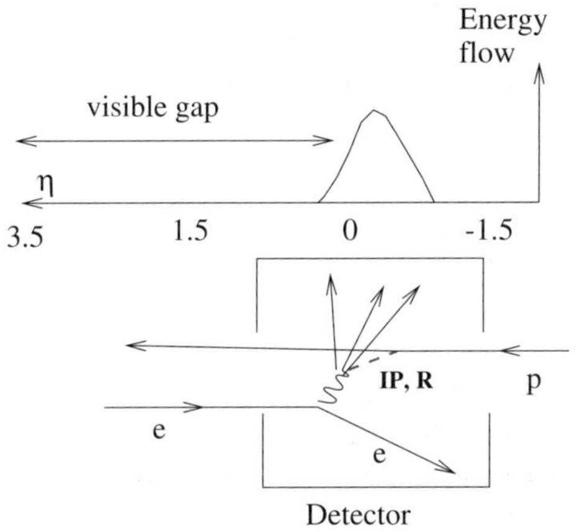


**Fig. 10.13.** Schematic view of *diffractive* DIS at HERA. From Cartiglia (1997).

the hadronization of the struck parton, typically at small or negative rapidity. The rapidity widths of these two systems are $\ln(W^2/m_N^2)$ and $\ln(xW^2/m_N^2)$, respectively. Thus, there is a rapidity interval $\sim \ln(1/x)$ between the proton remnant and the struck quark. Due to color interactions, this rapidity gap is actually completely filled by particles created in the hadronization process (Fig. 10.2b). Thus, no region devoid of activity is seen in the detectors. In the diffractive case, the proton remains almost intact and is lost in the beam pipe. No color interaction between its remnant and the jet in the photon fragmentation region takes place, and so a visible gap appears in the detector.

In the light of Regge theory one can easily understand the existence of large rapidity gaps in diffractive events. Using $\Delta\eta = \ln(1/x_{I\!P})$, and setting

$\alpha_{I\!P}(0) = 1$, $\alpha_{I\!R}(0) = 1/2$, Eqs. (10.22) and (10.38) lead to the following rapidity distributions of diffractive and non-diffractive events

$$\frac{\mathrm{d}N}{\mathrm{d}\Delta\eta} \sim \text{constant} \quad \text{(diffractive)}\,, \tag{10.46}$$

$$\frac{\mathrm{d}N}{\mathrm{d}\Delta\eta} \sim \mathrm{e}^{-\Delta\eta} \quad \text{(non − diffractive)}\,. \tag{10.47}$$

Therefore, rapidity gaps in non-diffractive events are exponentially suppressed.

The experimentalists have adopted various methods to select diffractive events from the full sample of DIS events at HERA. We briefly review some of these methods (see, for instance, Cartiglia 1997).

### 10.7.1 Large Rapidity Gaps

There are two selection methods based on the observation of rapidity gaps.

The *maximum pseudorapidity* method consists of setting a cut on the pseudorapidity $\eta_{\max}$ of the most forward energy deposit in the calorimeter. In the ZEUS experiment diffractive events are defined as those events with $\eta_{\max} \lesssim 1.5 - 2.0$. Since the smallest angle covered by the ZEUS detector corresponds to $\eta_{\max} \simeq 4.4$, the above cut amounts to requiring a rapidity gap in the forward direction of at least 2.5-3.0 units. The cut on $\eta_{\max}$, however, limits the pseudorapidity interval that can be covered by the $X$ system, which is in general given by $\Delta\eta \sim \ln(M^2/m_N^2)$. Therefore diffractive events with large $M$ ($M > 10 - 15$ GeV) are excluded. In 1998 a new forward calorimeter has been added to the ZEUS detector, which extends $\eta_{\max}$ to 5 and allows taking diffractive data with $M$ up to 40 GeV.

Alternatively, one can directly identify the *largest rapidity gap*. For instance, the H1 Collaboration selects diffractive events by requiring $x_{I\!P} < 0.05$ and a mass of the proton remnant $M_Y < 1.6$ GeV. This corresponds to observing no activity in the pseudorapidity range $3.4 < \eta < 7.5$.

### 10.7.2 $M$-Subtraction

This method is based on the different $M^2$ distributions of the diffractive and non-diffractive events. The diffractive contribution behaves as – see (10.22)

$$\frac{\mathrm{d}N}{\mathrm{d}\ln M^2} \sim \text{constant} \qquad \text{(diffractive)}\,, \tag{10.48}$$

whereas the non-diffractive contribution has an exponential dependence on $\ln M^2$, see (10.38)

$$\frac{\mathrm{d}N}{\mathrm{d}\ln M^2} \sim \exp(a \ln M^2)\,, \quad \text{with } a > 0 \quad \text{( non diffractive)}\,. \tag{10.49}$$

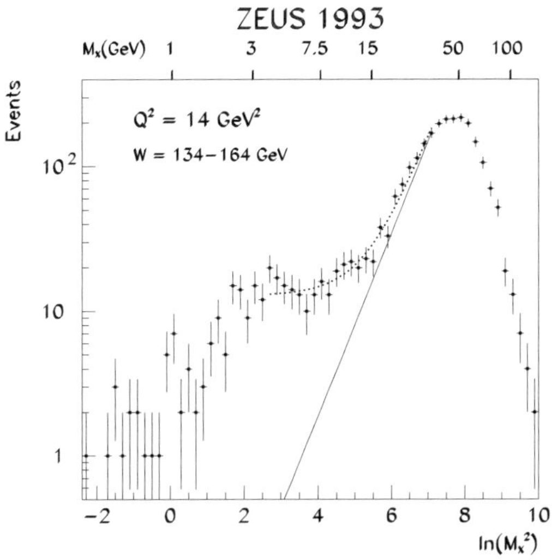


**Fig. 10.14.** The $\ln M^2$ distribution. From Cartiglia (1997).

The diffractive sample is defined as the excess contribution in the $\ln M^2$ distribution above the exponential fall-off of the non-diffractive peak (see Fig. 10.14).

### 10.7.3 Leading-Proton Detection

In principle, the cleanest way to identify diffractive events is to detect a leading proton (that is a proton, in the final state, carrying most of the momentum of the initial proton). Events with $x_F \gtrsim 0.97$ (which is the typical cut imposed by the experimentalists) lie within the diffractive peak (see Fig. 10.15). At $x_F \lesssim 0.9$ the continuum due to double dissociation, reggeon exchange and inclusive DIS, makes the distinction between diffractive and non-diffractive events quite difficult.

Since the leading protons have small transverse momenta and tend to stay very close to the beam line, their detection requires the use of to allow inserting high-precision detectors down to few millimeters from the beam. The main difficulty of this method comes from the limited acceptance of the detectors. As a consequence, the number of events with a tagged leading proton is quite small and there are relatively large statistical uncertainties.

## 10.8 Measurements of $F_2^D$

The discovery in 1993 of a relatively abundant production of diffractive events in DIS at HERA (Derrick et al. 1993; Ahmed et al. 1994) has been followed

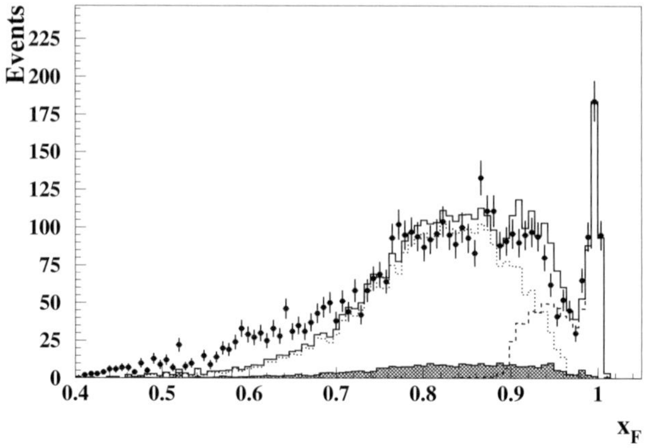


**Fig. 10.15.** The observed $x_F$ spectrum of protons (dots). Note the diffractive peak at $x_F \simeq 1$. The diffractive signal modelled by a Monte Carlo program is represented by the dashed line. The dotted line is the pion-exchange contribution. The shaded area corresponds to proton dissociation. (From Breitweg et al. 1998).

in the last years by more and more accurate measurements of various diffractive observables (for comprehensive reviews see, Cartiglia 1997, Marage 1999, Abramowicz and Caldwell 1999, Abramowicz 2000). In this section we review some of the experimental results on the diffractive structure function $F_2^D$.

The kinematic range covered by HERA is shown in Fig. 10.16. As one can see, the two experiments (ZEUS and H1) cover a wide range in $\beta, x_{I\!P}, Q^2$. The recent data at the extend the explored region to larger $x_{I\!P}$.

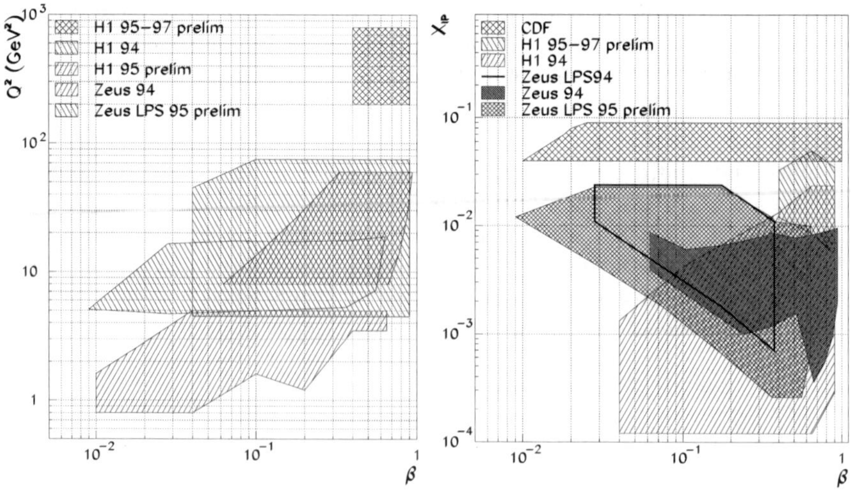


**Fig. 10.16.** The kinematic range in $Q^2$ and $x_{I\!P}$ covered by the present–day experiments

The starting point of most of the analyses is the Regge factorization formula (10.32) for $F_2^{D(4)}$. However, data on the diffractive structure function are usually integrated over $t$. What the experiments provide is actually

$$\widetilde{F}_2^{D(3)}(x_{I\!P}, \beta, Q^2) = \int_{|t|_{\min}}^{|t|_{\max}} \mathrm{d}|t| \, F_2^{D(4)}(x_{I\!P}, \beta, Q^2, t) \,, \tag{10.50}$$

where the tilde signals the fact that the integration is over a limited range of $t$. Although the use of leading proton detectors has recently made $F_2^{D(4)}$ available as well, most of the data are still of the type (10.50). Using Eq. (10.28) one gets for $\widetilde{F}_2^{D(3)}$

$$\begin{aligned} \widetilde{F}_2^{D(3)}(x_{I\!P}, \beta, Q^2) &\sim \frac{1}{b_0 + 2\alpha'_{I\!P} \ln(1/x_{I\!P})} \, x_{I\!P}^{1-2\alpha_{I\!P}(0)} \\ &\times \exp\left[\left(-(b_0 + 2\alpha'_{I\!P} \ln \frac{1}{x_{I\!P}}\right) |t|\right]_{|t|_{\min}}^{|t|_{\max}} F_2^{I\!P}(\beta, Q^2). \end{aligned} \tag{10.51}$$

To the extent that $|t|_{\min}$ and $|t|_{\max}$ are, respectively, small and large enough, one can set the exponential in (10.51) equal to unity, so that $\widetilde{F}_2^{D(3)} \simeq F_2^{D(3)}$. This is usually a legitimate procedure, which does not introduce significant errors. However, the limited range of the $t$ integration is one of the elements to be taken into account in determining the pomeron intercept from the $x_{I\!P}$ dependence of $\widetilde{F}_2^{D(3)}$. If we parametrize $\widetilde{F}_2^{D(3)}$ as

$$\widetilde{F}_2^{D(3)} \sim x_{I\!P}^{1-2a_{I\!P}} \,, \tag{10.52}$$

the effective pomeron intercept $a_{I\!P}$ differs from the real intercept $\alpha_{I\!P}(0)$ due to two effects:

- the denominator in the first line of (10.51) is a slowly rising function of $1/x_{I\!P}$ and causes $a_{I\!P}$ to be smaller than $\alpha_{I\!P}(0)$ (an effect which depends on the pomeron slope $\alpha'_{I\!P}$);
- since the $t$ range is limited, the exponential in (10.51) is a decreasing function of $1/x_{I\!P}$, and this again makes $a_{I\!P}$ smaller than $\alpha_{I\!P}(0)$ (by an amount which depends on $\alpha'_{I\!P}$, $b_0$ and the limits of integration over $|t|$).

The conclusion is that the real intercept $\alpha_{I\!P}(0)$ is larger than the effective one $a_{I\!P}$ by few percent. Note also that the determination of $\alpha_{I\!P}(0)$ from the $t$-integrated diffractive structure function $F_2^{D(3)}$ depends on the values of $\alpha'_{I\!P}$ and $b_0$.

The pomeron intercept obtained from measurements of diffractive and total $\gamma^* p$ and $\gamma p$ cross sections is shown in Fig. 10.17, for various $Q^2$ values, ranging from $Q^2 \simeq 0$ (the limit of real photoproduction) to $Q^2 \simeq 20 - 30$ GeV$^2$. Despite the large uncertainties, there appears to be a significant difference between the value at large $Q^2$ and the one obtained from photoproduction. The former is $\alpha_{I\!P}(0) \simeq 1.15 - 1.20$, the latter is $\alpha_{I\!P}(0) \simeq 1.10$, that

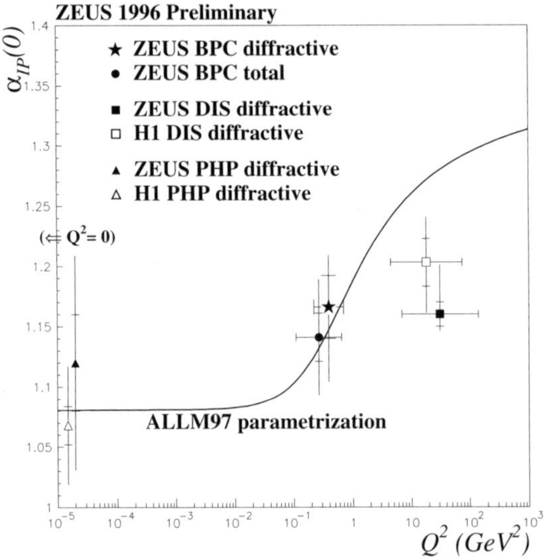


**Fig. 10.17.** The pomeron intercept for various $Q^2$ values. The curve is the parametrization of Abramowicz and Levy (1997), which updates Abramowicz, Levin, Levy and Maor (1991).

is the "soft" value. The curve in Fig. 10.17 comes from a parametrization of the total $\gamma^* p$ cross sections. It should be noted that the value of $\alpha_{I\!P}(0)$ extracted from diffractive DIS, while similar at low $Q^2$ to the value obtained from inclusive DIS (represented by the curve in Fig. 10.17), is definitely lower than it at large $Q^2$. This comes from the fact that, for $Q^2 \gtrsim 10\ \mathrm{GeV}^2$, the observed $W^2$-dependence of diffractive DIS is the same as for inclusive DIS, in contrast with Regge theory predictions which give different behaviors:

$$\sigma_{\mathrm{tot}}^{\gamma^* p} \sim (W^2)^{\alpha_{I\!P}(0)-1} \,, \qquad \frac{\mathrm{d}\sigma^D_{\gamma^* p}}{\mathrm{d}M^2\,\mathrm{d}t} \sim (W^2)^{2\,(\alpha_{I\!P}(0)-1)} \,. \tag{10.53}$$

The $x_{I\!P}$-dependence of $F_2^{D(3)}$ is shown in Fig. 10.18, for various $\beta$ and $Q^2$ values (Adloff et al. 1997). If the data are fitted with a function $F_2^{D(3)} = x_{I\!P}^{-a}\, A(\beta, Q^2)$, in the spirit of the Regge factorization formula (10.40), one finds a change in the exponent $a$ going from low to high values of $\beta$. This points to the existence of a non-pomeronic (i.e. non-diffractive) component in the measured $F_2^{D(3)}$. A fit with a pomeron ($\sim x_{I\!P}^{1-2\alpha_{I\!P}(t)}$) and a subleading reggeon ($\sim x_{I\!P}^{1-2\alpha_{I\!R}(t)}$) term, of the type

$$F_2^{D(4)}(\beta, Q^2, x_{I\!P}, t) = f_{I\!P}(x_{I\!P}, t)\, F_2^{I\!P}(\beta, Q^2) + f_{I\!R}(x_{I\!P}, t)\, F_2^{I\!R}(\beta, Q^2) \,, \tag{10.54}$$

gives

$$\alpha_{I\!P}(0) = 1.203 \pm 0.020\,(\mathrm{stat.}) \pm 0.013\,(\mathrm{syst.})^{+0.030}_{-0.035}\,(\mathrm{theor.}) \,, \tag{10.55}$$

$$\alpha_{I\!R}(0) = 0.50 \pm 0.11\,(\mathrm{stat.}) \pm 0.11\,(\mathrm{syst.})^{+0.09}_{-0.10}\,(\mathrm{theor.}) \,. \tag{10.56}$$

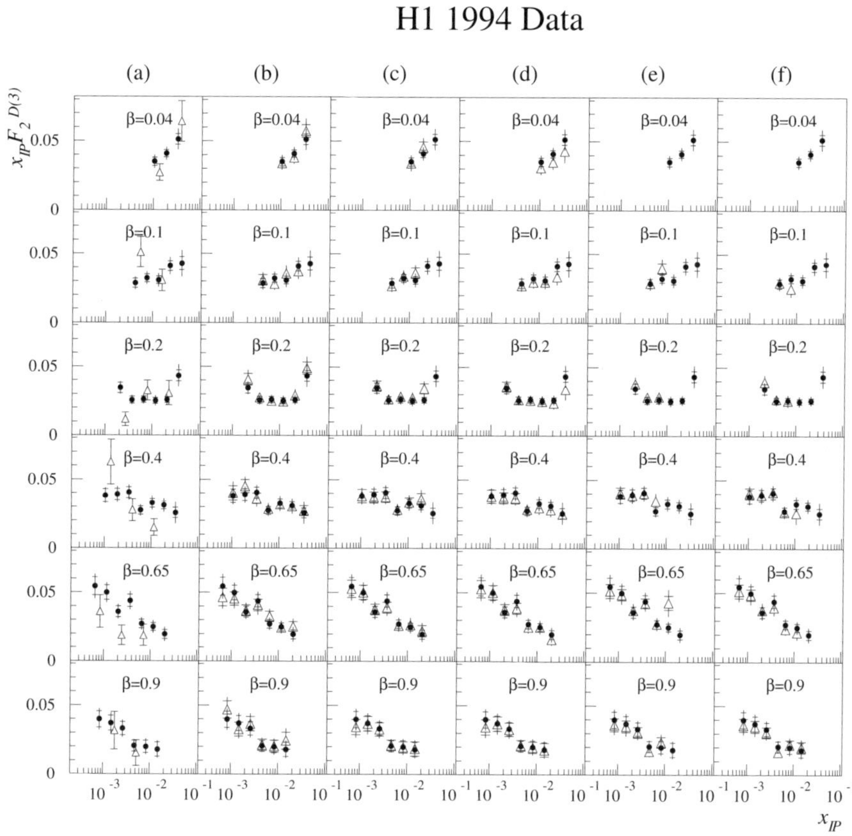


**Fig. 10.18a–f.** The diffractive structure function $x_{I\!P} F_2^{D(3)}$ (Adloff et al. 1997).

The $Q^2$ dependence of the diffractive structure function as measured by H1 (1998b) is shown in Fig. 10.19 for a range of $\beta$ values. This dependence is rather mild, as expected from the QCD factorization theorem, which predicts logarithmic scaling violations for $F_2^{D(3)}$. Going a little more into details, one notices a significant difference between $F_2$ and $F_2^{D(3)}$: the latter exhibits positive scaling violations even for relatively large values of $\beta$, up to $\beta = 0.4$, in contrast with the behavior of $F_2$. This is a clear indication of a different partonic content of the inclusive and diffractive structure functions.

As for the $\beta$-dependence of $F_2^{D(4)}$, this is rather flat over a wide range, as shown by the ZEUS measurements (ZEUS 1999) with the leading-proton spectrometer (Fig. 10.20). Again, this is at variance with the $x$-dependence of DIS structure functions.

The parton distributions of the pomeron (which are related to the diffractive parton distributions, see Sect. 10.6) have been extracted from a QCD fit to DDIS data by the H1 Collaboration (Adloff et al. 1997). The pomeron

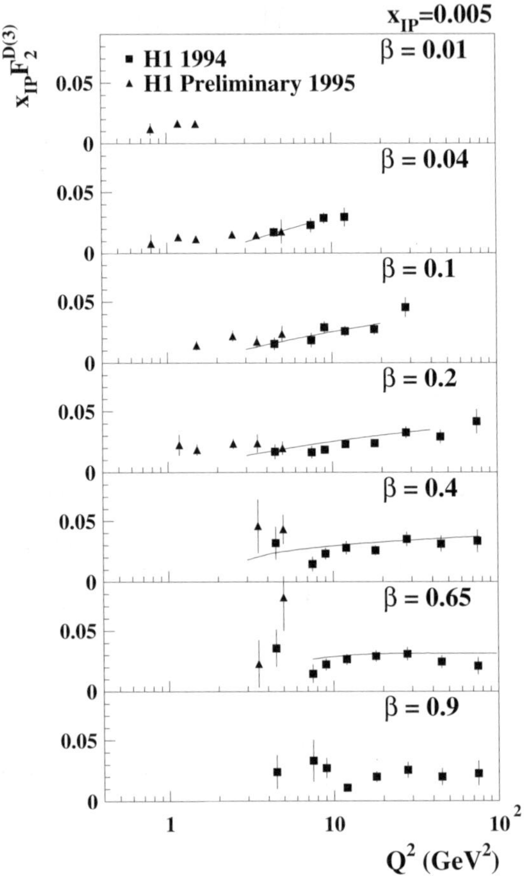


**Fig. 10.19.** $Q^2$ dependence of $x_{I\!P} F_2^{D(3)}$ (H1 1998b).

density functions $q_i(\beta, Q^2)$ and $g(\beta, Q^2)$ are given at some input scale $Q_0^2$ and then evolved by DGLAP equations (with the *caveat* that no energy-momentum sum rule holds). The result of this QCD analysis is shown in Fig. 10.21. Different solutions are found, due to the uncertainties of the data. In all cases, a large gluon component is found, which can be interpreted by saying that most of the momentum of the pomeron is carried by gluons. Note in Fig. 10.21 that one of the solutions of the fit corresponds to a gluon density peaked at large $\beta$. Once more, we recall that diffractive structure functions obey DGLAP evolution equations except for the higher-twist terms which are quite relevant at large $\beta$ values, $\beta \gtrsim 0.7$. Thus, the behavior of the extracted parton distributions in this $\beta$ region must be interpreted with great care.

Finally, the use of leading-proton detectors allows investigating directly the-$t$ dependence of the diffractive structure function. A measurement has been performed in the range $0.073 < |t| < 0.4$ GeV$^2$ by the ZEUS collabo-

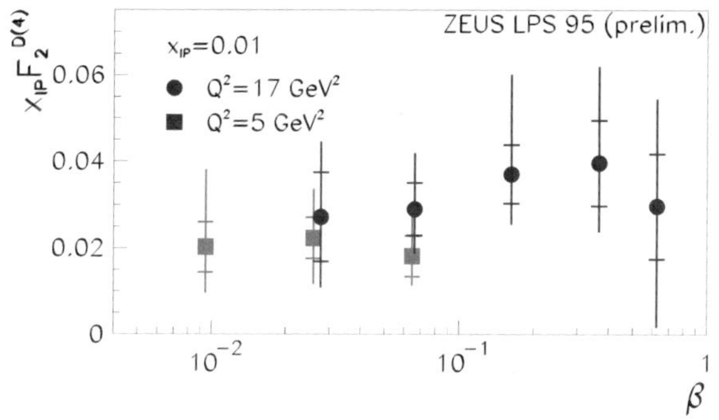


**Fig. 10.20.** $\beta$ dependence of $x_{I\!P}F_2^{D(4)}$ (figure from Abramowicz 2000)

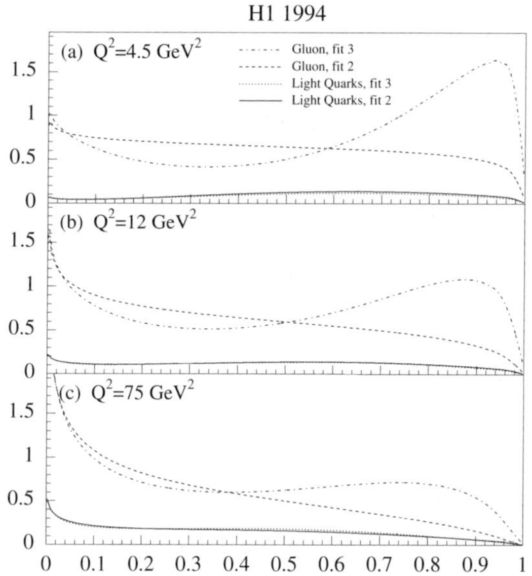


**Fig. 10.21a–c.** Parton distributions of the pomeron as a function of $\beta$ (Adloff et al. 1997).

ration (Breitweg et al. 1998). Assuming an exponential behavior $\mathrm{d}F_2^D/\mathrm{d}t \propto \exp(-b\,|t|)$, the slope parameter is found to be

$$b = 7.2 \pm 1.1\,(\text{stat.})\ ^{+0.7}_{-0.9}\,(\text{syst.})\ \text{GeV}^{-2} \tag{10.57}$$

This value is similar to those obtained in single diffractive $pp$ reactions.

## 10.9 Hadronic Final States in DDIS

Useful information on the internal quark-gluon dynamics of DDIS comes from other, less inclusive, diffractive $\gamma^* p$ processes, in which some particular hadronic final states are selected. One of such processes is high-$p_T$ dijet production

$$\gamma^* + p \to j\,j + X + p\,, \tag{10.58}$$

with a large rapidity gap between the proton remnant and the hadronic systems (Adloff et al. 1999, 2001; ZEUS 2000a). From the infinite-momentum-frame point of view, this reaction probes the gluon content of the pomeron via the photon–gluon fusion diagram (Fig. 10.10c). On the other hand, adopting the dipole picture of DIS, we interpret dijet production as the excitation of a higher Fock $q\bar{q}g$ fluctuation of the virtual photon (Fig. 10.10d). The observed rates (Adloff et al. 1999, 2001; ZEUS 2000a) are in agreement with calculations based on the diffractive parton distributions extracted from fits to $F_2^{D(3)}$. This can be seen in Fig. 10.22, where the $z_{I\!P}$-distribution is displayed (Adloff et al. 2001). The variable $z_{I\!P}$ is defined as

$$z_{I\!P} = \frac{q \cdot \kappa}{q \cdot (P - P')} \doteq \beta \left(1 + \frac{\hat{s}}{Q^2}\right) , \tag{10.59}$$

where $\kappa$ is the momentum of the gluon in Fig. 10.10c,d, and $\hat{s}$ is the invariant mass squared of the $q\bar{q}$ pair producing the dijet. Thus, $z_{I\!P}$ represents the fraction of the pomeron momentum carried by the gluon. The data are well reproduced by the flat gluon distribution extracted from the measurements of $F_2^{D(3)}$, while the peaked gluon distribution is slightly disfavored (Fig. 10.22).

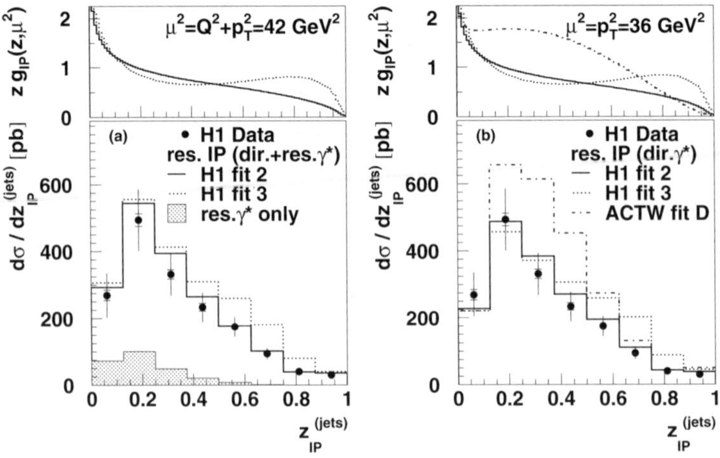


**Fig. 10.22.** The $z_{I\!P}$ distribution in diffractive jet production at HERA (Adloff et al. 2001).

Models based on the color dipole approach are able to reproduce the observed cross sections at low $x_{I\!P}$ values.

A related process studied at HERA (H1 1999; ZEUS 2000b) is diffractive open charm production, that is

$$\gamma^* + p \to c\bar{c} + X + p\,, \tag{10.60}$$

with hadronization of the charm quark (antiquark) into $D$ mesons. It is found that the ratio of diffractive to non-diffractive $D^{*\pm}$ production is about 6% (ZEUS 2000b). At the moment, however, the data are affected by large systematic and statistical errors, hence any comparison with the theoretical predictions is premature.

## 10.10 Vector Meson Production

An important class of diffractive processes is exclusive production of vector mesons in $\gamma p$ ($Q^2 \simeq 0$) or $\gamma^* p$ ($Q^2 > 0$) scattering,

$$\gamma\,(\gamma^*) + p \to V + p\,. \tag{10.61}$$

Both photoproduction and leptoproduction of vector mesons have been intensively studied at HERA for $\rho, \omega, \phi, \rho', J/\psi, \psi', \Upsilon$ (for an updated review see Adamczyk 2001; more references can be found in Marage 1999, Abramowicz 2000, Abramowicz 2001).

Let us start from photoproduction. In the spirit of the Vector Dominance Model (VDM) we can assume that the photon fluctuates into a vector meson which then interacts elastically with the proton via pomeron exchange (Fig. 10.23a). Since vector meson photoproduction represents the elastic part of $\sigma_{\rm tot}^{\gamma p}$, we can use the optical theorem to write

$$\left.\frac{d\sigma^{\gamma p \to V p}}{dt}\right|_{t=0} = \frac{1}{16\pi}\,(1+\rho^2)\,(\sigma_{\rm tot}^{\gamma p})^2\,, \tag{10.62}$$

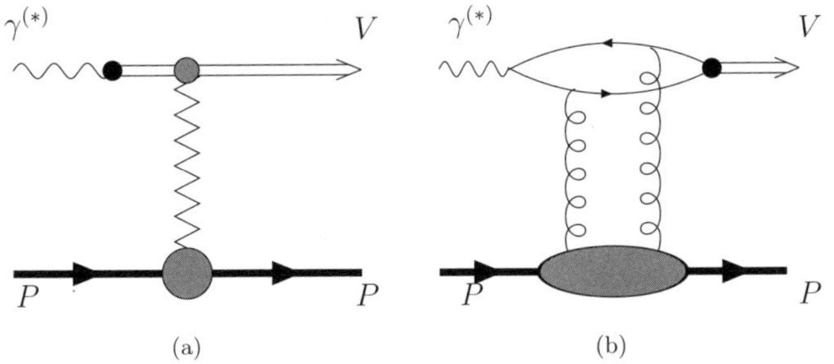


**Fig. 10.23a,b.** Vector meson production in VDM (**a**) and in the two-gluon exchange model (**b**).

where $\rho$ is the ratio of the real to imaginary part of the scattering amplitude. In case of pomeron dominance the amplitude is purely imaginary, so the $\rho$ parameter is zero and the whole $W^2$ dependence of the r.h.s of (10.62) is contained in $\sigma_{\text{tot}}^{\gamma p}$. Assuming the $t$-dependence of the elastic cross section as dictated by Regge theory

$$\frac{\mathrm{d}\sigma^{\gamma p\to Vp}}{\mathrm{d}t} = \left.\frac{\mathrm{d}\sigma^{\gamma p\to Vp}}{\mathrm{d}t}\right|_{t=0} \exp\left[\left(b_0 + 2\,\alpha'_{I\!P} \ln\frac{W^2}{W_0^2}\right) t\right] \tag{10.63}$$

and using (10.62) with $\rho = 0$, we get

$$\frac{\mathrm{d}\sigma^{\gamma p\to Vp}}{\mathrm{d}t} = A\,(\sigma_{\text{tot}}^{\gamma p})^2 \exp\left[\left(b_0 + 2\,\alpha'_{I\!P} \ln\frac{W^2}{W_0^2}\right) t\right] , \tag{10.64}$$

where $A$ is a constant. Integration of (10.64) over $|t|$, between $|t|_{\text{min}}$ and $|t|_{\text{max}}$ (with $|t|_{\text{min}} \simeq 0.7$ GeV$^2$ typically), gives, using $\sigma_{\text{tot}}^{\gamma p} \sim (W^2)^{\alpha_{I\!P}(0)-1}$

$$\begin{aligned}\sigma^{\gamma p\to Vp}(W^2) &\sim \frac{1}{b_0 + 2\alpha'_{I\!P}\,\ln(W^2/W_0^2)} \\ &\quad\times \left[\mathrm{e}^{-(b_0+2\alpha'_{I\!P}\,\ln(W^2/W_0^2)\,|t|)}\right]_{|t|_{\text{min}}}^{|t|_{\text{max}}} (W^2)^{2\,\alpha_{I\!P}(0)-2} \\ &\sim (W^2)^{2\,a_{I\!P}-2} .\end{aligned} \tag{10.65}$$

Here $a_{I\!P}$ is the average intercept which takes into account the $W^2$ dependence of the denominator and of the exponential in (10.65). As we said in Sect. 10.8, the difference between the real pomeron intercept $\alpha_{I\!P}(0)$ and $a_{I\!P}$ is few percent.

Experimentally, it is found that for $\rho, \omega$ and $\phi$ photoproduction the cross section goes approximately as (Fig. 10.10)

$$\sigma^{\gamma p\to Vp} \sim W^{0.2} , \tag{10.66}$$

which implies, according to (10.65), a typical "soft pomeron" intercept.

The cross section for $J/\psi$ photoproduction rises much faster than (10.66), and so does the electroproduction cross section of $\rho$ at high $Q^2$. In this case, a hard scale (the mass $m_{\text{V}}^2$ of the heavy meson, and/or the large virtuality of the photon $Q^2$) is present in the process, and perturbative QCD is suitable to describe it (Ryskin 1993; Nemchik, Nikolaev and Zakharov 1994; Brodsky et al. 1994). The photon fluctuates into a $q\bar{q}$ pair, which interacts, after a certain time, with the proton, by exchanging two gluons, or a gluonic ladder (see Fig. 10.23b). The fast increase of the cross section with $W^2$ is attributed to the rise of the gluon distribution at low $x$. The only component of the process which needs to be modelled, and thus is not completely under theoretical control, is the vector meson wave function.

Using a simple non-relativistic wave function for the vector meson, perturbative QCD in the two-gluon exchange approximation (see Sect. 11.7) predicts for the cross section of the process $\gamma^{(*)}p \to Vp$

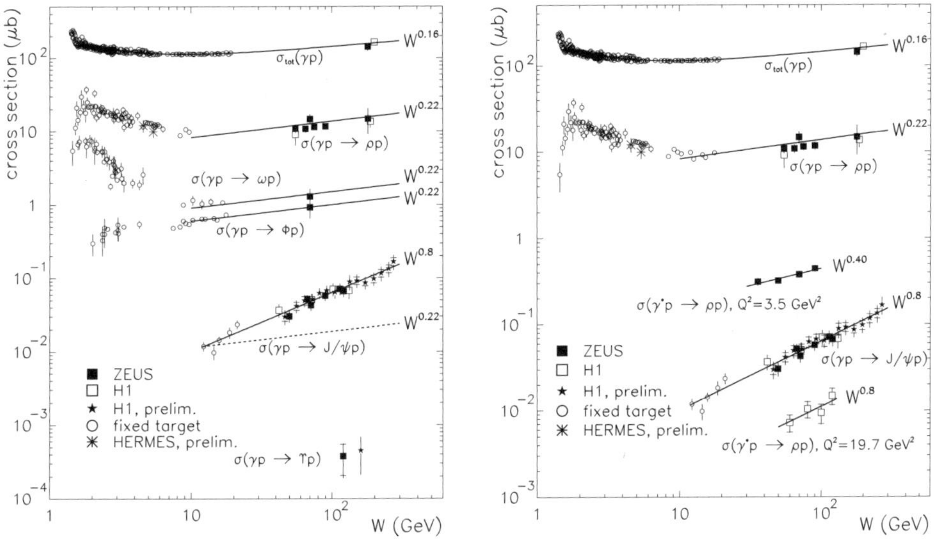


**Fig. 10.24.** Energy dependence of various vector-meson photoproduction and electroproduction processes. From Abramowicz (2000).

$$\left.\frac{\mathrm{d}\sigma^{\gamma^* p \to V p}}{\mathrm{d}t}\right|_{t=0} \propto \left[\alpha_s \, x_{I\!P} \, g\left(x_{I\!P}, \widetilde{Q}^2\right)\right]^2 , \tag{10.67}$$

where $g(x_{I\!P}, \widetilde{Q}^2)$ is the gluon distribution at an effective $Q^2$ value given by

$$\widetilde{Q}^2 \simeq \frac{Q^2 + M_V^2}{4} . \tag{10.68}$$

Note the dependence on the square of the gluon distribution, arising from the fact that two gluons are exchanged.

The ratio of the longitudinal to tranverse cross section goes as

$$\frac{\sigma_L^{\gamma^* p \to V p}}{\sigma_T^{\gamma^* p \to V p}} \sim \frac{Q^2}{M_V^2} , \tag{10.69}$$

and therefore electroproduction of (sufficiently heavy) vector mesons is dominated by the longitudinal contribution.

In the limit of photoproduction, one can use the above formulas, with $Q^2 = 0$, only insofar as $M_V^2$ is large enough to provide a hard scale for perturbation theory (this is the case in $J/\psi$ and $\Upsilon$ production). On the other hand, if $M_V^2$ is small, as it is for $\rho$ meson, but $Q^2$ is high, a perturbative treatment is still possible but the light-meson wave function is quite out of control and predictions are difficult.

Using the low-$x$ behavior of the gluon distribution $xg(x) \sim x^{-\lambda}$, the $W^2$ dependence expected on the basis of (10.67) is

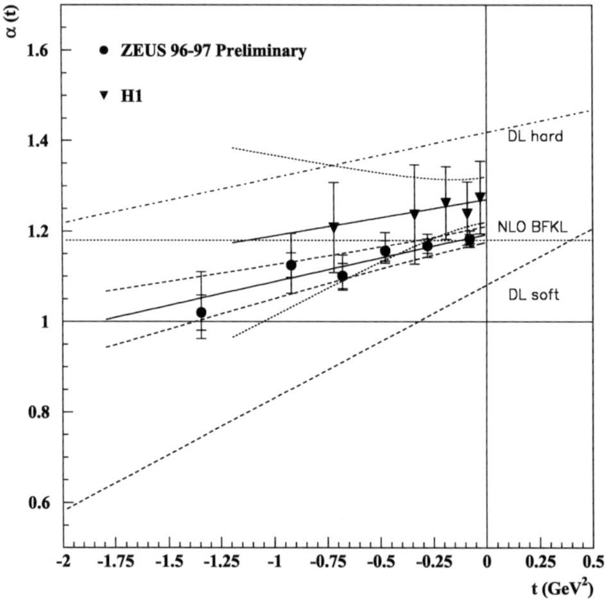


**Fig. 10.25.** The pomeron trajectory extracted from $J/\psi$ photoproduction. The solid lines are linear fits to the data (the one-standard-deviation contour is also indicated). Also shown are the Donnachie-Landshoff "soft' and "hard" pomerons (Donnachie and Landshoff 1999) and the intercept predicted by the NLO BFKL approach of Brodsky et al. 1999. Figure from ZEUS (2000c).

$$\sigma^{\gamma^{(*)}p\to Vp} \sim \left[x_{\mathbb{P}}\, g\left(x_{\mathbb{P}}, \tilde{Q}^2\right)\right]^2 \sim x_{\mathbb{P}}^{-2\lambda} \sim \left(W^2\right)^{2\lambda} \sim W^{0.8}\,, \qquad (10.70)$$

where $\lambda \approx 0.2$ is the approximate value emerging from the analyses of the inclusive DIS data. The rise (10.70) is in excellent agreement with the experimental findings for $\gamma\, p \to J/\psi\, p$, and $\gamma^* p \to \rho\, p$ at large $Q^2$ (see Fig. 10.10).

The pomeron trajectory as extracted from $J/\psi$ photoproduction data (Adloff et al. 2000b; ZEUS 2000c) is shown in Fig. 10.25 together with the results of linear fits of the form $\alpha_{\mathbb{P}}(t) = \alpha_{\mathbb{P}}(0) + \alpha'_{\mathbb{P}}t$. As an indication, we give the ZEUS findings

$$\begin{aligned}\alpha_{\mathbb{P}}(0) &= 1.193 \pm 0.011(\text{stat.})\,{}^{+0.015}_{-0.010}(\text{syst.})\\ \alpha'_{\mathbb{P}} &= 0.105 \pm 0.024(\text{stat.})\,{}^{+0.022}_{-0.020}(\text{syst.})\ \text{GeV}^{-2}\end{aligned}$$

The intercept turns out to be somewhat intermediate between those classically attributed to the "soft" and "hard" pomerons.

Let us discuss now the $Q^2$ dependence of vector meson production. If the cross sections at fixed $W^2$ are parametrized as

$$\sigma^{\gamma^* p\to Vp} \sim \frac{1}{(Q^2 + M_V^2)^n}\,, \qquad (10.71)$$

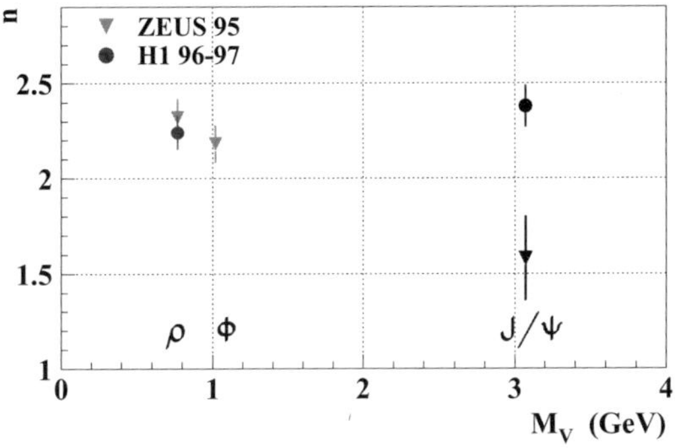


**Fig. 10.26.** The power $n$ in the $(Q^2 + M_V^2)^{-n}$ dependence of vector meson production.

one finds experimentally the power $n$ shown in Fig. 10.26. At large $Q^2$ the perturbative QCD approach predicts (see Sect. 11.7)

$$\sigma^{\gamma^* p \to Vp} \sim \frac{[x\,g(x, Q^2/4)]^2}{Q^6} , \tag{10.72}$$

and if we use the behavior

$$x\,g(x, Q^2) \sim (Q^2)^{\gamma}, \quad \text{with } \gamma = 0.3 \text{ for } 10^{-3} < x < 10^{-2} , \tag{10.73}$$

determined from global analyses of inclusive DIS data, we get

$$\sigma^{\gamma^* p \to Vp} \sim \frac{1}{Q^{4.8}} , \tag{10.74}$$

which is consistent with the data. However, as already mentioned, perturbative QCD, combined with a non-relativistic description of the meson wave function, predicts also that $\sigma_T^{\gamma^* p \to Vp}$ should be suppressed by a factor $1/Q^2$ with respect to $\sigma_L^{\gamma^* p \to Vp}$. As one can see from Fig. 10.27, such a behavior is not observed, at least in the case of $\rho$ production (Breitweg et al. 1999a, Adloff et al. 2000a). This is a clear indication that a better treatment of the recombination of the $q\bar{q}$ pair into the $\rho$ meson is needed (Nemchik et al. 1997; Martin, Ryskin and Teubner 1997; Royen and Cudell 1999). It should be also noticed that more recent data (ZEUS 2000d) seem to indicate that the exponent $n$ in (10.71) depends on $Q^2$, so that a fit of the form (10.71), with constant $n$, does not work over the full $Q^2$ range. This can be due to the variation with $Q^2$ of the exponent $\gamma$ in (10.73) or to some $Q^2$-dependent effects in the meson wave function.

The applicability of perturbative QCD at large $Q^2$, and/or for sufficiently heavy mesons, is supported by the measurements of the slope parameter $b$, which defines the diffractive peak according to the behavior $d\sigma/dt \sim \exp(-b\,|t|)$. It is found that $b$ gets smaller as $Q^2$ and $M_V^2$ increase

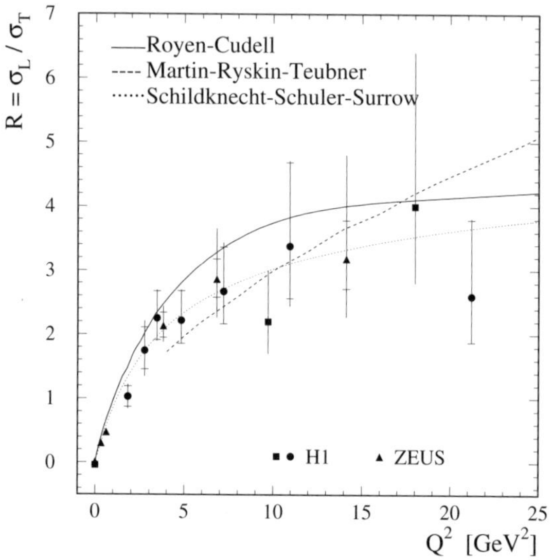


**Fig. 10.27.** The ratio of the longitudinal to transverse photon cross sections for elastic $\rho$ electroproduction as a function of $Q^2$. The predictions of some models are also displayed. From Adloff et al. (2000a).

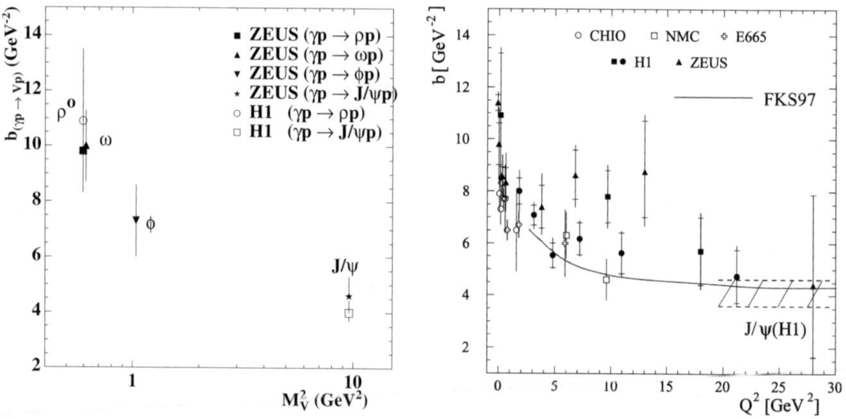


**Fig. 10.28.** *Left*: the slope $b$ of the $t$ distribution in $\gamma^* p \to V p$ at HERA as a function of the vector meson mass. *Right*: the slope $b$ as a function of $Q^2$ for $\rho$ production. From Abramowicz (2000).

(this behavior was predicted by Nikolaev, Zakharov and Zoller 1996; Nemchik et al. 1998). In particular, the value of $b$ for the $\rho$ meson decreases from $b \simeq 10$ GeV$^2$ at $Q^2 \simeq 0$ to $b \simeq 5$ GeV$^2$ for $Q^2 \gtrsim 10$ GeV$^2$. It also decreases by the same amount going from the $\rho$ and $\omega$ mesons to the $J/\psi$ (Fig. 10.28). Since $b = 5$ GeV$^2$ corresponds to an interaction radius which is close to the proton radius, we can conclude that, at large $Q^2$ and/or for heavy mesons, the size of the $\gamma^* \to V$ vertex, i.e. the size of the $q\bar{q}$ pair, is small, which justifies the use of perturbative QCD.

## 10.11 Diffraction in Hadron–Hadron Collisions

As already recalled, after the seminal paper by Ingelman and Schlein (1985), where high-$p_T$ jets produced via pomeron exchange were discussed for the first time, events of this type, containing two jets of high transverse energy and a leading proton (Fig. 10.29), were observed in $p\bar{p}$ interactions at $\sqrt{s} = 630$ GeV by the CERN UA8 experiment (Bonino et al. 1988). The rate of jet production in single-diffractive scattering was found to be $\sim 1-2$ % (Fig. 10.30) – in agreement with the predicted order of magnitude (Ingelman and Schlein 1985) –, but the number of events was quite limited.

Since then, the hard diffraction program in hadron–hadron scattering has been pursued by the UA8 Collaboration at the CERN SPS Collider, and by the CDF and D0 Collaborations at the Tevatron.

In 1992 the UA8 group (Brandt et al. 1992), using a larger sample of data, reported some evidence for a hard pomeron substructure, of the type $\beta(1-\beta)$, with a significant fraction of events exhibiting a peak near $\beta = 1$ (the final UA8 hard–diffraction results are presented in Brandt et al. 1998a).

The range of physical processes explored by the Tevatron experiments is broader. These experiments have been investigating various diffractive reactions at two center–of–mass energies, $\sqrt{s} = 1.8$ TeV and $\sqrt{s} = 630$ GeV. The first results were reported in 1994–1995 (Abachi et al. 1994; Abe et al. 1995; for a review of the latest observations see Goulianos 2001a, 2001b).

Three different classes of processes are investigated at the Tevatron (their topologies in the $\eta - \phi$ plane are shown in Fig. 10.31): double diffraction (Fig. 10.31a), single diffraction (Fig. 10.31b) and double pomeron exchange (Fig. 10.31c).

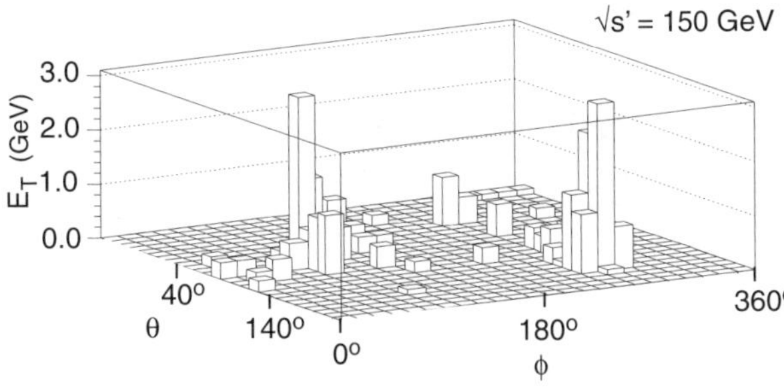


**Fig. 10.29.** A typical UA8 diffractive event with two high-$E_T$ jets and a detected leading proton ($s' \equiv s(1-x_F)$). From Brandt et al. (1998a).

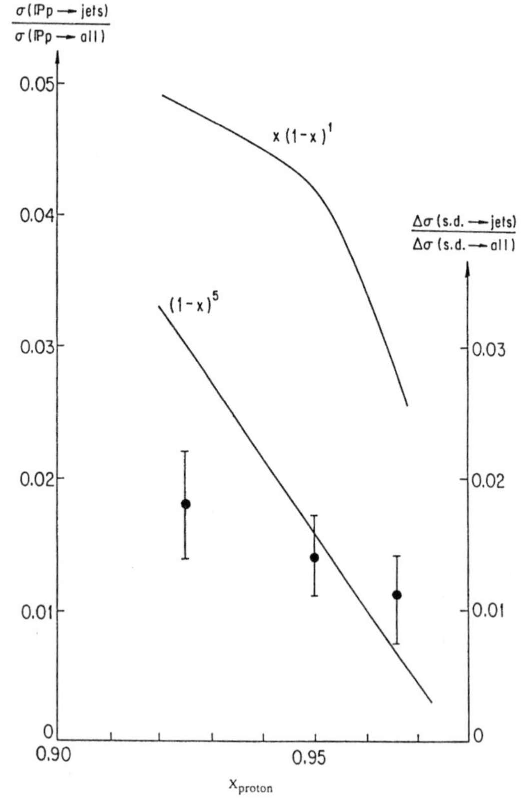


**Fig. 10.30.** The first evidence for transverse jets in diffractive interactions. The figure shows the rate of jet production in single-diffractive scattering as a function of $x_F$, measured by the UA8 experiment (from Bonino et al. 1988)

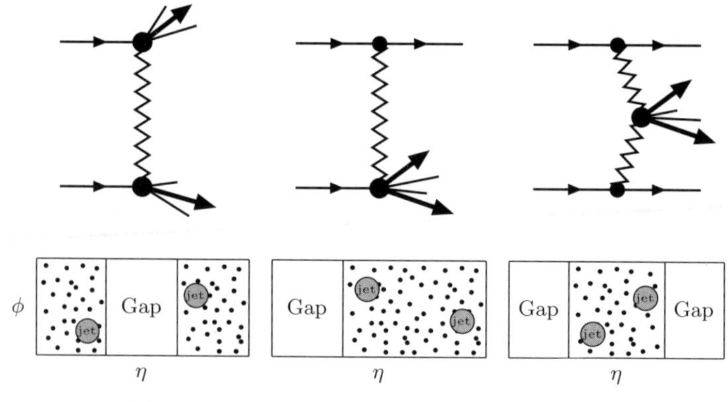


**Fig. 10.31.** The three classes of diffractive reactions studied at the Tevatron and their topologies ($\eta$ is the pseudorapidity, $\phi$ the azimuthal angle): (**a**) double diffraction; (**b**) single diffraction; (**c**) double pomeron exchange.

Both the CDF and the D0 detectors cover the pseudorapidity range $|\eta| \lesssim 4-5$, and the installed Roman Pots are able to tag leading antiprotons with $0.90 \lesssim x_F \lesssim 0.95$, that is[1] $0.05 \lesssim \xi \lesssim 0.10$.

### 10.11.1 Double Diffraction

Rapidity gaps between jets were proposed by Dokshitzer, Khoze and Troyan (1987) and Bjorken (1993) as a signature of color-singlet exchange. Events of this type are predominantly of diffractive nature, since the contribution from electroweak processes, which would give a similar configuration, is small.

The CDF and D0 experiments have collected dijet data with central rapidity gaps (Abachi et al. 1994, 1996; Abbott et al. 1998; Abe et al. 1995, 1998a, 1998b), and found the diffractive to non-diffractive production ratio $f$ to be about 1 % at $\sqrt{s} = 1.8$ GeV, that is 10 times smaller than the diffractive rate measured at HERA. A significant difference has been observed between the color–singlet rates at $\sqrt{s} = 1.8$ TeV and $\sqrt{s} = 630$ GeV: $f(630)/f(1800) \approx 2$.

The decrease of the double diffractive contribution with increasing energy can be explained by introducing the concept of gap survival. The survival probability of a rapidity gap, that we call $\langle S^2 \rangle$, is defined as the fraction of events for which spectator interactions do not fill the gap. In the eikonal picture $\langle S^2 \rangle$ is given by (Bjorken 1993; Gotsman, Levin and Maor 1993)

$$\langle S^2 \rangle = \frac{\int \mathrm{d}^2\boldsymbol{b}\, \Gamma_H(b)\, S^2(s,b)}{\int \mathrm{d}^2\boldsymbol{b}\, \Gamma_H(b)} , \tag{10.75}$$

where $\Gamma_H(b)$ is the profile function for the hard scattering process and $S^2(s,b)$ is the probability that no inelastic interaction occurs.

Bjorken's estimate for the survival probability is $\langle S^2 \rangle \approx 0.05 - 0.10$ at $\sqrt{s} = 1.8$ TeV. Similar values are found by Gotsman, Levin and Maor (1993) who use various phenomenological models.

It is reasonable to expect that $\langle S^2 \rangle$ varies with energy, in particular that it decreases with increasing $\sqrt{s}$, since the interactions between the particle remnants become stronger and tend to destroy the gap. The energy dependence of the survival probability may explain why $f(630)$ is larger than $f(1800)$. This has been confirmed by a calculation of the gap survival factor based on a perturbative QCD multiple interaction model (Cox, Forshaw and Lönnblad 1999).

As mentioned earlier, there is a significant difference between the diffractive rate observed at the Tevatron and the one measured at HERA ($f \sim 10\%$ at $\sqrt{s} \simeq 200$ GeV). The low yield of rapidity gap events at the Tevatron relative to HERA may be (in part, at least) explained in terms of survival probabilities as well. Since there are more spectator partons in $\bar{p}p$ collisions with respect to $\gamma^* p$, the rate of gap destruction is larger at the Tevatron than

[1] Note that in Tevatron papers $x_{I\!P}$ is called $\xi$.

at HERA (moreover, HERA energies are lower). However, the gap survival estimates of Cox, Forshaw and Lönnblad (1999) indicate a factor 3 difference between HERA and Tevatron, to be compared with the observed difference which is much larger, a factor 10 approximately.

### 10.11.2 Single Diffraction

The signature of hard single diffraction (SD) at the Tevatron (Fig. 10.31a) is two jets produced on the same side and either a forward rapidity gap along the direction of one of the initial particles (Abe et al. 1997b), or a leading particle with $\xi \lesssim 0.1$ (Affolder et al. 2000b). From a phenomenological point of view, the single dissociation process $\bar{p}p \to \bar{p}X$ is described by assuming that a pomeron is emitted by the incident antiproton and undergoes a hard scattering with the proton. This is an ideal reaction to study the partonic content of the pomeron, that is the $\beta$ dependence of the diffractive structure functions.

Comparing two samples of dijet events, a diffractive one, collected by tagging the leading antiproton with a Roman pot detector, and an inclusive one, collected with a minimum bias trigger, the CDF Collaboration has been able to extract the diffractive structure function of the antiproton (Affolder et al. 2000b). This is done by multiplying the measured ratio $R^{\mathrm{SD}}_{\mathrm{ND}}(x_{\bar{p}}, \xi_{\bar{p}})$ of single-diffractive to non-diffractive cross sections (Fig. 10.32) by the known inclusive structure function, which is

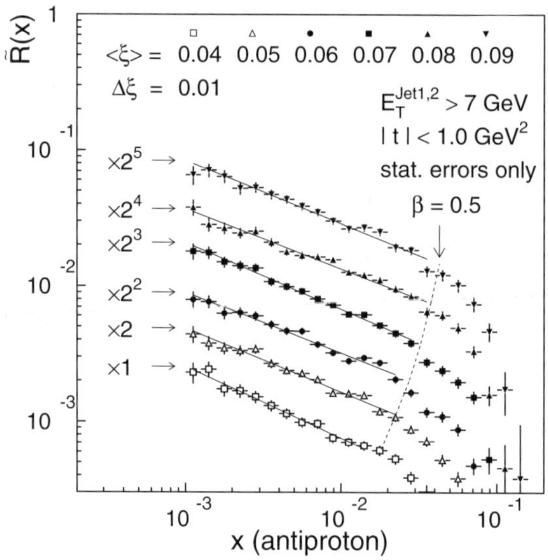


**Fig. 10.32.** Ratio of diffractive to non-diffractive event rates as a function of $x_{\bar{p}}$ (the momentum fraction of partons in antiproton). From Affolder et al. (2000b).

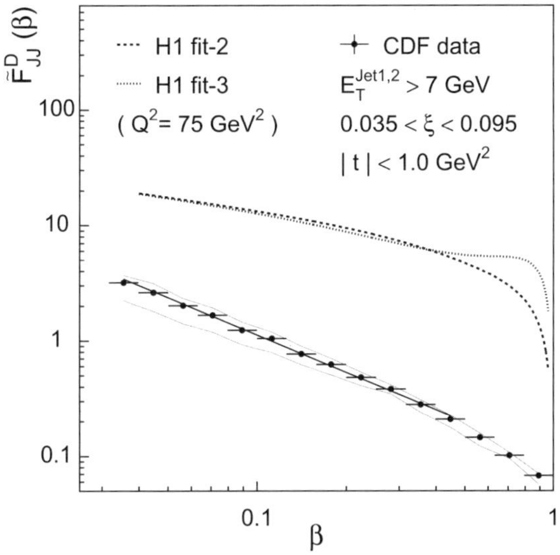


**Fig. 10.33.** The dijet diffractive structure function $F_{jj}^D$ as a function of $\beta$, measured by CDF (Affolder et al. 2000b). The dotted and dashed lines are the expectations based on the diffractive parton densities extracted from DDIS by H1.

$$F_{jj}(x_{\bar{p}}) = x_{\bar{p}} \left\{ g(x_{\bar{p}}) + \frac{4}{9} \sum_f [q_f(x_{\bar{p}}) + \bar{q}_f(x_{\bar{p}})] \right\} . \tag{10.76}$$

This procedure avoids the use of model–dependent Monte Carlo generators. From $F_{jj}^D(x_{\bar{p}}, \xi_{\bar{p}}) = R_{\rm ND}^{\rm SD}(x_{\bar{p}}, \xi_{\bar{p}})\, F_{jj}(x_{\bar{p}})$ one then gets $F_{jj}^D(\beta, \xi_{\bar{p}})$ by a change of variables, $\beta = x_{\bar{p}}/\xi_{\bar{p}}$. The resulting $\beta$ distribution is shown in Fig. 10.33 along with expectations based on diffractive parton densities extracted by H1 from DDIS measurements. A large discrepancy both in shape and normalization is found. As mentioned in Sect. 10.3, QCD factorization is violated in diffractive hadron–hadron scattering (Collins 1998), and therefore one cannot use diffractive parton distributions extracted from DDIS to predict hard diffractive processes in $p\bar{p}$ collisions. Still, the size of the discrepancy between DDIS-based expectations and dijet data is somewhat surprising.

Another evidence of the factorization breakdown comes from the CDF determination of the gluon fraction of the pomeron, based on a combined analysis of diffractive dijet, $W$ and $b$-quark measurements. The ratio $D$ of measured to predicted diffractive rates as a function of the gluon content of the pomeron is plotted in Fig. 10.34. As one can see, ZEUS and CDF agree on the value of the gluon fraction, which is about 50 %, but their $D$ ratios are largely different.

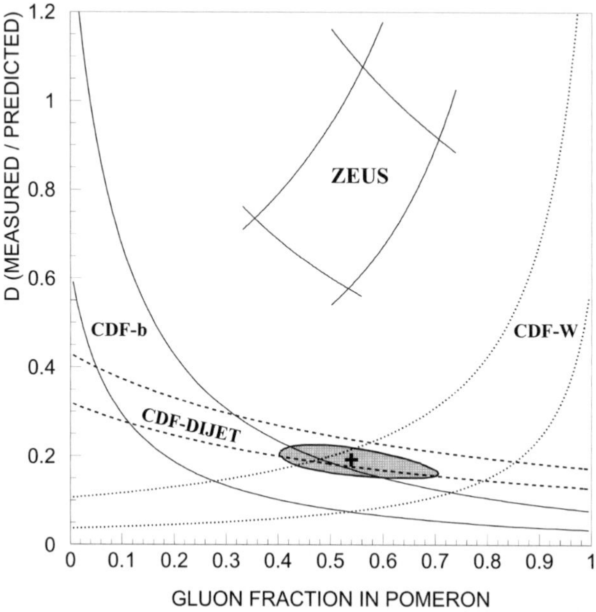


**Fig. 10.34.** The ratio of measured to predicted diffractive rates as a function of the gluon fraction of the pomeron. The black cross represents the best fit to the combined CDF data (the shaded ellipse is the corresponding $1\sigma$ contour). From Goulianos (2001b).

### 10.11.3 Double Pomeron Exchange

The first observation of dijet production via double pomeron exchange (DPE) in $p\bar{p}$ collisions was reported by the CDF Collaboration (Affolder et al. 2000c). The events are characterized by a leading antiproton, two jets in the central pseudorapidity region with transverse energy $E_T > 7$ GeV and a large rapidity gap on the outgoing proton side. The ratio of DPE to SD rates, $R_{\rm SD}^{\rm DPE}(x_p, \xi_p)$ was determined as a function of the proton's Bjorken variable $x_p$. In leading order QCD $R_{\rm SD}^{\rm DPE}(x_p, \xi_p)$ is equal to the ratio of the diffractive to non-diffractive color weighted structure functions of the proton, which have the form (10.76). Assuming factorization, one should have

$$R_{\rm SD}^{\rm DPE}(x_p, \xi_p) = R_{\rm ND}^{\rm SD}(x_p, \xi_p) \; . \tag{10.77}$$

The comparison is made with the available quantity $R_{\rm ND}^{\rm SD}(x_{\bar{p}}, \xi_{\bar{p}})$ and the result shown in Fig. 10.35 indicates that the equality of $R_{\rm SD}^{\rm DPE}$ and $R_{\rm ND}^{\rm SD}$ is not fulfilled: the two quantities are different from each other, the latter being about 20 % of the former.

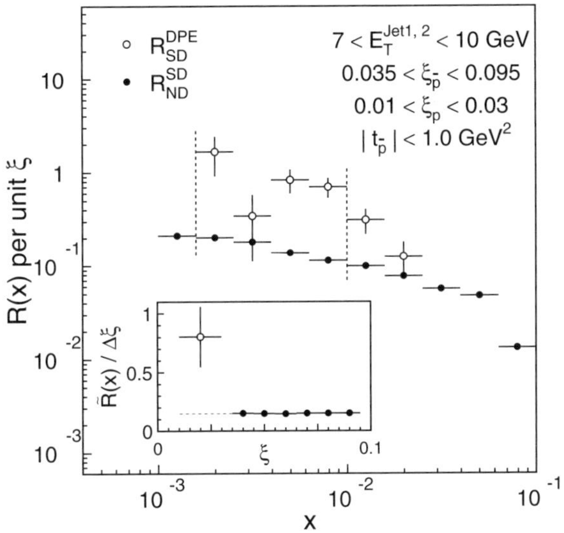


**Fig. 10.35.** Ratios of DPE to SD (SD to ND) dijet event rates per unit $\xi_p$ ($\xi_{\bar{p}}$) as a function of $x_p$ ($x_{\bar{p}}$). From Affolder et al. (2000c).

### 10.11.4 Other Diffractive Reactions

Other diffractive processes investigated at the Tevatron using the rapidity gap method include: *i)* $W$-boson production (the measured diffractive to non-diffractive rate is $\sim 1\%$, see Abe et al. 1997a); *ii)* $b$-quark production (the ratio of diffractive to total production rates is found to be $\lesssim 1\%$, see Affolder et al. 2000a; *iii)* $J/\psi$ production (Affolder et al. 2001b). For a recent general account of these (and other) measurements see Goulianos (2001a, 2001b). The perspectives of diffraction at Tevatron are reviewed by Santoro (2001).

# 11. Hard Diffraction in QCD

The BFKL picture of the pomeron, presented in Chap. 8, and the color dipole formalism, outlined in Sect. 9.9 (and discussed in detail in the next pages), are the main theoretical tools of the QCD approach to hard diffractive processes. These processes probe the transition regime from soft to hard physics. Therefore, the appropriate theoretical framework is perturbative QCD extrapolated to the "semihard" region. Such an extrapolation must be performed with great care, since non-perturbative dynamics plays an important rôle in most diffractive reactions. The full predictive power of the theory manifests itself in the study of those processes which are dominantly, or purely, perturbative, such as diffraction at large momentum transfer.

In what follows we are going to use QCD to describe a number of hard diffractive phenomena. We shall discuss both the achievements and the limits of the theory. The first part of the chapter is devoted to the two-gluon exchange description of diffractive DIS and related processes. In the second part, the focus will be on the BFKL phenomenology. The link between the two parts is represented by the color dipole picture of hard diffraction. Recent extensive reviews on the subject of this chapter are Wüsthoff and Martin (1999) and Hebecker (2000).

## 11.1 Quantum Mechanics of Diffractive Scattering

What is surprising about diffraction is that some old ideas, developed long time ago, when our understanding of strong interactions was – to say the least – rather incomplete, have proven to be valid also in the light of modern theories of hadronic phenomena. One of the reasons of such longevity resides in the fact that diffraction is essentially a quantum mechanical effect. As we shall see, the new, crucial, ingredient that QCD has brought into the field is the identification of the relevant degrees of freedom.

Let us start from the quantum mechanical picture of diffraction developed by Good and Walker in their pioneering paper (Good and Walker 1960; see also Kopeliovich and Lapidus 1978; Miettinen and Pumplin 1978; Nikolaev 1981). Let $T$ be the transition matrix describing elastic and diffractive scattering off a definite hadron (say $N$). We assume that the scattering amplitudes are purely imaginary, hence we set $T = \mathrm{i}\,\mathcal{D}$, where $\mathcal{D}$ is real. Consider

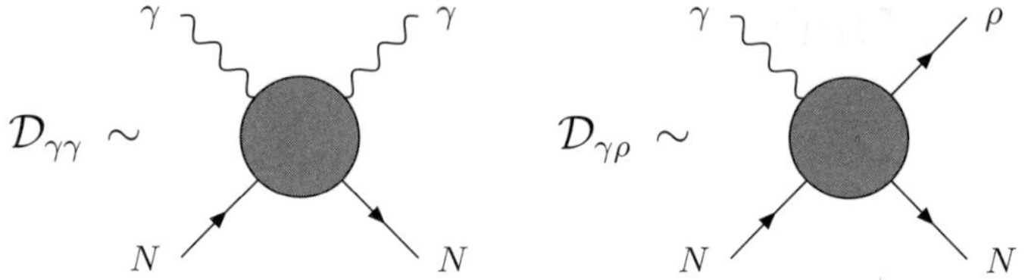


**Fig. 11.1.** Two of the $\mathcal{D}$ matrix elements for the set of states described in the text.

now a basis of physical hadronic states $|i\rangle$ with the same quantum numbers. Diffractive scattering takes one of these states into another; elastic scattering takes each state into itself. Thus $\mathcal{D}_{ik} \equiv \langle k|\mathcal{D}|i\rangle$ is the amplitude for the diffractive transition $|k\rangle \to |i\rangle$. The diagonal elements $\mathcal{D}_{ii} \equiv \langle i|\mathcal{D}|i\rangle$ are the elastic amplitudes. As an example, consider the following set of states

$$\{|i\rangle\} = \{|\gamma\rangle,\ |\rho\rangle,\ |\omega\rangle,\ |\phi\rangle\}\ . \tag{11.1}$$

The $\mathcal{D}$ matrix elements are then the amplitudes of the processes shown in Fig. 11.1.

We introduce a complete set of eigenstates of $\mathcal{D}$

$$\mathcal{D}\,|\alpha\rangle = d_\alpha\,|\alpha\rangle\ . \tag{11.2}$$

The eigenvalue $d_\alpha$ is proportional to the total cross section for $\alpha N$ scattering, that we call $\sigma_\alpha$

$$\sigma_\alpha \equiv \sigma_{\text{tot}}^{\alpha N} = \frac{1}{s}\,\text{Im}\,\langle\alpha|\mathrm{i}\,\mathcal{D}|\alpha\rangle = \frac{1}{s}\,d_\alpha\ . \tag{11.3}$$

Expanding the physical states $|i\rangle$ in terms of $|\alpha\rangle$

$$|i\rangle = \sum_\alpha c_{i\alpha}\,|\alpha\rangle\ , \tag{11.4}$$

the $\mathcal{D}$ matrix elements are

$$\begin{aligned}\mathcal{D}_{ik} &= \langle k|\mathcal{D}|i\rangle \\ &= \sum_\alpha\sum_\beta c_{k\alpha}^* c_{i\beta}\,\langle\alpha|\mathcal{D}|\beta\rangle \\ &= \sum_\alpha c_{k\alpha}^* c_{i\alpha} d_\alpha\ ,\end{aligned} \tag{11.5}$$

All physical observables can be calculated starting from $\sigma_\alpha$ and the coefficients $c_{i\alpha}$. In particular, the elastic amplitude

$$\mathcal{D}_{ii} = \langle i|\mathcal{D}|i\rangle = \sum_\alpha |c_{i\alpha}|^2\,d_\alpha\ , \tag{11.6}$$

gives immediately the total cross section for $iN$ scattering, via optical theorem

$$\sigma_{\text{tot}}^{iN} = \frac{1}{s}\,\mathcal{D}_{ii} = \sum_{\alpha} |c_{i\alpha}|^2\,\sigma_\alpha\,. \tag{11.7}$$

In general, given an operator $O$, its quantum mechanical expectation value in the $|i\rangle$ state is

$$\begin{aligned}\langle O\rangle \equiv \langle i|O|i\rangle &= \sum_{\alpha\beta} \langle i|\alpha\rangle\langle\alpha|O|\beta\rangle\langle\beta|i\rangle \\ &= \sum_{\alpha\beta} c_{i\alpha}c_{i\beta}^*\,\langle\alpha|O|\beta\rangle\,. \end{aligned} \tag{11.8}$$

If $O$ is diagonal in the $|\alpha\rangle$ basis, then (11.8) becomes

$$\langle O\rangle = \sum_{\alpha} |c_{i\alpha}|^2\,O_\alpha\,, \tag{11.9}$$

where $O_\alpha \equiv \langle\alpha|O|\alpha\rangle$. Thus we can rewrite (11.7) as

$$\sigma_{\text{tot}}^{iN} = \langle\sigma_\alpha\rangle\,. \tag{11.10}$$

By definition, the diffractive cross section at $t = 0$ is

$$\begin{aligned}\left.\frac{\mathrm{d}\sigma_{iN}^D}{\mathrm{d}t}\right|_{t=0} &= \frac{1}{16\pi s^2}\sum_{k\neq i}\mathcal{D}_{ik}^2 \\ &= \frac{1}{16\pi s^2}\left(\sum_k \mathcal{D}_{ik}^2 - \mathcal{D}_{ii}^2\right)\,. \end{aligned} \tag{11.11}$$

Using the completeness of the $|i\rangle$ states, Eq. (11.11) becomes

$$\left.\frac{\mathrm{d}\sigma_{iN}^D}{\mathrm{d}t}\right|_{t=0} = \frac{1}{16\pi s^2}\left(\langle i|\mathcal{D}^2|i\rangle - \langle i|\mathcal{D}|i\rangle^2\right)\,, \tag{11.12}$$

and expanding $|i\rangle$ as in (11.4), we finally get

$$\left.\frac{\mathrm{d}\sigma_{iN}^D}{\mathrm{d}t}\right|_{t=0} = \frac{1}{16\pi}\left(\langle\sigma_\alpha^2\rangle - \langle\sigma_\alpha\rangle^2\right)\,. \tag{11.13}$$

This is a remarkable formula which expresses the cross section of diffraction dissociation in terms of quantum mechanical expectation values. It tells us that, once the scattering eigenstates $|\alpha\rangle$ have been found, the calculation of $\mathrm{d}\sigma^D/\mathrm{d}t$ is a rather straightforward task. One of the main achievements of the theory in the past decade has been the identification of the hard diffraction eigenstates in the framework of quantum chromodynamics.

## 11.2 Diffractive DIS in the Impact-Parameter Representation

We now apply the quantum mechanical formalism developed in the previous section to a specific case of hard diffraction: diffractive DIS (DDIS).

In DDIS, the diffractive eigenstates, i.e. the states which diagonalize the diffraction matrix, are the $q\bar{q}$ color dipoles[1] described in Sect. 9.9. As we have seen, in fact, color dipoles have a lifetime much longer than the interaction time and keep their transverse size frozen during the scattering process. Thus the states $|\alpha\rangle$ defined in the previous section are to be identified with the $|q\bar{q}\rangle$ Fock states into which the virtual photon fluctuates before encountering the target. The cross section for dipole–proton scattering is $\sigma(x,\rho)$ and corresponds to the $\sigma_\alpha$ introduced in our previous discussion. In conclusion, the quantum mechanical formula (11.13) translates into the following expression for the DDIS cross section

$$\left.\frac{\mathrm{d}\sigma^D_{L,T}}{\mathrm{d}t}\right|_{t=0} = \frac{1}{16\pi}\left(\langle\sigma^2(x,\rho)\rangle_{L,T} - \langle\sigma(x,\rho)\rangle^2_{L,T}\right) , \tag{11.14}$$

where the expectation values are defined as

$$\langle\sigma(x,\rho)\rangle_{L,T} \equiv \int_0^1 \mathrm{d}z \int \mathrm{d}^2\boldsymbol{\rho}\, |\Psi_{L,T}(z,\rho)|^2\, \sigma(x,\rho) . \tag{11.15}$$

Since $\langle\sigma(x,\rho)\rangle_{L,T} \equiv \sigma^{\gamma^* p}_{L,T}(x,Q^2) = \mathcal{O}(\alpha_{\mathrm{em}})$, we can neglect $\langle\sigma(x,\rho)\rangle^2$ in (11.14) and hence we obtain

$$\begin{aligned}\left.\frac{\mathrm{d}\sigma^D_{L,T}}{\mathrm{d}t}\right|_{t=0} &= \frac{1}{16\pi}\langle\sigma(x,\rho)^2\rangle_{L,T} \\ &= \frac{1}{16\pi}\int_0^1 \mathrm{d}z \int \mathrm{d}^2\rho\, |\Psi_{L,T}(z,\rho)|^2\, \sigma^2(x,\rho) .\end{aligned} \tag{11.16}$$

Despite the semi–intuitive derivation we have just offered, Eq. (11.16) is exact, as one can check by an explicit calculation of Feynman diagrams (see below, Sect. 11.4.1).

We next make a back-of-the-envelope evaluation of the various color dipole contributions to transverse and longitudinal diffractive DIS cross sections. We use the results for $\Psi_{L,T}$ obtained in Sect. 9.9, and proceed as we did there for the case of inclusive DIS cross sections.

Large-size pairs give the following contributions

[1] The contribution of the higher Fock fluctuations $|q\bar{q}g\cdots\rangle$ of the photon will be considered in Sect. 11.4.2. For these states a quantum mechanical approach is possible as well (see Sect. 9.10 and Sect. 11.10.2).

$$\left.\frac{\mathrm{d}\sigma_T^D}{\mathrm{d}t}\right|_{t=0} \sim \int \mathrm{d}z\,[z^2+(1-z)^2] \int_{1/\mu^2}^{\infty} \mathrm{d}\rho^2\,\varepsilon^2\,\frac{1}{\varepsilon^2\rho^2}\,\sigma^2(\rho)$$

$$\sim \frac{\mu^2}{Q^2}\,\frac{1}{\mu^4} = \frac{1}{\mu^2 Q^2}\,, \tag{11.17}$$

$$\left.\frac{\mathrm{d}\sigma_L^D}{\mathrm{d}t}\right|_{t=0} \sim Q^2 \int \mathrm{d}z\,z^2(1-z)^2 \int_{1/\mu^2}^{\infty} \mathrm{d}\rho^2\,\sigma^2(\rho)$$

$$\sim Q^2\,\frac{\mu^6}{Q^6}\,\frac{1}{\mu^6} = \frac{1}{Q^4}\,, \tag{11.18}$$

where $\mu \sim m_q \sim 1/R$ is a soft parameter. Similarly, for small-size pairs we have the contributions

$$\left.\frac{\mathrm{d}\sigma_T^D}{\mathrm{d}t}\right|_{t=0} \sim \int \mathrm{d}z\,[z^2+(1-z)^2] \int_0^{1/Q^2} \mathrm{d}\rho^2\,\varepsilon^2\,\frac{1}{\varepsilon^2\rho^2}\,\sigma^2(\rho) \sim \frac{1}{Q^4} \tag{11.19}$$

$$\left.\frac{\mathrm{d}\sigma_L^D}{\mathrm{d}t}\right|_{t=0} \sim Q^2 \int \mathrm{d}z\,z^2(1-z)^2 \int_0^{1/Q^2} \mathrm{d}\rho^2\,\sigma^2(\rho) \sim Q^2\,\frac{1}{Q^6} = \frac{1}{Q^4} \tag{11.20}$$

To summarize the above estimates, let us rewrite formally (11.15) and (11.16) as (Kopeliovich and Povh 1997)

$$\sigma_{L,T}^{\gamma^* p} \sim W_{L,T}^{\alpha}\,\sigma_\alpha\,, \qquad \left.\frac{\mathrm{d}\sigma_{L,T}^D}{\mathrm{d}t}\right|_{t=0} \sim W_{L,T}^{\alpha}\,\sigma_\alpha^2\,, \tag{11.21}$$

where $\alpha$ denotes the color dipoles and we introduced the weights $W_{L,T}^{\alpha}$ which incorporate the photon wave functions and the integrations. The dipole cross sections $\sigma_\alpha$ are universal quantities behaving as

$$\sigma_\alpha \sim \frac{1}{Q^2} \qquad \text{(small–size dipoles)}\,, \tag{11.22}$$

$$\sigma_\alpha \sim \frac{1}{\mu^2} \qquad \text{(large–size dipoles)}\,. \tag{11.23}$$

The weights $W_{L,T}^{\alpha}$ depend on the polarization state of the virtual photon. We have (setting conventionally to 1 the weight corresponding to small–size pairs)

$$W_T^{\alpha} \sim 1\,, \quad W_L^{\alpha} \sim 1 \qquad \text{(small–size dipoles)}\,, \tag{11.24}$$

$$W_T^{\alpha} \sim \frac{\mu^2}{Q^2}\,, \quad W_L^{\alpha} \sim \frac{\mu^4}{Q^4} \qquad \text{(large–size dipoles)}\,. \tag{11.25}$$

The results for the physical cross sections are collected in Table 11.1 for the transverse case, and in Table 11.2 for the longitudinal case.

The first conclusion we can draw from these results is that diffractive DIS is dominated by *large-size asymmetric pairs* (the so-called "aligned jet configuration") (Nikolaev and Zakharov 1991, 1992, 1994; Bjorken 1994), with the

**Table 11.1.** Contributions of 'hard' and 'soft' dipole configurations to the DIS and DDIS transverse cross sections.

| dipole | $W_T^\alpha$ | $\sigma_\alpha$ | $W_T^\alpha\,\sigma_\alpha$ | $W_T^\alpha\,\sigma_\alpha^2$ |
|---|---|---|---|---|
| small size ('hard') | $\sim 1$ | $\sim 1/Q^2$ | $\sim 1/Q^2$ | $\sim 1/Q^4$ |
| large size ('soft') | $\sim \mu^2/Q^2$ | $\sim 1/\mu^2$ | $\sim 1/Q^2$ | $\sim 1/\mu^2 Q^2$ |

**Table 11.2.** Contributions of 'hard' and 'soft' dipole configurations to the DIS and DDIS longitudinal cross sections.

| dipole | $W_L^\alpha$ | $\sigma_\alpha$ | $W_L^\alpha\,\sigma_\alpha$ | $W_L^\alpha\,\sigma_\alpha^2$ |
|---|---|---|---|---|
| small size ('hard') | $\sim 1$ | $\sim 1/Q^2$ | $\sim 1/Q^2$ | $\sim 1/Q^4$ |
| large size ('soft') | $\sim \mu^4/Q^4$ | $\sim 1/\mu^2$ | $\sim \mu^2/Q^4$ | $\sim 1/Q^4$ |

dipole cross section close to the saturation limit. By contrast, inclusive DIS receives comparable contributions from both large and small-size pairs. So, even though $Q^2$ is a hard scale, DDIS is largely non-perturbative. To make it more perturbative, one must select special final states: e.g., large $p_T$ jets, charm, longitudinally polarized vector mesons. For these "hard" configurations, the dipole is constrained to have a small transverse size so that perturbation theory is applicable. In the inclusive case, instead, short-distance (i.e., hard) and long-distance (i.e., soft) phenomena coexist. One may hope to discriminate between these two classes of contributions by looking at the energy dependence of the diffractive cross section, since the hard component is expected to grow faster with energy (at fixed $M^2$, the invariant mass of the diffracted system) than the soft one.

Even in DDIS with no hard final states, however, some information can be obtained in a purely perturbative way. Indeed, the light-cone wave functions of the virtual photon (which are calculated perturbatively) determine the main features of the $M^2$ spectrum of the process, i.e., its $\beta$ dependence (see Sect. 11.3). The $x_{I\!P}$ behavior, on the contrary, is genuinely non-perturbative, and is only open to a phenomenological parametrization.

Another observation concerning the results (11.17–11.20) is that the longitudinal contribution is higher-twist and so is generally suppressed with respect to the transverse one. However, a more careful study of the $M^2$ spectrum shows that $\frac{\mathrm{d}\sigma_L}{\mathrm{d}M^2\mathrm{d}t}$ is comparable to (or even larger than) $\frac{\mathrm{d}\sigma_T}{\mathrm{d}M^2\mathrm{d}t}$ at small $M^2$ (i.e. $\beta \to 1$) and small $Q^2$.

## 11.3 The $M^2$ Spectrum of DDIS: a Preliminary Evaluation

By undoing the $z$ integration in the dipole formula (11.16) we can crudely evaluate the $M^2$ spectrum of DDIS (for a more detailed study a description of the process in momentum space is mandatory, see Sect. 11.4). This is done as follows (Nikolaev and Zakharov 1991). Retaining the quark mass $m_q$, which acts as a regulator for the invariant mass $M$ of the dipole state (i.e., the mass of the diffracted state), we have[2]

$$M^2 = \frac{\boldsymbol{\kappa}^2 + m_q^2}{z(1-z)} . \tag{11.26}$$

For large-size asymmetric pairs (that give the main contribution to DDIS, as seen in Sect. 11.2) one has $\boldsymbol{\kappa}^2 \sim 0$ and $z \sim 0$, and (11.26) becomes

$$M^2 \simeq \frac{m_q^2}{z} . \tag{11.27}$$

Thus the $z$-integral in (11.16) can be reinterpreted as an integral over $M^2$ with

$$\mathrm{d}z = \frac{m_q^2}{M^2}\,\mathrm{d}M^2 . \tag{11.28}$$

This allows us to extract $(\mathrm{d}\sigma^D/\mathrm{d}M^2\mathrm{d}t)_{t=0}$ from the integrated cross section $(\mathrm{d}\sigma^D/\mathrm{d}t)_{t=0}$, given by (11.16). Focusing on the dominant transverse contribution, we find, with the wave function (9.239) and $\varepsilon^2 = Q^2 z(1-z) + m_q^2 \simeq m_q^2(M^2+Q^2)/M^2$

$$\begin{aligned}\left.\frac{\mathrm{d}\sigma_T^D}{\mathrm{d}M^2\mathrm{d}t}\right|_{t=0} &\sim \frac{m_q^2}{M^4}\int_{1/\varepsilon^2} \mathrm{d}\rho^2\,\varepsilon^2\,K_1^2(\varepsilon\rho)\,\sigma^2(\rho) \\ &\sim \frac{m_q^2}{M^4}\,\frac{1}{\varepsilon^4} \sim \frac{1}{m_q^2}\,\frac{1}{(M^2+Q^2)^2} .\end{aligned} \tag{11.29}$$

This is approximately the mass spectrum of DDIS when $M^2 > Q^2$. At smaller $M^2$ we cannot restrict the $z$-integral to the endpoint contribution and hence the procedure we applied above is inapplicable.

For very large masses, $M^2 \gg Q^2$, DDIS is dominated by the $q\bar{q}g$ fluctuations of the virtual photon, which yield a spectrum

$$\left.\frac{\mathrm{d}\sigma_T^D}{\mathrm{d}M^2\mathrm{d}t}\right|_{t=0} \sim \frac{1}{M^2} , \tag{11.30}$$

corresponding, in the Regge theory language, to the triple pomeron component. We shall discuss this contribution in Sect. 11.4.2.

[2] In this section boldface vectors are transverse vectors.

Let us now translate the $M^2$ dependence of $\left.\frac{d\sigma_T^D}{dM^2 dt}\right|_{t=0}$ into the $\beta$ dependence of the diffractive structure functions. Using the definitions (10.6, 10.7) and (10.11), and factorizing $F_T^{D(4)}(x_{I\!P}, \beta, Q^2, t = 0)$ as the product of the pomeron flux $f_{I\!P}(x_{I\!P})$ and the pomeron structure function $F_T^{I\!P}(\beta, Q^2)$ – see (10.32) –, we obtain from (11.29)

$$F_T^{I\!P}(\beta, Q^2) \sim \beta \,. \tag{11.31}$$

Since the diffractive cross section must vanish for $M^2 < 4m_q^2$, and for light flavors $m_q^2 \ll Q^2$, $F_T^{D(4)}$, or equivalently $F_T^{I\!P}$, is expected to vanish for $\beta = \frac{Q^2}{M^2+Q^2} \to 1$.

This is all we can learn about the $M^2$ spectrum of DDIS from the impact parameter picture. In order to get more information, we must work in momentum space, that is calculate Feynman diagrams.

## 11.4 Diffractive Cross-sections in the Two-Gluon Exchange Approximation

In first approximation, the diffractively excited states of the photon (the color dipoles) interact with the target proton by exchanging a color-singlet two-gluon pair, which represents the pomeron. The perturbatively calculable dipole interaction can be factorized from the non-perturbative gluon distribution, which must be somehow parametrized. Let us begin by considering the $q\bar{q}$ contribution.

### 11.4.1 The $q\bar{q}$ Contribution

Two representative Feynman diagrams for the interaction of a quark-antiquark dipole with the proton, via two-gluon exchange, are shown in Fig. 11.2. The gluons couple to the $q\bar{q}$ pair in all possible ways.

The kinematics of the process is as follows. The Sudakov decomposition of the final quark and antiquark momenta (see Fig. 11.2 for notations) is

$$\kappa_0 = (1-z)q' + \frac{\boldsymbol{\kappa}^2}{(1-z)W^2}P - \kappa_\perp \,, \tag{11.32}$$

$$\kappa_1 = zq' + \frac{\boldsymbol{\kappa}^2}{zW^2}P + \kappa_\perp \,, \tag{11.33}$$

where $q' = q + xP$, the on-shellness condition for massless particles has been used. The mass of the diffractively produced state is therefore

$$M^2 = (\kappa_0 + \kappa_1)^2 = \frac{\boldsymbol{\kappa}^2}{z(1-z)} \,. \tag{11.34}$$

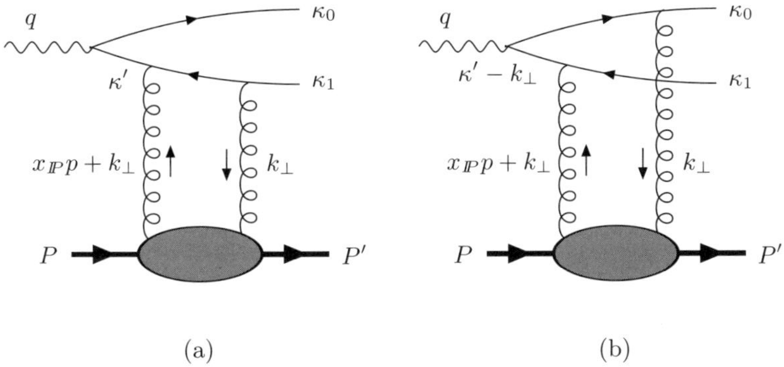


**Fig. 11.2a,b.** Two of the two-gluon exchange diagrams for the $q\bar{q}$ contribution to diffractive DIS.

(Recall that $M^2 = W^2 x_{I\!P}(1-\beta)$). The momentum of the off-shell (anti)quark in the $q\bar{q}$ box (see Fig. 11.2a) is

$$\kappa' = \kappa_1 - x_{I\!P} p = z q' + \left( \frac{\boldsymbol{\kappa}^2}{z W^2} - x_{I\!P} \right) P + \kappa_\perp \, , \tag{11.35}$$

and hence

$$\kappa'^2 = -\frac{\boldsymbol{\kappa}^2}{(1-z)(1-\beta)} = -\frac{\boldsymbol{\kappa}^2 + z(1-z)Q^2}{1-z} \, , \tag{11.36}$$

where the second equality follows from using $\beta = Q^2/(Q^2+M^2)$ and (11.34).

The calculation of $\left.\frac{\mathrm{d}\sigma^D_{q\bar{q}}}{\mathrm{d}t}\right|_{t=0}$ in the two-gluon exchange approximation was first performed by Nikolaev and Zakharov (1991a, 1991b, 1994c) (for subsequent related work see Diehl 1995, Bartels, Lotter and Wüsthoff 1996, Wüsthoff 1997). We refer the reader to the original papers for the details of the calculation. The result is

$$\begin{aligned} \left.\frac{\mathrm{d}\sigma^D_{T,q\bar{q}}}{\mathrm{d}t}\right|_{t=0} &= \frac{\alpha_{\mathrm{em}}}{6\pi} \sum_a e_a^2 \int_0^1 \mathrm{d}z \, [z^2 + (1-z)^2] \\ &\times \int \mathrm{d}^2\boldsymbol{\kappa} \, \mathrm{d}^2\boldsymbol{k} \, \mathrm{d}^2\boldsymbol{k}' \, \alpha_s^2 \, \frac{1}{\boldsymbol{k}^4} f(x_{I\!P}, \boldsymbol{k}^2) \, \frac{1}{\boldsymbol{k}'^4} f(x_{I\!P}, \boldsymbol{k}'^2) \\ &\times \left\{ \frac{\boldsymbol{\kappa}^2}{(\boldsymbol{\kappa}^2 + \varepsilon^2)} - \frac{\boldsymbol{\kappa} \cdot (\boldsymbol{\kappa} + \boldsymbol{k})}{(\boldsymbol{\kappa}^2 + \varepsilon^2)[(\boldsymbol{\kappa} + \boldsymbol{k})^2 + \varepsilon^2]} - \frac{\boldsymbol{\kappa} \cdot (\boldsymbol{\kappa} - \boldsymbol{k}')}{(\boldsymbol{\kappa}^2 + \varepsilon^2)[(\boldsymbol{\kappa} - \boldsymbol{k}')^2 + \varepsilon^2]} \right. \\ &+ \left. \frac{(\boldsymbol{\kappa} + \boldsymbol{k}) \cdot (\boldsymbol{\kappa} - \boldsymbol{k}')}{[(\boldsymbol{\kappa} + \boldsymbol{k})^2 + \varepsilon^2][(\boldsymbol{\kappa} - \boldsymbol{k}')^2 + \varepsilon^2]} \right\} , \end{aligned} \tag{11.37}$$

for the transverse cross section, and

$$\left.\frac{\mathrm{d}\sigma^D_{L,q\bar{q}}}{\mathrm{d}t}\right|_{t=0} = \frac{\alpha_{\mathrm{em}}}{6\pi} \sum_a e_a^2 \int_0^1 \mathrm{d}z \, 4Q^2 \, z^2 (1-z)^2$$

$$\times \int \mathrm{d}^2\boldsymbol{\kappa}\, \mathrm{d}^2\boldsymbol{k}\, \mathrm{d}^2\boldsymbol{k}'\, \alpha_s^2 \, \frac{1}{\boldsymbol{k}^4} f(x_{I\!P}, \boldsymbol{k}^2) \frac{1}{\boldsymbol{k}'^4} f(x_{I\!P}, \boldsymbol{k}'^2)$$
$$\times \left\{ \frac{1}{(\boldsymbol{\kappa}^2+\varepsilon^2)} - \frac{1}{(\boldsymbol{\kappa}^2+\varepsilon^2)[(\boldsymbol{\kappa}+\boldsymbol{k})^2+\varepsilon^2]} - \frac{1}{(\boldsymbol{\kappa}^2+\varepsilon^2)[(\boldsymbol{\kappa}-\boldsymbol{k}')^2+\varepsilon^2]} \right.$$
$$\left. + \frac{1}{[(\boldsymbol{\kappa}+\boldsymbol{k})^2+\varepsilon^2][(\boldsymbol{\kappa}-\boldsymbol{k}')^2+\varepsilon^2]} \right\}, \tag{11.38}$$

for the longitudinal cross section.

In Eqs. (11.37, 11.38) $f(x_{I\!P}, \boldsymbol{k}^2)$ is the unintegrated gluon distribution defined in (9.144). The strong coupling $\alpha_s$ is to be taken at some value $\widetilde{\boldsymbol{\kappa}}^2$, to be specified later. An important *caveat* is in order. In general, we cannot expect $\widetilde{\boldsymbol{\kappa}}^2$ to be so large that perturbation theory may be applied in a totally reliable way, unless we require high-$p_T$ jets in the final state (or consider heavy-quark excitations). For the moment, we ignore this difficulty and simply assume that there is some infrared hadronic scale $\mu^2$, hidden in $f(x_{I\!P}, \boldsymbol{k}^2)$, which separates the regions of soft and hard dynamics (in high-$p_T$ jet production $\mu$ is of the order of the lowest transverse momentum of jets, whereas in open charm production $\mu$ is essentially the charm mass $m_c$). It should be kept in mind that the perturbative treatment of DDIS is somewhat uncertain, if hard final-state configurations are not selected.

By means of the identities (9.233) and (9.234) it is not difficult to show that Eqs. (11.37) and (11.38) are equivalent to the impact parameter formula (11.16), with the dipole cross section given by (9.237). Therefore, as far as $q\bar{q}$ fluctuations are considered, there is an exact equivalence between the Feynman diagram results and the quantum mechanical picture of Sect. 11.2.

Although less appealing than (11.16), Eqs. (11.37, 11.38) carry much more information and allow us to fully reconstruct the $M^2$ spectrum of DDIS. The $x_{I\!P}$ dependence, by contrast, is contained in the non-perturbative quantity $f(x_{I\!P}, \boldsymbol{k}^2)$ that we are unable to predict.

From the invariant mass (11.34) of the $q\bar{q}$ pair we get

$$\mathrm{d}z\, \mathrm{d}\boldsymbol{\kappa}^2 = \frac{\boldsymbol{\kappa}^2}{M^4} \left(1 - 4\frac{\boldsymbol{\kappa}^2}{M^2}\right)^{-\frac{1}{2}} \mathrm{d}M^2\, \mathrm{d}\boldsymbol{\kappa}^2 \,, \tag{11.39}$$

and we can rewrite the cross sections (11.37) and (11.38) in a more convenient form, upon integrations over the azymuthal angles of $\boldsymbol{k}$ and $\boldsymbol{k}'$,

$$\left.\frac{\mathrm{d}\sigma^D_{T,q\bar{q}}}{\mathrm{d}M^2 \mathrm{d}\boldsymbol{\kappa}^2 \mathrm{d}t}\right|_{t=0} = \frac{\pi^2\, \alpha_{\mathrm{em}}}{24} \sum_q e_q^2 \frac{1}{M^4} \left(1 - 4\frac{\boldsymbol{\kappa}^2}{M^2}\right)^{-\frac{1}{2}}$$
$$\times \left(1 - 2\frac{\boldsymbol{\kappa}^2}{M^2}\right) I_T^2, \tag{11.40}$$

$$\left.\frac{\mathrm{d}\sigma^D_{L,q\bar{q}}}{\mathrm{d}M^2 \mathrm{d}\boldsymbol{\kappa}^2 \mathrm{d}t}\right|_{t=0} = \frac{2\pi^2\, \alpha_{\mathrm{em}}}{3} \sum_q e_q^2 \frac{Q^2}{M^4(M^2+Q^2)^2}$$

$$\times \boldsymbol{\kappa}^2 \left(1 - 4\frac{\boldsymbol{\kappa}^2}{M^2}\right)^{-\frac{1}{2}} I_L^2 \, , \, . \tag{11.41}$$

where

$$I_T = \int \frac{\mathrm{d}\boldsymbol{k}^2}{\boldsymbol{k}^4} \alpha_s \, f(x_{I\!P}, \boldsymbol{k}^2) \times \left[\frac{\boldsymbol{\kappa}^2 - \varepsilon^2}{\boldsymbol{\kappa}^2 + \varepsilon^2} + \frac{\boldsymbol{k}^2 - \boldsymbol{\kappa}^2 + \varepsilon^2}{\sqrt{(\boldsymbol{k}^2 + \boldsymbol{\kappa}^2 + \varepsilon^2)^2 - 4\boldsymbol{k}^2\boldsymbol{\kappa}^2}}\right] , \tag{11.42}$$

$$I_L = \int \frac{\mathrm{d}\boldsymbol{k}^2}{\boldsymbol{k}^4} \alpha_s \, f(x_{I\!P}, \boldsymbol{k}^2) \left[1 - \frac{\boldsymbol{\kappa}^2 + \varepsilon^2}{\sqrt{(\boldsymbol{k}^2 + \boldsymbol{\kappa}^2 + \varepsilon^2)^2 - 4\boldsymbol{k}^2\boldsymbol{\kappa}^2}}\right] , \tag{11.43}$$

with $\varepsilon^2 = z(1-z)Q^2 = \boldsymbol{\kappa}^2 Q^2/M^2$.

It is instructive to rederive the above results in the framework of the wave-function formalism, applied to momentum space. The wave functions $\Psi_{T,L}^{\lambda\lambda'}$ of a virtual photon fluctuating into a quark of helicity $\lambda$ and an antiquark of helicity $\lambda'$ are (see, e.g., Wüsthoff 1997; Ivanov and Wüsthoff 1999)

$$\Psi_T^{+-}(z, \boldsymbol{\kappa}) = \frac{\sqrt{2}\, e_q z \, \boldsymbol{\kappa} \cdot \boldsymbol{\varepsilon}^{(+)}}{\boldsymbol{\kappa}^2 + z(1-z)Q^2} \, , \tag{11.44}$$

$$\Psi_T^{-+}(z, \boldsymbol{\kappa}) = -\frac{\sqrt{2}\, e_q (1-z) \, \boldsymbol{\kappa} \cdot \boldsymbol{\varepsilon}^{(+)}}{\boldsymbol{\kappa}^2 + z(1-z)Q^2} \, , \tag{11.45}$$

for a transverse photon of helicity $\lambda_\gamma = +1$ described by the polarization vector $\boldsymbol{\varepsilon}^{(+)}$, and

$$\Psi_L^{+-}(z, \boldsymbol{\kappa}) = \Psi_L^{-+}(z, \boldsymbol{\kappa}) = \frac{2 e_q z(1-z) Q}{\boldsymbol{\kappa}^2 + z(1-z)Q^2} \, , \tag{11.46}$$

for a longitudinal photon. The two-dimensional Fourier transforms of (11.44–11.46), i.e.

$$\Psi(z, \boldsymbol{\rho}) = \int \frac{\mathrm{d}^2\boldsymbol{\kappa}}{(2\pi)^2} \mathrm{e}^{\mathrm{i}\boldsymbol{\kappa}\cdot\boldsymbol{\rho}} \, \Psi(z, \boldsymbol{\kappa}) \, , \tag{11.47}$$

are the wave functions in the impact parameter space already encountered in Sect. 9.9. In detail one has

$$\Psi_T^{+-}(z, \boldsymbol{\rho}) = \frac{\sqrt{2}\mathrm{i}e_q}{2\pi} z^{\frac{3}{2}} (1-z)^{\frac{1}{2}} Q \, K_1(\varepsilon\rho) \, (\hat{\rho}_x + \mathrm{i}\hat{\rho}_y) \, , \tag{11.48}$$

$$\Psi_T^{-+}(z, \boldsymbol{\rho}) = -\frac{\sqrt{2}\mathrm{i}e_q}{2\pi} z^{\frac{1}{2}} (1-z)^{\frac{3}{2}} Q \, K_1(\varepsilon\rho) \, (\hat{\rho}_x + \mathrm{i}\hat{\rho}_y) \, , \tag{11.49}$$

$$\Psi_L^{+-}(z, \boldsymbol{\rho}) = \Psi_L^{-+}(z, \boldsymbol{\rho}) = \frac{e_q}{\pi} z(1-z) Q \, K_0(\varepsilon\rho) \, , \tag{11.50}$$

which lead to (9.238, 9.239), except for a different normalization.

In terms of the photon wave functions (11.44–11.46) the diffractive structure functions are given by a $\boldsymbol{k}_\perp$-factorization formula, namely

$$F_{T,L}^{D(3)} \sim \int \mathrm{d}t \int \mathrm{d}^2\boldsymbol{\kappa} \left| \int \frac{\mathrm{d}^2\boldsymbol{k}}{\boldsymbol{k}^4} D\Psi(z,\boldsymbol{\kappa},\boldsymbol{k}) f(x_{I\!P},\boldsymbol{k}^2) \right|^2 , \tag{11.51}$$

where the quantity

$$D\Psi(z,\boldsymbol{\kappa},\boldsymbol{k}) \equiv 2\,\Psi(z,\boldsymbol{\kappa}) - \Psi(z,\boldsymbol{\kappa}+\boldsymbol{k}) - \Psi(z,\boldsymbol{\kappa}-\boldsymbol{k}) , \tag{11.52}$$

takes into account all possible couplings of the gluons to the $q\bar{q}$ pair.

We can distinguish a "hard" limit ($\boldsymbol{k}^2 \ll \boldsymbol{\kappa}^2$) and a "soft" limit ($\boldsymbol{k}^2 \gg \boldsymbol{\kappa}^2$) for $D\Psi$. In these two extreme situations we have

$$D\Psi(z,\boldsymbol{\kappa},\boldsymbol{k}) = \begin{cases} -k^i k^j \frac{\partial^2 \Psi}{\partial \kappa^i \, \partial \kappa^j} & \text{for } \boldsymbol{k}^2 \ll \boldsymbol{\kappa}^2 , \\ 2\,\Psi(z,\boldsymbol{\kappa}) & \text{for } \boldsymbol{k}^2 \gg \boldsymbol{\kappa}^2 , \end{cases} \tag{11.53}$$

In the hard limit, the two gluons interact with the whole $q\bar{q}$ dipole and the picture of a pomeron structure function is questionable. In the opposite, soft, limit, the two gluons couple to the same (anti)quark line and the quark (or antiquark) of the pair can be considered as a valence constituent of the pomeron. The latter case is also consistent with the picture of the pomeron as a $C = +1$ photon advocated by Landshoff and collaborators (see Sect. 8.1).

Let us come back to the diffractive cross sections (11.40, 11.41). If we assume that their $t$-dependence of is exponential with a slope $b$, we can derive from (11.40, 11.41) the diffractive structure functions $F_{L,T}^{D(3)}$ defined by (10.14) and (10.11), obtaining

$$\begin{aligned} x_{I\!P} F_{T,q\bar{q}}^{D(3)} &= \frac{1}{96\,b} \sum_q e_q^2 \frac{\beta}{1-\beta} \int \mathrm{d}\widetilde{\boldsymbol{\kappa}}^2 \left(1 - 4\beta\frac{\widetilde{\boldsymbol{\kappa}}^2}{Q^2}\right)^{-\frac{1}{2}} \\ &\quad \times \left(1 - 2\beta\frac{\widetilde{\boldsymbol{\kappa}}^2}{Q^2}\right) I_T^2 , \end{aligned} \tag{11.54}$$

$$x_{I\!P} F_{L,q\bar{q}}^{D(3)} = \frac{1}{6\,b} \sum_q e_q^2 \beta^3 \int \mathrm{d}\widetilde{\boldsymbol{\kappa}}^2 \frac{\widetilde{\boldsymbol{\kappa}}^2}{Q^2} \left(1 - 4\beta\frac{\widetilde{\boldsymbol{\kappa}}^2}{Q^2}\right)^{-\frac{1}{2}} I_L^2 . \tag{11.55}$$

In (11.54, 11.55) we have introduced the scale

$$\widetilde{\boldsymbol{\kappa}}^2 = \frac{\boldsymbol{\kappa}^2}{1-\beta} = \boldsymbol{\kappa}^2 + \varepsilon^2 . \tag{11.56}$$

In the limit $z \to 0$, which gives the main contribution to DDIS, one sees from (11.36) and (11.56) that

$$\widetilde{\boldsymbol{\kappa}}^2 \simeq -\kappa'^2 . \tag{11.57}$$

This is the virtuality of the (anti)quark of the dipole, and represents the scale at which $\alpha_s$ is to be taken. So, although Eqs. (11.55, 11.56) are infrared safe, since a cutoff $\mu \sim 1/R$ ($R$ being the confinement radius) is hidden in the unintegrated gluon distribution $f(x_{I\!P}, \boldsymbol{k}^2)$, the perturbative treatment is justified only if $\widetilde{\boldsymbol{\kappa}}^2$ is large enough. This does not generally happen in DDIS, unless one tags jets or heavy quarks in final states.

### 11.4.2 The $q\bar{q}g$ Contribution

We discuss now the $q\bar{q}g$ contribution to diffractive DIS (Fig. 11.3). Although this contribution might seem, at first sight, subleading with respect to the $q\bar{q}$ contribution due to the extra $\alpha_s$ factor, there is a kinematical region, namely large $M^2$ (i.e., small $\beta$), where $q\bar{q}g$ production dominates. This can be understood by noticing that the $q\bar{q}g$ diagrams (Fig. 11.3) have an extra $t$-channel spin-1 gluon in contrast to the $t$-channel spin-$\frac{1}{2}$ quark of the $q\bar{q}$ diagrams, which are consequently suppressed by a power in $M^2$. In the language of Regge theory the $q\bar{q}g$ contribution is identified with the triple-pomeron term, which is known to be dominant at large $M^2$.

The Sudakov parametrization of the final quark, antiquark and gluon momenta is (see Fig. 11.3a)

$$\kappa_0 = (1 - z_1 - z_2)q' + \frac{(\boldsymbol{\kappa}_1 + \boldsymbol{\kappa}_2)^2}{(1 - z_1 - z_2)W^2} P - (\kappa_{1\perp} + \kappa_{2\perp}) \, , \quad (11.58)$$

$$\kappa_1 = z_1 q' + \frac{\boldsymbol{\kappa}_1^2}{z_1 W^2} P + \kappa_{1\perp} \, , \quad (11.59)$$

$$\kappa_2 = z_2 q' + \frac{\boldsymbol{\kappa}_2^2}{z_2 W^2} P + \kappa_{2\perp} \, . \quad (11.60)$$

The invariant mass of the $q\bar{q}g$ state is

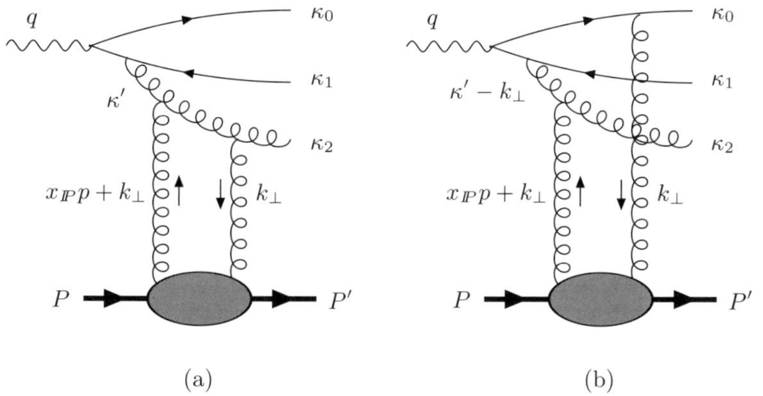


**Fig. 11.3a,b.** Two of the diagrams representing the $q\bar{q}g$ contribution to diffractive DIS via two-gluon exchange.

$$M^2 = \frac{\boldsymbol{\kappa}_1^2}{z_1} + \frac{\boldsymbol{\kappa}_2^2}{z_2} + \frac{(\boldsymbol{\kappa}_1 + \boldsymbol{\kappa}_2)^2}{1 - z_1 - z_2} . \tag{11.61}$$

In terms of the invariant mass of the $q\bar{q}$ pair, $m_{q\bar{q}}^2 = (\kappa_0 + \kappa_1)^2$, $M^2$ is given by

$$M^2 = \frac{z_2 m_{q\bar{q}}^2 + \boldsymbol{\kappa}_2^2}{z_2(1 - z_2)} . \tag{11.62}$$

The momentum of the gluon emitted by the (antiquark) of the primary $q\bar{q}$ dipole is

$$\kappa' = \kappa_2 - x_{I\!P} P = z_2 q' - \left(1 - \frac{\boldsymbol{\kappa}_2^2}{z_2 W^2 x_{I\!P}}\right) x_{I\!P} P + \kappa_{2\perp} . \tag{11.63}$$

If we call $y$ the fraction of the pomeron momentum $x_{I\!P} P$ carried by this gluon, we have

$$y = 1 - \frac{\boldsymbol{\kappa}_2^2}{z_2 W^2 x_{I\!P}} , \tag{11.64}$$

and the virtuality of the gluon is

$$\kappa'^2 = -\frac{\boldsymbol{\kappa}_2^2}{1 - y} . \tag{11.65}$$

Note that $y$ coincides with the $z_{I\!P}$ variable encountered in Sect. 10.9 and used by the experimentalists.

The diffractive cross sections arising from the $q\bar{q}g$ Fock component of the photon have been investigated in various kinematic regimes (Nikolaev and Zakharov 1994b; Levin and Wüsthoff 1994; Wüsthoff 1997; Bartels, Jung and Wüsthoff 1999). In leading twist and leading $\ln Q^2$ approximation, the transverse separation between the quark and the antiquark is much smaller than the distance between the quarks and the gluon. thus, the $q\bar{q}$ pair and the gluon form an effective color dipole, as shown in Fig. 11.4. The corresponding diffractive structure function is (Wüsthoff 1997)

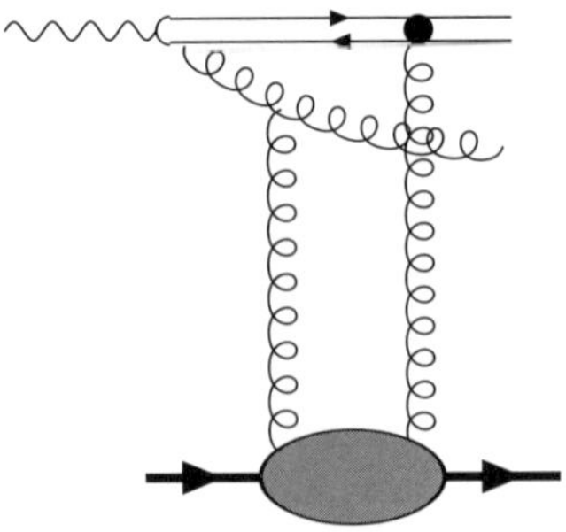

**Fig. 11.4.** The effective gluon dipole contributing to DDIS in leading $\ln Q^2$ approximation.

$$x_{I\!P} F^D_{q\bar{q}g} = \frac{9}{64\,b} \sum_q e_q^2 \beta \int d\widetilde{\boldsymbol{\kappa}}^2 \, \frac{\alpha_s}{2\pi} \, \ln \frac{Q^2}{\widetilde{\boldsymbol{\kappa}}^2} \times \int_\beta^1 \frac{dy}{y^2(1-y)^2} \left[ \left(1 - \frac{\beta}{y}\right)^2 + \left(\frac{\beta}{y}\right)^2 \right] I_g^2 \,, \tag{11.66}$$

where

$$I_g = \int \frac{d\boldsymbol{k}^2}{\boldsymbol{k}^4} \, \alpha_s \, f(x_{I\!P}, \boldsymbol{k}^2) \times \left[ y^2 + (1-y)^2 + \frac{\boldsymbol{k}^2}{\widetilde{\boldsymbol{\kappa}}^2} - \frac{[(1-2y)\widetilde{\boldsymbol{\kappa}}^2 - \boldsymbol{k}^2]^2 + 2y(1-y)\widetilde{\boldsymbol{\kappa}}^4}{\widetilde{\boldsymbol{\kappa}}^2 \sqrt{(\boldsymbol{k}^2 + \widetilde{\boldsymbol{\kappa}}^2)^2 - 4(1-y)\boldsymbol{k}^2 \widetilde{\boldsymbol{\kappa}}^2}} \right] \tag{11.67}$$

Few remarks are in order. First of all, the variable $\widetilde{\boldsymbol{\kappa}}^2$ used in (11.66, 11.67) is now (we set $\boldsymbol{\kappa} \equiv \boldsymbol{\kappa}_2$)

$$\widetilde{\boldsymbol{\kappa}}^2 = \frac{\boldsymbol{\kappa}^2}{1-y} \,, \tag{11.68}$$

and is related to the virtuality of the gluon in the effective dipole, see (11.65). Second, the extra integral over $y$ comes from the convolution with the quark box, which is signaled by the presence of the DGLAP kernel (the factor in square brackets in (11.67)).

Let us now try to derive some features of the diffractive structure functions from the results of the last two sections.

### 11.4.3 The $\beta$ Dependence

The main properties of the $\beta$ spectrum of DDIS in the two-gluon exchange approximation can be inferred from (11.54, 11.55) and (11.42, 11.43), by exploring the limits $\beta \to 1$ and $\beta \to 0$. To do so, it is convenient to reexpress the integrals $I_T$ and $I_L$ in terms of $\beta$. We find

$$I_T = \int \frac{d\boldsymbol{k}^2}{\boldsymbol{k}^4} \, \alpha_s \, f(x_{I\!P}, \boldsymbol{k}^2) \times \left[ 1 - 2\beta + \frac{\boldsymbol{k}^2 - (1-2\beta)\widetilde{\boldsymbol{\kappa}}^2}{\sqrt{(\boldsymbol{k}^2 + \widetilde{\boldsymbol{\kappa}}^2)^2 - 4(1-\beta)\boldsymbol{k}^2 \widetilde{\boldsymbol{\kappa}}^2}} \right] , \tag{11.69}$$

$$I_L = \int \frac{d\boldsymbol{k}^2}{\boldsymbol{k}^4} \, \alpha_s \, f(x_{I\!P}, \boldsymbol{k}^2) \left[ 1 - \frac{\widetilde{\boldsymbol{\kappa}}^2}{\sqrt{(\boldsymbol{k}^2 + \widetilde{\boldsymbol{\kappa}}^2)^2 - 4(1-\beta)\boldsymbol{k}^2 \widetilde{\boldsymbol{\kappa}}^2}} \right] , \tag{11.70}$$

For $\beta \to 1$, Eq. (11.54) vanishes as $(1-\beta)$, whereas (11.55) gets a finite contribution. Thus, at small diffractive masses (i.e., large $\beta$) the longitudinal contribution is larger than the transverse one. One has to recall, however, that the longitudinal structure function is higher twist and hence is suppressed at high $Q^2$. In the limit $\beta \to 0$ (large diffractive masses) one finds that the transverse contribution goes as $\beta$, whereas the longitudinal contribution behaves as $\beta^3$.

A similar study can be carried out for the $q\bar{q}g$ term (11.66). For small $\beta$ the region of small $y$ dominates and, setting $y = 0$ in (11.66), one finds, upon integration over $y$, that the structure function is finite, $F^D_{q\bar{q}g} \sim \beta^0$. This corresponds to the familiar triple-pomeron behavior, $\mathrm{d}\sigma^D/\mathrm{d}M^2 \sim (Q^2 + M^2)^{-1}$. In the opposite limit, $\beta \to 1$, one finds that the structure function vanishes as $(1-\beta)^3$. This result is obtained by taking the limit $y \to 1$ and integrating over $y$.

We conclude that the three contributions to DDIS, transverse $q\bar{q}$ and $q\bar{q}g$ and longitudinal $q\bar{q}$, are important in distinct regions in $\beta$, namely medium, small and large $\beta$, respectively. The precise form of $f(x_{I\!P}, \boldsymbol{k}^2)$ has a little influence on the $\beta$ spectrum. The situation is shown in Fig. 11.5, where the results of Golec-Biernat and Wüsthoff (1999b) are shown .These authors use a specific model for the unintegrated gluon distribution and set the coupling constant to a fixed value (as already discussed, this is actually unjustified).

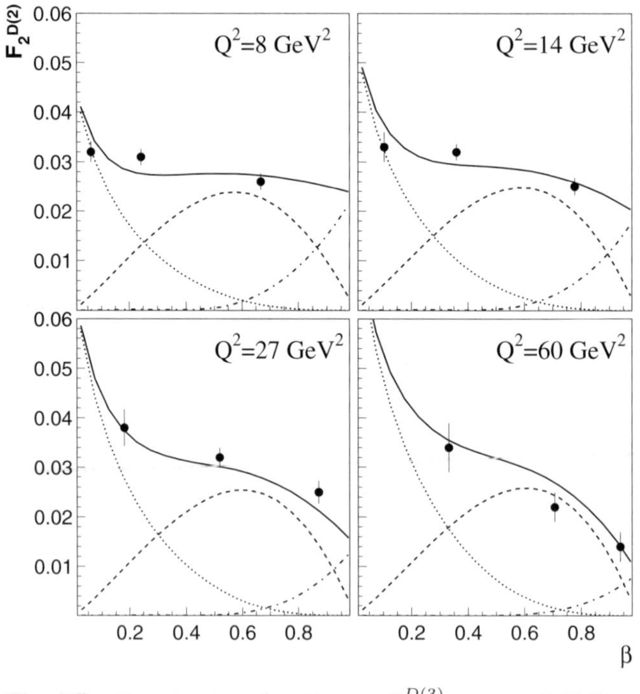


**Fig. 11.5.** The diffractive structure function $x_{I\!P} F_2^{D(3)}$ at $x_{I\!P} = 0.0042$, as computed by Golec-Biernat and Wüsthoff (1999b). The three contributions corresponding to (11.54), (11.55) and (11.66) are separately shown. Dashed line: transverse $q\bar{q}$; dashed-dotted line: longitudinal $q\bar{q}$; dotted line: $q\bar{q}g$. The data are from ZEUS.

Note that as $Q^2$ increases, the longitudinal contribution gets suppressed, whereas the rise of the $q\bar{q}g$ term at small $\beta$ is due to the presence of a $\ln(Q^2/Q_0^2)$ arising from the phase-space integration of the quark box.

A fit of the ZEUS and H1 measurements based on the semi-quantitative analysis of the $\beta$ spectrum presented above has been performed by Bartels et al. (1999). They assume

$$F^D_{T,q\bar{q}} \sim \beta(1-\beta) , \tag{11.71}$$

$$F^D_{L,q\bar{q}} \sim \beta^3(1-2\beta)^2 , \tag{11.72}$$

$$F^D_{q\bar{q}g} \sim (1-\beta)^\gamma . \tag{11.73}$$

Here $\gamma$ is left as a free parameter. The ZEUS data are reproduced with $\gamma = 4.3$, close the value 3 derived above. The H1 data admit two solutions: a large $\gamma$ value, 8.55, and a small one, 0.28. The latter is consistent with the singular gluon distribution at $\beta = 1$ found by H1 (see Sect. 10.8). The quality of one of these fits is shown in Fig. 11.6.

A different result is obtained by Nikolaev and Zakharov (1994b) and Genovese, Nikolaev and Zakharov (1995), who find that glue and charged partons carry approximately the same momentum fraction.

Note, in conclusion, that a very small $\beta$ (i.e., very large diffractive masses $M^2$) the color dipole picture becomes insufficient, due to the presence of large logarithms in $1/\beta$ ($M^2/Q^2$) which need to be resummed.

## 11.5 Jets in Diffractive DIS

In order to circumvent the difficulties mentioned in Sects. 11.2 and 11.4.1, that is the lack of an explicit infrared cutoff in diffractive DDIS which makes the process strongly non-perturbative, one can search for diffractive final states containing high-$p_T$ jets. In this way, the partonic transverse momenta in the color dipole are bounded from below, and large enough to let us apply perturbative QCD safely.

The production of quark-antiquark jets with large transverse momenta in DDIS has been studied by several authors (Nikolaev and Zakharov 1992, 1994c; Bartels, Lotter and Wüsthoff 1996; Bartels et al. 1996; Gotsman, Levin and Maor 1997). Since $\widetilde{\boldsymbol{\kappa}}^2$ is large, in leading $\ln \widetilde{\boldsymbol{\kappa}}^2$ approximation we can expand (11.69) and (11.70) in powers of $\boldsymbol{k}^2/\widetilde{\boldsymbol{\kappa}}^2$, thus obtaining

$$\begin{aligned} I_T &\simeq \int \frac{\mathrm{d}\boldsymbol{k}^2}{\boldsymbol{k}^4}\, \alpha_s\, f(x_{I\!P}, \boldsymbol{k}^2)\, \frac{\boldsymbol{k}^2}{\widetilde{\boldsymbol{\kappa}}^2}\, 4\beta(1-\beta) \\ &= \frac{4\alpha_s\, \beta(1-\beta)}{\widetilde{\boldsymbol{\kappa}}^2}\, x_{I\!P}\, g(x_{I\!P}, \widetilde{\boldsymbol{\kappa}}^2) , \end{aligned} \tag{11.74}$$

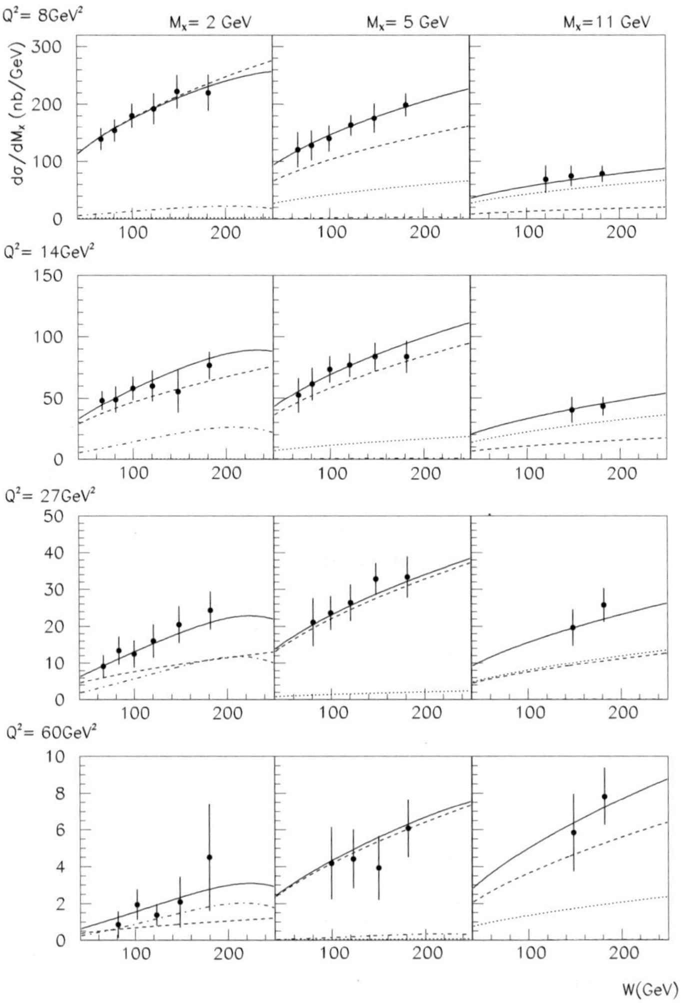


**Fig. 11.6.** The fit of Bartels et al. (1999) compared to the ZEUS data. Upper solid line: total result; dashed line: transverse $q\bar{q}$; dashed-dotted line: longitudinal $q\bar{q}$; dotted line: $q\bar{q}g$.

$$I_L \simeq -\int \frac{\mathrm{d}\boldsymbol{k}^2}{\boldsymbol{k}^4}\, \alpha_s\, f(x_{I\!P}, \boldsymbol{k}^2)\, \frac{\boldsymbol{k}^2}{\widetilde{\boldsymbol{\kappa}}^2}\, (1-2\beta)$$

$$= -\frac{\alpha_s\,(1-2\beta)}{\widetilde{\boldsymbol{\kappa}}^2}\, x_{I\!P} g(x_{I\!P}, \widetilde{\boldsymbol{\kappa}}^2)\,. \qquad (11.75)$$

Therefore the transverse, dominant, cross section for diffractive production of jets is given by

$$d\sigma_T \sim \left[ \frac{\alpha_s}{\boldsymbol{\kappa}^2} x_{I\!P} g \left( x_{I\!P}, \frac{\boldsymbol{\kappa}^2}{1-\beta} \right) \right]^2 . \tag{11.76}$$

For $\beta$ near 1, i.e. small $M^2$, the reaction is indeed dominated by the hard pomeron. As for the azimuthal dependence of the cross section, one finds that in the photon–pomeron c.m. frame the two jets prefer to lie perpendicular to the scattering plane (defined by the $\gamma^* I\!P$ direction and the outgoing lepton momentum). By contrast, in the photon–gluon fusion process of DIS, the produced quark and antiquark tend to lie inside the scattering plane. As seen in Sect. 10.9, the experimental results on jet production in DDIS are still quite uncertain and no conclusive comparison with the theoretical expectations are possible (although the color dipole picture seems to do well in the low-$x_{I\!P}$ region).

When $\beta$ is away from 1, i.e. $M^2$ is not so small, the diffractive dissociation of the virtual photon into $q\bar{q}g$ becomes important. The production of $q\bar{q}g$ jets with large transverse momenta has been investigated, among others, by Wüsthoff (1997) and Bartels, Jung and Wüsthoff (1999). Working in the leading $\ln(1/\beta)$ approximation, which is clearly adequate to describe the low-$\beta$ region (i.e., large diffractive masses), Bartels, Jung and Wüsthoff have estimated that the jet contribution to DDIS amount to $\sim 8$ % of the whole diffractive cross section. It also turns out that a large fraction of the $q\bar{q}g$ final states appear as dijet configurations, with the quark-antiquark pair forming a single jet opposite to the gluon jet.

## 11.6 Diffractive Production of Open Charm

Another process which offers the possibility of testing the perturbative QCD description of DDIS is diffractive leptoproduction of charm quarks and antiquarks fragmenting into $D$ mesons (Genovese, Nikolaev and Zakharov 1996; Levin et al. 1997). Using the notations of Sect. 11.4.1, the invariant mass of the produced $c\bar{c}$ pair is

$$M^2 = \frac{\boldsymbol{\kappa}^2 + m_c^2}{z(1-z)} , \tag{11.77}$$

and the scale $\widetilde{\boldsymbol{\kappa}}^2$ is

$$\widetilde{\boldsymbol{\kappa}}^2 = \frac{\boldsymbol{\kappa}^2 + m_c^2}{1-\beta} . \tag{11.78}$$

The mass of the charm quark sets a lower limit on $\widetilde{\boldsymbol{\kappa}}^2$, namely

$$\widetilde{\boldsymbol{\kappa}}_0^2 = \frac{m_c^2}{1-\beta} , \tag{11.79}$$

which allows a relatively safe perturbative treatment of the process, at least when $\beta$ is not very large. Translated in the color dipole language, this means that the produced $c\bar{c}$ pair has a small transverse size $\rho$ and interacts with a cross section $\sigma(\rho) \sim \rho^2$. The obvious drawback is that one expects a sensible suppression fr this process. Note that the diffractive mass (11.77) is bounded from below, being always $\geq 4m_c^2$.

The cross sections for $c\bar{c}$ production in diffractive DDIS are a generalization of Eqs. (11.40, 11.41) for the case of massive quarks (Nikolaev and Zakharov 1994c; Genovese, Nikolaev and Zakharov 1996):

$$\left.\frac{\mathrm{d}\sigma^D_{T,c\bar{c}}}{\mathrm{d}M^2\mathrm{d}\boldsymbol{\kappa}^2\mathrm{d}t}\right|_{t=0} = \frac{\pi^2}{12}\frac{\alpha_{\mathrm{em}}e_c^2}{}\frac{\boldsymbol{\kappa}^2+m_c^2}{\boldsymbol{\kappa}^2 M^4}\left(1-4\frac{\boldsymbol{\kappa}^2+m_c^2}{M^2}\right)^{-\frac{1}{2}}$$
$$\times\left\{\left(1-2\frac{\boldsymbol{\kappa}^2+m_c^2}{M^2}\right)I_T^2+\frac{4\boldsymbol{\kappa}^2 m_c^2}{(\boldsymbol{\kappa}^2+\varepsilon^2)^2}I_L^2\right\}, \quad (11.80)$$

$$\left.\frac{\mathrm{d}\sigma^D_{L,c\bar{c}}}{\mathrm{d}M^2\mathrm{d}\boldsymbol{\kappa}^2\mathrm{d}t}\right|_{t=0} = \frac{\pi^2}{3}\frac{\alpha_{\mathrm{em}}e_c^2}{}\frac{Q^2}{M^4(M^2+Q^2)^2}(\boldsymbol{\kappa}^2+m_c^2)$$
$$\times\left(1-4\frac{\boldsymbol{\kappa}^2+m_c^2}{M^2}\right)^{-\frac{1}{2}} I_L^2\,, \quad (11.81)$$

where now

$$I_T = \int\frac{\mathrm{d}\boldsymbol{k}^2}{\boldsymbol{k}^4}\,\alpha_s\, f(x_{I\!P},\boldsymbol{k}^2)$$
$$\times\left\{1-2\beta-\frac{2m_c^2}{\widetilde{\boldsymbol{\kappa}}^2}+\frac{\boldsymbol{k}^2-(1-2\beta)\widetilde{\boldsymbol{\kappa}}^2+2m_c^2}{\sqrt{(\boldsymbol{k}^2+\widetilde{\boldsymbol{\kappa}}^2)^2-4[(1-\beta)\widetilde{\boldsymbol{\kappa}}^2-m_c^2]\boldsymbol{k}^2}}\right\}, \quad (11.82)$$

$$I_L = \int\frac{\mathrm{d}\boldsymbol{k}^2}{\boldsymbol{k}^4}\,\alpha_s\, f(x_{I\!P},\boldsymbol{k}^2)\left\{1-\frac{\widetilde{\boldsymbol{\kappa}}^2}{\sqrt{(\boldsymbol{k}^2+\widetilde{\boldsymbol{\kappa}}^2)^2-4[(1-\beta)\widetilde{\boldsymbol{\kappa}}^2-m_c^2]\boldsymbol{k}^2}}\right\}. \quad (11.83)$$

The second term in (11.80), proportional to $I_L^2$, results from a spin flip on the (anti)quark line, which is possible in the massive case.

From (11.80, 11.81) we can derive the diffractive structure functions, which read

$$x_{I\!P}F^{D(3)}_{T,c\bar{c}} = \frac{e_c^2}{48\,b}\frac{\beta}{(1-\beta)^2}\int\frac{\mathrm{d}\boldsymbol{\kappa}^2}{\boldsymbol{\kappa}^2}(\boldsymbol{\kappa}^2+m_c^2)\left(1-4\beta\frac{\widetilde{\boldsymbol{\kappa}}^2}{Q^2}\right)^{-\frac{1}{2}}$$
$$\times\left[\left(1-2\beta\frac{\widetilde{\boldsymbol{\kappa}}^2}{Q^2}\right)I_T^2+\frac{4\boldsymbol{\kappa}^2 m_c^2}{\widetilde{\boldsymbol{\kappa}}^4}I_L^2\right], \quad (11.84)$$

$$x_{I\!P}F^{D(3)}_{L,c\bar{c}} = \frac{e_c^2}{3\,b\,Q^2}\frac{\beta^3}{1-\beta}\int\mathrm{d}\boldsymbol{\kappa}^2\,\frac{\widetilde{\boldsymbol{\kappa}}^2}{Q^2}\left(1-4\beta\frac{\widetilde{\boldsymbol{\kappa}}^2}{Q^2}\right)^{-\frac{1}{2}} I_L^2\,. \quad (11.85)$$

In the leading $\ln \widetilde{\boldsymbol{\kappa}}^2$ approximation – which is meaningful since $\widetilde{\boldsymbol{\kappa}}^2$ is always large regardless the value of $\boldsymbol{\kappa}^2$, due to (11.78) –, we can expand $I_T$ and $I_L$ in powers of $\boldsymbol{k}^2/\widetilde{\boldsymbol{\kappa}}^2$. We thus get

$$
\begin{aligned}
I_T &\simeq \int \frac{\mathrm{d}\boldsymbol{k}^2}{\boldsymbol{k}^4}\, \alpha_s\, f(x_{I\!P}, \boldsymbol{k}^2)\, \frac{\boldsymbol{k}^2}{\widetilde{\boldsymbol{\kappa}}^2} \left[1 - \left(1 - 2\beta - \frac{2m_c^2}{\widetilde{\boldsymbol{\kappa}}^2}\right)^2\right] \\
&= \frac{\alpha_s\, x_{I\!P} g(x_{I\!P}, \widetilde{\boldsymbol{\kappa}}^2)}{\widetilde{\boldsymbol{\kappa}}^2} \left[1 - \left(1 - \frac{2\boldsymbol{\kappa}^2}{\widetilde{\boldsymbol{\kappa}}^2}\right)^2\right] , \qquad (11.86)
\end{aligned}
$$

$$
\begin{aligned}
I_L &\simeq - \int \frac{\mathrm{d}\boldsymbol{k}^2}{\boldsymbol{k}^4}\, \alpha_s\, f(x_{I\!P}, \boldsymbol{k}^2)\, \frac{\boldsymbol{k}^2}{\widetilde{\boldsymbol{\kappa}}^2} \left(1 - 2\beta - \frac{2m_c^2}{\widetilde{\boldsymbol{\kappa}}^2}\right) \\
&= \frac{\alpha_s\, x_{I\!P} g(x_{I\!P}, \widetilde{\boldsymbol{\kappa}}^2)}{\widetilde{\boldsymbol{\kappa}}^2} \left(1 - \frac{2\boldsymbol{\kappa}^2}{\widetilde{\boldsymbol{\kappa}}^2}\right) . \qquad (11.87)
\end{aligned}
$$

The structure functions for diffractive production of open charm are therefore proportional to the square of the gluon distribution at the scale $m_c^2/(1-\beta)$,

$$
F_{T,c\bar{c}}^{D(3)} \propto \left[\alpha_s x_{I\!P} g\left(x_{I\!P}, \frac{m_c^2}{1-\beta}\right)\right]^2 , \qquad (11.88)
$$

and

$$
F_{L,c\bar{c}}^{D(3)} \propto \frac{1}{Q^2} \ln\left(\frac{(1-\beta)Q^2}{4\beta m_c^2}\right) \left[\alpha_s x_{I\!P} g\left(x_{I\!P}, \frac{Q^2}{4\beta}\right)\right]^2 . \qquad (11.89)
$$

The logarithm appearing in the longitudinal structure function comes from the integration over $\boldsymbol{\kappa}^2$.

The measurements of diffractive charm production, briefly discussed in Sect. 10.9, are still at a preliminary level.

## 11.7 Diffractive Vector Meson Production at $t = 0$

An important class of diffractive reactions open – to some extent – to a perturbative treatment is vector meson production in DDIS: $\gamma^* p \to V p$. This is a quasi-elastic reaction in the sense that the photon and the meson $V$ have the same quantum numbers. The process has been intensively investigated from an experimental viewpoint, and a large body of data (on $\rho, \omega, \phi, J/\psi, \psi', \Upsilon$ production) is by now available, thanks to the HERA measurements (that we reviewed in Sect. 10.10). On the theoretical side, a lot of work has been devoted to the subject, with a variety of approaches. Here we shall focus on the two-gluon exchange description. For reference to other work in the field see Wüsthoff and Martin (1999).

For a perturbative QCD study a hard scale is necessary (though sometimes not sufficient): this scale can be the photon virtuality $Q^2$, the heavy quark mass in the case of $J/\psi, \psi', \Upsilon$ production, or the momentum transfer $t$. In this section we consider the forward case ($t = 0$) postponing the discussion of diffractive meson production at large $t$, which involves the BFKL dynamics, to Sect. 11.12.1.

Two-gluon exchange diagrams for vector meson leptoproduction are shown in Fig. 11.7. Since at high energy (i.e., low $x$) the $\gamma^* \to q\bar{q}$ fluctuation time and the $q\bar{q} \to V$ formation time are both much longer than the interaction time of the $q\bar{q}$ pair with the proton, the $\gamma^* p \to Vp$ scattering amplitude can be factorized into the product of the $\gamma^* \to q\bar{q}$ transition, the scattering of the $q\bar{q}$ system with the target via a color singlet two-gluon exchange and the recombination of $q\bar{q}$ into the observed meson. Thus, the only extra ingredient with respect to open $q\bar{q}$ production is the vector meson wave function.

As already anticipated in Sect. 10.10, in leading $\ln Q^2$ the cross section for $\gamma^* p \to Vp$ is proportional to the square of the gluon density of the proton

$$\left.\frac{\mathrm{d}\sigma^{\gamma^* p \to Vp}}{\mathrm{d}t}\right|_{t=0} \propto \left[x_{I\!P}\, g\left(x_{I\!P}, \widetilde{Q}^2\right)\right]^2 , \qquad (11.90)$$

at the scale $\widetilde{Q}^2 = (Q^2 + M_V^2)/4$. The derivation of this result will be sketched in Sect. 11.7.1.

There is an important subtlety to be noticed. The two gluons exchanged in the diagrams of Fig. 11.7 carry different fractions of the proton longitudinal momentum. Thus, what is probed in vector meson production is not the usual gluon distribution but rather the *skewed* (or *off-forward*) gluon distribution $g(x_1, x_2)$ (Ji 1997, Radyushkin 1997, Martin and Ryskin 1998). Equation (11.90) should actually be rewritten as

$$\left.\frac{\mathrm{d}\sigma^{\gamma^* p \to Vp}}{\mathrm{d}t}\right|_{t=0} \propto \left[x_2 g\left(x_1, x_2, \widetilde{Q}^2\right)\right]^2 , \qquad (11.91)$$

with

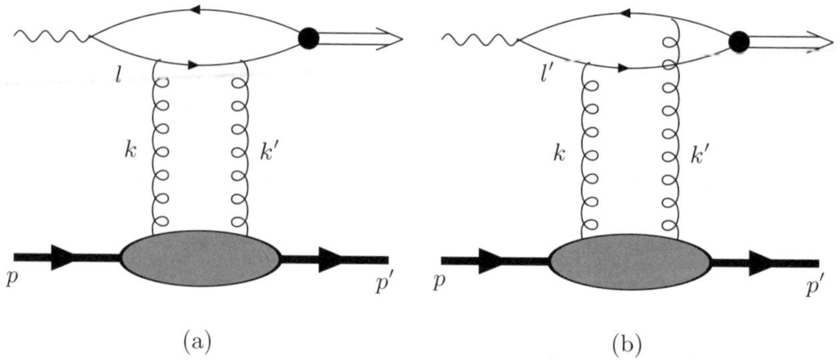


**Fig. 11.7a,b.** Vector meson leptoproduction via two-gluon exchange.

$$x_1 = x_{I\!P} = \frac{Q^2 + M^2}{W^2}, \quad x_2 = \frac{M^2 - M_V^2}{W^2} \ll x_1 . \qquad (11.92)$$

Here $M^2$ is the invariant mass of the $q\bar{q}$ system. It has been shown (Shuvaev et al. 1999) that at small $x_{I\!P}$ the skewed distributions are completely determined by the conventional diagonal distributions. The effect of skewedness can be incorporated into a constant enhancement factor $\mathcal{R}$, so that (11.91) becomes

$$\left. \frac{\mathrm{d}\sigma^{\gamma^* p \to V p}}{\mathrm{d}t} \right|_{t=0} \propto \left[ \mathcal{R}\, x_{I\!P} g \left( x_{I\!P}, \widetilde{Q}^2 \right) \right]^2 . \qquad (11.93)$$

The factor $\mathcal{R}$ affects only the normalization of cross sections, not their energy dependence. In the case of $J/\psi$ production it gives a 30% correction. The effect is larger for $\rho$ production. From now on, for simplicity, we shall ignore the enhancement due to skewedness.

### 11.7.1 $J/\psi$ Production at $t = 0$

As previously seen (Sect. 11.6), when the virtual photon dissociates into a heavy quark-antiquark pair (for instance $c\bar{c}$), the quark mass acts as an infrared cutoff which keeps the process away from the soft regime. Thus, one can study in a perturbative manner not only the leptoproduction of heavy vector mesons (i.e., large $Q^2$), but also their photoproduction ($Q^2 \simeq 0$). A QCD analysis of these processes has been carried out by Ryskin (1993), Nemchik, Nikolaev and Zakharov (1994) and Brodsky et al. (1994).

The recombination of the $c\bar{c}$ pair into the $J/\psi$ (or $\psi'$) depends on the meson wave function. This introduces a model dependence and consequently some theoretical uncertainty into the calculation. In the case of a heavy quark-antiquark bound state such as the $J/\psi$, a simplified approach, based on a non-relativistic static picture of the $c\bar{c}$ pair, is sufficient to understand the main features of the process (Ryskin 1993). A more accurate study, on the other hand, requires a careful consideration of the $J/\psi$ wave function and of possible relativistic effects (Nemchik, Nikolaev and Zakharov 1994; Brodsky et al. 1994; Frankfurt, Koepf and Strikman 1996, 1998; Ryskin et al. 1997).

Let us assume for simplicity that the quark and the antiquark have the same longitudinal momentum fraction (i.e., $z = 1/2$) and no relative transverse momentum (i.e., $\boldsymbol{\kappa}_\perp = 0$). The $J/\psi$ wave function is therefore (in the momentum representation)

$$\Psi(z, \boldsymbol{\kappa}_\perp) \sim \delta \left( z - \frac{1}{2} \right) \delta^2(\boldsymbol{\kappa}_\perp) . \qquad (11.94)$$

If we enforce these constraints in the formulas for open charm production, Eqs. (11.84, 11.85), we find after some manipulations

$$x_{I\!P} F^D_{T,\psi}(x_{I\!P}, Q^2) \propto \frac{Q^4 M_\psi^2}{(Q^2 + M_\psi^2)^5} \left[ \alpha_s x_{I\!P} g(x_{I\!P}, \widetilde{Q}^2) \right]^2 , \qquad (11.95)$$

$$x_{I\!P} F_{L,\psi}^{D}(x_{I\!P}, Q^2) \propto \frac{Q^6}{(Q^2 + M_\psi^2)^5} \left[\alpha_s x_{I\!P} g(x_{I\!P}, \widetilde{Q}^2)\right]^2 , \qquad (11.96)$$

where

$$x_{I\!P} = \frac{Q^2 + M_\psi^2}{Q^2 + W^2} , \quad \widetilde{Q}^2 = \frac{1}{4}(Q^2 + M_\psi^2) . \qquad (11.97)$$

The longitudinal production of $J/\psi$ is therefore enhanced with respect to the transverse production by a factor $Q^2/M_\psi^2$.

In order to obtain the cross section with its correct normalization we have to start directly from the scattering amplitude of the process. Referring to Fig. 11.7 for notations, we find that the forward amplitude for diffractive $J/\psi$ production from a transversely polarized photon is (Ryskin 1993; Ryskin et al. 1997)

$$A_T(W^2, t=0) = -4\pi^2 \mathrm{i}\, \alpha_s W^2 \int \frac{\mathrm{d}\boldsymbol{k}^2}{\boldsymbol{k}^4} \left( \frac{1}{l^2 - m_c^2} - \frac{1}{l'^2 - m_c^2} \right) \times f(x_{I\!P}, \boldsymbol{k}^2)\, e_c g_\psi M_\psi . \qquad (11.98)$$

The cross sections is

$$\frac{\mathrm{d}\sigma_T^{\gamma^{(*)}p \to \psi\, p}}{\mathrm{d}t} = \frac{1}{16\pi W^4} |A_T|^2 . \qquad (11.99)$$

The constant $g_\psi$ specifies the $c\bar{c}$ coupling to $J/\psi$ and may be determined from the width $\Gamma_{e^+e^-}^{\psi}$ of the $J/\psi \to e^+e^-$ decay. One finds

$$e_c^2 g_\psi^2 = \frac{\Gamma_{e^+e^-}^{\psi} M_\psi}{12 \alpha_{\mathrm{em}}} . \qquad (11.100)$$

The two terms in bracket in Eq. (11.98) correspond to the diagrams of Fig. 11.7a and Fig. 11.7b, respectively. Color factors give rise to the opposite sign of the two contributions. The denominators of the quark propagators appearing in (11.98) are

$$l^2 - m_c^2 = -2\widetilde{Q}^2 - 2\boldsymbol{k}^2 , \qquad (11.101)$$

$$l'^2 - m_c^2 = -2\widetilde{Q}^2 . \qquad (11.102)$$

In the leading $\ln \widetilde{Q}^2$ approximation the amplitude (11.98) reads

$$A_T \simeq 2\pi^2 \mathrm{i}\, e_c g_\psi M_\psi\, \alpha_s(\widetilde{Q}^2) W^2 \frac{x_{I\!P} g(x_{I\!P}, \widetilde{Q}^2)}{\widetilde{Q}^4} , \qquad (11.103)$$

and hence the transverse cross section is

$$\left.\frac{\mathrm{d}\sigma_T^{\gamma^{(*)}p \to \psi\, p}}{\mathrm{d}t}\right|_{t=0} = \frac{16\, \Gamma_{e^+e^-}^{\psi} M_\psi^3 \pi^3}{3\, \alpha_{\mathrm{em}} (Q^2 + M_\psi^2)^4} \left[\alpha_s(\widetilde{Q}^2) x_{I\!P} g(x_{I\!P}, \widetilde{Q}^2)\right]^2 . \qquad (11.104)$$

As for the longitudinal contribution, we found above that $\sigma_L/\sigma_T = Q^2/M_\psi^2$, and therefore the full (longitudinal plus transverse) cross section in leading $\ln \widetilde{Q}^2$ approximation is

$$\left.\frac{\mathrm{d}\sigma^{\gamma^{(*)} p \to \psi\, p}}{\mathrm{d}t}\right|_{t=0} = \frac{16\, \Gamma^{\psi}_{e^+e^-} M_\psi^3 \pi^3}{3\, \alpha_{\mathrm{em}} (Q^2 + M_\psi^2)^4} \times \left[\alpha_s(\widetilde{Q}^2) x_{I\!P}\, g(x_{I\!P}, \widetilde{Q}^2)\right]^2 \left(1 + \frac{Q^2}{M_\psi^2}\right). \quad (11.105)$$

The effect on the cross section of the relativistic corrections to the $J/\psi$ wave function is controversial (see, e.g., Frankfurt, Koepf and Strikman 1996, 1998; Ryskin et al. 1997).

In the limit of photoproduction we simply set $Q^2 = 0$ in (11.105). In this case the longitudinal contribution vanishes. The perturbative treatment is still legitimate since the charm mass ensures a hard enough scale.

The $W$ dependence of the cross section is determined by the behavior in $x_{I\!P}$ of the gluon distribution. As we saw in Sect. 10.10, with $xg(x) \sim x^{-\lambda}$ and the experimentally determined value of $\lambda \approx 0.2$ one finds the behavior $\sigma^{\gamma p \to \psi\, p} \sim W^{0.8}$, in good agreement with the data.

### 11.7.2 Light-Meson Production at $t = 0$

Most of the available data on vector meson production concern light mesons, in particular the $\rho$. In this case, quark masses do not provide a hard scale and a perturbative treatment is possible only for electroproduction. Unfortunately, light-meson wave functions are not well understood from a theoretical viewpoint and therefore it is difficult to produce reliable predictions (Brodsky et al. 1994; Nemchik et al. 1997; Martin, Ryskin and Teubner 1997). Some results were already discussed in Sect. 10.10.

To avoid the problems arising from meson wave functions it has been proposed to look for photons in the final state. At $t = 0$ this situation corresponds to deeply virtual Compton scattering (DVCS) – see, e.g. Frankfurt, Freund and Strikman (1998). The two-gluon exchange diagram for DVCS is shown in Fig. 11.8. The first signals of DVCS in DIS events with $Q^2 > 6$

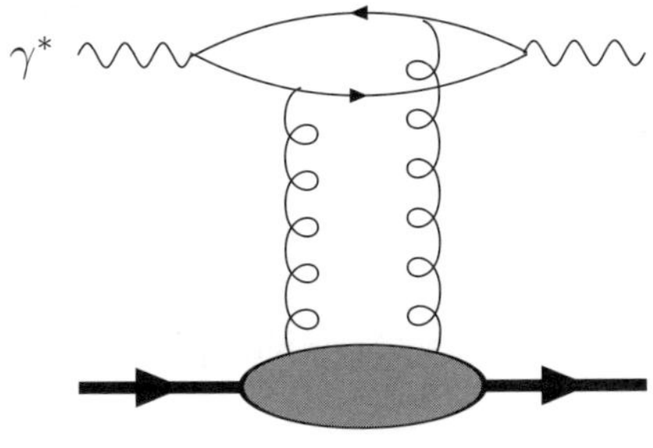


**Fig. 11.8.** Deeply virtual Compton scattering via two-gluon exchange.

$\mathrm{GeV}^2$ and $5 \cdot 10^{-4} < x < 10^{-2}$ have been recently found at HERA (ZEUS 1999b, H1 2000).

## 11.8 Other QCD Approaches to DDIS

Besides the perturbative QCD approach to DDIS based on two-gluon exchange, other QCD-based pictures of DDIS have been developed in the past few years (for a comprehensive review, see Hebecker 2000).

It has been suggested that *soft color interactions* might be responsible for the large diffractive cross sections measured at HERA (Buchmüller and Hebecker 1995; Edin, Ingelman and Rathsman 1996). Support to this idea comes from the observation of a similarity between the $x$ and $Q^2$ dependence of diffractive and inclusive DIS at low $x$ (Buchmüller 1995). This induces one to think that the hard hadronic processes should be the same in both reactions, whereas the difference might be due to non-perturbative soft color exchange in the final state. In particular, Buchmüller and Hebecker (1995) proposed photon–gluon fusion as the dominant partonic process, with a soft color neutralization of the produced $q\bar{q}$ pair (an idea which dates back to Nachtmann and Reiter 1984). Final-state soft color interactions were implemented in a Monte Carlo event generator by Edin, Ingelman and Rathsman (1996, 1997). These models are phenomenologically quite successful but the soft color exchange is introduced in an ad hoc way and do not properly account for some perturbative effects which are known to be at work in certain kinematical regimes, such as color transparency.

An attempt to combine perturbative QCD with the idea of soft color interactions is the so-called *semiclassical approach* (Nachtmann 1991; Buchmüller and Hebecker 1996; Buchmüller, McDermott and Hebecker 1997). We cannot enter into the details of this model, which are rather technical, and hence we limit ourselves to a brief discussion (for a more extensive treatment see Hebecker 2000). In the semiclassical approach the proton is modelled by a soft color field and the interaction of partons with this field is calculated in the eikonal approximation (Fig. 11.9). The ingredients of the model are therefore the wave functions of the virtual photon and the eikonal amplitudes for the scattering of partons off the target color field.

The $n$-gluon exchange contribution to the scattering amplitude for the interaction of a quark with the target color source (Fig. 11.9) turns out to be

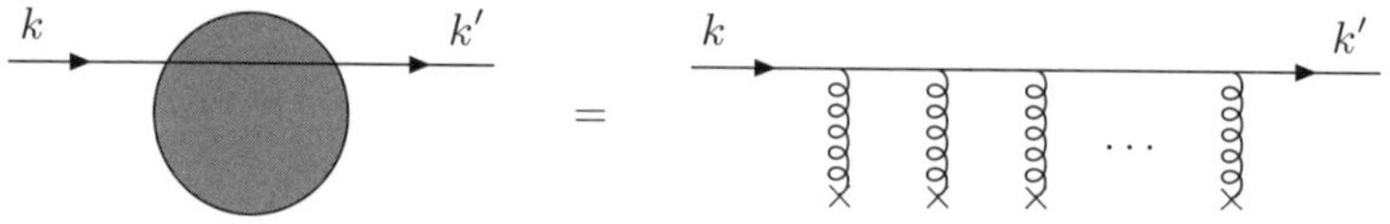


**Fig. 11.9.** Interaction of a parton with the target color field as a multiple scattering off a superposition of classical external fields.

$$A^{(n)}(k,k') \sim -2\mathrm{i}k^+ \int \mathrm{d}^2\boldsymbol{x}\, \mathrm{e}^{\mathrm{i}\boldsymbol{x}\cdot(\boldsymbol{k}-\boldsymbol{k}')} \times \frac{1}{n!}\mathcal{P}\left(-\frac{\mathrm{i}g}{2}\int \mathrm{d}x^+\,\mathcal{A}^-(x^+,0,\boldsymbol{x})\right)^n , \quad (11.106)$$

where $\mathcal{A}^\mu$ is the color field and $\mathcal{P}$ denotes path ordering. Note that in the high-energy limit $k^+$ is the large component of the momentum of the interacting parton. Summing over $n$ yields the eikonal formula for the scattering amplitude. This reads, in the impact-parameter space

$$A(\boldsymbol{x}) \sim \mathbb{1} - U(\boldsymbol{x}) , \quad (11.107)$$

where $\mathbb{1}$ is the unit matrix of $SU(N_c)$ and

$$U(\boldsymbol{x}) = \mathcal{P}\exp\left(-\frac{\mathrm{i}g}{2}\int_{\infty}^{+\infty}\mathrm{d}x^+\,\mathcal{A}^-(x^+,0,\boldsymbol{x})\right) , \quad (11.108)$$

is the non-abelian eikonal factor.

Let us consider now the interaction of a $q\bar{q}$ pair with the target color field. The corresponding amplitude contains terms proportional to $UU^\dagger$, $U$ and $U^\dagger$. Introducing the quantity

$$W_{\boldsymbol{x}}(\boldsymbol{y}) = U(\boldsymbol{x})U^\dagger(\boldsymbol{x}+\boldsymbol{y}) - \mathbb{1} , \quad (11.109)$$

which incorporates all the information about the external field, the virtual photon cross sections are expressed in terms of the $W$ matrix. In particular, the inclusive DIS cross sections are linear functionals of

$$\mathrm{Tr}\,\left(W_{\boldsymbol{x}}(\boldsymbol{y})W_{\boldsymbol{x}}^\dagger(\boldsymbol{y}')\right) . \quad (11.110)$$

The trace appears because of the summation over all colors of the $q\bar{q}$ pair in the final state.

Diffractive DIS, requiring a color singlet configuration, implies the replacement

$$\mathrm{Tr}\,\left(W_{\boldsymbol{x}}(\boldsymbol{y})W_{\boldsymbol{x}}^\dagger(\boldsymbol{y}')\right) \to \frac{1}{N_c}\,\mathrm{Tr}\,W_{\boldsymbol{x}}(\boldsymbol{y})\,\mathrm{Tr}\,W_{\boldsymbol{x}}^\dagger(\boldsymbol{y}') . \quad (11.111)$$

This change in the color structure has crucial consequences on the resulting cross sections. What one ultimately finds is that the leading-twist inclusive DIS cross section receives contributions from both small- and large-size $q\bar{q}$ pairs, the latter corresponding to the aligned jet configuration. The requirement of color neutrality in the final state suppresses the small-size contributions, and the leading-twist diffractive DIS cross section turns out to be dominated by the production of $q\bar{q}$ pairs with large transverse size. One therefore reaches the same conclusions of the color dipole model (Sect. 11.2). Note however that in the semiclassical approach the production of $q\bar{q}$ pairs both in color singlet and color octet state is described in a unique way. In other terms, diffractive and non-diffractive events are pictured in the same manner, the only difference being the color configuration of the final state, which is assumed to be the result of non-perturbative interactions with the proton color field, in the spirit of the soft color neutralization model. The production of $q\bar{q}g$ states in the semiclassical model has been considered by Buchmüller, McDermott and Hebecker (1997).

## 11.9 Nuclear Shadowing and Diffractive Dissociation

Another, indirect, manifestation of diffraction in deep inelastic scattering is the phenomenon of *nuclear shadowing*. We call "nuclear shadowing" (NS) the low-$x$ depletion of bound-nucleon structure functions with respect to free-nucleon structure functions (see Arneodo 1994 for an overview of the subject),

$$\frac{F_2^A}{F_2^N} < 1, \quad \text{for } x < 10^{-1}\,. \tag{11.112}$$

Here $F_2^A$ is the structure function per nucleon of a nucleus of atomic weight $A$ and $F_2^N$ is the free-nucleon structure function.

At small $Q^2$ (below 1 GeV$^2$), nuclear effects were already studied in the 70's by low-energy electron experiments. In the limit of photoproduction ($Q^2 \to 0$), shadowing can be understood by assuming that the photon fluctuates into a superposition of vector mesons, like the $\rho$, the $\omega$, or the $\phi$, which then interact strongly with the nucleons on the surface of the target nucleus. These nucleons absorb most of the hadronic flux and hence cast a shadow onto the inner ones. This is, in essence, the explanation of NS at $Q^2 \approx 0$, based on the Vector Dominance Model (for the application of this model to shadowing see Donnachie and Shaw 1978; Grammer and Sullivan 1978; Piller, Ratzka and Weise 1995).

The common belief that NS would quickly disappear at $Q^2 > 1$ GeV$^2$ was first challenged by the EMC NA28 experiment at CERN (Arneodo et al. 1988, 1990), which explored the range $0.003 < x < 0.1$ and $0.3 < Q^2 < 3.2$ GeV$^2$, and reported some evidence for shadowing[3]. This result was later confirmed by a more precise measurement carried out by the New Muon Collaboration (Amaudruz et al. 1991), which found that for intermediate–mass nuclei the ratio (11.112) differs from unity by as much as $\sim 20$ %, in the $x \approx 0.003 - 0.01$ range and for $Q^2 \approx 2 - 10$ GeV$^2$. For light nuclei (helium, carbon), shadowing is a $\sim 10$ % effect. DIS off a heavier nuclear target (Xe) has been investigated by the E665 experiment at Fermilab in the range $0.001 < x < 0.25$ and $0.1 < Q^2 < 20$ GeV$^2$ (Adams et al. 1992). The results of various measurements are collected in Fig. 11.10.

[3] From a theoretical viewpoint, NS was *predicted* to survive at large $Q^2$ by Nikolaev and Zakharov (1975), long time before its experimental discovery.

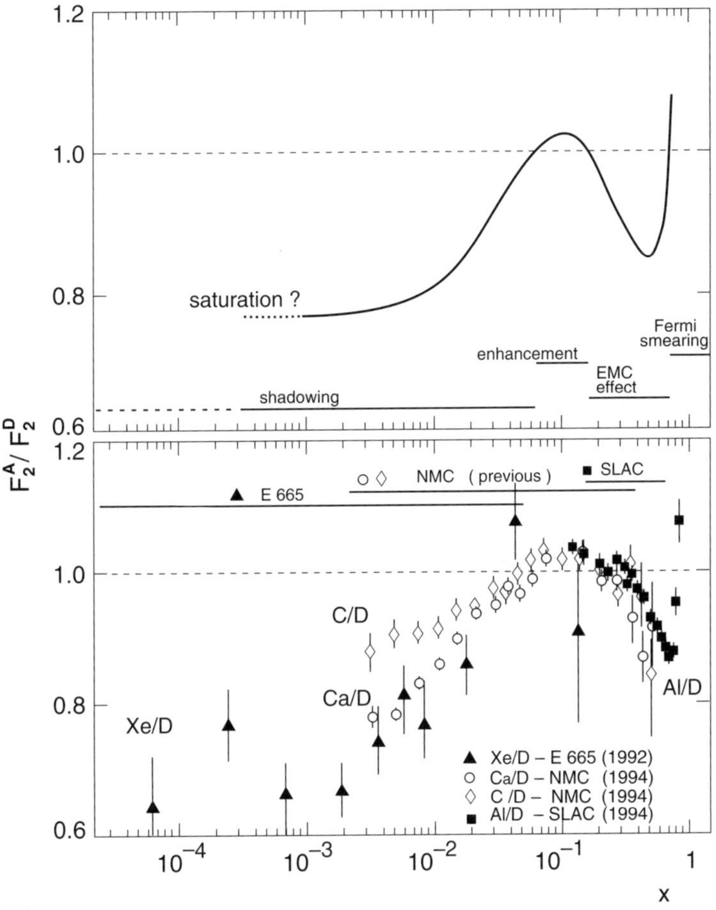


**Fig. 11.10.** A compilation of data and interpretations of DIS off nuclear targets.

What emerges clearly from experiments is that shadowing persists at large $Q^2$ and exhibits a very weak $Q^2$ dependence. In other terms, it appears to be a leading–twist effect, as it was suggested long time ago by Nikolaev and Zakharov (1975). More recently, a QCD picture of nuclear shadowing based on the color dipole approach has been presented by Nikolaev and Zakharov (1991), and Barone et al. (1993).

As shown by Gribov (1969), nuclear shadowing (of hadrons, but the analysis can be extended to hadronic fluctuations of photons) is connected to diffractive dissociation. For this reason, we can say that the observation of nuclear shadowing at large $Q^2$ in the early 90's represented, in a certain sense, the first (indirect) evidence of diffraction in DIS.

To see how the relation between nuclear shadowing and diffractive dissociation arises we start by summarizing the main elements of the Glauber–Gribov multiple scattering theory (Glauber 1958, Gribov 1969).

### 11.9.1 Glauber Theory

Let us consider the scattering of a hadron (or of a hadronic fluctuation) $h$ off a nucleus $A$. When a high-energy particle collides with a composite system, it may undergo many rescatterings with the individual components of the target (Fig. 11.11). A general treatment of this reaction is obviously very complicated, and one has to resort to some approximations. For fast incident particles, which tend to emerge at small angles after being scattered, we can use the eikonal approach (see Sect. 2.3). The elastic scattering amplitude reads

$$f_{hA}(\boldsymbol{q}) = \frac{\mathrm{i}k}{2\pi} \int \mathrm{d}^2\boldsymbol{b}\, \mathrm{e}^{-\mathrm{i}\boldsymbol{q}\cdot\boldsymbol{b}} \langle A|\Gamma_A(\boldsymbol{b}, \boldsymbol{b}_1, \ldots, \boldsymbol{b}_A)|A\rangle \,, \tag{11.113}$$

where $\boldsymbol{b}$ is the impact parameter of the incident wave, and $\boldsymbol{b}_1, \ldots, \boldsymbol{b}_A$ are the positions of the nucleons relative to the axis of collision. $\Gamma_A$ is the profile function of the nucleus, related to the overall phase shift $\chi_A$ by

$$\Gamma_A(\boldsymbol{b}, \boldsymbol{b}_1, \ldots, \boldsymbol{b}_A) = 1 - \mathrm{e}^{\mathrm{i}\chi_A(\boldsymbol{b}, \boldsymbol{b}_1, \ldots, \boldsymbol{b}_A)} \,. \tag{11.114}$$

According to (2.142) the total cross section for $hA$ scattering is

$$\sigma_{hA} = 2 \int \mathrm{d}^2\boldsymbol{b}\, \mathrm{Re}\, \langle A|\Gamma_A(\boldsymbol{b}, \boldsymbol{b}_1, \ldots, \boldsymbol{b}_A)|A\rangle \,. \tag{11.115}$$

In the Glauber approach (Glauber 1958) the nucleus is assumed to be a system of uncorrelated nucleons, and the phase shift of the wave after it has passed the entire nucleus is written as a sum of phase shifts for each nucleon, i.e.

$$\chi_A(\boldsymbol{b}, \boldsymbol{b}_1, \ldots, \boldsymbol{b}_A) = \sum_{i=1}^{A} \chi(\boldsymbol{b} - \boldsymbol{b}_i) \,. \tag{11.116}$$

This amounts to requiring that the potential which determines $\chi_A$ is a sum of single-nucleon potentials. The profile function $\Gamma_A$ then becomes (to simplify the formula we adopt the notation $\Gamma_i \equiv \Gamma(\boldsymbol{b} - \boldsymbol{b}_i)$)

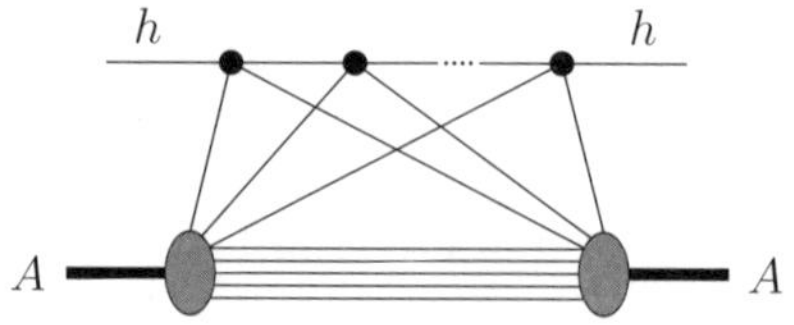


**Fig. 11.11.** Contribution to the multiple scattering series in $hA$ collisions.

$$\begin{aligned}\Gamma_A(\boldsymbol{b}, \boldsymbol{b}_1, \ldots, \boldsymbol{b}_A) &= 1 - \prod_{i=1}^{A}[1 - \Gamma(\boldsymbol{b} - \boldsymbol{b}_i)] \\ &= \sum_i \Gamma_i - \sum_{i<j} \Gamma_i \Gamma_j + \sum_{i<j<k} \Gamma_i \Gamma_j \Gamma_k \\ &\quad - \ldots + (-1)^{A-1} \prod_i \Gamma_i \,, \end{aligned} \tag{11.117}$$

Note the alternance of signs in the expansion of $\Gamma_A$.

Using (11.117) and ignoring the center-of-mass motion of the nucleus, we obtain for $\langle A|\Gamma_A|A\rangle$, that is the scattering amplitude in the impact parameter space,

$$\begin{aligned} f_{hA}(\boldsymbol{b}) &\equiv \langle A|\Gamma_A|A\rangle \\ &= 1 - [1 - \langle N|\Gamma_N(\boldsymbol{b} - \boldsymbol{c})|N\rangle]^A \\ &= 1 - \left[1 - \frac{1}{A} \int \mathrm{d}^2\boldsymbol{c}\, \Gamma(\boldsymbol{b} - \boldsymbol{c}) \int_{-\infty}^{+\infty} \mathrm{d}z\, n_A(\boldsymbol{c}, z)\right]^A , \end{aligned} \tag{11.118}$$

where $|N\rangle$ is a single-nucleon state and $n_A(\boldsymbol{b}, z)$ is the nuclear density normalized as

$$\int_{-\infty}^{+\infty} \mathrm{d}z \int \mathrm{d}^2\boldsymbol{b}\, n_A(\boldsymbol{r}) = A \,. \tag{11.119}$$

Let us now parametrize the profile function $\Gamma(\boldsymbol{b} - \boldsymbol{c})$ as a Gaussian with a slope $B$

$$\Gamma(\boldsymbol{b} - \boldsymbol{c}) = \frac{\sigma_{hN}\,(1 - \mathrm{i}\,\rho_{hN})}{4\pi B} \exp\left[-\frac{(\boldsymbol{b} - \boldsymbol{c})^2}{2B}\right] , \tag{11.120}$$

where $\sigma_{hN}$ is the total cross section for $hN$ scattering and $\rho_{hN}$ is the ratio of the real to the imaginary part of the $hN$ scattering amplitude. Introducing the nuclear optical thickness

$$T(\boldsymbol{c}) = \int_{-\infty}^{+\infty} \mathrm{d}z\, n_A(\boldsymbol{c}, z) \,, \tag{11.121}$$

one obtains, by means of (11.120),

$$f_{hA}(\boldsymbol{b}) = 1 - \left[1 - \frac{1}{2A} \sigma_{hN}(1 - \mathrm{i}\rho_{hN})\, \widetilde{T}(\boldsymbol{b})\right]^A , \tag{11.122}$$

where $\widetilde{T}(\boldsymbol{b})$ is defined as

$$\widetilde{T}(\boldsymbol{b}) = \frac{1}{2\pi B} \int \mathrm{d}^2\boldsymbol{c}\, T(\boldsymbol{c}) \exp\left[-\frac{(\boldsymbol{b} - \boldsymbol{c})^2}{2B}\right] . \tag{11.123}$$

Since the slope $B$ is much smaller than the nuclear radius squared, one has $\widetilde{T}(\boldsymbol{b}) \simeq T(\boldsymbol{b})$. From now on, we shall exploit this approximate equality and replace in our formulas $\widetilde{T}(\boldsymbol{b})$ by $T(\boldsymbol{b})$.

Inserting (11.122) into (11.115) gives for the hadron–nucleus total cross section

$$\sigma_{hA} = 2 \int \mathrm{d}^2\boldsymbol{b} \left\{ 1 - \left[ 1 - \frac{1}{2A}\, \sigma_{hN} (1 - \mathrm{i}\rho_{hN})\, \widetilde{T}(\boldsymbol{b}) \right]^A \right\} . \tag{11.124}$$

If we neglect the (small) real part of the scattering amplitude, we get

$$\begin{aligned} \sigma_{hA} &= 2 \int \mathrm{d}^2\boldsymbol{b} \left\{ 1 - \left[ 1 - \frac{1}{2A}\, \sigma_{hN}\, \widetilde{T}(\boldsymbol{b}) \right]^A \right\} \\ &= 2 \sum_{n=1}^{A} \frac{A!}{n!\,(A-n)!} (-1)^{n-1} \left( \frac{\sigma_{hN}}{2A} \right)^n \int \mathrm{d}^2\boldsymbol{b}\, T^n(\boldsymbol{b}) \, . \end{aligned} \tag{11.125}$$

In the limit of large $A$, the power in (11.122) can be approximated by an exponential

$$f_{hA}(\boldsymbol{b}) \simeq 1 - \exp\left[ -\frac{1}{2}\, \sigma_{hN}\, T(\boldsymbol{b}) \right] , \tag{11.126}$$

and (11.125) becomes

$$\sigma_{hA} \simeq 2 \int \mathrm{d}^2\boldsymbol{b} \left\{ 1 - \exp\left[ -\frac{1}{2}\, \sigma_{hN}\, T(\boldsymbol{b}) \right] \right\} . \tag{11.127}$$

Equation (11.125), or (11.127), is the main result of Glauber theory.

The first non vanishing term in the expansion of (11.127) gives $\sigma_{hA} \simeq A\,\sigma_{hN}$ (the impulse approximation). The second term, which corresponds to double scattering, adds a negative contribution, which gives rise to nuclear shadowing. Explicitly we have

$$\sigma_{hA} = A\,\sigma_{hN} \left( 1 - \frac{A-1}{4\,A^2}\, \sigma_{hN} \int \mathrm{d}^2\boldsymbol{b}\, T^2(\boldsymbol{b}) + \ldots \right) . \tag{11.128}$$

The remaining terms in (11.127), represented by the dots, can be approximately resummed by introducing an exponential $\exp(-\frac{1}{2}\sigma_{hN}\, T(\boldsymbol{b}))$, that is

$$\sigma_{hA} \simeq A\,\sigma_{hN} \left( 1 - \frac{A-1}{4\,A^2}\, \sigma_{hN} \int \mathrm{d}^2\boldsymbol{b}\, T^2(\boldsymbol{b})\, \mathrm{e}^{-\frac{1}{2}\sigma_{hN} T(\boldsymbol{b})} \right) . \tag{11.129}$$

For illustration, if we take a Gaussian nuclear density of the form

$$n_A(\boldsymbol{r}) = \frac{1}{\pi^{3/2} R^3} \exp\left( - \frac{\boldsymbol{r}^2}{A^{2/3} R^2} \right) , \tag{11.130}$$

with a radius $\langle r^2 \rangle^{1/2} = \sqrt{3/2}\, A^{1/3} R$, we get from (11.128)

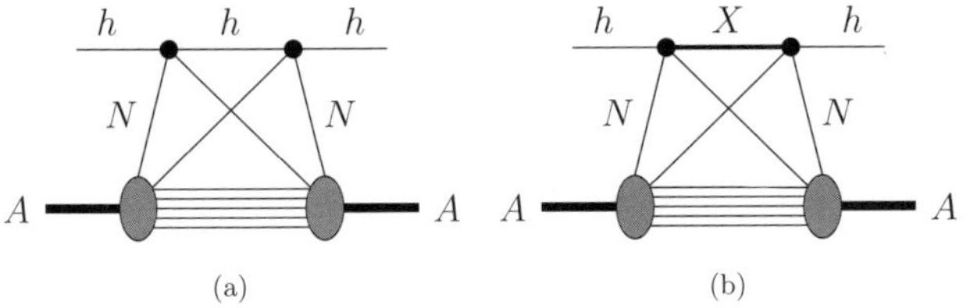


**Fig. 11.12.** Double scattering diagrams: (**a**) elastic rescattering; (**b**) inelastic rescattering.

$$\sigma_{hA} = A\,\sigma_{hN}\left(1 - \frac{A-1}{A}\,A^{1/3}\,\frac{\sigma_{hN}}{8\pi R^2} + \dots\right), \tag{11.131}$$

which displays the typical $A^{1/3}$ dependence of shadowing in $hA$ collisions.

In field–theoretical terms the results discussed above can be reobtained by calculating the multiple scattering diagrams of Fig. 11.11. In particular, Glauber theory is recovered if one considers only diagonal (i.e. elastic) rescatterings, in which case the first correction to the impulse approximation comes from the diagram of Fig. 11.12a, and neglects the longitudinal momenta transferred to the interacting nucleons.

### 11.9.2 Gribov Inelastic Corrections

The Glauber approximation would be exact for the scattering of an eigenstate of the interaction Hamiltonian. The incoming hadron $h$, however, is not such an eigenstate and thus it can be excited into a new state as a result of its interaction with the nucleus. The off–diagonal (i.e. inelastic) corrections to Glauber theory (Fig. 11.12b), that correspond to diffractive dissociation, were first studied by Gribov (1969). Computing these corrections is a difficult task, and again some approximations are needed. Assuming that the inelastic contributions are small compared to the elastic one[4], so that we can consider only one diffractive excitation, Karmanov and Kondratyuk (1973) derived the following formula for the inelastic correction to the $hA$ total cross section

$$\Delta_{\text{in}}\,\sigma_{hA} = -4\pi\,\frac{A-1}{A}\int \mathrm{d}^2\boldsymbol{b}\int \mathrm{d}M^2\;\frac{\mathrm{d}\sigma^D}{\mathrm{d}t\,\mathrm{d}M^2}\bigg|_{t=0} \times |\mathcal{F}(k_L,\boldsymbol{b})|^2\,\mathrm{e}^{-\frac{1}{2}\sigma_{hN}T(\boldsymbol{b})}, \tag{11.132}$$

where $\mathrm{d}\sigma^D/\mathrm{d}t\,\mathrm{d}M^2$ is the cross section for diffractive dissociation of the hadron $h$ into a state of invariant mass $M^2$, and $\mathcal{F}(k_L,\boldsymbol{b})$ is the longitudinal form factor of the nucleus

$$\mathcal{F}(k_L,\boldsymbol{b}) = \int_{-\infty}^{+\infty} \mathrm{d}z\,n_A(\boldsymbol{b},z)\,\mathrm{e}^{\mathrm{i}k_L z}. \tag{11.133}$$

---

[4] As shown by Nikolaev (1981), this is true in the case of high-energy hadron–nucleus scattering.

The phase shift $\mathrm{e}^{\mathrm{i}k_L z}$ arises from the longitudinal momentum transfer $k_L$ at the diffractive dissociation vertex, which is given by

$$k_L = \frac{M^2 - m_h^2}{2E} \,, \tag{11.134}$$

where $m_h$ is the mass of hadron $h$ and $E$ is the incident energy in the target rest frame. At high energies, if $k_L R_A \ll 1$, one has $\mathcal{F}(k_L, \boldsymbol{b}) \simeq T(\boldsymbol{b})$.

The easiest way to derive the Karmanov–Kondratyuk formula (11.132) is to use the eigenstate formalism presented in Sect. 11.1. We start by expanding the physical state $|h\rangle$ of the incident hadron in the basis of the eigenstates $|\alpha\rangle$ of the interaction Hamiltonian

$$|h\rangle = \sum_\alpha c_{h\alpha} |\alpha\rangle \tag{11.135}$$

The scattering amplitude in the impact parameter space is then

$$f_{hA}(\boldsymbol{b}) = 1 - \left\langle \mathrm{e}^{-\frac{1}{2}\sigma_\alpha T(\boldsymbol{b})} \right\rangle \,, \tag{11.136}$$

where $\langle O_\alpha \rangle \equiv \sum_\alpha |c_{h\alpha}|^2 \, O_\alpha$, and the resulting cross section reads

$$\sigma_{hA} \simeq 2 \int \mathrm{d}^2\boldsymbol{b} \, \left\{ 1 - \left\langle \mathrm{e}^{-\frac{1}{2}\sigma_\alpha T(\boldsymbol{b})} \right\rangle \right\} \,. \tag{11.137}$$

The assumption underlying (11.136) is that the states $|\alpha\rangle$ do not mix with each other during their propagation through the nucleus, in spite of the fact that they are not eigenstates of the *vacuum* Hamiltonian. This is true if the fluctuation time of the states $|\alpha\rangle$ is much larger than the nuclear radius $R_A$, that is if

$$\frac{2E}{m_{h*}^2 - m_h^2} \gg R_A \,, \tag{11.138}$$

where $h^*$ is the first excited state of $h$.

The inelastic correction to Glauber formula is the difference between (11.137) and the elastic contribution (11.127), that, recalling $\sigma_{hN} = \langle \sigma_\alpha \rangle$, we rewrite as

$$\sigma_{hA} \simeq 2 \int \mathrm{d}^2\boldsymbol{b} \, \left\{ 1 - \mathrm{e}^{-\frac{1}{2}\langle\sigma_\alpha\rangle T(\boldsymbol{b})} \right\} \,. \tag{11.139}$$

Thus we get

$$\begin{aligned} \Delta_{\mathrm{in}} \sigma_{hA} &= \int \mathrm{d}^2\boldsymbol{b} \, \left\{ \left\langle \mathrm{e}^{-\frac{1}{2}\sigma_\alpha T(\boldsymbol{b})} \right\rangle - \mathrm{e}^{-\frac{1}{2}\langle\sigma_\alpha\rangle T(\boldsymbol{b})} \right\} \\ &\simeq -\frac{1}{4} \int \mathrm{d}^2\boldsymbol{b} \, T^2(\boldsymbol{b}) \, \mathrm{e}^{-\frac{1}{2}\sigma_{hN} T(\boldsymbol{b})} \, \left\{ \langle\sigma_\alpha^2\rangle - \langle\sigma_\alpha\rangle^2 \right\} \,. \end{aligned} \tag{11.140}$$

In the second line of (11.140) we neglected all terms of order higher than $\sigma_\alpha^2$. In view of the expression (11.13) for the diffractive cross section, (11.140) becomes

$$\Delta_{\rm in}\,\sigma_{hA} \simeq -4\pi \int \mathrm{d}^2\boldsymbol{b}\, \left.\frac{\mathrm{d}\sigma^D}{\mathrm{d}t}\right|_{t=0} T^2(\boldsymbol{b})\,\mathrm{e}^{-\frac{1}{2}\sigma_{hN}\,T(\boldsymbol{b})}\,, \tag{11.141}$$

which coincides with (11.132) in the limit (11.138) (and for heavy nuclei). Adding (11.141) to (11.129) gives the full (elastic + inelastic) shadowing correction.

### 11.9.3 Color Transparency

As discussed in Sect. 11.1, the rôle of the quantum number $\alpha$ is played in QCD by the transverse size of the color dipole fluctuation of the projectile. Thus the quantum mechanical average $\langle O_\alpha \rangle$ is replaced by $\int \mathrm{d}^2\boldsymbol{\rho}\, |\Psi(\rho)|^2\, O(\rho)$, where $\Psi(\rho)$ is the dipole wave function. This has important consequences on the interaction between the projectile and the target nucleus, as we shall now show (Zamolodchikov, Kopeliovich and Lapidus 1981; Bertsch et al. 1981; for a review see Nikolaev 1994, 2001; an introduction to the subject can be found in the appendix of Kopeliovich and Marage 1993).

According to Glauber theory, the attenuation amplitude $A(\boldsymbol{b})$ of a hadron $h$ scattering off a nucleus $A$ is exponential – see (11.127),

$$A_{hA}(\boldsymbol{b}) \sim \mathrm{e}^{-\frac{1}{2}\sigma_{hN}T(\boldsymbol{b})}\,. \tag{11.142}$$

Analogously, since a color dipole of transverse size $\rho$ is an eigenstate of the interaction Hamiltonian, its attenuation amplitude has the form

$$A_{dA}(\boldsymbol{b}) \sim \mathrm{e}^{-\frac{1}{2}\sigma(\rho)T(\boldsymbol{b})}\,. \tag{11.143}$$

If we account for inelastic corrections the attenuation amplitude for the scattering of a hadron $h$ off a nucleus $A$ is obtained by averaging over its wave function. Thus (11.142) is replaced by – see (11.136)

$$A_{hA}(\boldsymbol{b}) = \int \mathrm{d}^2\boldsymbol{\rho}\, |\Psi(\rho)|^2\, \mathrm{e}^{-\frac{1}{2}\sigma(\rho)T(\boldsymbol{b})}\,. \tag{11.144}$$

Now, consider a hadron whose wave function is dominated by a small-size $q\bar{q}$ Fock state. This is the case, for instance, of the charmonium. For small $\rho$, as we saw in Sect. 9.9, the dipole cross section has the following (perturbative) behavior (the so–called *color transparency* property)

$$\sigma(\rho) \sim \frac{\rho^2}{R^2}\,\sigma_{\psi N}\,, \tag{11.145}$$

where $R$ is the charmonium size. With a simple Gaussian wave function $\Psi(\rho) \sim \mathrm{e}^{-\rho^2/2\,R^2}$, Eq. 11.144 gives

$$A_{hA}(\boldsymbol{b}) \sim \frac{1}{1 + \frac{\sigma_{\psi N}\,T(\boldsymbol{b})}{2}}\,. \tag{11.146}$$

We discover the important result that the quantum mechanical average over the hadronic wave function leads from an exponential attenuation – Eq. (11.142) –, to a non-exponential attenuation – Eq. (11.146). In other terms, the nucleus becomes more transparent than in the eikonal approximation. This is true for charmonium–nucleus scattering, but is also valid more generally for the scattering of any hadron off a sufficiently heavy nucleus. In fact, in the limit of large $T(\boldsymbol{b})$, the exponential in (11.144) selects small $\rho$ values, so that, irrespective of the details of the hadron wave function, we find

$$A_{hA}(\boldsymbol{b}) \sim \int \mathrm{d}^2\boldsymbol{\rho}\, |\Psi(0)|^2 \,\mathrm{e}^{-\frac{\rho^2}{2R^2}\sigma_{hN}\,T(\boldsymbol{b})} \sim \frac{1}{\sigma_{hN}T(\boldsymbol{b})} \,. \tag{11.147}$$

Again, a non-exponential attenuation amplitude is found. Thus, heavy nuclei filters the small-size dipole fluctuations of the projectiles.

### 11.9.4 Nuclear Shadowing in DIS

Let us now specialize to the case of nuclear shadowing in deep inelastic scattering. According to the theory developed above, at large $Q^2$, where vector meson contributions are negligible, the shadowing correction to $\sigma_{\gamma^* A}$ is

$$\begin{aligned} \Delta_{\text{shad}}\,\sigma_{\gamma^* A} = -4\pi\,\frac{A-1}{A} \int \mathrm{d}^2\boldsymbol{b} \int \mathrm{d}M^2 \, \left.\frac{\mathrm{d}\sigma^D}{\mathrm{d}t\,\mathrm{d}M^2}\right|_{t=0} \\ \times |\mathcal{F}(k_L,\boldsymbol{b})|^2\,\mathrm{e}^{-\frac{1}{2}\sigma_{\text{eff}}T(\boldsymbol{b})} \,, \end{aligned} \tag{11.148}$$

where

$$k_L = \frac{M^2+Q^2}{2\nu} \,, \tag{11.149}$$

and $\sigma_{\text{eff}}$ is the effective cross section for the interaction of the diffracted states with a nucleon. An estimate of $\sigma_{\text{eff}}$ is

$$\sigma_{\text{eff}} \approx 16\pi\,\frac{1}{\sigma_{\gamma^* N}} \left.\frac{\mathrm{d}\sigma^D}{\mathrm{d}t}\right|_{t=0} . \tag{11.150}$$

In terms of structure functions, and using the usual notations of diffractive DIS (see Sect. 10.3), we have ($x_{I\!P} = k_L/m_N$)

$$\begin{aligned} F_2^A(x,Q^2) = F_2^N(x,Q^2) - 4\pi\,\frac{A-1}{A^2} \int \mathrm{d}x_{I\!P}\, F_2^{D(4)}(\beta,Q^2,x_{I\!P},t=0) \\ \times \int \mathrm{d}^2\boldsymbol{b}\, |\mathcal{F}(x_{I\!P},\boldsymbol{b})|^2\,\mathrm{e}^{-\frac{1}{2}\sigma_{\text{eff}}T(\boldsymbol{b})} \,. \end{aligned} \tag{11.151}$$

Thus we have established a link (valid at large $Q^2$) between diffractive structure functions and nuclear structure functions. Using the data on $F_2^{D(4)}$, one can predict nuclear shadowing by means of (11.151). In Fig. 11.13 we show

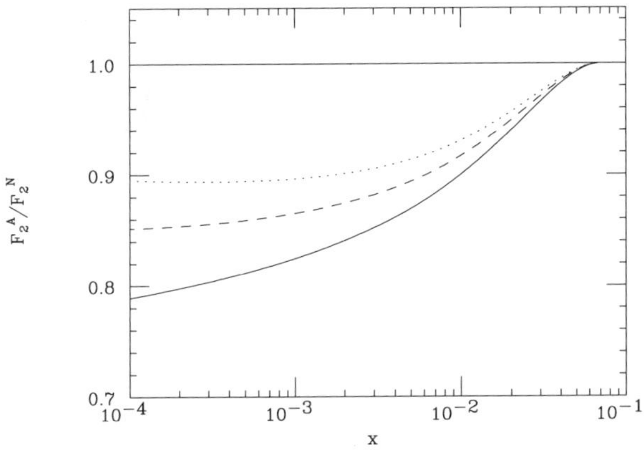


**Fig. 11.13.** The ratio $F_2^A/F_2^N$ for calcium in the $Q^2$ range $10 - 30$ GeV$^2$. The predictions are based on the Glauber-Gribov theory and on the ZEUS measurement of the diffractive structure function $F_2^{D(3)}$ which is parametrized as $F_2^{D(3)} \sim x_{I\!P}^{-a}$. Dotted line: $a = 1.15$; dashed line: $a = 1.20$; solid line: $a = 1.25$. From Barone and Genovese (1997).

the result of such a calculation (Barone and Genovese 1997; for a similar study see Capella et al. 1998). As one can see, the harder the pomeron the more pronounced is nuclear shadowing. The investigation of nuclear DIS in the region $x < 10^{-3}$ for $Q^2 \gtrsim 10$ GeV$^2$ would be an important task for a future collider.

## 11.10 Dipole Scattering

The second part of this chapter is devoted to the BFKL phenomenology. We introduce this subject by studying an ideal process, the scattering of two color dipoles. In the previous sections we identified such dipoles as the relevant degrees of freedom in hard diffractive reactions. Dipoles interact by exchanging two gluons (in Born approximation), or a BFKL ladder (in leading $\ln s$ approximation). The interest of dipole-dipole scattering is at least threefold: *i)* it allows a derivation of the BFKL equation in the impact parameter space, in terms of cross sections; *ii)* it has some phenomenological applications, e.g. to $\gamma^*\gamma^*$ scattering; *iii)* last but no least, it sheds light on the unitarization of cross sections in QCD.

### 11.10.1 The Cross-section for the Scattering of Two Dipoles

Let us consider the high-energy scattering of two colorless hadronic states A and B, both composed of a quark–antiquark pair (for instance, two quarkonia, or two photons in their leading Fock configuration). We suppose that

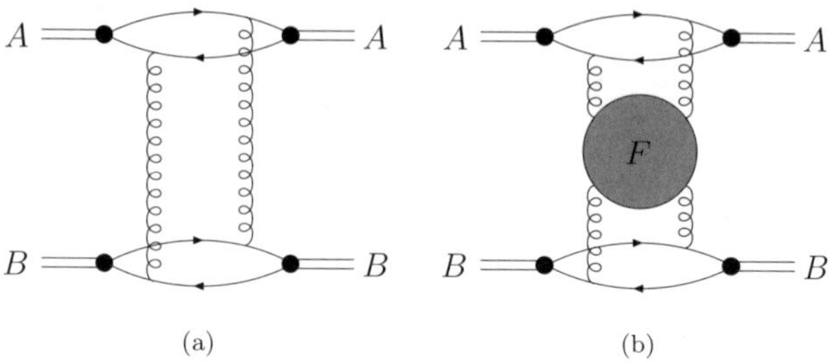


**Fig. 11.14.** Diagrams for the elastic scattering of two quarkonia. (**a**) Born approximation (two-gluon exchange). (**b**) BFKL dynamics (exchange of a gluon ladder).

these states have very small transverse sizes, so that perturbation theory is applicable.

According to Eq. (8.266), the total cross section $\sigma_{AB}$ for the scattering of two quarkonia (which is related to the imaginary part of the diagrams in Fig. 11.14) can be written as

$$\sigma_{AB} = \frac{N_c^2 - 1}{(2\pi)^4} \int \frac{\mathrm{d}^2\boldsymbol{k}}{\boldsymbol{k}^2} \int \frac{\mathrm{d}^2\boldsymbol{k}'}{\boldsymbol{k}'^2} \, \Phi_A(\boldsymbol{k}) \, \Phi_B(\boldsymbol{k}') \, F(x, \boldsymbol{k}, \boldsymbol{k}') \;, \tag{11.152}$$

where the impact factors $\Phi_{A,B}$ are related to the quarkonium wave functions $\Psi_{A,B}(z, r)$ by (see (9.240)

$$\Phi(\boldsymbol{k}) = \frac{8\,\pi^2\,\alpha_s}{3} \int_0^1 \mathrm{d}z \int \mathrm{d}^2\boldsymbol{r} \, |\Psi(z,r)|^2 \, (1 - \mathrm{e}^{\mathrm{i}\,\boldsymbol{k}\cdot\boldsymbol{r}}) \;. \tag{11.153}$$

We denote by $z(z')$ the longitudinal momentum fraction of the quark in $A(B)$, and by $\boldsymbol{r}(\boldsymbol{r}')$ the respective quark–antiquark transverse separation.

We start from the Born approximation, which corresponds to two–gluon exchange (Fig. 11.14a). In this case the BFKL kernel $F(x, \boldsymbol{k}, \boldsymbol{k}')$ reduces to $\delta^2(\boldsymbol{k} - \boldsymbol{k}')$, and inserting (11.153) into (11.152) we find (with $N_c = 3$)

$$\begin{aligned} \sigma_{AB} = \frac{32}{9}\, \alpha_s^2 & \int \mathrm{d}z \, \mathrm{d}^2\boldsymbol{r} \int \mathrm{d}z' \, \mathrm{d}^2\boldsymbol{r}' \, |\Psi_A(z,r)|^2 \, |\Psi_B(z',r')|^2 \\ & \times \int \frac{\mathrm{d}^2\boldsymbol{k}}{\boldsymbol{k}^4} \, (1 - \mathrm{e}^{\mathrm{i}\,\boldsymbol{k}\cdot\boldsymbol{r}}) \, (1 - \mathrm{e}^{\mathrm{i}\,\boldsymbol{k}\cdot\boldsymbol{r}'}) \;. \end{aligned} \tag{11.154}$$

This is a factorizable expression which can be recast in the form

$$\sigma_{AB} = \int \mathrm{d}z \, \mathrm{d}^2\boldsymbol{r} \int \mathrm{d}z' \, \mathrm{d}^2\boldsymbol{r}' \, |\Psi_A(z,r)|^2 \, |\Psi_B(z',r')|^2 \, \hat{\sigma}^{(0)}(\boldsymbol{r}, \boldsymbol{r}') \;, \tag{11.155}$$

where $\hat{\sigma}^{(0)}$, given by

$$\hat{\sigma}^{(0)}(\boldsymbol{r}, \boldsymbol{r}') = \frac{32}{9} \alpha_s^2 \int \frac{\mathrm{d}^2\boldsymbol{k}}{\boldsymbol{k}^4} \, (1 - \mathrm{e}^{\mathrm{i}\,\boldsymbol{k}\cdot\boldsymbol{r}}) \, (1 - \mathrm{e}^{\mathrm{i}\,\boldsymbol{k}\cdot\boldsymbol{r}'}) \;, \tag{11.156}$$

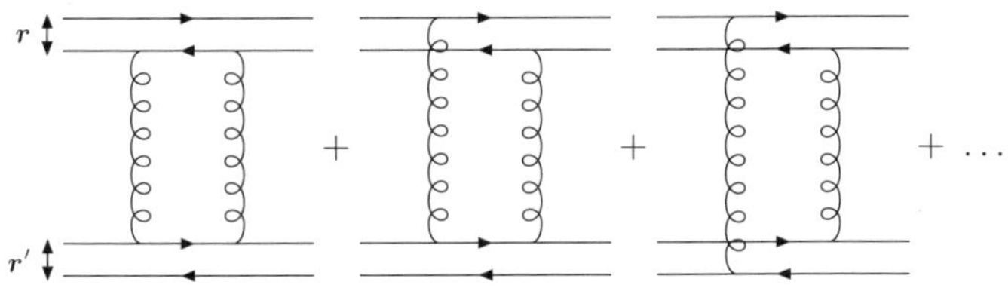


**Fig. 11.15.** Two-gluon exchange diagrams contributing to dipole-dipole scattering.

is the cross section for the scattering of two $q\bar{q}$ pairs ("dipoles") of transverse sizes $\boldsymbol{r}$ and $\boldsymbol{r}'$, via two–gluon exchange (Fig. 11.15). Note that this is a purely perturbative quantity. The non-perturbative dynamics of the scattering is incorporated into the wave functions $\Psi_{A,B}$ of the two quarkonia.

Let us recall the origin of the factors $(1 - \mathrm{e}^{\mathrm{i}\,\boldsymbol{k}\cdot\boldsymbol{r}})$ and $(1 - \mathrm{e}^{\mathrm{i}\,\boldsymbol{k}\cdot\boldsymbol{r}'})$. The unity arises from the diagrams where the gluons couple to the same quark or antiquark line. The exponentials arise from the coupling to both the quark and the antiquark of the dipole.

The factorization formula (11.155) is valid only if $\alpha_s$ is small enough. In the spirit of leading-order BFKL, in (11.156) we used a fixed strong coupling and we factorized it out of the integral. However, computing the diagrams of Fig. 11.15, one finds, instead of $\alpha_s^2$, the product

$$\alpha_s(\max\{\boldsymbol{k}^2, 1/\boldsymbol{r}^2\})\,\alpha_s(\max\{\boldsymbol{k}^2, 1/\boldsymbol{r}'^2\})\,. \tag{11.157}$$

Therefore, as anticipated, we can use perturbative QCD, and apply Eq. (11.155), only if the $q\bar{q}$ Fock components of the interacting quarkonia have small sizes. With this important *caveat* in mind, we proceed to consider higher-order corrections to Eq. (11.155).

As discussed in Sect. 9.10, the primary dipoles, that is the $q\bar{q}$ pairs of the lowest Fock components of $A$ and $B$, start emitting soft gluons, and consequently produce a cascade of interacting secondary dipoles. So, the effect of the BFKL ladder (Fig. 11.14b) is to modify Eq. (11.155) as follows

$$\begin{aligned}\sigma_{AB} = &\int \mathrm{d}z\,\mathrm{d}^2\boldsymbol{r} \int \mathrm{d}z'\,\mathrm{d}^2\boldsymbol{r}'\,|\Psi_A(z,r)|^2\,|\Psi_B(z,r')|^2 \\ &\times \int \frac{\mathrm{d}^2\boldsymbol{\rho}}{2\pi\rho^2} \int \frac{\mathrm{d}^2\boldsymbol{\rho}'}{2\pi\rho'^2}\, n_d(r,\rho,y')\, n_d(r',\rho',y-y')\,\hat{\sigma}^{(0)}(\boldsymbol{\rho},\boldsymbol{\rho}')\,. \end{aligned} \tag{11.158}$$

Here $n_d(r,\rho,y')$ is the number density of secondary dipoles of transverse size $\rho$ inside a primary dipole of transverse size $r$, separated from this by $y'$ units of rapidity ($y \equiv \ln(s/\boldsymbol{k}^2)$). Note that $y'$ is arbitrary. Choosing $y'$ to be equal to 0 ($y$) is equivalent to assuming all secondary dipoles to be radiated from quarkonium B (A). Any intermediate choice ($0 < y' < y$) corresponds to sharing the emitted dipoles between A and B. The normalization of $n_d(r,\rho,y')$ is such that the integral

$$N_d = \int \frac{\mathrm{d}^2\boldsymbol{\rho}}{2\pi\rho^2}\, n_d(r,\rho,y') \tag{11.159}$$

is the total number of dipoles inside the initial quarkonium.

Equation (11.158) is still of the form (11.155), but with the Born cross section $\hat{\sigma}^{(0)}$ replaced by

$$\hat{\sigma}(y,r,r') = \int \frac{\mathrm{d}^2\boldsymbol{\rho}}{2\pi\rho^2} \int \frac{\mathrm{d}^2\boldsymbol{\rho}'}{2\pi\rho'^2}\, n_d(r,\rho,y')\, n_d(r',\rho',y)\, \hat{\sigma}^{(0)}(\boldsymbol{\rho},\boldsymbol{\rho}')\,, \tag{11.160}$$

which is the dipole-dipole cross section incorporating the gluon ladder. Note that, beyond the Born approximation, the dipole cross section acquires a dependence on on the rapidity $y$ (i.e., on $s$).

If we assume that all dipoles are emitted from one of the two quarkonia (say $A$), then we have $n_d(r',\rho',0) = r'\,\delta(r'-\rho')$, and (11.158) becomes

$$\sigma_{AB} = \int \mathrm{d}z\, \mathrm{d}^2\boldsymbol{r} \int \mathrm{d}z'\, \mathrm{d}^2\boldsymbol{r}'\, |\Psi_A(z,r)|^2\, |\Psi_B(z,r')|^2\, \hat{\sigma}(y,r,r')\,, \tag{11.161}$$

with the dipole cross section

$$\begin{aligned}\hat{\sigma}(y,r,r') &= \int \frac{\mathrm{d}^2\boldsymbol{\rho}}{2\pi\rho^2}\, n_d(r,\rho,y)\, \hat{\sigma}^{(0)}(\boldsymbol{\rho},\boldsymbol{r}') \\ &= \frac{32}{9}\alpha_s^2 \int \frac{\mathrm{d}^2\boldsymbol{\rho}}{2\pi\rho^2} \int \frac{\mathrm{d}^2\boldsymbol{k}}{\boldsymbol{k}^4}\, n_d(r,\rho,y) \\ &\quad \times (1-\mathrm{e}^{\mathrm{i}\,\boldsymbol{k}\cdot\boldsymbol{\rho}})\,(1-\mathrm{e}^{\mathrm{i}\,\boldsymbol{k}\cdot\boldsymbol{r}'})\,.\end{aligned} \tag{11.162}$$

The BFKL theory gives the following expression for $n_d(r,\rho,y)$

$$n_d(r,\rho,y) = \sqrt{\frac{2}{\pi\lambda'}}\,\frac{r}{\rho}\,\frac{\mathrm{e}^{\lambda y}}{\sqrt{y}}\,\exp\left[-\frac{\ln^2(r^2/\rho^2)}{2\lambda' y}\right]\,, \tag{11.163}$$

where $\lambda$ and $\lambda'$ are defined in (8.205). In Sect. 11.10.2 we shall present a proof of (11.162) and (11.163). For an alternative derivation, based on the BFKL equation (8.178), see Forshaw and Ross (1997).

The evaluation of (11.160), with (11.163), yields the total cross section for the scattering of two dipoles, which is (for dipoles with the same transverse size $r$)

$$\hat{\sigma}(y,r) == 8\pi\,\alpha_s^2\, r^2 \sqrt{\frac{2}{\pi\lambda'}}\,\frac{\mathrm{e}^{\lambda y}}{\sqrt{y}} \sim \frac{r^2}{\sqrt{\ln(s/s_0)}}\left(\frac{s}{s_0}\right)^{\lambda}\,. \tag{11.164}$$

This is evidently the familiar BFKL behavior.

### 11.10.2 The BFKL Equation for the Dipole Cross-section

Let us now see how Eq. (11.162), and in particular the explicit form of $n_d(r,b,y)$, Eq. (11.163), are derived in the color dipole formalism (Nikolaev and Zakharov 1994b; Nikolaev, Zakharov and Zoller 1994a, 1994b; see also Mueller 1994). By doing so, we shall also get a new form of the BFKL equation, in terms of the dipole cross section. The idea underlying this approach is that $\hat{\sigma}(y,r,r')$ is obtained by summing the contributions of all higher Fock components of the dipole, i.e., of the states with any number of soft gluons.

To start with, consider a $q\bar{q}g$ state, where $g$ is a soft gluon. We call $\boldsymbol{\rho}_1$ and $\boldsymbol{\rho}_2$ the transverse separations of the gluon from the quark and the antiquark respectively ($\boldsymbol{\rho}_2 = \boldsymbol{\rho}_1 - \boldsymbol{r}$). The transverse coordinates of the three partons are shown in Fig. 11.16. The cross section for the interaction of the $q\bar{q}g$ state with a $q\bar{q}$ dipole of size $r'$ is (from now on, we suppress the dependence on the target variable $r'$ in the cross sections)

$$\hat{\sigma}_{q\bar{q}g}(r,\rho_1) = \frac{9}{8}\left[\hat{\sigma}^{(0)}(\rho_1) + \hat{\sigma}^{(0)}(\rho_2)\right] - \frac{1}{8}\,\hat{\sigma}^{(0)}(r)\,. \tag{11.165}$$

This form can be heuristically inferred from gauge invariance considerations. In fact, when the $q\bar{q}$ separation $r$ is small with respect to $\rho_1 \simeq \rho_2$, the $q\bar{q}$ pair is indistinguishable from the gluon, and (11.165) correctly reduces to (recall that $\hat{\sigma}^{(0)}(r) \to 0$ as $r \to 0$)

$$\hat{\sigma}_{q\bar{q}g}(0,\rho_1) = \frac{9}{4}\,\hat{\sigma}^{(0)}(\rho_1)\,, \tag{11.166}$$

where $9/4$ is the familiar ratio of the octet and triplet color factors, i.e. $C_A/C_F$. On the other hand, in the limit of vanishing $\rho_1(\rho_2)$, the $qg(\bar{q}g)$ system is indistinguishable from the (anti)quark, hence

$$\hat{\sigma}_{q\bar{q}g}(r,0) = \hat{\sigma}^{(0)}(r)\,, \tag{11.167}$$

as expected. A formal derivation of (11.165) is based on the calculation of diagrams like those of Fig. 11.15, but with the upper system consisting of a quark, an antiquark and a gluon.

If we call $\Psi^{(0)}(z,r)$ the wave function of the $q\bar{q}$ bound state, i.e. of the lowest Fock component of the meson, and by $\Psi^{(1)}(z,r,z_1,\rho_1)$ the wave function of the $q\bar{q}g$ state, where $z_1$ is the longitudinal momentum fraction of the quarkonium momentum carried by the gluon, we have the following relation

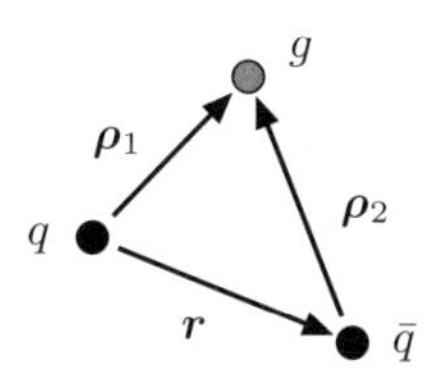


**Fig. 11.16.** Transverse coordinates of the $q\bar{q}g$ state.

$$|\Psi^{(1)}(z,r,z_1,\rho_1)|^2 = \frac{C_F\,\alpha_s}{\pi^2}\,\frac{r^2}{\rho_1^2\,\rho_2^2}\,|\Psi^{(0)}(z,r)|^2\,. \tag{11.168}$$

This relation was derived in Sect. 9.10, where the quantity

$$\Pi(r,\rho_1) = \frac{C_F\,\alpha_s}{\pi^2}\,\frac{r^2}{\rho_1^2\,\rho_2^2} \tag{11.169}$$

was identified as the (differential) probability to find, in the $q\bar{q}g$ state, a gluon with longitudinal momentum fraction $z_1$ at a transverse distance $\rho_1$ from the quark. The number of such gluons is (we introduce a lower cutoff $z_0$ in the integration over $z_1$)

$$n_g(z,r) = \int_{z_0}^{z}\frac{\mathrm{d}z_1}{z_1}\int \mathrm{d}^2\boldsymbol{\rho}_1\,\frac{C_F\,\alpha_s}{\pi^2}\,\frac{r^2}{\rho_1^2\,\rho_2^2}\,. \tag{11.170}$$

We are now able to write down the dipole cross section taking into account the emission of one soft gluon. It reads

$$\begin{aligned}\hat{\sigma}(z,r) &= [1-n_g(z,r)]\,\hat{\sigma}^{(0)}(r)\\ &\quad+\int_{z_0}^{z}\frac{\mathrm{d}z_1}{z_1}\int \mathrm{d}^2\boldsymbol{\rho}_1\,\Xi(r,\rho_1)\,\sigma_{q\bar{q}g}(r,\rho_1)\,.\end{aligned} \tag{11.171}$$

The first term is the lowest order contribution renormalized for the gluon emission, whereas the second term is the contribution of the $q\bar{q}g$ Fock state. We rearrange the two terms, hence obtaining

$$\begin{aligned}\hat{\sigma}(z,r) &= \hat{\sigma}^{(0)}(r) + \int\frac{\mathrm{d}z_1}{z_1}\int \mathrm{d}^2\boldsymbol{\rho}_1\,\Pi(r,\rho_1)\\ &\quad\times\frac{9}{8}\left[\hat{\sigma}^{(0)}(\rho_1)+\hat{\sigma}^{(0)}(\rho_2)-\hat{\sigma}^{(0)}(r)\right]\,.\end{aligned} \tag{11.172}$$

This can be put in the form

$$\hat{\sigma}(z,r) = \hat{\sigma}^{(0)}(r) + \hat{\sigma}^{(1)}(r)\,\ln\frac{z}{z_0}\,, \tag{11.173}$$

where $\hat{\sigma}^{(1)}$ is given by (the coefficient 9/8 in (11.172) is $C_A/2C_F$)

$$\begin{aligned}\hat{\sigma}^{(1)}(r) &= \mathcal{K}\otimes\hat{\sigma}^{(0)}(r)\\ &= \frac{N_c\alpha_s}{2\pi^2}\int \mathrm{d}^2\boldsymbol{\rho}_1\,\frac{r^2}{\rho_1^2\,\rho_2^2}\left[\hat{\sigma}^{(0)}(\rho_1)+\hat{\sigma}^{(0)}(\rho_2)-\hat{\sigma}^{(0)}(r)\right]\,.\end{aligned} \tag{11.174}$$

This is a general recurrence relation which connects the cross section of a $(n+1)$-gluon state to the cross section of a $n$-gluon state, that is

$$\hat{\sigma}^{(n+1)} = \mathcal{K}\otimes\hat{\sigma}^{(n)}\,. \tag{11.175}$$

The $n$-gluon contribution to the dipole cross section is proportional to $\ln^n(z/z_0)$ and the generalization of Eq. (11.173) to all orders (i.e., taking into account an arbitrary number of soft gluons) is

$$\hat{\sigma}(y,r) = \sum_{n=0}^{\infty} \frac{1}{n!}\, \sigma^{(n)}(r)\, y^n \,, \tag{11.176}$$

written in terms of the rapidity $y = \ln(z_0/z)$. From (11.174–11.176) we get the BFKL equation for the dipole cross section, which reads

$$\frac{\partial \hat{\sigma}(y,r)}{\partial y} = \frac{N_c \alpha_s}{2\pi^2} \int \mathrm{d}^2 \boldsymbol{\rho}_1 \, \frac{r^2}{\rho_1^2 \rho_2^2} \left[\hat{\sigma}(y,\rho_1) + \hat{\sigma}(y,\rho_2) - \hat{\sigma}(y,r)\right] , \tag{11.177}$$

The solution of this equation is found by using the methods already applied in Sects. 8.7 and 9.10. At leading $\ln s$ we have

$$\begin{aligned}
\hat{\sigma}(y,r) &= \frac{1}{2\pi} \int_{-\infty}^{+\infty} \mathrm{d}\nu \left(\frac{r^2}{r_0^2}\right)^{\frac{1}{2}+\mathrm{i}\nu} \mathrm{e}^{\omega_0(\nu) y} \int \frac{\mathrm{d}\rho^2}{r_0^2} \left(\frac{\rho^2}{r_0^2}\right)^{-\frac{3}{2}-\mathrm{i}\nu} \hat{\sigma}(0,r) \\
&= \frac{1}{2\pi} \int \frac{\mathrm{d}\rho^2}{\rho^2} \frac{r}{\rho} \int_{-\infty}^{+\infty} \mathrm{d}\nu \, \exp\left[\mathrm{i}\nu \ln\left(\frac{r^2}{\rho^2}\right) + \lambda y - \frac{1}{2}\lambda' \nu^2\right] \hat{\sigma}^{(0)}(r) \\
&= \frac{\mathrm{e}^{\lambda y}}{\sqrt{2\pi \lambda' y}} \int \frac{\mathrm{d}\rho^2}{\rho^2} \frac{r}{\rho} \exp\left[-\frac{\ln^2(r^2/\rho^2)}{2\lambda' y}\right] \hat{\sigma}^{(0)}(\rho) \,,
\end{aligned} \tag{11.178}$$

Comparing this result with (11.162) we get the explicit expression of the dipole number density $n_d$, that is (11.163).

### 11.10.3 Multiple Scattering of Dipoles

The exchange of a BFKL pomeron (in the leading logarithm approximation) leads to total cross sections which grow very rapidly with $s$, and eventually violate unitarity. This was seen explicitly in the case of parton–parton scattering (Sect. 8.15), DIS at low-$x$ (Sect. 9.11), and dipole scattering (Sect. 11.10.1). In BFKL theory the rise of total cross sections with increasing energy is attributed to the proliferation of soft gluon emissions. At large $s$ some mechanism must set on to tame such a growth.

The color dipole formalism allows one to investigate the unitarization of cross sections in a very elegant way. Back to Sect. 9.11, we unitarized the DIS cross sections by a simple phenomenological procedure, using standard eikonalization methods. We now adopt a perturbative QCD point of view. Unitarity corrections in dipole-dipole interactions will be related to *multiple scattering effects*.

The BFKL formulas (11.158) and (11.182) for the scattering of two quarkonia correspond to one-pomeron exchange. The crucial underlying assumption is that only a single pair of dipoles is involved in the scattering process. As the energy gets larger and larger, this may not be true any longer. With increasing $s$, in fact, the number of emitted dipoles rises exponentially (see (11.163)) and the probability of multiple scattering becomes non negligible. It turns out that, while one-pomeron exchange violates the Froissart-Martin bound, the whole multiple scattering series (corresponding to multi-pomeron exchange) restores unitarity. Multiple scattering of dipoles has been investigated both formally and numerically by Mueller (1995), Mueller and Patel (1994), Salam (1996) and Mueller and Salam (1996).

Consider the elastic scattering of two quarkonia $A$ and $B$ at momentum transfer $t \equiv -\boldsymbol{q}^2$. In the framework of BFKL theory the scattering amplitude is

$$A(s,t) = \mathrm{i}s \, \frac{N_c^2 - 1}{(2\pi)^4} \int \mathrm{d}^2\boldsymbol{k} \int \mathrm{d}^2\boldsymbol{k}' \, \Phi_A(\boldsymbol{k},\boldsymbol{q}) \, \Phi_B(\boldsymbol{k}',\boldsymbol{q}) \, \frac{F(x,\boldsymbol{k},\boldsymbol{k}')}{\boldsymbol{k}'^2(\boldsymbol{k}-\boldsymbol{q})^2} \, , \tag{11.179}$$

where the impact factors $\Phi_{A,B}$ are

$$\Phi_{A,B}(\boldsymbol{k},\boldsymbol{q}) = \frac{4\,\pi^2\,\alpha_s}{3} \int_0^1 \mathrm{d}z \int \mathrm{d}^2\boldsymbol{r} \, |\Psi_{A,B}(z,r)|^2 \\ \times (1 - \mathrm{e}^{\mathrm{i}\,\boldsymbol{k}\cdot\boldsymbol{r}/2} - \mathrm{e}^{-\mathrm{i}\,\boldsymbol{k}\cdot\boldsymbol{r}/2})(1 - \mathrm{e}^{\mathrm{i}\,(\boldsymbol{q}-\boldsymbol{k})\cdot\boldsymbol{r}/2} - \mathrm{e}^{-\mathrm{i}\,(\boldsymbol{q}-\boldsymbol{k})\cdot\boldsymbol{r}/2}) \, . \tag{11.180}$$

The Born approximation (i.e., two-gluon exchange) for $A(s,t)$ is

$$A(s,t) = \mathrm{i}s \, \frac{8}{9} \, \alpha_s^2 \int \mathrm{d}z \, \mathrm{d}^2\boldsymbol{r} \int \mathrm{d}z' \, \mathrm{d}^2\boldsymbol{r}' \, |\Psi_A(z,r)|^2 \, |\Psi_B(z',r')|^2 \\ \times \int \frac{\mathrm{d}^2\boldsymbol{k}}{\boldsymbol{k}^2(\boldsymbol{k}-\boldsymbol{q})^2} \left(\mathrm{e}^{\mathrm{i}\,\boldsymbol{k}\cdot\boldsymbol{r}/2} - \mathrm{e}^{-\mathrm{i}\,\boldsymbol{k}\cdot\boldsymbol{r}/2}\right) \left(\mathrm{e}^{\mathrm{i}\,(\boldsymbol{q}-\boldsymbol{k})\cdot\boldsymbol{r}/2} - \mathrm{e}^{-\mathrm{i}\,(\boldsymbol{q}-\boldsymbol{k})\cdot\boldsymbol{r}/2}\right) \\ \times \left(\mathrm{e}^{\mathrm{i}\,\boldsymbol{k}\cdot\boldsymbol{r}'/2} - \mathrm{e}^{-\mathrm{i}\,\boldsymbol{k}\cdot\boldsymbol{r}'/2}\right) \left(\mathrm{e}^{\mathrm{i}\,(\boldsymbol{q}-\boldsymbol{k})\cdot\boldsymbol{r}'/2} - \mathrm{e}^{-\mathrm{i}\,(\boldsymbol{q}-\boldsymbol{k})\cdot\boldsymbol{r}'/2}\right) \tag{11.181}$$

Resummation of the whole BFKL series yields the following generalization of (11.181)

$$A(s,t) = \frac{8}{9} \mathrm{i}s \, \alpha_s^2 \int \mathrm{d}z \, \mathrm{d}^2\boldsymbol{r} \int \mathrm{d}z' \, \mathrm{d}^2\boldsymbol{r}' \, |\Psi_A(z,r)|^2 \, |\Psi_B(z',r')|^2 \\ \times \int \frac{\mathrm{d}^2\boldsymbol{\rho}}{2\pi\rho^2} \int \frac{\mathrm{d}^2\boldsymbol{\rho}'}{2\pi\rho'^2} \int \frac{\mathrm{d}^2\boldsymbol{k}}{\boldsymbol{k}^2(\boldsymbol{k}-\boldsymbol{q})^2} \, n_d(r,\rho,y,q) \, n_d(r',\rho',y-y',q) \\ \times \left(\mathrm{e}^{\mathrm{i}\,\boldsymbol{k}\cdot\boldsymbol{r}/2} - \mathrm{e}^{-\mathrm{i}\,\boldsymbol{k}\cdot\boldsymbol{r}/2}\right) \left(\mathrm{e}^{\mathrm{i}\,(\boldsymbol{q}-\boldsymbol{k})\cdot\boldsymbol{r}/2} - \mathrm{e}^{-\mathrm{i}\,(\boldsymbol{q}-\boldsymbol{k})\cdot\boldsymbol{r}/2}\right) \\ \times \left(\mathrm{e}^{\mathrm{i}\,\boldsymbol{k}\cdot\boldsymbol{r}'/2} - \mathrm{e}^{-\mathrm{i}\,\boldsymbol{k}\cdot\boldsymbol{r}'/2}\right) \left(\mathrm{e}^{\mathrm{i}\,(\boldsymbol{q}-\boldsymbol{k})\cdot\boldsymbol{r}'/2} - \mathrm{e}^{-\mathrm{i}\,(\boldsymbol{q}-\boldsymbol{k})\cdot\boldsymbol{r}'/2}\right) \, , \tag{11.182}$$

where $n_d(r,\rho,y,q)$ is the number density of dipoles at $t = -\boldsymbol{q}^2$.

Introducing the profile function $\Gamma(s,b)$ of quarkonium-quarkonium scattering

$$\Gamma(y,b) = \frac{1}{2\mathrm{i}s} \int \frac{\mathrm{d}^2\boldsymbol{q}}{(2\pi)^2} \, \mathrm{e}^{\mathrm{i}\boldsymbol{q}\cdot\boldsymbol{b}} \, A(s,t) \, , \tag{11.183}$$

which is the scattering amplitude in the impact-parameter space, Eq. (11.182) gives

$$\Gamma(y,b) = \frac{8}{9} \int \mathrm{d}z\, \mathrm{d}^2\boldsymbol{r} \int \mathrm{d}z'\, \mathrm{d}^2\boldsymbol{r}' \\ \times |\Psi_A(z,r)|^2\, |\Psi_B(z',r')|^2\, \hat{\Gamma}(y,r,r',b)\;. \qquad (11.184)$$

Here $\hat{\Gamma}(y,r,r',b)$ is the profile function for the scattering of two dipoles of sizes $r$ and $r'$ at an impact parameter $b$. It is given by

$$\hat{\Gamma}(y,r,r',b) = \frac{\alpha_s^2}{2} \int \frac{\mathrm{d}^2\boldsymbol{\rho}}{2\pi\rho^2} \int \frac{\mathrm{d}^2\boldsymbol{\rho}'}{2\pi\rho'^2} \int \mathrm{d}^2\boldsymbol{R} \int \mathrm{d}^2\boldsymbol{R}' \\ \times n_d(r,\rho,R,y')\, n_d(r',\rho',|\boldsymbol{R}'-\boldsymbol{b}|,y-y') \\ \times \int \frac{\mathrm{d}^2\boldsymbol{q}}{(2\pi)^2}\, \mathrm{e}^{\mathrm{i}\,\boldsymbol{q}\cdot(\boldsymbol{R}-\boldsymbol{R}')} \int \frac{\mathrm{d}^2\boldsymbol{k}}{\boldsymbol{k}^2\,(\boldsymbol{k}-\boldsymbol{q})^2} \\ \times \left[\mathrm{e}^{\mathrm{i}\,(\boldsymbol{q}-\boldsymbol{k})\cdot\boldsymbol{\rho}/2} - \mathrm{e}^{-\mathrm{i}\,(\boldsymbol{q}-\boldsymbol{k})\cdot\boldsymbol{\rho}/2}\right] \left[\mathrm{e}^{\mathrm{i}\,\boldsymbol{k}\cdot\boldsymbol{\rho}/2} - \mathrm{e}^{-\mathrm{i}\,\boldsymbol{k}\cdot\boldsymbol{\rho}/2}\right] \\ \times \left[\mathrm{e}^{\mathrm{i}\,(\boldsymbol{q}-\boldsymbol{k})\cdot\boldsymbol{\rho}'/2} - \mathrm{e}^{-\mathrm{i}\,(\boldsymbol{q}-\boldsymbol{k})\cdot\boldsymbol{\rho}'/2}\right] \left[\mathrm{e}^{\mathrm{i}\,\boldsymbol{k}\cdot\boldsymbol{\rho}'/2} - \mathrm{e}^{-\mathrm{i}\,\boldsymbol{k}\cdot\boldsymbol{\rho}'/2}\right]\,, \qquad (11.185)$$

where $n_d(r,\rho,R,y)$, namely the number density of dipoles of size $\rho$ inside a parent dipole of size $r$ at a transverse distance $R$ from it, is the two-dimensional Fourier tranform of $n_d(r,\rho,y,q)$, i.e.

$$n_d(r,\rho,R,y) = \int \frac{\mathrm{d}^2\boldsymbol{q}}{(2\pi)^2}\, \mathrm{e}^{\mathrm{i}\boldsymbol{q}\cdot\boldsymbol{r}}\, n_d(r,\rho,y,q)\;. \qquad (11.186)$$

Using (11.163) gives, at leading $\ln s$ (we assume that the two primary dipoles have the same transverse size $r$)

$$\hat{\Gamma}(y,r,b) = 8\pi\alpha_s^2 \left(\frac{2}{\pi\lambda'}\right)^{3/2} \frac{r^2}{b^2} \ln\left(\frac{b^2}{r^2}\right) \\ \times \frac{\mathrm{e}^{\lambda y}}{y^{3/2}} \exp\left[-\frac{2\,\ln^2(b^2/r^2)}{\lambda' y}\right]\;. \qquad (11.187)$$

Equation (11.187) is the one-pomeron exchange contribution to the $S$-matrix describing the interaction of two dipoles in the impact-parameter representation. If only one pomeron is exchanged, then we have simply

$$\hat{S}(y,r,b) = 1 + \hat{\Gamma}(y,r,b)\;, \qquad (11.188)$$

Taking into account multi-dipole interactions, $\hat{S}(y,r,b)$ is given by a multiple scattering series, which can be written as (Mueller 1995)

$$\hat{S}(y,r,b) = \sum_{m,n} P_m P_n\, \exp\left(-\langle n,m|\hat{f}|n,m\rangle\right)\;, \qquad (11.189)$$

where $|n,m\rangle$ is a state of $n$ dipoles in one of the two quarkonia and $m$ dipoles in the other, $P_n$ is the probability of a $n$-dipole configuration, and $\hat{f}$ is the

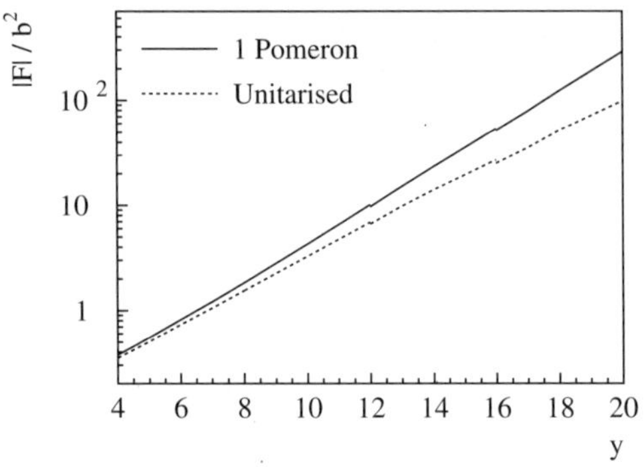


**Fig. 11.17.** The total cross section for the scattering of two dipoles of size $R$, as a function of the rapidity $y$. Solid line: one-pomeron exchange; dashed line: the result of the summation of the whole multiple scattering series. From Salam (1996).

dipole-dipole scattering operator (for more details about this formalism, see Mueller 1995).

Equation (11.188) corresponds to retaining only the first-order term of (11.189). When $\hat{\Gamma} \sim 1$ this approximation breaks down and one has to consider the successive terms in (11.189). Using a Monte Carlo program, G. Salam (1996) evaluated the effect of multiple scattering corrections on the total cross section of dipole scattering. He found that, for large enough $y$, the many-pomeron exchange contributions exceed the one-pomeron exchange contribution. Summing the whole series gives the result shown in Fig. 11.17. The unitarization corrections are rather small for $y \lesssim 10$ but become more and more relevant as the rapidity increases.

## 11.11 The $\gamma^*\gamma^*$ Total Cross-section

A physical situation in which a dipole-dipole interaction is concretely realized, and the BFKL dynamics is at work, is $\gamma^*\gamma^*$ scattering (Bartels, De Roeck and Lotter 1996; Brodsky, Hautmann and Soper 1997). Photons with large virtualities fluctuate in fact into small-size dipoles, and therefore perturbative theory is appropriate.

Total cross sections of $\gamma^*\gamma^*$ scattering can be measured at $e^+e^-$ colliders by tagging both outgoing leptons close to the forward direction. The main practical limitation is that, because of the photon propagators, the event rate for the process falls off very rapidly with increasing photon virtualities. Thus, one cannot reach very large values of $Q^2$.

A representative Feynman diagram contributing to the process $e^+e^- \to e^+e^-$ + 'anything' in the high-energy limit is shown in Fig. 11.18. The relevant kinematic variables are

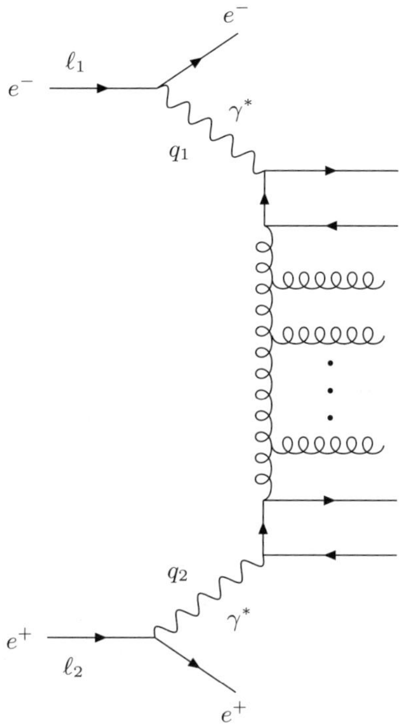


**Fig. 11.18.** Feynman diagram for the process $e^+e^- \to e^+e^-$ + anything.

$$y_1 = \frac{q_1 \cdot \ell_2}{\ell_1 \cdot \ell_2}\,, \quad y_2 = \frac{q_2 \cdot \ell_1}{\ell_1 \cdot \ell_2}\,, \tag{11.190}$$

and

$$x_1 = \frac{Q_1^2}{2q_1 \cdot \ell_2}\,, \quad x_2 = \frac{Q_2^2}{2q_2 \cdot \ell_1}\,, \tag{11.191}$$

where the photon virtualities are $Q_i^2 = -q_i^2$, $i = 1, 2$. The $e^+e^-$ c.m. energy is $s = (\ell_1 + \ell_2)^2$. The $\gamma^*\gamma^*$ c.m. energy is $\hat{s} = (q_1 + q_2)^2 \simeq y_1 y_2 s$. We have $Q_i^2 = x_i y_i s$, $i = 1, 2$.

The expression for the cross section of the process under discussion can be found in Bartels, De Roeck and Lotter (1996). Here we are mainly interested in the subprocess, i.e. $\gamma^*\gamma^*$ scattering. In leading-order BFKL one finds for the total $\gamma^*\gamma^*$ cross section (cfr. (11.164))

$$\begin{aligned} \sigma_{\text{tot}}^{\gamma^*\gamma^*} &= \frac{\sigma_0}{\sqrt{Q_1^2 Q_2^2 Y}} \left(\frac{s}{s_0}\right)^{\lambda} \\ &\simeq \frac{\sigma_0}{\sqrt{Q_1^2 Q_2^2 Y}} \left(\frac{\hat{s}}{\sqrt{Q_1^2 Q_2^2}}\right)^{\lambda}\,, \end{aligned} \tag{11.192}$$

where $\sigma_0$ is a constant and

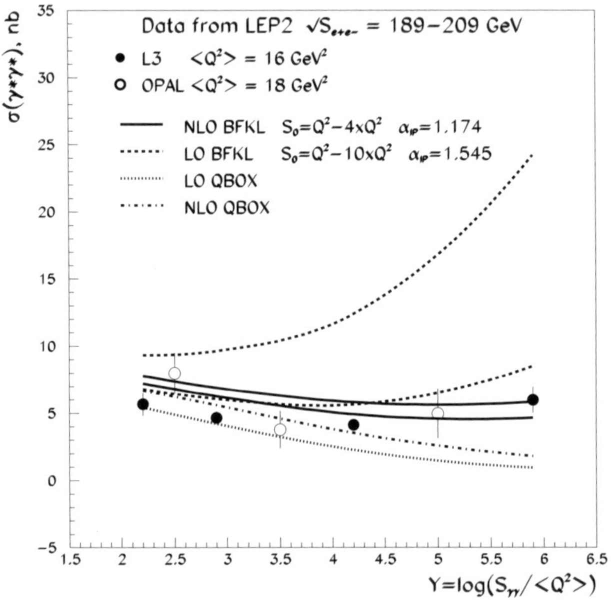


**Fig. 11.19.** The $\gamma^*\gamma^*$ total cross section (from Brodsky et al. 2001).

$$Y = \ln \frac{s}{s_0}\,, \qquad s_0 = \frac{\sqrt{Q_1^2 Q_2^2}}{y_1 y_2}\,. \tag{11.193}$$

In (11.192) the exponent $\lambda \equiv \alpha_{I\!P}^{\mathrm{BFKL}} - 1$ is the familiar BFKL eigenvalue $N_c \alpha_s 4 \ln 2/\pi$.

A measurement of $\sigma_{\mathrm{tot}}^{\gamma^*\gamma^*}$ has been performed by the L3 and OPAL Collaborations at LEP (Acciarri et al. 1999; Achard et al. 2001; Abbiendi et al. 2001). The data are collected in Fig. 11.19. The leading-order BFKL result is represented by the dashed lines. The spread in the curves reflect the uncertainty in the choice of the Regge scale parameter, which defines the onset of the asymptotic regime. As one can see, the LO BFKL prediction overestimates $\sigma_{\mathrm{tot}}^{\gamma^*\gamma^*}$. Also shown in figure is the NLO QCD prediction of Cacciari et al. (2001) which apparently fails at large $Y$. A recent NLO BFKL calculation (Brodsky et al. 2001) seems to improve the agreement with the data (see the solid curves in Fig. 11.19) but it is an incomplete estimate since it does not take into account the contribution of the (presently unknown) NLO photon impact factor.

## 11.12 Diffractive Photoproduction at High $|t|$

Diffractive real or virtual photoproduction at high $|t|$ (with $|t| \ll s$) represents an excellent testing ground of BFKL dynamics. Due to the largeness of $|t|$, in fact, one can legitimately use perturbative QCD, and apply the non-forward

BFKL equation, that we solved in Sect. 8.14. Although one still have to face possible complications arising from the internal structure of the particles involved, QCD factorization properties allow separating the non-perturbative dynamics from the hard subprocesses.

Two relevant examples of high-$|t|$ quasi-elastic processes which have been studied in BFKL theory (Forshaw and Ryskin 1995; Bartels et al. 1996; Ivanov 1996; Ginzburg and Ivanov 1996; Ivanov and Wüsthoff 1999) are $\gamma^{(*)} p \to \gamma p$ and and $\gamma^{(*)} p \to J/\psi\ p$. (The outgoing proton may be replaced by an unresolved hadronic system $Y$ with the same quantum numbers of the proton.) The initial photon is either real or virtual. In the former case one has a single-scale problem, with only one type of powers to be resummed, i.e. $\alpha_s^n \ln^n(s/|t|)$, and, consequently, BFKL dynamics does not coexist with any other evolution mechanism. By contrast, if the photon is (highly) virtual, another hard scale ($Q^2$) appears, and one has to take care of logarithmic terms of the type $\alpha_s^n \ln^n(Q^2/\mu^2)$.

$J/\psi$ diffractive production at high $|t|$ has been investigated experimentally at HERA (H1 1997; ZEUS 1997), but the rate is low. From a theoretical point of view, this process is also affected by uncertainties coming from our ignorance of the meson wave function. Requiring a photon, instead of a meson, in the final state yields a suppression by an extra $\alpha_{\mathrm{em}}$, but the process has the advantage of being completely calculable.

A reaction which has been proposed in order to avoid the drawbacks of high-$|t|$ vector meson production is double diffraction dissociation in photoproduction at large $|t|$ (Cox and Forshaw 1998), $\gamma p \to XY$. The produced hadronic systems $X$ and $Y$ are separated by a large rapidity gap. The first experimental study of this reaction has been reported by the H1 Collaboration (1998a).

We shall now present the BFKL description of diffractive processes at high $|t|$, focusing on meson production and double diffractive photoproduction.

### 11.12.1 Vector Meson Production at High $|t|$

In the BFKL framework, vector meson photoproduction at high-$|t|$, $\gamma p \to V p$ is described by the diagram in Fig. 11.20a. A BFKL ladder in a color-singlet configuration (that is, a hard pomeron) is exchanged between the photon and the proton. According to Eq. (8.267), the amplitude for this process is ($t = -\boldsymbol{q}^2$)

$$A(s,t) = \frac{\mathrm{i}s}{(2\pi)^6}\, \mathcal{C} \int_{-\infty}^{+\infty} \mathrm{d}\nu\, \frac{\nu^2}{\left(\nu^2 + \frac{1}{4}\right)^2}\, \mathrm{e}^{\omega_0(\nu) y}\, I_\nu^V(\boldsymbol{q})\, I_\nu^{p*}(\boldsymbol{q})\,, \tag{11.194}$$

with

$$I_\nu^{V,p}(\boldsymbol{q}) = \int \frac{\mathrm{d}^2\boldsymbol{k}}{(2\pi)^2}\, \Phi_{V,p}(\boldsymbol{k},\boldsymbol{q}) \int \mathrm{d}^2\boldsymbol{r}_1\, \mathrm{d}^2\boldsymbol{r}_2\, \mathrm{e}^{-\mathrm{i}\boldsymbol{k}\cdot\boldsymbol{r}_1 - \mathrm{i}(\boldsymbol{q}-\boldsymbol{k})\cdot\boldsymbol{r}_2}\, \phi_\nu(\boldsymbol{r}_1,\boldsymbol{r}_2)\,. \tag{11.195}$$

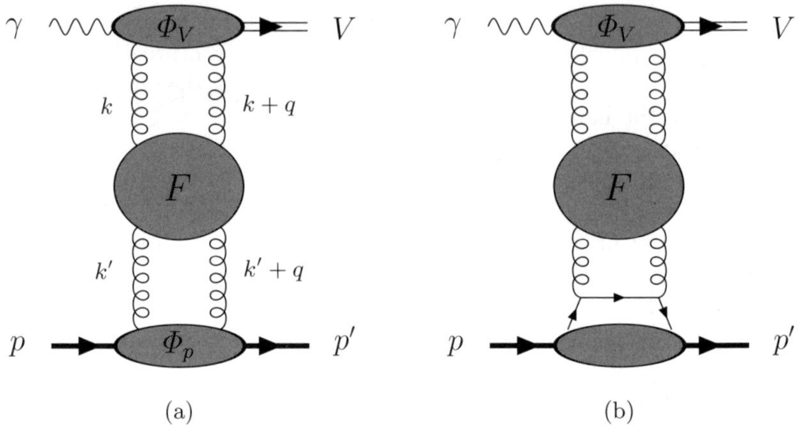


**Fig. 11.20a,b.** The BFKL diagram for diffractive vector meson photoproduction

In (11.195) $\Phi_V$ and $\Phi_p$ are the meson and proton impact factors (see Fig. 11.20a) and $\phi_\nu$ are the BFKL eigenfunctions defined in (8.251). The rapidity variable is

$$y = \ln\left(\frac{s}{|t| + M_V^2}\right) , \tag{11.196}$$

where $M_V$ is the meson mass.

Since in the high-$|t|$ regime the size of the pomeron ($\sim 1/\sqrt{|t|}$) is smaller than the size of the proton, the pomeron couples to a single parton. This factorization property is discussed in Forshaw and Ryskin (1995) and Bartels et al. (1995). The process is then pictured as in Fig. 11.20b and its cross section is the sum of the cross sections for the partonic subprocesses $\gamma\, a \to V\, a$, multiplied each by the parton distribution $f_a(x)$, namely

$$\frac{\mathrm{d}\sigma^{\gamma p \to V p}}{\mathrm{d}x'\,\mathrm{d}t} = \sum_a f_a(x', t)\, \frac{\mathrm{d}\hat{\sigma}^{\gamma\, a \to V\, a}}{\mathrm{d}t} . \tag{11.197}$$

Here $x'$ is the fraction of the proton momentum carried by the parton which couples to the BFKL ladder. Note that the factorization (11.197) corresponds to setting

$$\Phi_p = 8\pi^2 \alpha_s \sum_a f_a(x', t) \tag{11.198}$$

in (11.195), cfr. (8.269).

Taking into account the ratio $(C_A/C_F)^2$ between the coupling of gluons and quarks to a color-singlet BFKL ladder (recall the discussion at the end of Sect. 8.7), Eq. (11.197) becomes

$$\frac{\mathrm{d}\sigma^{\gamma p \to V p}}{\mathrm{d}x'\,\mathrm{d}t} = f_{\mathrm{eff}}(x', t)\, \frac{\mathrm{d}\hat{\sigma}^{\gamma q \to V q}}{\mathrm{d}t} , \tag{11.199}$$

where the effective parton density $f_{\rm eff}(x', t)$ is defined as the combination

$$f_{\rm eff}(x', t) = \left(\frac{9}{4}\right)^2 g(x', t) + \sum_f \left[q_f(x', t) + \bar{q}_f(x', t)\right] \tag{11.200}$$

The cross section for the process $\gamma q \to V q$ is given by

$$\frac{{\rm d}\hat{\sigma}^{\gamma q \to V q}}{{\rm d}t} = \frac{1}{16\pi \hat{s}^2} \left|\hat{A}(\hat{s}, t)\right|^2 , \tag{11.201}$$

where the $\gamma q$ scattering amplitude $\hat{A}(\hat{s}, t)$ is (4/9 is a color factor)

$$\hat{A}(s, t) = \frac{4}{9} \frac{{\rm i}\hat{s}}{(2\pi)^6} \int_{-\infty}^{+\infty} {\rm d}\nu \, \frac{\nu^2}{\left(\nu^2 + \frac{1}{4}\right)^2} \, {\rm e}^{\omega_0(\nu) y} \, I_\nu^V(\boldsymbol{q}) \, I_\nu^{q*}(\boldsymbol{q}) , \tag{11.202}$$

with

$$I_\nu^q(\boldsymbol{q}) = 8\pi^2 \alpha_s \int \frac{{\rm d}^2\boldsymbol{k}}{(2\pi)^2} \int {\rm d}^2\boldsymbol{r}_1 \, {\rm d}^2\boldsymbol{r}_2 \, {\rm e}^{-{\rm i}\boldsymbol{k}\cdot\boldsymbol{r}_1 - {\rm i}(\boldsymbol{q}-\boldsymbol{k})\cdot\boldsymbol{r}_2} \, \phi_\nu^{MT}(\boldsymbol{r}_1, \boldsymbol{r}_2) . \tag{11.203}$$

In (11.203) we used the Mueller-Tang prescription (8.258) for the BFKL eigenfunctions $\phi_\nu^{MT}(\boldsymbol{r}_1, \boldsymbol{r}_2)$.

As for the impact factor $\Phi_V$, the simplest way to model it is to assume that the meson is a non-relativistic quark-antiquark bound state. This is a reasonable assumption for heavy mesons, such as the $J/\psi$. In Bartels et al. (1996) the following form is used

$$\Phi_V(\boldsymbol{k}, \boldsymbol{q}) \sim C \left[\frac{1}{\widetilde{Q}^2 + (\boldsymbol{k} - \boldsymbol{q}/2)^2} - \frac{1}{\widetilde{Q}^2 + (\boldsymbol{q}/2)^2}\right] , \tag{11.204}$$

where $C$ is a normalization constant and $\widetilde{Q}^2 = (Q^2 + M_V^2)/4$.

Evaluating (11.201) with the ansatz (11.204) in the large-$s$ limit at fixed (but high) $|t|$ gives

$$\frac{{\rm d}\hat{\sigma}^{\gamma q \to V q}}{{\rm d}t} \sim \frac{\alpha_s^4}{t^4} f(\tau) \frac{{\rm e}^{2\lambda y}}{y^3} . \tag{11.205}$$

Here $\lambda = N_c \alpha_s 4 \ln 2/\pi$ and $f(\tau)$ is a function of $\tau = |t|/(Q^2 + M_V^2)$ whose form depends on the region of $\tau$ explored.

### 11.12.2 Double Diffractive Dissociation at High $|t|$

Cox and Forshaw (1998) proposed to measure double diffractive dissociation in $\gamma p$ scattering at high values of momentum transfer as as a further way to explore the BFKL pomeron. The process, shown in Fig. 11.21, is analogous to vector meson production, but is more inclusive. The experimental signature is a large rapidity gap between the photon and proton remnants.

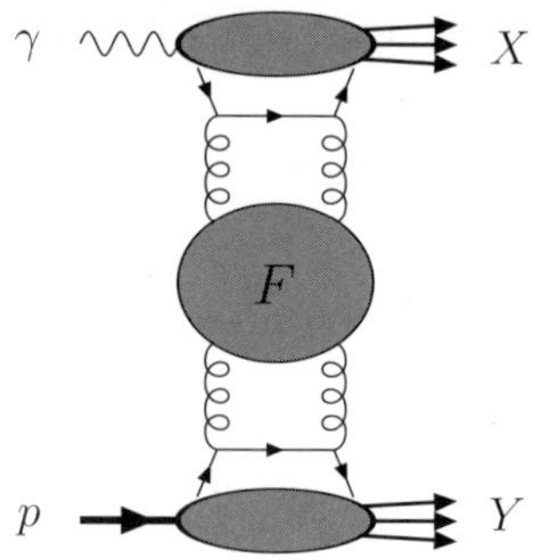


**Fig. 11.21.** Double diffractive dissociation at high $|t|$ in the BFKL picture.

In the large-$|t|$ limit we can factorize the long-distance physics (which is embodied in parton distrbution functions) from the hard subprocess which is parton–parton elastic scattering via color-singlet exchange. Thus, we can write

$$\frac{\mathrm{d}\sigma^{\gamma p \to XY}}{\mathrm{d}x_\gamma \mathrm{d}x_p \mathrm{d}t} = f_{\mathrm{eff}}(x_\gamma, t)\, f_{\mathrm{eff}}(x_p, t)\, \frac{\mathrm{d}\hat{\sigma}^{qq}_{\mathrm{sing}}}{\mathrm{d}t}\,, \tag{11.206}$$

where $x_\gamma$ ($x_p$) is the fraction of the photon (proton) momentum carried by the partons, and $\mathrm{d}\hat{\sigma}/\mathrm{d}t$ is the quark–quark elastic scattering cross section via pomeron exchange given in (8.261), with the rapidity $y$ defined, in the present context, as

$$y = \ln\left(\frac{x_\gamma x_p W^2}{|t|}\right)\,. \tag{11.207}$$

$W^2$ is the c.m. energy squared for $\gamma p$ scattering.

Working out the kinematics of the process in the Regge limit, one finds that the variables $x_\gamma$ and $x_p$ are related to the masses of the hadronic systems $X$ and $Y$ by

$$x_\gamma = \frac{|t|}{M_X^2 + |t|}\,, \quad x_p = \frac{|t|}{M_Y^2 + |t|}\,. \tag{11.208}$$

The rapidity gap between $X$ and $Y$ is $\ln(x_\gamma x_p W^2/2\,|t|)$.

In terms of $x_{I\!P} = (M_X^2 + |t|)/W^2$ the cross section (11.206) becomes

$$\frac{\mathrm{d}\sigma^{\gamma p \to XY}}{\mathrm{d}x_{I\!P}\mathrm{d}t} = \frac{1}{t^2}\int \mathrm{d}M_Y^2\,(x_\gamma x_p W)^2 \times f_{\mathrm{eff}}(x_\gamma, t)\, f_{\mathrm{eff}}(x_p, t)\, \frac{\mathrm{d}\hat{\sigma}^{qq}_{\mathrm{sing}}}{\mathrm{d}t}\,. \tag{11.209}$$

Keeping $M_X^2$ and $t$ fixed, the $x_{I\!P}$ dependence is determined completely by the hard partonic scattering. With the BFKL asymptotic behavior for the elastic quark–quark cross section, i.e. $\mathrm{d}\hat{\sigma}/\mathrm{d}t \sim \exp(2\lambda y)$, we get

$$\frac{\mathrm{d}\sigma^{\gamma p \to XY}}{\mathrm{d}x_{I\!P}\mathrm{d}t} \sim \left(\frac{1}{x_{I\!P}}\right)^{2\lambda+1}\,. \tag{11.210}$$

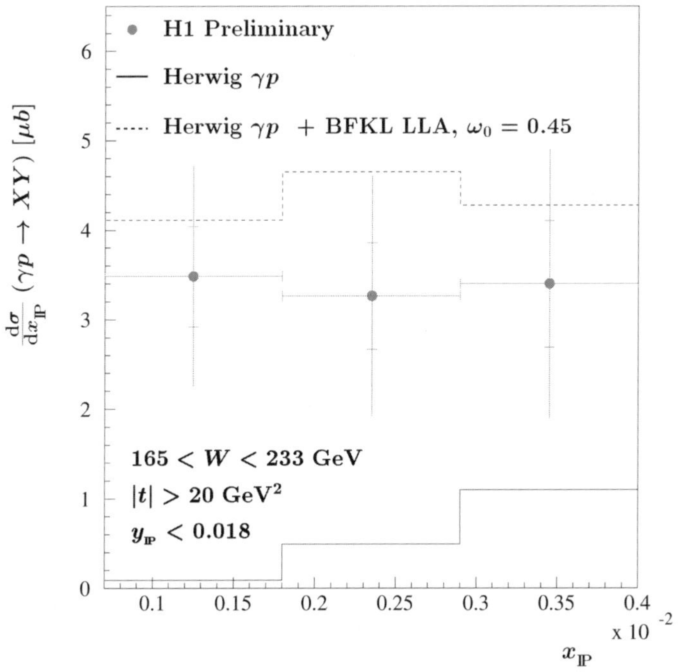


**Fig. 11.22.** The double diffractive cross section $\mathrm{d}\sigma^{\gamma p \to XY}/\mathrm{d}x_{I\!P}$. From H1 (1998a).

Alternatively, if $W^2$ and $t$ are fixed, the $x_{I\!P}$ dependence is also driven by the photon parton densities (note that $x_\gamma = |t|/(x_{I\!P} W^2)$), and we have

$$\frac{\mathrm{d}\sigma^{\gamma p \to XY}}{\mathrm{d}x_{I\!P}\mathrm{d}t} \sim \left(\frac{1}{x_{I\!P}}\right)^{2\lambda+2} f_{\mathrm{eff}}(x_\gamma, t) \; . \tag{11.211}$$

The BFKL results, implemented in a Monte Carlo program, are in fair agreement – within the experimental uncertainties – with the measurement of the double diffractive cross section at $|t| > 20$ GeV$^2$ performed by H1 (1998a). The theoretical expectations are compared to the data in Fig. 11.22.

## 11.13 BFKL Dynamics in High-$p_T$ Jet Production

High-$p_T$ jet production at the Tevatron and at future colliders probes the semi-hard regime of QCD, where the center-of-mass energy $\sqrt{s}$ is much larger than the jet transverse momenta. While the high values of $p_T$ guarantee the applicability of perturbative QCD, the process involves also large logarithms of the type $\ln(s/p_T^2)$ which must be taken care of. To have an idea of the underlying dynamics, let us look at the partonic subprocesses. If $x_1$ and $x_2$

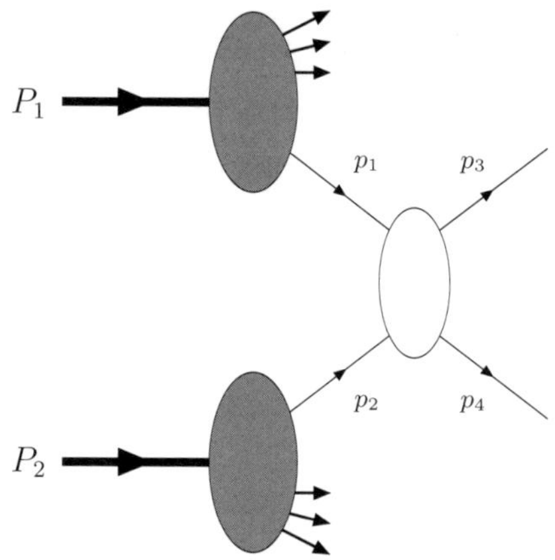


**Fig. 11.23.** Two-jet production.

are the momentum fractions of the partons in the two colliding hadrons (1 and 2), the subprocess c.m. energy is $\sqrt{\hat{s}} = \sqrt{x_1 x_2 s}$. Thus

$$\ln \frac{s}{p_T^2} = \ln \frac{1}{x_1} + \ln \frac{\hat{s}}{p_T^2} + \ln \frac{1}{x_2} \,. \tag{11.212}$$

The logarithms $\ln(1/x_1)$ and $\ln(1/x_2)$ appear in the evolution of the parton densities. Considering (relatively) large $x_{1,2}$ values, these logarithms are small and do not raise any problem: normal DGLAP evolution applies. On the contrary, logarithms of the form $\ln(\hat{s}/p_T^2)$ are large and must be resummed. This is done by the BFKL technique, as seen in Chap. 8 (for an excellent review on the BFKL analysis of jet cross sections see Del Duca 1995).

In order to analyze jet production in a way that resembles the BFKL configuration, one can select all the jets with transverse momentum above some minimum value and tag the two jets with the largest and smallest rapidity (Fig. 11.23), measuring the distributions as a function of these tagging jets (Mueller and Navelet 1987). Keeping the partonic momentum fractions fixed and varying the rapidity interval between the jets, one can disentangle the BFKL dynamics from the DGLAP evolution.

In the following, we shall briefly discuss two-jet production in the framework of BFKL theory (for a more detailed account see Del Duca 1995).

### 11.13.1 Two-Jet Production with a Rapidity Gap

To start with, consider the production of two jets with nearly balancing transverse momenta higher than some minimum value $\boldsymbol{p}^2_{\min}$, and a rapidity gap devoid of any other jets above a transverse momentum scale $\mu$ (Mueller and Tang 1992). Under such conditions, the hard subprocess is the elastic scattering of a parton $a$ in hadron 1 with a parton $b$ in hadron 2 (in brackets the four-momenta)

$$a(p_1) + b(p_2) \to c(p_3) + d(p_4) \tag{11.213}$$

The outgoing partons $c$ and $d$ give rise to the two observed jets. The factorization formula for the two-jet production cross section, expressed in terms of the jet rapidities, is[5]

$$\frac{\mathrm{d}\sigma^{jj}}{\mathrm{d}^2\boldsymbol{p}_3\mathrm{d}^2\boldsymbol{p}_4\mathrm{d}y_3\mathrm{d}y_4} = \sum_{ab}\int \mathrm{d}x_1 \int \mathrm{d}x_2\, f_{a/1}(x_1, \boldsymbol{p}^2_{\mathrm{min}})\, f_{b/2}(x_2, \boldsymbol{p}^2_{\mathrm{min}}) \times \frac{\mathrm{d}\hat{\sigma}^{ab}_{\mathrm{el}}}{\mathrm{d}^2\boldsymbol{p}_3\mathrm{d}^2\boldsymbol{p}_4\mathrm{d}y_3\mathrm{d}y_4} . \tag{11.214}$$

Here $\sum_{ab}$ is a sum over parton species and flavors, $f_{a/1}$ ($f_{b/2}$) is the probability to find a parton $a$ ($b$) in hadron 1 (2) with longitudinal momentum fraction $x_1$ ($x_2$), and $\mathrm{d}\hat{\sigma}^{ab}_{\mathrm{el}}$ is the cross section of parton–parton elastic scattering. The factorization scale has been identified with $\boldsymbol{p}^2_{\mathrm{min}}$.

In the hadron c.m. frame the parton momenta are

$$p_1 = \left(\sqrt{\frac{s}{2}}\, x_1, 0, \boldsymbol{0}\right) , \tag{11.215}$$

$$p_2 = \left(0, \sqrt{\frac{s}{2}}\, x_2, \boldsymbol{0}\right) , \tag{11.216}$$

$$p_3 = \left(\frac{|\boldsymbol{p}_3|}{\sqrt{2}}\, \mathrm{e}^{y_3}, \frac{|\boldsymbol{p}_3|}{\sqrt{2}}\, \mathrm{e}^{-y_3}, \boldsymbol{p}_3\right) , \tag{11.217}$$

$$p_4 = \left(\frac{|\boldsymbol{p}_4|}{\sqrt{2}}\, \mathrm{e}^{y_4}, \frac{|\boldsymbol{p}_4|}{\sqrt{2}}\, \mathrm{e}^{-y_4}, \boldsymbol{p}_4\right) . \tag{11.218}$$

Energy-momentum conservation implies $\boldsymbol{p}_3 = -\boldsymbol{p}_4 \equiv \boldsymbol{p}$ and

$$x_1 = 2\, \frac{|\boldsymbol{p}|}{\sqrt{s}}\, \mathrm{e}^{\overline{y}} \cosh\frac{y}{2} , \tag{11.219}$$

$$x_2 = 2\, \frac{|\boldsymbol{p}|}{\sqrt{s}}\, \mathrm{e}^{-\overline{y}} \cosh\frac{y}{2} , \tag{11.220}$$

where we introduced the combinations

$$\overline{y} = \frac{1}{2}\,(y_3 + y_4) , \tag{11.221}$$

$$y = y_3 - y_4 . \tag{11.222}$$

Here $\overline{y}$ is the rapidity of the parton c.m. frame with respect to the hadron c.m. frame, and $y$ is the rapidity difference between the produced partons (i.e., between the jets).

The large-$\hat{s}$ limit with fixed $\hat{t}$ corresponds to large $y$. In this limit one has $\hat{t} \simeq -\boldsymbol{p}^2$, and the parton momentum fractions become

---

[5] Recall that boldface vectors are transverse vectors.

$$x_1 \simeq \frac{|\boldsymbol{p}|}{\sqrt{s}}\,\mathrm{e}^{y_3}\,, \quad x_2 \simeq \frac{|\boldsymbol{p}|}{\sqrt{s}}\,\mathrm{e}^{-y_4}\,. \tag{11.223}$$

The partonic cross section in (11.214) is

$$\mathrm{d}\hat{\sigma}_{\mathrm{el}}^{ab} = \frac{1}{2\hat{s}}\,|A_{ab}|^2\,\mathrm{d}\Pi_2\,, \tag{11.224}$$

with

$$\mathrm{d}\Pi_2 = \frac{\mathrm{d}y_3\mathrm{d}^2\boldsymbol{p}_3}{4\pi\,(2\pi)^2}\,\frac{\mathrm{d}y_4\mathrm{d}^2\boldsymbol{p}_4}{4\pi\,(2\pi)^2}\,(2\pi)^4\,\delta^4(p_1+p_2-p_3-p_4)\,. \tag{11.225}$$

Using the delta function of momentum conservation and Eqs. (11.223), the factorization formula (11.214) for the two-jet production cross section becomes

$$\frac{\mathrm{d}\sigma^{jj}}{\mathrm{d}\boldsymbol{p}^2\,\mathrm{d}x_1\mathrm{d}x_2} = \sum_{ab} f_{a/1}(x_1,\boldsymbol{p}^2_{\mathrm{min}})\,f_{b/2}(x_2,\boldsymbol{p}^2_{\mathrm{min}})\,\frac{\mathrm{d}\hat{\sigma}_{\mathrm{el}}^{ab}}{\mathrm{d}\hat{t}}\,, \tag{11.226}$$

where

$$\frac{\mathrm{d}\hat{\sigma}_{\mathrm{el}}^{ab}}{\mathrm{d}\hat{t}} = \frac{1}{16\pi\hat{s}^2}\,|A_{ab}|^2\,. \tag{11.227}$$

It is important to notice that, while the parton distributions depend on both $y$ and $\overline{y}$, the partonic cross section depends only on the rapidity interval $y$. Thus, if our principal aim is to study the hard dynamics, it is convenient to fix $\overline{y}$, or to integrate it out.

In the Regge limit ($\hat{s} \to \infty$ with fixed $|\hat{t}| \gg \mu^2$) the dominant hard subprocess is parton–parton scattering via exchange of a BFKL ladder in a color-singlet configuration, i.e., via (hard) pomeron exchange (Fig. 11.24). Taking into account the relation (8.264) between quark–quark and gluon–gluon elastic scattering, the two-jet production cross section (11.226) becomes in the case at hand (we assume, for simplicity, that 1 and 2 are alike hadrons)

$$\frac{\mathrm{d}\sigma^{jj}_{\mathrm{sing}}}{\mathrm{d}\boldsymbol{p}^2\,\mathrm{d}x_1\mathrm{d}x_2} = f_{\mathrm{eff}}(x_1,\boldsymbol{p}^2_{\mathrm{min}})\,f_{\mathrm{eff}}(x_2,\boldsymbol{p}^2_{\mathrm{min}})\,\frac{\mathrm{d}\hat{\sigma}^{qq}_{\mathrm{sing}}}{\mathrm{d}\hat{t}}\,, \tag{11.228}$$

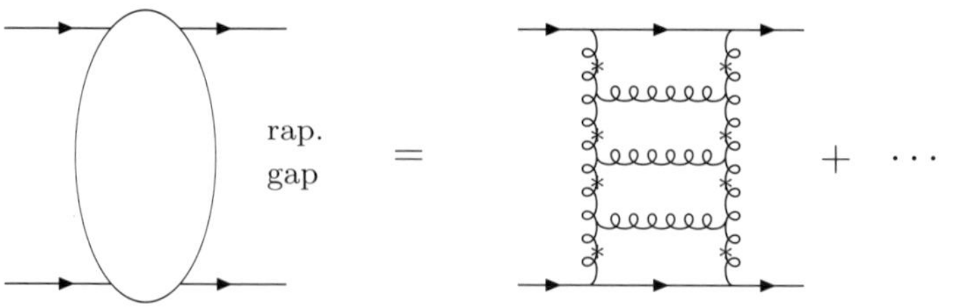


**Fig. 11.24.** Hard subprocesses contributing to two-jet production with a rapidity gap.

where $f_{\rm eff}(x,t)$ is the effective parton density defined in (11.200), and $d\hat{\sigma}^{qq}_{\rm sing}$ is the BFKL quark–quark elastic cross section, that is – see (8.261)

$$\frac{d\hat{\sigma}^{qq}_{\rm sing}}{d\hat{t}} = \left(\frac{8}{9}\right)^2 \pi^3 \frac{\alpha_s^4}{\hat{t}^2} \frac{e^{2\lambda y}}{(\pi\lambda' y/8)^3} . \tag{11.229}$$

The background to the color-singlet exchange comes from the exchange of a reggeized gluon. The color-octet contribution to the jet cross section is

$$\frac{d\sigma^{jj}_{\rm oct}}{d\boldsymbol{p}^2\, dx_1 dx_2} = f'_{\rm eff}(x_1, \boldsymbol{p}^2_{\rm min})\, f'_{\rm eff}(x_2, \boldsymbol{p}^2_{\rm min}) \frac{d\hat{\sigma}^{qq}_{\rm oct}}{d\hat{t}} . \tag{11.230}$$

Since only one reggeized gluon is exchanged the effective parton distribution is

$$f'_{\rm eff}(x,t) = \frac{9}{4} g(x,t) + \sum_f \left[q_f(x,t) + \overline{q}_f(x,t)\right] . \tag{11.231}$$

The quark–quark elastic scattering cross section via color-octet exchange was given in (8.262),

$$\frac{d\hat{\sigma}^{qq}_{\rm oct}}{d\hat{t}} = \frac{8}{9}\pi \frac{\alpha_s^2}{\hat{t}^2} e^{2\epsilon(\hat{t})y} . \tag{11.232}$$

We are requiring that no additional jets with transverse momenta above some scale $\mu^2$ be found in the rapidity interval between the tagging jets. Thus $\mu^2$ is the infrared cutoff to be used in the expression for $\epsilon(\hat{t})$, Eq. (8.45). Therefore we have – see Eq. (8.47),

$$\epsilon(\hat{t}) = -\frac{N_c \alpha_s}{2\pi} \ln \frac{|\hat{t}|}{\mu^2} . \tag{11.233}$$

for $|\hat{t}| \gg \mu^2$. What Eqs. (11.233) tells us is that we are allowing gluons to be radiated only up to the scale $\mu^2$. Note that if $\mu \to \Lambda_{\rm QCD}$, which corresponds to requiring a *large* rapidity gap devoid of particles between the tagging jets, the color-octet exchange is strongly suppressed and the process is governed by pomeron exchange. However, in such a case QCD factorization is lost and the use of (11.226) is questionable.

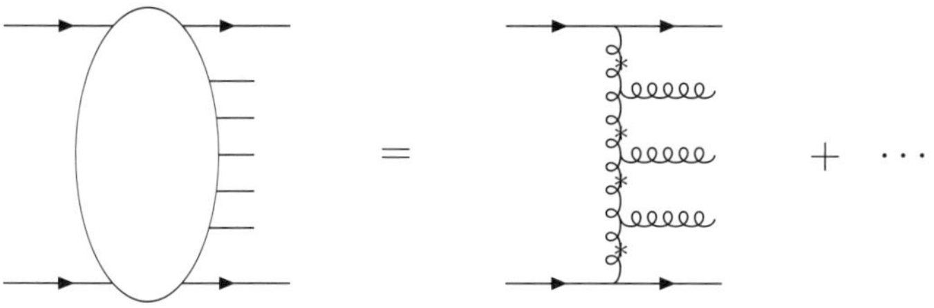

**Fig. 11.25.** Hard subprocesses contributing to Mueller-Navelet jets.

### 11.13.2 Total Cross-section for Tagging Jets

Let us now suppose that the rapidity interval between the tagging jets is not empty, but rather is filled by extra jets (the so-called "minijets"). In this case, the relevant partonic subprocess is parton–parton scattering with multi-gluon production (Fig. 11.25) and what one measures is the total cross section for tagging jets (Mueller and Navelet 1987; see also, Del Duca, Peskin and Tang 1993; Del Duca and Tang 1993 Del Duca and Schmidt 1994a, 1994b; Stirling 1994).

As discussed in Sect. 8.3.5, in the large-$\hat{s}$ limit the leading $\ln(\hat{s}/|\hat{t}|)$ contribution comes from production of partons with a strong ordering in rapidity, the multi-Regge kinematics. Let $p_3$ and $p_4$ be the momenta of the partons with the largest and smallest rapidities ($y_3$ and $y_4$, respectively), and $\kappa_i$ ($i = 1, \dots, n$) the momenta of the other partons (we call their rapidities $y_i'$). The multi-Regge kinematics is

$$y_3 \gg y_1' \gg y_2' \gg \cdots \gg y_n' \gg y_4 \, , \tag{11.234}$$

$$p_{3\perp}^2 \simeq \kappa_{1\perp}^2 \simeq \kappa_{2\perp}^2 \simeq \cdots \simeq \kappa_{n\perp}^2 \simeq p_{4\perp}^2 \, . \tag{11.235}$$

It is easy to check that in this regime the parton momentum fractions $x_1$ and $x_2$ are given by

$$x_1 \simeq \frac{|\boldsymbol{p}_3|}{\sqrt{s}} \, \mathrm{e}^{y_3} \, , \quad x_2 \simeq \frac{|\boldsymbol{p}_4|}{\sqrt{s}} \, \mathrm{e}^{-y_4} \, . \tag{11.236}$$

The two-jet production cross section is still given by the factorization formula (11.214), but the partonic cross section is now

$$\mathrm{d}\hat{\sigma}^{ab} = \frac{1}{2\hat{s}} \sum_{n=0}^{\infty} |A_{ab\to n+2}|^2 \, \mathrm{d}\Pi_{n+2} \, , \tag{11.237}$$

where $A_{ab\to n+2}$ is the amplitude for the production of $n+2$ particles and $\mathrm{d}\Pi_{n+2}$ is the corresponding phase space, that is

$$\begin{aligned} \mathrm{d}\Pi_{n+2} &= \int \frac{\mathrm{d}y_3 \, \mathrm{d}^2\boldsymbol{p}_3}{4\pi \, (2\pi)^2} \int \frac{\mathrm{d}y_4 \, \mathrm{d}^2\boldsymbol{p}_4}{4\pi \, (2\pi)^2} \prod_{i=1}^{n} \int \frac{\mathrm{d}y_i' \, \mathrm{d}^2\boldsymbol{\kappa}_i}{4\pi \, (2\pi)^2} \\ &\times (2\pi)^4 \, \delta^4 \left( p_1 + p_2 - p_3 - p_4 - \sum_{i=1}^{n} \kappa_i \right) \, . \end{aligned} \tag{11.238}$$

Integrating over all tranverse momenta in multi-Regge kinematics, and taking into account the different couplings of quarks and gluons to the exchanged reggeized gluon, we get the total cross section for Mueller-Navelet jets,

$$\frac{\mathrm{d}\sigma_{\mathrm{tot}}^{jj}}{\mathrm{d}x_1 \mathrm{d}x_2} = f_{\mathrm{eff}}'(x_1, \boldsymbol{p}_{\mathrm{min}}^2) \, f_{\mathrm{eff}}'(x_2, \boldsymbol{p}_{\mathrm{min}}^2) \, \hat{\sigma}_{\mathrm{tot}}^{qq}(\hat{s}) \, . \tag{11.239}$$

Here $\hat{\sigma}_{\mathrm{tot}}^{qq}$ is the quark–quark total cross section – see Eq. (8.218),

$$\hat{\sigma}_{\text{tot}}^{qq} = \frac{8\pi}{9} \frac{\alpha_s^2}{\boldsymbol{p}_{\min}^2} \frac{\mathrm{e}^{\lambda y}}{\sqrt{\pi \lambda' y/8}} . \qquad (11.240)$$

The probability that two-jet production be driven by an elastic partonic subprocess is quantified by the ratio (Mueller and Tang 1992, Del Duca and Tang 1993)

$$R(\mu) = \frac{\sigma_{\text{sing}}^{jj} + \sigma_{\text{oct}}^{jj}}{\sigma_{\text{tot}}^{jj}} , \qquad (11.241)$$

where $\sigma_{\text{sing}}^{jj}$ are $\sigma_{\text{oct}}^{jj}$ are the cross sections derived in Sect. 11.13.1 – Eqs. (11.228), (11.230), respectively –, integrated over $\boldsymbol{p}^2$ above the minimum value $\boldsymbol{p}_{\min}^2$, whereas $\sigma_{\text{tot}}^{jj}$ is the Mueller-Navelet total cross section (11.240).

$R(\mu)$ is not exactly the probability of observing a rapidity gap in the final state, because at the hadron level there is an additional complication. Accompanying any hard reaction in hadron–hadron collisions, in fact, there are spectator partons. These may produce particles which tend to fill the rapidity gap arising from elastic parton scattering. Thus, the probability to have two-jet production events with a large rapidity gap is given by $R(\mu)$ multiplied by the gap survival factor $\langle S^2 \rangle$ introduced in Sect. 10.11.1.

# A. Conventions

We collect here some conventions and definitions of general use. Throughout the book we adopt natural units: $\hbar = c = 1$ (unless otherwise specified).

## A.1 Four-Vectors

The metric tensor is

$$g^{\mu\nu} = g_{\mu\nu} = \text{diag}\,(+1, -1, -1, -1)\,. \tag{A.1}$$

A generic four-vector $A^\mu$ is written, in Cartesian contravariant components, as

$$A^\mu = (A^0, A^1, A^2, A^3) = (A^0, \boldsymbol{A}_\perp, A^3) = (A^0, \boldsymbol{A})\,, \tag{A.2}$$

The light-cone components of $A^\mu$ are defined as

$$A^\pm = \frac{1}{\sqrt{2}}\,(A^0 \pm A^3)\,. \tag{A.3}$$

and in these components $A^\mu$ is written as

$$A^\mu = (A^+, A^-, \boldsymbol{A}_\perp)\,. \tag{A.4}$$

The norm of $A^\mu$ is given by

$$A^2 = \left(A^0\right)^2 - \boldsymbol{A}^2 = 2A^+A^- - \boldsymbol{A}_\perp^2\,. \tag{A.5}$$

The scalar product of two four-vectors $A^\mu$ and $B^\mu$ is

$$A \cdot B = A^0 B^0 - \boldsymbol{A} \cdot \boldsymbol{B} = A^+B^- + A^-B^+ - \boldsymbol{A}_\perp \cdot \boldsymbol{B}_\perp\,. \tag{A.6}$$

## A.2 Sudakov Parametrization

We introduce two light-like vectors $p$ and $n$ (the Sudakov vectors), defined as

$$p^\mu = \frac{1}{\sqrt{2}}\,(\Lambda, 0, 0, \Lambda), \tag{A.7}$$

$$n^\mu = \frac{1}{\sqrt{2}}\,(\Lambda^{-1}, 0, 0, -\Lambda^{-1})\,, \tag{A.8}$$

where $\Lambda$ is an arbitrary number. These vectors satisfy

$$p^2 = n^2 = 0, \quad p \cdot n = 1, \quad n^+ = p^- = 0\,. \tag{A.9}$$

In light-cone components they read

$$p^\mu = (\Lambda, 0, \mathbf{0}_\perp), \tag{A.10}$$

$$n^\mu = (0, \Lambda^{-1}, \mathbf{0}_\perp). \tag{A.11}$$

A generic vector $A^\mu$ can be parametrised as (*Sudakov decomposition*)

$$A^\mu = \alpha\, p^\mu + \beta\, n^\mu + A_\perp^\mu \tag{A.12}$$

$$= (A \cdot n)\, p^\mu + (A \cdot p)\, n^\mu + A_\perp^\mu, \tag{A.13}$$

with $A_\perp^\mu = (0, \boldsymbol{A}_\perp, 0)$. The square of $A^\mu$ is

$$A^2 = 2\,\alpha\beta - \boldsymbol{A}_\perp^2. \tag{A.14}$$

By means of the Sudakov vectors we can construct the *perpendicular* metric tensor $g_\perp^{\mu\nu}$ which projects onto the plane perpendicular to $p$ and $n$,

$$g_\perp^{\mu\nu} = g^{\mu\nu} - (p^\mu n^\nu + p^\nu n^\mu)\,. \tag{A.15}$$

## A.3 Reference Frames

In DIS processes, a natural reference frame is the one where the virtual photon and the target nucleon move collinearly (say, along the $z$ axis). We can represent the nucleon momentum $P$ and the photon momentum $q$ in terms of the Sudakov vectors $p$ and $n$ as

$$P^\mu = p^\mu + \frac{M^2}{2} n^\mu \simeq p^\mu\,, \tag{A.16}$$

$$q^\mu \simeq (P \cdot q)\, n^\mu - x p^\mu = M\nu\, n^\mu - x p^\mu\,, \tag{A.17}$$

where the approximate equality sign means that we are neglecting $M^2$ with respect to large scales, such as $Q^2$, or $(P^+)^2$ in the infinite momentum frame. Conventionally, we always take the nucleon to be directed in the positive $z$ direction.

With the identification (A.16) the parameter $\Lambda$ appearing in the definition of the Sudakov vectors (A.7, A.8) coincides with $P^+$ and fixes the specific frame. In particular:

- in the *target rest frame* one has

$$P^\mu = (M, 0, 0, 0)\ , \tag{A.18}$$

$$q^\mu = (\nu, 0, 0, -\sqrt{\nu^2 + Q^2})\ , \tag{A.19}$$

  and $\Lambda \equiv P^+ = M/\sqrt{2}$. The Bjorken limit in this frame corresponds to $q^- = \sqrt{2}\,\nu \to \infty$ with $q^+ = -Mx/\sqrt{2}$ fixed.
- in the *infinite momentum frame* the momenta are

$$P^\mu \simeq \frac{1}{\sqrt{2}}\,(P^+, 0, 0, P^+)\ , \tag{A.20}$$

$$q^\mu \simeq \frac{1}{\sqrt{2}}\left(\frac{M\nu}{P^+} - xP^+, 0, 0, -\frac{M\nu}{P^+} - xP^+\right)\ , \tag{A.21}$$

  Here we have $P^- \to 0$ and $\Lambda \equiv P^+ \to \infty$. In this frame the vector $n^\mu$ is suppressed by a factor of $(1/P^+)^2$ with respect to $p^\mu$.

## A.4 Fermionic States

Our fermionic states are normalized as

$$\begin{aligned}\langle p|p'\rangle &= (2\pi)^3\, 2E\, \delta^3(\boldsymbol{p} - \boldsymbol{p}') \\ &= (2\pi)^3\, 2p^+\, \delta(p^+ - p'^+)\, \delta(\boldsymbol{p}_\perp - \boldsymbol{p}'_\perp)\ ,\end{aligned} \tag{A.22}$$

with $E = (\boldsymbol{p}^2 + m^2)^{1/2}$. The Dirac spinors obey

$$\bar{u}(p,s)\gamma^\mu u(p,s') = 2p^\mu\, \delta_{ss'}\ . \tag{A.23}$$

The creation and annihilation operators satisfy the anticommutators

$$\{b(p,s), b^\dagger(p',s')\} = \{d(p,s), d^\dagger(p',s')\} = (2\pi)^3\, 2E\, \delta_{ss'}\, \delta^3(\boldsymbol{p} - \boldsymbol{p}')\ . \tag{A.24}$$

# B. Mellin Transforms

The Mellin transform $\widetilde{f}(\omega)$ of a function $f(s)$ is defined as

$$\widetilde{f}(\omega) = \int_1^\infty \mathrm{d}\left(\frac{s}{s_0}\right)\left(\frac{s}{s_0}\right)^{-\omega-1} f(s) \,, \tag{B.1}$$

where $s_0$ is a scale introduced for dimensional reasons.

The inverse Mellin transform is

$$f(s) = \frac{1}{2\pi \mathrm{i}} \int_C \mathrm{d}\omega \left(\frac{s}{s_0}\right)^{\omega} \widetilde{f}(\omega) \,. \tag{B.2}$$

The integration contour $C$ in (B.2) is located to the right of all singularities of $\widetilde{f}(\omega)$ in the complex $\omega$-plane.

A useful example of Mellin transform is

$$f(s) = s^\alpha \ln^p s \,, \quad \widetilde{f}(\omega) = s_0^\alpha \frac{\Gamma(p+1)}{(\omega-\alpha)^{p+1}} \,. \tag{B.3}$$

**Convolution property.** Suppose that $f(s)$ is the convolution of $n$ functions $g_i$, that is

$$f(s) = \prod_{i=1}^{n} \int_{\alpha_{i+1}}^{1} \frac{\mathrm{d}\alpha_i}{\alpha_i}\, g_i\left(\frac{\alpha_{i-1}}{\alpha_i}\right) s_0\, \delta(\alpha_n s - s_0) \,, \tag{B.4}$$

with $\alpha_0 = 1$ and $\alpha_{n+1} = 0$. To calculate the Mellin transform of (B.4) we first perform the integration over $s$ using the delta function, and then we change variables from $\alpha_i$ to $\rho_i$, where

$$\rho_i = \frac{\alpha_i}{\alpha_{i-1}} \,. \tag{B.5}$$

The final result for $\widetilde{f}(\omega)$ is

$$\widetilde{f}(\omega) = \prod_{i=1}^{n} \int_0^1 \mathrm{d}\rho_i\, \rho_i^{\omega-1} g_i\left(\frac{1}{\rho_i}\right) = \prod_{i=1}^{n} \widetilde{g}_i(\omega) \,. \tag{B.6}$$

Therefore, the Mellin transform of the convolution (B.4) is the product of the Mellin transforms of the functions $g_i$.

# C. QCD Formulas

We use the following notations:

- Greek indices $(\alpha, \beta, \ldots, \mu, \nu, \ldots)$ are Lorentz tensor indices.
- $i, j, k, l$ are quark color indices $(i, j, k, l = 1, 2, 3)$.
- $a, b, c, d, e$ are gluon color indices $(a, b, c, d, e = 1, \ldots, 8)$.
- $g_s$ is the strong coupling constant. It is related to $\alpha_s$ by $\alpha_s = g_s^2/4\pi$.
- $n_f$ is the number of flavors. $N_c$ is the number of colors.

Repeated indices are summed.

## C.1 QCD Lagrangian and *SU*(3) Matrices

The QCD (classical) Lagrangian, describing the interaction of quarks and gluons, is (as usual, repeated indices are summed)

$$\mathcal{L} = -\frac{1}{4} F^a_{\mu\nu} F^{\mu\nu,a} + \sum_{\text{flavors}} \overline{\psi}_i \, (\mathrm{i}\not{D} - m))_{ij} \psi_j \; . \tag{C.1}$$

Here $\psi$ is the quark field, $D^\mu$ is the covariant derivative (see later) and $F^a_{\mu\nu}$ is the field strength tensor constructed from the gluon field $A^a_\mu$ as

$$F^a_{\mu\nu} = \partial_\mu A^a_\nu - \partial_\nu A^a_\mu - g_s \, f^{abc} A^b_\mu A^c_\nu \; . \tag{C.2}$$

The $f^{abc}$ (with $a, b, c = 1, \ldots, 8$) are the structure constants of the $SU(3)$ color group. They are antisymmetric under the interchange of any two indices and satisfy the Jacobi identity

$$f_{abe} f_{ecd} + f_{cbe} f_{aed} + f_{dbe} f_{ace} = 0 \; . \tag{C.3}$$

The covariant derivative $D$ has the form

$$(D_\mu)_{ij} = \partial_\mu \delta_{ij} + \mathrm{i} \, g_s \, (t^a A^a_\mu)_{ij} \; , \tag{C.4}$$

when acting on quark fields, and

$$(D_\mu)_{ab} = \partial_\mu \delta_{ab} + \mathrm{i} \, g_s \, (T^c A^c_\mu)_{ab} \; , \tag{C.5}$$

when acting on gluon fields.

We call $t^a$ and $T^a$ the generators in the fundamental and adjoint representations, respectively, of $SU(3)$. Their commutators are

$$[t^a, t^b] = \mathrm{i}\, f^{abc}\, t^c\,, \tag{C.6}$$

$$[T^a, T^b] = \mathrm{i}\, f^{abc}\, T^c\,, \quad (T^a)_{bc} = -\mathrm{i}\, f^{abc}\,. \tag{C.7}$$

The $t^a$ are normalized as

$$\mathrm{Tr}\,(t^a t^b) = T_R\, \delta^{ab}, \quad T_R = \frac{1}{2}\,, \tag{C.8}$$

and obey the relations

$$t^a_{ij} t^a_{kl} = \frac{1}{2}\left[\delta_{il}\delta_{jk} - \frac{1}{N_c}\,\delta_{ij}\delta_{kl}\right]\,, \tag{C.9}$$

$$t^a_{ij} t^a_{jl} = C_F\, \delta_{il}\,, \quad C_F = \frac{N_c^2 - 1}{2\,N_c}\,. \tag{C.10}$$

For the $T^a$ we have

$$\mathrm{Tr}\,(T^a T^b) = f^{acd} f^{bcd} = C_A\, \delta^{ab}\,, \quad C_A = N_c\,. \tag{C.11}$$

The anticommutator of the $t^a$ is

$$\{t^a, t^b\} = \frac{1}{N_c}\,\delta^{ab}\mathbb{1} + d^{abc} t^c\,, \tag{C.12}$$

where the $d^{abc}$ are symmetric under the interchange of any two indices, and satisfy the relations

$$d_{abb} = 0\,, \quad d_{acd} d_{bcd} = \frac{N_c^2 - 4}{N_c}\,\delta_{ab}\,, \tag{C.13}$$

$$f_{abe} d_{ecd} + f_{cbe} d_{aed} + f_{dbe} d_{ace} = 0\,. \tag{C.14}$$

The trace of three $t$ matrices is

$$\mathrm{Tr}\,(t^a t^b t^c) = \frac{1}{4}\,(d^{abc} + \mathrm{i} f^{abc})\,. \tag{C.15}$$

## C.2 Feynman Rules for QCD

The Feynman rules for the QCD vertices are summarized in Fig. C.1.

The form of the gluon propagator (Fig. C.2) depends on the choice of gauge. In the *covariant gauges* the gauge fixing term in the Lagrangian is

$$\mathcal{L}_{\mathrm{gf}} = -\frac{1}{2\eta}\,(\partial_\mu A_a^\mu)^2\,, \tag{C.16}$$

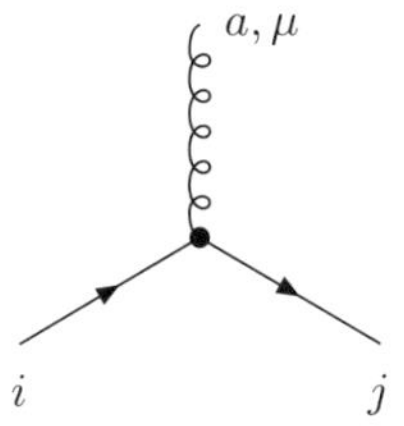


$$-\mathrm{i}g_s(t^a)_{ji}\gamma^\mu$$

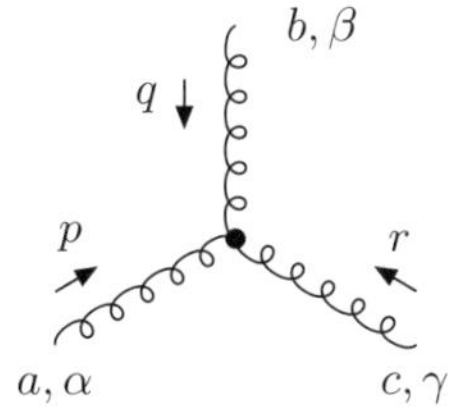


$$-g_s f_{abc}[(p-q)^\gamma g^{\alpha\beta} + (q-r)^\alpha g^{\beta\gamma} + (r-p)^\beta g^{\gamma\alpha}]$$
$$(p+q+r=0)$$

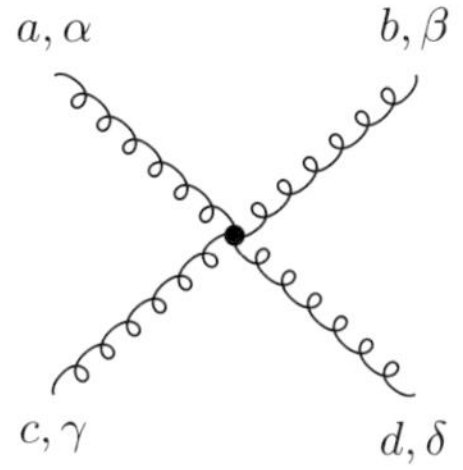


$$\begin{aligned} &-\mathrm{i}g_s^2\{f_{eac}f_{ebd}[g^{\alpha\beta}g^{\gamma\delta} - g^{\alpha\delta}g^{\beta\gamma}] \\ &+f_{ead}f_{ebc}[g^{\alpha\beta}g^{\gamma\delta} - g^{\alpha\gamma}g^{\beta\delta}] \\ &+f_{eab}f_{ecd}[g^{\alpha\gamma}g^{\beta\delta} - g^{\alpha\delta}g^{\beta\gamma}]\} \end{aligned}$$

**Fig. C.1.** Feynman rules for QCD vertices

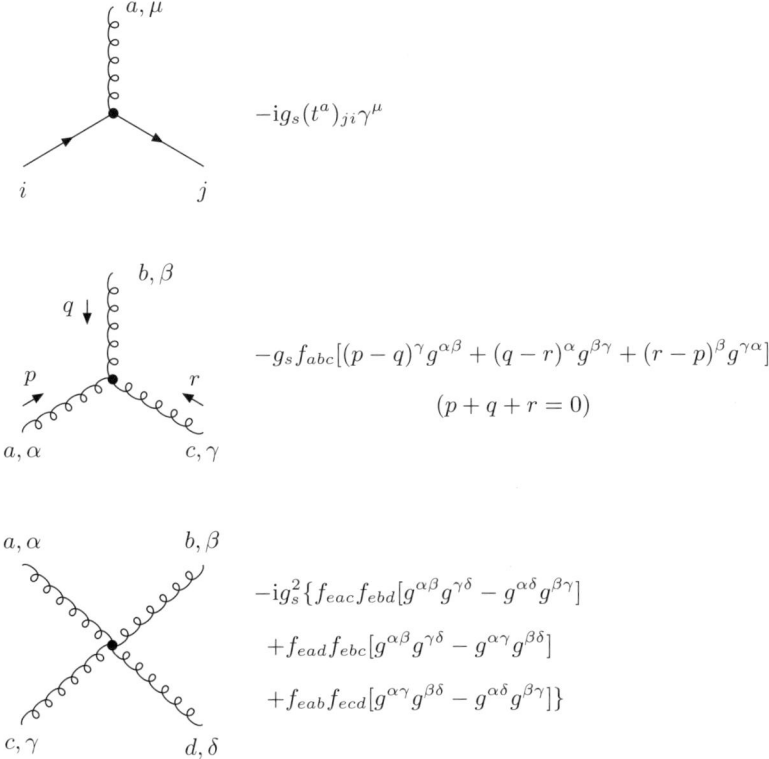

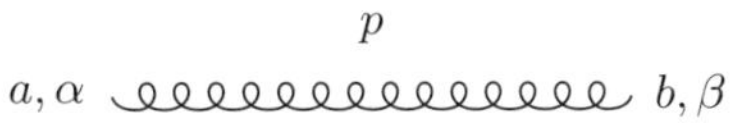


**Fig. C.2.** The gluon propagator

where $\eta$ is an arbitrary parameter. This class of gauges requires the introduction of ghost fields, but the propagator is quite simple

$$\Gamma^{ab}_{\alpha\beta}(p) = \delta^{ab}\left[-g_{\alpha\beta} + (1-\eta)\,\frac{p_\alpha p_\beta}{p^2+\mathrm{i}\epsilon}\right]\frac{\mathrm{i}}{p^2+\mathrm{i}\epsilon}\,. \tag{C.17}$$

In particular, for $\eta = 1$ (*Feynman gauge*), it becomes identical to the photon propagator (except for the color factor $\delta^{ab}$). Another special choice is $\eta = 0$ (*Landau gauge*).

The class of *axial gauges* is defined by a gauge fixing Lagrangian which reads

$$\mathcal{L}_{\mathrm{gf}} = -\frac{1}{2\eta}\,(n_\mu A^\mu_a)^2\,, \tag{C.18}$$

where $n$ is an arbitrary four-vector. These gauges do not require ghosts, but the gluon propagator is more complicated

$$\Gamma_{\alpha\beta}^{ab}(p) = \delta^{ab}\left[-g_{\alpha\beta} + \frac{n_\alpha p_\beta + p_\alpha n_\beta}{n \cdot p} - \frac{(n^2 + \eta\, p^2)\, p_\alpha p_\beta}{(n \cdot p)^2}\right] \frac{\mathrm{i}}{p^2 + \mathrm{i}\epsilon} \,. \tag{C.19}$$

The choice $\lambda = 0$ and $n^2 = 0$ corresponds to the so-called *light-cone gauges.* In this case the propagator reduces to

$$\Gamma_{\alpha\beta}^{ab}(p) = \delta^{ab}\left[-g_{\alpha\beta} + \frac{n_\alpha p_\beta + p_\alpha n_\beta}{n \cdot p}\right] \frac{\mathrm{i}}{p^2 + \mathrm{i}\epsilon} \,. \tag{C.20}$$

In the limit $p^2 \to 0$ one finds

$$n^\alpha\, \Gamma_{\alpha\beta}^{ab} = p^\alpha\, \Gamma_{\alpha\beta}^{ab} = 0 \,, \tag{C.21}$$

and thus only two physical polarization states, orthogonal to $p$ and $n$, propagate. These gauges are called *physical gauges.* The sum over the two gluon polarizations $\lambda$ is

$$\sum_\lambda \varepsilon_\lambda^\alpha\, \varepsilon_\lambda^{\beta *} = -g^{\alpha\beta} + \frac{n^\alpha p^\beta + p^\alpha n^\beta}{n \cdot p} = -g_\perp^{\alpha\beta} \,. \tag{C.22}$$

# References


- Abachi, S. et al., D0 (1994), Phys. Rev. Lett. **72**, 2332.
- Abachi, S. et al., D0 (1996), Phys. Rev. Lett. **76**, 734.
- Abatzis, S. et al., WA91 (1994), Phys. Lett. **B324**, 509
- Abbiendi, G. et al., OPAL (2001), "Measurement of the Hadronic Cross-Section for the Scattering of Two Virtual Photons at LEP", e-print `hep-ex/0110006`.
- Abbott, B. et al., D0 (1998), Phys. Lett **B440**, 189.
- Abe, F. et al., CDF (1994), Phys. Rev. **D50**, 5518; 5535; 5550.
- Abe, F. et al., CDF (1995), Phys. Rev. Lett. **74**, 855.
- Abe, F. et al., CDF (1997a), Phys. Rev. Lett. **78**, 2698.
- Abe, F. et al., CDF (1997b), Phys. Rev. Lett. **79**, 2636.
- Abe, F. et al., CDF (1998a), Phys. Rev. Lett. **80**, 1156.
- Abe, F. et al., CDF (1998b), Phys. Rev. Lett. **81**, 5278.
- Abramowicz, H. (2000), Int. J. Mod. Phys. **A15** Suppl. 1, 495.
- Abramowicz, H. (2001), Nucl. Phys. B (Proc. Suppl.) **99**, 79.
- Abramowicz, H., Bartels, J., Frankfurt, L. and Jung, H. (1996), "Diffractive hard scattering", Summary report of the Working Group on Diffractive Hard Scattering, in *Future Physics at HERA*, vol. 2, DESY, Hamburg.
- Abramowicz, H. and Caldwell, A. (1999), Rev. Mod. Phys. **71**, 1275.
- Abramowicz, H., Levin, E., Levy, A. and Maor, U. (1991), Phys. Lett. **B269**, 465.
- Abramowicz, H. and Levy, A. (1997), "The ALLM parametrization of $\sigma_{\text{tot}}^{\gamma^* p}$: an update", preprint DESY 97-251, `hep-ph/9712415`.
- Abramowitz, M. and Stegun, I.A., eds. (1965), *Handbook of Mathematical Functions*, Dover, New York.
- Abramowsky, V.A., Gribov, V.N. and Kancheli, O.V. (1974), Sov. J. Nucl. Phys. **18**, 308.
- Acciarri, M. et al., L3 (1999), Phys. Lett. **B453**, 333.
- Achard, P. et al., L3 (2001), "Double-Tag Events in Two-Photon Collisions at LEP", e-print `hep-ex/0111012`.
- Adamczyk, L. (2001), Nucl. Phys. B (Proc. Suppl.) **99**, 97.
- Adams, M.R. et al., E665 (1992a), Phys. Rev. Lett. **68**, 3266.
- Adams, M.R. et al., E665 (1992b), Phys. Lett. **B287**, 375.
- Adloff, C. et al., H1 (1997), Z. Phys. **C76**, 613.

- Adloff, C. et al., H1 (1999), Eur. Phys. J. **C6**, 421.
- Adloff, C. et al., H1 (2000a), Eur. Phys. J. **C13**, 371.
- Adloff, C. et al., H1 (2000b), Phys. Lett. **B483**, 23.
- Adloff, C. et al., H1 (2001), Eur. Phys. J. **C20**, 29.
- Affolder, T. et al., CDF (2000a), Phys. Rev. Lett. **84**, 232.
- Affolder, T. et al., CDF (2000b), Phys. Rev. Lett. **84**, 5043.
- Affolder, T. et al., CDF (2000c), Phys. Rev. Lett. **85**, 4215.
- Affolder, T. et al., CDF (2001a), Phys. Rev. Lett. **87**, 141802.
- Affolder, T. et al., CDF (2001b), Phys. Rev. Lett. **87**, 241802.
- Ahmed, T. et al., H1 (1994), Nucl. Phys. **B429**, 477.
- Akhiezer, A.I. and Pomeranchuk, I.Ya. (1958), Usp. Fiz. Nauk. **65**, 593.
- Alberi, G. and Goggi, G. (1981), Phys. Rep. **74**, 1.
- Albrow, M. et al. (1976), Nucl. Phys. **B108**, 1.
- de Alfaro, V. and Regge, T. (1965), *Potential Scattering*, North-Holland, Amsterdam.
- Alner, G.J. et al., UA5 (1986), Z. Phys. **C32**, 133.
- Altarelli, G. and Parisi, G. (1977), Nucl. Phys. **B26**, 298.
- Amaldi, U. et al. (1973), Phys. Lett. **B44**, 112.
- Amaldi, U., Jacob, M. and Matthiae, G. (1976), Ann. Rev. Nucl. Sci. **26**, 385.
- Amaldi, U. and Schubert, K.R. (1980), Nucl. Phys. **B166**, 301.
- Amati, D., Fubini, S. and Stanghellini, A. (1962), Phys. Lett. **1**, 29.
- Amaudruz, P. et al., NMC (1991), Z. Phys. **C51**, 387.
- Amos, N.A. et al. (1990a), Phys. Lett. **B243**, 158.
- Amos, N.A. et al. (1990b), Phys. Lett. **B247**, 127.
- Amos, N.A. et al. (1992), Phys. Rev. Lett. **68**, 2433.
- Arneodo, M. et al., EMC (1988), Phys. Lett. **B211**, 493.
- Arneodo, M. et al., EMC (1990), Nucl. Phys. **B333**, 1.
- Arneodo, M. et al., NMC (1994), Phys. Rev **D50**, R1.
- Arneodo, M. (1994), Phys. Rep. **240**, 301.
- Askew, A.J., Kwieciński, J., Martin, A.D. and Sutton, P.J. (1994), Phys. Rev. **D49**, 4402.
- Auberson, G., Kinoshita, T. and Martin, A. (1971), Phys. Rev. **D3**, 3185.
- Augier, C. et al. (1993), Phys. Lett. **B315**, 503.
- Baker, M. and Ter-Martirosyan, K. (1976), Phys. Rep. **28**, 1.
- Balitsky, Y.Y. and Lipatov, L.N. (1978), Sov. J. Nucl. Phys. **28**, 822.
- Ball, R.D. and Forte, S. (1994a), Phys. Lett. **B335**, 77.
- Ball, R.D. and Forte, S. (1994b), Phys. Lett. **B336**, 77.
- Baltrusaitis, R.M. et al. (1984), Phys. Rev. Lett. **52**, 1380.
- Barbiellini, G. et al. (1972), Phys. Lett. **B39**, 663.
- Barone, V. and Genovese, M. (1997), Phys. Lett. **B412** 143.
- Barone, V., Genovese, M., Nikolaev, N.N., Predazzi, E. and Zakharov, B.G. (1993), Z. Phys. **C58**, 541.

- Barone, V., Genovese, M., Nikolaev, N.N., Predazzi, E. and Zakharov, B.G. (1994), Phys. Lett. **B326**, 161.
- Barone, V., Pascaud, C. and Zomer, F. (2000), Eur. Phys. J. **C12**, 243.
- Bartels, J. (1975), Phys. Rev. **D11**, 2977; 2989.
- Bartels, J. (1979), Nucl. Phys. **B151**, 293.
- Bartels, J. (1980), Nucl. Phys. **B175**, 365.
- Bartels, J., De Roeck, A. and Lotter, H. (1996), Phys. Lett. **B389**, 742.
- Bartels, J., Ellis, J., Kowalski, H. and Wüsthoff, M. (1999), Eur. Phys. J. **C7**, 443.
- Bartels, J., Ewerz, C., Lotter, H. and Wüsthoff, M. (1996), Phys. Lett. **B386**, 389.
- Bartels, J., Forshaw, J.R., Lotter, H., Lipatov, L.N., Ryskin, M.G. and Wüsthoff, M. (1995), Phys. Lett. **B348**, 589.
- Bartels, J., Forshaw, J.R., Lotter, H. and Wüsthoff, M. (1996), Phys. Lett. **B375**, 301.
- Bartels, J., Jung, H. and Wüsthoff, M. (1999), Eur. Phys. J. **C11**, 111.
- Bartels, J., Lipatov, L.N. and Vacca, G.P. (2000), Phys. Lett. **B477**, 178.
- Bartels, J., Lotter, H. and Wüsthoff, M. (1996), Phys. Lett. **B379**, 239. Erratum, *ibid.* **B382** (1996), 449.
- Basile, M. et al. (1980), Phys. Lett. **92B**, 367.
- Bateman Manuscript Project (1953), *Higher Transcendental Functions*, A. Erdelyi, W. Magnus, F. Oberhettinger and F.G. Tricomi eds., McGraw-Hill, New York.
- Bateman Manuscript Project (1954), *Tables of Integral Transforms*, A. Erdelyi, W. Magnus, F. Oberhettinger and F.G. Tricomi eds., McGraw-Hill, New York.
- Berera, A. and Soper, D.E. (1996), Phys. Rev. **D53**, 6162.
- Berger, E.L., Collins, J.C., Soper, D.E. and Sterman, G. (1987), Nucl. Phys. **B286**, 704.
- Bertocchi, L., Fubini, S. and Tonin, M. (1962), Nuovo Cimento **25**, 626.
- Bertsch, G., Brodsky, S.J., Goldhaber, A.S. and Gunion, J.R. (1981), Phys. Rev. Lett. **47**, 267.
- Bethe, H. (1958), Ann. Phys. **3**, 190.
- Bjorken, J.D. (1969), Phys. Rev. **179**, 1547.
- Bjorken, J.D. (1993), Phys. Rev. **D47**, 101.
- Bjorken, J.D. (1994), "Hard Diffraction and Deep Inelastic Scattering", in Proc. of Int. Workshop on DIS and Related Subjects, Eilat (Israel), A. Levy ed., World Scientific, Singapore, p. 151.
- Bjorken, J.D., Kogut, J. and Soper, D.E. (1971), Phys. Rev. **D3**, 1382.
- Block, M.M., Fletcher, R., Halzen, F., Margolis, B. and Valin, P. (1990), Phys. Rev. **D41**, 978.
- Blümlein, J., Ravindran, V. and van Neerven, W.L. (1998), Phys. Rev. **D58**, 091502.
- Bojak, I. and Ernst, M. (1996), Phys. Rev. **D52**, 80.

- Bonino, R. et al., UA8 (1988), Phys. Lett. **B211**, 239.
- Born, M. and Wolf, E. (1959), *Principles of Optics*, Pergamon Press, London.
- Bottino, A., Longoni, A.M. and Regge, T. (1962), Nuovo Cimento **23**, 954.
- Bourrely, C., Soffer, J. and Wu, T.T. (1984), Nucl. Phys. **B247**, 15.
- Bourrely, C., Soffer, J. and Wu, T.T. (1988), Z. Phys. **C37**, 369.
- Bourrely, C., Soffer, J. and Wu, T.T. (1990), Phys. Lett. **B252**, 287.
- Bozzo, M. et al., UA4 (1984a), Phys. Lett. **136B**, 217.
- Bozzo, M. et al., UA4 (1984b), Phys. Lett. **147B**, 392.
- Brandt, A. et al., UA8 (1992), Phys. Lett. **B297**, 417.
- Brandt, A. et al., UA8 (1998a), Phys. Lett. **B421**, 395.
- Brandt, A. et al., UA8 (1998b), Nucl. Phys. **B514**, 3.
- Breakstone, A. et al. (1984), Nucl. Phys. **B248**, 253.
- Breitweg, J. et al., ZEUS (1998), Eur. Phys. J. **C1**, 81.
- Breitweg, J. et al., ZEUS (1999a), Eur. Phys. J. **C6**, 603.
- Breitweg, J. et al., ZEUS (1999b), Eur. Phys. J. **C7**, 609.
- Brodsky, S.J., Frankfurt, L., Gunion, J.F., Mueller, A.H. and Strikman, M. (1994), Phys. Rev. **D50**, 3134.
- Brodsky, S.J., Hautmann, F. and Soper, D.E. (1997), Phys. Rev. **D56**, 6957.
- Brodsky, S.J., Fadin, V.S., Kim, V.T., Lipatov, L.N. and Pivovarov, G.B. (2001), "High-Energy Asymptotics of Photon-Photon Collisions in QCD", e-print `hep-ph/0111390`.
- Bronzan, J.B, Kane, G.L. and Sukhatme, U. P. (1974), Phys. Lett. **49B**, 272.
- Brower, R.C., DeTar, C.E. and Weis, J.H. (1974), Phys. Rep. **14C**, 257.
- Buchmüller, W. (1995), Phys. Lett. **B353**, 335.
- Buchmüller, W. and Hebecker, A. (1995), Phys. Lett. **B355**, 573.
- Buchmüller, W. and Hebecker, A. (1996), Nucl. Phys. **B476**, 203.
- Buchmüller, W., McDermott, M.F. and Hebecker, A. (1997), Nucl. Phys. **B487**, 283; Erratum, *ibid.* **B500** (1997), 621.
- Buttimore, N.H., Gotsman, E. and Leader, E. (1978), Phys. Rev. **D18**, 694.
- Byckling, E. and Kajantie, K. (1973), *Relativistic Kinematics*, Wiley, New York.
- Byers, N. and Yang, C.N. (1966), Phys. Rev. **142**, 976.
- Cacciari, M., Del Duca, V., Frixione, S. and Trocsanyi, Z. (2001), JHEP **0102**, 029.
- Cahn, R. (1982), Z. Phys. **C15**, 253.
- Camici, G. and Ciafaloni, M. (1998), Phys. Lett. **B430**, 349.
- Caneschi, L., ed. (1989), *Regge Theory of Low-$p_T$ Hadronic Interaction*, North-Holland, Amsterdam.
- Capella, A. and Kaplan, J. (1974), Phys. Lett. **B52**, 448.

- Capella, A., Kaidalov, A., Merino, C. and Tran Thanh Van, J. (1994), Phys. Lett. **B337**, 358.
- Capella, A., Kaidalov, A., Merino, C. and Tran Thanh Van, J. (1995), Phys. Lett. **B343**, 403.
- Capella, A., Kaidalov, A., Merino, C., Pertermann, D. and Tran Thanh Van, J. (1996), Phys. Rev. **D53**, 2309.
- Capella, A., Kaidalov, A., Merino, C., Pertermann, D. and Tran Thanh Van, J. (1998), Eur. Phys. J. **C5**, 111.
- Cartiglia, N. (1997), "Diffraction at HERA", talk at 24th Annual SLAC Summer Inst. on Particle Physics "The Strong Interaction, From Hadrons to Protons", Stanford, CA, e-print `hep-ph/9703245`.
- Catani, S., Ciafaloni, M. and Hautmann, F. (1990), Phys. Lett. **B242**, 97.
- Catani, S., Ciafaloni, M. and Hautmann, F. (1991), Nucl. Phys. **B366**, 135.
- Catani, S., Fiorani, F. and Marchesini, G. (1990a), Phys. Lett. **B234**, 339.
- Catani, S., Fiorani, F. and Marchesini, G. (1990b), Nucl. Phys. **B336**, 18.
- Cheng, H. and Lo, C.Y. (1976), Phys. Rev. **D13**, 1131.
- Cheng, H. and Lo, C.Y. (1977), Phys. Rev. **D15**, 2959.
- Cheng, H. and Wu, T.T. (1969), Phys. Rev. **182**, 1852, 1868, 1873.
- Cheng, H. and Wu, T.T. (1970a), Phys. Rev. Lett. **24**, 1456.
- Cheng, H. and Wu, T.T. (1970b), Phys. Rev. **D1**, 1069, 1083.
- Cheng, H. and Wu, T.T. (1987), *Expanding Protons. Scattering at High Energies*, MIT Press, Cambridge, MA.
- Chew, G.F.P. (1961), *Theory of Strong Interactions*, Benjamin, New York.
- Chew, G.F.P. and Frautschi, S.C. (1961), Phys. Rev. Lett. **7**, 394.
- Chou, T.T. and Yang, C.N. (1968), Phys. Rev. Lett. **20**, 1213.
- Chou, T.T. and Yang, C.N. (1970), Phys. Rev. Lett. **25**, 1072.
- Chou, T.T. and Yang, C.N. (1979), Phys. Rev. **D19**, 3268.
- Ciafaloni, M. (1988), Nucl. Phys. **B296**, 49.
- Collins, J.C. (1989), "Sudakov Form Factors", in *Perturbative Quantum Chromodynamics*, ed. A.H. Mueller, World Scientific, Singapore.
- Collins, J.C. (1998), Phys. Rev. **D57**, 3051. Erratum, *ibid.* **D61** (2000), 019902.
- Collins, J.C. and Kwieciński, J. (1989), Nucl. Phys. **B316**, 307.
- Collins, J.C., Soper, D.E. (1982), Nucl. Phys. **B194**, 445.
- Collins, J.C., Soper, D.E. and Sterman, G. (1989), "Factorization of hard processes in QCD", in *Perturbative Quantum Chromodynamics*, A.H. Mueller ed., World Scientific, Singapore.
- Collins, P.D.B. (1977), *An Introduction to Regge Theory and High Energy Physics*, Cambridge University Press, Cambridge, UK.
- Collins, P.D.B., Gault, F.D. and Martin, A. (1974), Nucl. Phys. **B80**, 135.
- Collins, P.D.B. and Martin, A.D. (1984), *Hadron Interactions*, Adam Hilger, Bristol.

- Collins, P.D.B. and Squires, E.J. (1968), *Regge Poles in Particle Physics*, Springer, Berlin.
- Combridge, B.L., Kripfganz, J. and Ranft, J. (1977), Phys. Lett. **70B**, 234.
- Cool, R. L. et al. (1981), Phys. Rev. Lett. **47**, 701.
- Corianò, C. and White, A.R. (1996), Nucl. Phys. **B468**, 175.
- Cornille, H. and Martin, A. (1972), Nucl. Phys. **B48**, 104.
- Covolan, R.M.J., Montanha, J. and Goulianos, K. (1996), Phys. Lett. **B389**, 176.
- Cox, B.E. and Forshaw, J.R. (1998), Phys. Lett. **B434**, 133.
- Cox, B., Forshaw, J. and Lönnblad, L. (1999), JHEP **9910**, 023.
- Cudell, J.-R., Kang, K. and Kim S.K. (1997), Phys. Lett. **B395**, 311.
- Cutkosky, R.E. (1960), J. Math. Phys. **1**, 429.
- Cutler, R. and Sivers, D. (1977), Phys. Rev. **D16**, 679.
- Cutler, R. and Sivers, D. (1978), Phys. Rev. **D17**, 196.
- Czyżewski, J., Kwieciński, J., Motyka, L. and Sazikowski, M. (1997), Phys. Lett. **B398**, 400. Erratum, *ibid.* **B411** (1997), 402.
- Del Duca, V. (1995), Scientifica Acta **10**, 91 (e-print hep-ph/9503226).
- Del Duca, V., Peskin, M.E. and Tang, W.-K. (1993), Phys. Lett. **B306**, 151.
- Del Duca, V. and Schmidt, C.R. (1994a), Phys. Rev. **D49**, 177.
- Del Duca, V. and Schmidt, C.R. (1994b), Phys. Rev. **D49**, 177.
- Del Duca, V. and Tang, W.-K. (1993), Phys. Lett. **B312**, 227.
- Del Duca, V. (1996a), Phys. Rev. **D54**, 989.
- Del Duca, V. (1996b), Phys. Rev. **D54**, 4474.
- Del Duca, V. and Glover, N. (2001), JHEP **0110**, 035.
- Del Duca, V. and Schmidt, C.R. (1999), Phys. Rev. **D59**, 074004.
- De Roeck, A. (2000), Nucl. Phys. **A666&667**, 129c.
- Derrick, M. et al., ZEUS (1993), Phys. Lett. **B315**, 481.
- De Rùjula, A., Glashow, S.L., Politzer, H.D., Treiman, S.B., Wilczek, F. and Zee, A. (1974), Phys. Rev. **D10**, 1649.
- Desgrolard, P., Giffon, M., Martynov, E. and Predazzi, E. (2000), Eur. Phys. J. **C14**, 683.
- Dias de Deus, J. and Kroll, P. (1978), Acta Phys. Pol. **B9**, 159.
- Dias de Deus, J. and Kroll, P. (1983), J. Phys. **G9**, L81.
- Diehl, M. (1995), Z. Phys. **C66**, 181.
- Dokshitzer, Yu.L. (1977), Sov. Phys. JETP **46**, 6451.
- Dokshitzer, Yu.L., Khoze, V.A. and Troyan, S.I. (1987), in *Proc. of the 6th Int. Conf. on Physics in Collisions (1986)*, M. Derrick ed., World Scientific, Singapore.
- Dolen, R., Horn, D. and Schmid, C. (1968), Phys. Rev. **166**, 1768.
- Donnachie, A. and Shaw, G. (1978), in *Electromagnetic Interactions of Hadrons*, eds. A. Donnachie and G. Shaw, Plenum, New York.
- Donnachie, A. and Landshoff, P.V. (1979), Z. Phys. **C2**, 55.
- Donnachie, A. and Landshoff, P.V. (1984), Nucl. Phys. **B244**, 322.

- Donnachie, A. and Landshoff, P.V. (1987), Phys. Lett. **B191**, 309.
- Donnachie, A. and Landshoff, P.V. (1988), Nucl. Phys. **B303**, 634.
- Donnachie, A. and Landshoff, P.V. (1988/89), Nucl. Phys. **B311**, 509.
- Donnachie, A. and Landshoff, P.V. (1991), Nucl. Phys. **B348**, 297.
- Donnachie, A. and Landshoff, P.V. (1992), Phys. Lett. **B296**, 227.
- Donnachie, A. and Landshoff, P.V (1999), Phys. Lett. **B470**, 243.
- Dosch, H.G., Ferreira, E. and Krämer, A. (1992), Phys. Lett. **B289**, 153.
- Dosch, H.G., Ferreira, E. and Krämer, A. (1994), Phys. Rev. **D50**, 1992.
- Dosch, H.G. and Rueter, M. (1996), Phys. Lett. **B380**, 177.
- Dubovikov, M.S., Kopeliovich, B.Z., Lapidus, L.I. and Ter-Martirosyan, K.A. (1977), Nucl. Phys. **B123**, 147.
- Durand, L. and Lipes, R. (1968), Phys. Rev. Lett. **20**, 637.
- Eden, R.J. (1967), *High Energy Collisions of Elementary Particles*, Cambridge University Press, Cambridge, UK.
- Eden, R.J., Landshoff, P.V., Olive, D.I. and Polkinghorne, J.C. (1966), *The Analytic S-Matrix*, Cambridge University Press, Cambridge, UK.
- Edin, A., Ingelman, G. and Rathsman, J. (1996), Phys. Lett. **B366**, 371.
- Edin, A., Ingelman, G. and Rathsman, J. (1997), Z. Phys. **C75**, 57.
- Ellis, R.K., Stirling, W.J. and Webber, B.R. (1996), *QCD and Collider Physics*, Cambridge University Press, Cambridge, UK.
- Engel, R., Gaisser, T.K. and Lipari, P. (1998), Phys. Rev. **D58**, 014019.
- Engel, R., Ivanov, D.Y., Kirschner, R. and Szymanowski, L. (1998), Eur. Phys. J. **C4**, 93.
- Fadin, V.S. (2000), Nucl. Phys. **A666&667**, 155c.
- Fadin, V.S. (2001), Nucl. Phys. B (Proc. Suppl.) **99**, 204.
- Fadin, V.S. and Fiore, R. (1992), Phys. Lett. **B294**, 286.
- Fadin, V.S. and Fiore, R. (1998), Phys. Lett. **B440**, 359.
- Fadin, V.S., Fiore, R., Flachi, A. and Kotsky, M.I. (1998), Phys. Lett. **B422**, 287.
- Fadin, V.S., Fiore, R. and Kotsky, M.I. (1995), Phys. Lett. **B359**, 181.
- Fadin, V.S., Fiore, R. and Kotsky, M.I. (1996a), Phys. Lett. **B387**, 593.
- Fadin, V.S., Fiore, R. and Kotsky, M.I. (1996b), Phys. Lett. **B389**, 737.
- Fadin, V.S., Fiore, R.. Kotsky, M.I. and Papa, A. (2000a), Phys. Rev. **D61**, 094005.
- Fadin, V.S., Fiore, R.. Kotsky, M.I. and Papa, A. (2000b), Phys. Rev. **D61**, 094006.
- Fadin, V.S., Fiore, R. and Quartarolo, A. (1994), Phys. Rev. **D50**, 5893.
- Fadin, V.S., Fiore, R. and Quartarolo, A. (1996), Phys. Rev. **D53** (1996) 2729.
- Fadin, V.S. and Gorbachev, D.A. (2000), JETP Lett. **71**, 222.
- Fadin, V.S., Kotsky, M.I. and Lipatov, L.N. (1997), Phys. Lett. **B415**, 97.
- Fadin, V.S. and Lipatov, L.N. (1989), Sov. J. Nucl. Phys. **50**, 712.
- Fadin, V.S. and Lipatov, L.N. (1993), Nucl. Phys. **B406**, 259.
- Fadin, V.S. and Lipatov, L.N. (1996), Nucl. Phys. **B477**, 767.

- Fadin, V.S. and Lipatov, L.N. (1998), Phys. Lett. **B429**, 127.
- Feinberg, E. and Pomeranchuk, I.Ya. (1956), Nuovo Cim. Suppl. **3**, 652.
- Ferreira, E. and Pereira, F. (1997), Phys. Rev. **D55**, 130.
- Ferreira, E. and Pereira, F. (1999), Phys. Rev. **D59**, 014008.
- Field, R.D. (1989), *Applications of Perturbative QCD*, Addison-Wesley, Reading, MA.
- Forshaw, J.R. and Ryskin, M.G. (1995), Z. Phys. **C68**, 137.
- Forshaw, J.R. and Ross, D.A. (1997), *Quantum Chromodynamics and the Pomeron*, Cambridge University Press, Cambridge, UK.
- Forte, S. (1997), Nucl. Phys. B (Proc. Suppl.) **54A**, 163.
- Forte, S. (2000), Nucl. Phys. **A666&667**, 113c.
- Frankfurt, L., Freund, A. and Strikman, M. (1998), Phys. Rev. **D58**, 114001; Erratum, *ibid.* **D59** (1999), 119901.
- Frankfurt, L., Koepf, W. and Strikman, M. (1996), Phys. Rev. **D54**, 319.
- Frankfurt, L., Koepf, W. and Strikman, M. (1998), Phys. Rev. **D57**, 513.
- Frankfurt, L.L. and Sherman, V.E. (1976), Sov. J. Nucl. Phys. **23**, 581.
- Frautschi, S.C. (1963), *Regge Poles and S-Matrix Theory*, Benjamin, New York.
- Fried, H.M. (1990), *Functional Methods and Eikonal Models*, Editions Frontières, Paris.
- Fritzsch, H. and Streng, K.-H. (1985), Phys. Lett. **164B**, 391.
- Froissart, M. (1961), Phys. Rev. **123**, 1053.
- Frolov, G.V., Gribov, V.N. and Lipatov, L.N. (1970a), Phys. Lett. **31B**, 34.
- Frolov, G.V., Gribov, V.N. and Lipatov, L.N. (1970b), Yad. Fiz. **12**, 994 [transl. Sov. J. Nucl. Phys. **12** (1971), 543].
- Ganguli, S.N. and Roy, D.P. (1980), Phys. Rep. **67**, 201.
- Genovese, M., Nikolaev, N.N. and Zakharov, B.G. (1995), JETP **81**, 625.
- Genovese, M., Nikolaev, N.N. and Zakharov, B.G. (1996), Phys. Lett. **B378**, 347.
- Giffon, M., Nahabetian, R.S. and Predazzi, E. (1987), Z. Phys. **C36**, 67.
- Giffon, M., Nahabetian, R.S. and Predazzi, E. (1988), Phys. Lett. **B205**, 363.
- Giffon, M. and Predazzi, E. (1980), Lett. Nuovo Cim. **29**, 110.
- Giffon, M., Predazzi, E. and Samokhin, A. (1996), Phys. Lett. **B375**, 315.
- Ginzburg, I.F. and Ivanov, D.Yu. (1996), Phys. Rev. **D54**, 5523.
- Glauber, R.J. (1958), "High-energy collision theory", in *Lectures in Theoretical Physics*, vol. 1, Interscience, New York.
- Glauber, R.J. and Velasco, J. (1984), Phys. Lett. **B147**, 380.
- Goldberger, M.L. and Watson, K.M. (1964), *Collision Theory*, Wiley, New York.
- Golec-Biernat, K. and Wüsthoff, M. (1999a), Phys. Rev. **D59**, 014017.
- Golec-Biernat, K. and Wüsthoff, M. (1999b), Phys. Rev. **D60**, 114023.
- Good, M.L. and Walker, W.D. (1960), Phys. Rev. **120**, 1857.

- Gotsman, E., Levin, E. and Maor, U. (1993), Phys. Lett. **B390**, 199.
- Gotsman, E., Levin, E. and Maor, U. (1997), Nucl. Phys. **B493**, 354.
- Goulianos, K. (1983), Phys. Rep. **101**, 169.
- Goulianos, K. (1995), Phys. Lett. **B358**, 379. Erratum, *ibid.* **B363** (1995), 268.
- Goulianos, K. (2001a), Nucl. Phys. B (Proc. Suppl.) **99**, 9.
- Goulianos, K. (2001b), Nucl. Phys. B (Proc. Suppl.) **99**, 37.
- Goulianos, K. and Montanha, J. (1999), Phys. Rev. **D59**, 114017.
- Grammer Jr., G. and Sullivan, J.D. (1978), in *Electromagnetic Interactions of Hadrons*, eds. A. Donnachie and G. Shaw, Plenum, New York.
- Grazzini, M., Trentadue, L. and Veneziano, G. (1998), Nucl. Phys. **B519**, 394.
- Green, M.B., Schwarz, J.H. and Witten, E. (1987), *Superstring Theory*, Vol. 1, Cambridge University Press, Cambridge, UK.
- Gribov, V.N. (1961), Zh. Eksp. Teor. Fiz. **41**, 667 [transl. Sov. Phys. JETP **14** (1962) 478].
- Gribov, V.N. (1968), Sov. Phys. JETP **26**, 414.
- Gribov, V.N. (1969), Sov. Phys. JETP **29**, 483.
- Gribov, V.N. and Lipatov, L.N. (1972), Sov. J. Nucl. Phys. **15**, 78.
- Gribov, L.V., Levin, E.M. and Ryskin, M.G. (1983), Phys. Rep. **100**, 1.
- Gribov, V.N., Mur V.D., Kobzarev, I.Yu., Okun, L.B. and Popov, V.S. (1971), Sov. J. Nucl. Phys. **12**, 699.
- Gribov, V.N. and Pomeranchuk, I.Ya. (1962), Phys. Rev. Lett. **9**, 238.
- Gribov, V.N., Pomeranchuk, I.Ya. and Ter-Martirosyan, K.A. (1965), Phys. Rev. **139**, B184.
- Grunberg, G. and Truong, T.N. (1973), Phys. Rev. Lett. **31**, 63.
- Guryn, W. (2001), Nucl. Phys. B (Proc. Suppl.) **99**, 299.
- H1 Collaboration (1997), "Production of $J/\Psi$ mesons with large $|t|$ at HERA", contribution to the Int. Europhys. Conf. on High Energy Physics, Jerusalem, Israel, August 1997.
- H1 Collaboration (1998a), "Double diffraction dissociation at large $|t|$ in photoproduction at HERA", contribution to the 29th Int. Conf. on High-Energy Physics, Vancouver, Canada, July 1998.
- H1 Collaboration (1998b), "Measurements of the diffractive structure function $F_2^{D(3)}$ at low and high $Q^2$ at HERA", contribution to the 29th Int. Conf. on High-Energy Physics, Vancouver, Canada, July 1998.
- H1 Collaboration (1999), "Measurement of the production of $D^{*\pm}$ mesons in deep-inelastic diffractive interactions at HERA", contribution to the Int. Europhys. Conf. on High-Energy Physics, Tampere, Finland, July 1999.
- H1 Collaboration (2000), "Measurement of the deeply virtual Compton scattering at HERA", contribution to the 30th Int. Conf. on High-Energy Physics, Osaka, Japan, July 2000.
- Halzen, F. (1993), Summary talk at the Int. Conf. on Elastic and Diffractive Scattering (5th Blois Workshop), Providence, RI, 1993, hep-ph/9307237.

- Hebecker, A. (2000), Phys. Rep. **331**, 1.
- Henzi, R. and Valin, P. (1983), Phys. Lett. **B132**, 443.
- Henzi, R. and Valin, P. (1985), Phys. Lett. **B160**, 167.
- Honda, M. et al. (1993), Phys. Rev. Lett. **70**, 525.
- Ingelman, G. and Schlein, P.E. (1985), Phys. Lett. **152B**, 256.
- Irving, A.C. and Worden, R.P. (1977), Phys. Rep. **34**, 117.
- Ivanov, D.Yu. (1996), Phys. Rev. **D53**, 3564.
- Ivanov, D.Yu. and Wüsthoff, M. (1999), Eur. Phys. J. **C8**, 107.
- Janik, R.A. and Wosiek, J. (1999), Phys. Rev. Lett. **82**, 1092.
- Jaroszkiewicz, G.A. and Landshoff, P.V. (1974), Phys. Rev. **D10**, 170.
- Jenkovszky, L.L., Paccanoni, F. and Predazzi, E. (1992), Nucl. Phys. Proc. Suppl. **25B**, 80.
- Ji, X. (1997), Phys. Rev. **D55**, 7114.
- Joynson, D., Leader, E., Lopez, C. and Nicolescu, B. (1975), Nuovo Cimento **30A**, 345.
- Kaidalov, A.B. (1979), Phys. Rep. **50**, 157.
- Kaidalov, A.B. and Ter-Martirosyan, K.A. (1974), Nucl. Phys. **B75**, 471.
- Karmanov, V.A. and Kondratyuk, L.A. (1973), JETP Lett. **18**, 266.
- Khuri, N.N. and Kinoshita, T. (1965), Phys. Rev. **137**, B720.
- Kibble, T.W.B. (1960), Phys. Rev. **117**, 1159.
- King, R.W.P. and Wu, T.T. (1959), *Scattering and Diffraction of Waves*, Harvard University Press, Cambridge, MA.
- Kokkedee, J.J.J. and Van Hove, L. (1966), Nuovo Cimento **42**, 711.
- Kopeliovich, B.Z. and Lapidus, L.I. (1978), JETP Lett. **28**, 614.
- Kopeliovich, B.Z. and Marage, P. (1993), Int. J. Mod. Phys. **A8**, 1513.
- Kopeliovich, B.Z., Nikolaev, N.N. and Potashnikova, I.K. (1989), Phys. Rev. **D39**, 769.
- Kopeliovich, B.Z. and Povh, B. (1997), Z. Phys. **A356**, 467.
- Kopeliovich, B.Z., Povh, B. and Predazzi, E. (1997), Phys. Lett. **B 405**, 361.
- Kopeliovich, B.Z., Potashnikova, I., Povh, B. and Predazzi, E. (2000), Phys. Rev. Lett. **85**, 507.
- Kopeliovich, B.Z. and Tarasov, A.V. (2001), Phys. Lett. **B497**, 44.
- Kuraev, E.A., Lipatov, L.N. and Fadin, V.S. (1976), Sov. Phys. JETP **44**, 443.
- Kuraev, E.A., Lipatov, L.N. and Fadin, V.S. (1977), Sov. Phys. JETP **45**, 199.
- Kwieciński, J., Martin, A.D. and Sutton, P.J. (1995), Phys. Rev. **D52**, 1445.
- Kwieciński, J. and Praszałowicz (1980), Phys. Lett. **B94**, 413.
- Landau, L.D. and Pomeranchuk, I.Ya. (1953), Zh. Eksp. Teor. Fiz. **24**, 505.
- Landshoff, P.V. and Nachtmann, O. (1987), Z. Phys. **C35**, 405.
- Landshoff, P.V. and Nachtmann, O. (1998), "Some Remarks on the Pomeron and the Odderon in theory and Experiment", e-print `hep-ph/9808233`.

- Landshoff, P.V. and Polkinghorne, J.C. (1971), Nucl. Phys. **B32**, 541.
- Leader, E. (2001), *Spin in Particle Physics*, Cambridge University Press, Cambridge, UK.
- Leader, E. and Predazzi, E. (1996), *An introduction to gauge theories and modern particle physics* , Cambridge University Press, Cambridge, UK.
- Lehmann, H. (1958), Nuovo Cimento **10**, 579.
- Levin, E. (1997), "Everything about Reggeons. Part I: Reggeons in 'Soft' Interaction", report DESY 97-213, `hep-ph/9710546`.
- Levin, E.M. and Frankfurt, L.L. (1965), JETP Lett. **2**, 65.
- Levin, E.M., Martin, A.D., Ryskin, M.G. and Teubner, T. (1997), Z. Phys. **C74**, 671.
- Levin, E.M. and Wüsthoff (1994), Phys. Rev. **D50**, 4306.
- Lipatov, L.N. (1976), Sov. J. Nucl. Phys. **23**, 338.
- Lipatov, L.N. (1986), Sov. Phys. JETP **63**, 904.
- Lipatov, L.N. (1989), "Pomeron in Quantum Chromodynamics", in *Perturbative Quantum Chromodynamics*, ed. A.H. Mueller, World Scientific, Singapore.
- Lipatov, L.N. (1997), Phys. Rep. **286**, 131.
- Lipatov, L.N. and Fadin, V.S. (1989), Sov. J. Nucl. Phys. **50**, 712.
- Lipkin, H.J. and Scheck, F. (1966), Phys. Rev. Lett. **16**, 71.
- Low, F.E. (1975), Phys. Rev. **D12**, 163.
- Łukaszuk, L. and Nicolescu, B. (1973), Lett. Nuovo Cim. **8**, 405.
- MacDowell, S.W. and Martin, A. (1964), Phys. Rev. **135**, B960.
- Mandelstam, S. (1958), Phys. Rev. **112**, 1344.
- Mandelstam, S. (1963), Nuovo Cimento **30**, 1113, 1127, 1148.
- Marage, P. (1999), "Hadronic structure, low-$x$ physics and diffraction", Plenary report at the Int. Europhys. Conf. on High-Energy Physics, Tampere, Finland, `hep-ph/9911426`.
- Margolis, B. et al. (1988), Phys. Lett. **B213**, 221.
- Martin, A. (1963), Phys. Rev. **129**, 1432.
- Martin, A.D., Ryskin, M.G. (1998), Phys. Rev. **D57**, 6692.
- Martin, A.D., Ryskin, M.G. and Teubner, T. (1997), Phys. Rev. **D55**, 4329.
- Matthiae, G. (1994), Rep. Prog. Phys. **57**, 743.
- Matthiae, G. (2001), Nucl. Phys. B (Proc. Suppl.) **99**, 281.
- Miettinen, H.I. and Pumplin, J. (1978), Phys. Rev. **D18**, 1696.
- Molière, G. (1947), Zeit. Natur. **2A**, 133.
- Mueller, A.H. (1970), Phys. Rev. **D2**, 2963.
- Mueller, A.H. (1994), Nucl. Phys. **B415**, 373.
- Mueller, A.H. (1995), Nucl. Phys. **B437**, 107.
- Mueller, A.H. and Navelet, H. (1987), Nucl. Phys. **B282**, 727.
- Mueller, A.H. and Patel, B. (1994), Nucl. Phys. **B425**, 471.
- Mueller, A.H. and Salam, G.P. (1996), Nucl. Phys. **B475**, 293.
- Mueller, A.H. and Tang, W.-K. (1992), Phys. Lett. **B284**, 123.
- Nachtmann, O. (1991), Ann. Phys. **209**, 436.

- Nachtmann, O. and Reiter, A. (1984), Z. Phys. **C24**, 283.
- Nemchik, J., Nikolaev, N.N. and Zakharov, B.G. (1994), Phys. Lett. **B374**, 199.
- Nemchik, J., Nikolaev, N.N., Predazzi, E. and Zakharov, B.G. (1997), Z. Phys. **C75**, 71.
- Nemchik, J., Nikolaev, N.N., Predazzi, E., Zakharov, B.G. and Zoller, V.R. (1998), JETP **86**, 1054.
- Newton, R.G. (1964), *The Complex j-Plane*, Benjamin, New York.
- Newton, R.G. (1966), *Scattering Theory of Waves and Particles*, McGraw-Hill, New York.
- Nicolescu, B. (1999), "Recent Advances in Odderon Physics", e-print hep-ph/9911334.
- Nikolaev, N.N. (1981), Sov. Phys. JETP **54**, 434.
- Nikolaev, N.N. (1993), Phys. Rev. **D48**, 1904.
- Nikolaev, N.N. (1994), Int. J. Mod. Phys. **E3**, 1. Erratum, *ibid.* **E3** (1994), 995.
- Nikolaev, N.N. (2001), Nucl. Phys. B (Proc. Suppl.) **99**, 249.
- Nikolaev, N.N. and Zakharov, B.G. (1991a), Phys. Lett. **B260**, 414.
- Nikolaev, N.N. and Zakharov, B.G. (1991b), Z. Phys. **C49**, 607.
- Nikolaev, N.N. and Zakharov, B.G. (1992), Z. Phys. **C53**, 331.
- Nikolaev, N.N. and Zakharov, B.G. (1994a), Phys. Lett. **B327**, 149.
- Nikolaev, N.N. and Zakharov, B.G. (1994b), Z. Phys. **C64**, 631.
- Nikolaev, N.N. and Zakharov, B.G. (1994c), Phys. Lett. **B332**, 177.
- Nikolaev, N.N. and Zakharov, V.I (1975), Phys. Lett. **B55**, 397.
- Nikolaev, N.N., Zakharov, B.G. and Zoller, V.R. (1994a), JETP Lett. **60**, 694.
- Nikolaev, N.N., Zakharov, B.G. and Zoller, V.R. (1994b), Phys. Lett. **B328**, 486.
- Nikolaev, N.N., Zakharov, B.G. and Zoller, V.R. (1996), Phys. Lett. **B366**, 1996.
- Nussinov, S. (1975), Phys. Rev. Lett. **34**, 1286.
- Nussinov, S. (1976), Phys. Rev. **D14**, 246.
- Okun, L.B. and Pomeranchuk, I.Ya. (1956), Sov. Phys. JETP **3**, 307.
- Pilkuhn, H.M. (1979), *Relativistic Particle Physics*, Springer, Berlin.
- Piller, G., Ratzka, W. and Weise, W. (1995), Z. Phys. **A352**, 427.
- Pomeranchuk, I.Ya. (1956), Sov. Phys. JETP **3**, 306.
- Pomeranchuk, I.Ya. (1958), Sov. Phys. JETP **7**, 499.
- Predazzi, E. (1966) Ann. Phys. **36**, 228; 250.
- Predazzi, E. (1976), Riv. Nuovo Cim. **6**, 217.
- Radyushkin, A.V. (1997), Phys. Rev. **D56**, 5524.
- Regge, T. (1959), Nuovo Cimento **14**, 951.
- Regge, T. (1960), Nuovo Cimento **18**, 947.
- Roberts, R.G. (1990), *The Structure of the Proton*, Cambridge University Press, Cambridge, UK.

- Roy, S.M. (1972), Phys. Rep. **5C**, 125.
- Royen, I. and Cudell, J.-R. (1999), Nucl. Phys. **B545**, 505.
- Rueter, M., Dosch, H.G. and Nachtmann, O. (1999), Phys. Rev. **D59**, 014018.
- Ryskin, M.G. (1993), Z. Phys. **C57**, 89.
- Ryskin, M.G., Roberts, R.G., Martin, A.D. and Levin, E.M. (1997), Z. Phys. **C76**, 231.
- Salam, G.P. (1996), Nucl. Phys. **B461**, 512.
- Santoro, A. (2001), Nucl. Phys. B (Proc. Suppl.) **99**, 289.
- Shuvaev, A.G., Golec-Biernat, K., Martin, A.D. and Ryskin, M.G. (1999), Phys. Rev. **D60**, 014015.
- Singh, V. and Roy, S.M. (1970), Phys. Rev. **D1**, 2638.
- Sitenko, A. (1959), Usp. Fiz. Nauk. **67**, 377.
- Sommerfeld, A. (1949), *Partial Differential Equations in Physics*, Academic Press, New York.
- Squires, E.J. (1963), *Complex Angular Momentum and Particle Physics*, Benjamin, New York.
- Stirling, W.J. (1994), Nucl. Phys. **B423**, 56.
- Titchmarsh, E.C. (1939), *The Theory of Functions*, 2nd edition, Oxford University Press, Oxford.
- TOTEM Collaboration (1997), Letter of Intent, CERN/LHCC 97-49.
- Trentadue, L. and Veneziano, G. (1994), Phys. Lett. **B323**, 201.
- Tyburski, L. (1976), Phys. Rev. **D13**, 1107.
- Vacca, G.P. (2001), "Odderon in QCD", contribution to the 9th Int. Workshop on DIS and QCD, Bologna, Italy, April 2001, `hep-ph/0106224`.
- Van de Hulst, H.C. (1957), *Light Scattering by Small Particles*, Wiley, New York.
- Van Hove, L. (1964), Rev. Mod. Phys. **36**, 655.
- Veneziano, G. (1968), Nuovo Cimento **57A**, 190.
- Watson, G.N. (1918), Proc. Roy. Soc. **95**, 83.
- West, G. and Yennie, D.R. (1968), Phys. Rev. **172**, 1413.
- Wu, T.T. (1995), "Scattering by Spheres at Short Wavelengths – From Maxwell to HERA", preprint CERN-TH/95-238.
- Wu, T.T. (2000), "Scattering and Production Processes at High Energies", preprint CERN-TH/2000-274.
- Wu, T.T. and Yang, C.N. (1965), Phys. Rev. **137**, B708.
- Wüsthoff, M. (1997), Phys. Rev. **D56**, 4311.
- Wüsthoff, M. and Martin, A.D. (1999), J. Phys. G **25**, R309.
- Zakharov, B.G. (1989), Sov. J. Nucl. Phys. **49**, 860.
- Zamolodchikov, A.B., Kopeliovich, B.Z. and Lapidus, L.I. (1981), JETP Lett. **33**, 595.
- ZEUS Collaboration (1997), "Study of vector meson production at large $|t|$ at HERA", contribution to the Int. Europhys. Conf. on High Energy Physics, Jerusalem, Israel, August 1997.

- ZEUS Collaboration (1999a), "Diffractive dissociation of virtual photons in *ep* scattering at HERA", contribution to the Int. Europhys. Conf. on High Energy Physics, Tampere, Finland, July 1999.
- ZEUS Collaboration (1999b), "Observation of deeply virtual Compton scattering in $e^+p$ interactions at HERA", contribution to the Int. Europhys. Conf. on High Energy Physics, Tampere, Finland, July 1999.
- ZEUS Collaboration (2000a), "Three-jet production in diffractive deep inelastic scattering at HERA", contribution to the 30th Int. Conf. on High Energy Physics, Osaka, Japan, July 2000.
- ZEUS Collaboration (2000b), "Diffractive $D^{*\pm}(2010)$ production in deep inelastic scattering at HERA", contribution to the 30th Int. Conf. on High Energy Physics, Osaka, Japan, July 2000.
- ZEUS Collaboration (2000c), "Exclusive photoproduction of $J/\psi$ mesons", contribution to the 30th Int. Conf. on High Energy Physics, Osaka, Japan, July 2000.
- ZEUS Collaboration (2000d), "Exclusive electroproduction of $\rho^0$ mesons with the ZEUS detector at HERA", contribution to the 30th Int. Conf. on High Energy Physics, Osaka, Japan, July 2000.

# Subject Index

# Texts and Monographs in Physics



**Essential Relativity** Special, General, and Cosmological Revised 2nd edition By W. Rindler

**The Elements of Mechanics** By G. Gallavotti

**Generalized Coherent States and Their Applications** By A. Perelomov

**Quantum Mechanics II** By A. Galindo and P. Pascual

**Geometry of the Standard Model of Elementary Particles** By. A. Derdzinski

**From Electrostatics to Optics** A Concise Electrodynamics Course By G. Scharf

**Finite Quantum Electrodynamics** The Causal Approach 2nd edition By G. Scharf

**Path Integral Approach to Quantum Physics** An Introduction 2nd printing By G. Roepstorff

**Supersymmetric Methods in Quantum and Statistical Physics** By G. Junker

**Relativistic Quantum Mechanics and Introduction to Field Theory** By F. J. Ynduráin

**Local Quantum Physics** Fields, Particles, Algebras 2nd revised and enlarged edition By R. Haag

**The Mechanics and Thermodynamics of Continuous Media** By M. Šilhavý

**Quantum Relativity** A Synthesis of the Ideas of Einstein and Heisenberg By D. R. Finkelstein

**Scattering Theory of Classical and Quantum *N*-Particle Systems** By. J. Derezinski and C. Gérard

**Effective Lagrangians for the Standard Model** By A. Dobado, A. Gómez-Nicola, A. L. Maroto and J. R. Peláez

**Quantum** The Quantum Theory of Particles, Fields, and Cosmology By E. Elbaz

**Quantum Groups and Their Representations** By A. Klimyk and K. Schmüdgen

**Multi-Hamiltonian Theory of Dynamical Systems** By M. Błaszak

**Renormalization** An Introduction By M. Salmhofer

**Fields, Symmetries, and Quarks** 2nd, revised and enlarged edition By U. Mosel

**Statistical Mechanics of Lattice Systems** Volume 1: Closed-Form and Exact Solutions 2nd, revised and enlarged edition By D. A. Lavis and G. M. Bell

**Statistical Mechanics of Lattice Systems** Volume 2: Exact, Series and Renormalization Group Methods By D. A. Lavis and G. M. Bell

**Conformal Invariance and Critical Phenomena** By M. Henkel

**The Theory of Quark and Gluon Interactions** 3rd revised and enlarged edition By F. J. Ynduráin

**Quantum Field Theory in Condensed Matter Physics** By N. Nagaosa

**Quantum Field Theory in Strongly Correlated Electronic Systems** By N. Nagaosa

**Information Theory and Quantum Physics** Physical Foundations for Understanding the Conscious Process By H.S. Green

**Magnetism and Superconductivity** By L.-P. Lévy

**The Nuclear Many-Body Problem** By P. Ring and P. Schuck

**Perturbative Quantum Electrodynamics and Axiomatic Field Theory** By O. Steinmann

**Quantum Non-linear Sigma Models** From Quantum Field Theory to Supersymmetry, Conformal Field Theory, Black Holes and Strings By S. V. Ketov

Series homepage – http://www.springer.de/phys/books/tmp